D1321197

Systems & Control: Foundations & Applications

Systems and Control in the Twenty-First Century

Christopher I. Byrnes
Biswa N. Datta
Clyde F. Martin
David S. Gilliam
Editors

Birkhäuser
Boston • Basel • Berlin

Christopher I. Byrnes
School of Engineering and
Applied Science
Washington University
St. Louis, MO 63130-4899

Biswa N. Datta
Dept. of Mathematical Sciences
Northern Illinois University
DeKalb, IL 60115

David S. Gilliam
Clyde F. Martin
Dept. of Mathematics
Texas Tech University
Lubbock, TX 79409

Library of Congress Cataloging-in-Publication Data
Systems and control in the twenty-first century / [edited by]
Christopher Byrnes . . . [et al.].
p. cm. -- (Progress in systems and control theory ; v. 22)
"[Papers] presented at the 12th International Symposium on the Mathematical Theory of Networks and Systems, held in St. Louis, Missouri, from June 24-28, 1996" -- Pref.
Includes bibliographical references and index.
ISBN 0-8176-3881-4 (alk. paper) ISBN 3-7643-3881-4 (alk. paper)
1. System analysis--Congresses. 2. Control Theory--Congresses.
I. Byrnes, Christopher I., 1949- . II. International Symposium on the Mathematical Theory of Networks and Systems (12th : 1996 : Saint Louis, Mo.) III. Series.
QA402.S9694 1997
003--dc21 96-45612
CIP

Printed on acid-free paper

ISBN 0-8176-3881-4
ISBN 3-7643-3881-4
Camera-ready copy prepared by the editors in LaTeX.
Printed and bound by Quinn-Woodbine, Woodbine, NJ.
Printed in the U.S.A.

9 8 7 6 5 4 3 2 1

Contents

Preface

The mathematical theory of networks and systems has a long, and rich history, with antecedents in circuit synthesis and the analysis, design and synthesis of actuators, sensors and active elements in both electrical and mechanical systems. Fundamental paradigms such as the state-space realization of an input/output system, or the use of feedback to prescribe the behavior of a closed-loop system have proved to be as resilient to change as were the practitioners who used them.

This volume celebrates the resiliency to change of the fundamental concepts underlying the mathematical theory of networks and systems. The articles presented here are among those presented as plenary addresses, invited addresses and minisymposia presented at the 12th International Symposium on the Mathematical Theory of Networks and Systems, held in St. Louis, Missouri from June 24 - 28, 1996. Incorporating models and methods drawn from biology, computing, materials science and mathematics, these articles have been written by leading researchers who are on the vanguard of the development of systems, control and estimation for the next century, as evidenced by the application of new methodologies in distributed parameter systems, linear nonlinear systems and stochastic systems for solving problems in areas such as aircraft design, circuit simulation, imaging, speech synthesis and visionics.

We wish to thank these authors, and all the contributors to MTNS-96, for making this conference an outstanding intellectual celebration of this area and its ability to embrace and lead paradigm shifts, a celebration which will continue to grow in importance as we enter the next century. We also take great pleasure in thanking Rose Brower, Bijoy Ghosh, Michalina Karina, Susan McLaughlin, Giorgio Picci, Beth Scnettler, Elizabeth SoRelle and Sue Schenker for years of outstanding service to MTNS-96. This endeavor could not have succeeded without their dedication.

Chris Byrnes	Biswa Datta	David Gilliam	Clyde Martin
St. Louis, MO	Dekalb, Il	Lubbock, TX	Lubbock, TX

Contributors List

D. Alpay Department of Mathematics, Ben-Gurion University of the Negev, Israel

B.D.O. Anderson Department of Systems Engineering and Cooperative Research, Australian National University, Canberra, Australia

H.T. Banks Center for Research in Scientific Computation, North Carolina State University, USA

D. Boley Dept. of Computer Science, University of Minnesota

R. Brockett Division of Engineering and Appied Sciences, Harvard University, USA

C.I. Byrnes Department of Systems Science and Mathematics, Washington University, USA

S. Coraluppi Electrical Engineering Department and Institute for Systems Research, University of Maryland, USA

B.N. Datta Dept. of Mathematical Sciences, Northern Illinois University, USA

P. Fard Electrical Engineering Department and Institute for Systems Research, University of Maryland, USA

E. Fernández-Gaucherand Systems and Industrial Engineering Department, University of Arizona, USA

M. Fliess Laboratoire des Signaux et Systèmes, CNRS-Supélec, Plateau de Moulon, France

H. Frankowska CNRS URA 749, CEREMADE, Université Paris-Dauphine, France

R.W. Freund Bell Laboratories, Lucent Technologies, USA

B.K. Ghosh Department of Systems Science and Mathematics, Washington University, St. Louis, USA

I. Gohberg School of Mathematical Sciences, Tel-Aviv University, Tel-Aiv, Israel

U. Grenander Division of Applied Mathematics, Brown University, USA

U. Helmke Department of Mathematics, University of Würzburg, Germany

D. Hernández-Hernández Department of Mathematics, CINVESTAV-IPN, Mexico

K. Hüper Department of Mathematics, University of Würzburg, Germany

M. Janković Ford Motor Company, Scientific Research Laboratories, USA

Petar V. Kokotović Center for Control Engineering and Computation, Dept. of Electrical and Computer Engineering, University of California, USA

A.B. Kurzhanski Moscow State University, Russia

I. Lasiecka Department of Applied Mathematics, University of Virginia, USA

J. Levine Centre Automatique et Systèmes, École des Mines de Paris, France

A. Lindquist Department of Optimization and Systems Theory, Royal Institute of Technology, Sweden

E.P. Loucks Chiron Corporation, St. Louis, Missouri, USA

N. Lybeck Center for Research in Scientific Computation, North Carolina State University, USA

C.F. Martin Department of Mathematics, Texas Tech University, USA

P. Martin Centre Automatique et Systèmes, École des Mines de Paris, France

S.I. Marcus Electrical Engineering Department and Institute for Systems Research, University of Maryland, USA

M.I. Miller Department of Electrical Engineering, Washington University, St. Louis, USA

F. Ollivier GAGE-CNRS, Centre de Mathématiques, École polytechnique, France

Y.M. Ram Department of Mechanical Engineering, University of Adelaid, Australia

J. Rosenthal Department of Mathematics, University of Notre Dame, Notre Dame, USA

P. Rouchon Centre Automatique et Systémes, École des Mines de Paris, France

L. Schovanec Department of Mathematics, Texas Tech University, USA

R. Sepulchre Center for Systems Engineering and Applied Mechanics, Université Catholique de Louvain, Belgium

A. Srivastava Department of Electrical Engineering, Washington University, St. Louis, USA

H.J. Sussmann Department of Mathematics, Rutgers University, USA

A.R. Teel Electrical Engineering Department, University of Minnesota, USA

X.A. Wang Department of Mathematics, Texas Tech University, USA

G. Weiss Center for Systems and Control Engineering, School of Engineering, University of Exeter, United Kingdom

Jan C. Willems University of Groningen, The Netherlands

K.A. Wise McDonnel Douglas Aerospace, St. Louis, USA

State Space Method for Inverse Spectral Problems

D. Alpay and I. Gohberg

1 Introduction

Let H denote a differential operator of the form

$$(Hf)(t) = -iJ\frac{\mathrm{d}\,f}{\mathrm{d}\,t}(t) - V(t)f(t), \quad t \geq 0, \tag{1.1}$$

where

$$J = \begin{pmatrix} I_m & 0 \\ 0 & -I_m \end{pmatrix} \quad \text{and} \quad V(t) = \begin{pmatrix} 0 & k(t) \\ k(t)^* & 0 \end{pmatrix}. \tag{1.2}$$

Here, $k(t)$ is a $\mathbb{C}^{m\times m}$–valued function with entries in $L_1(0,\infty)$. It is sometimes called the potential of the differential operator, or the local reflexivity coefficient function (see [10] for this latter interpretation). Associated to the operator H are two important functions: the scattering function and the spectral function.

To define the scattering function, consider for real λ the $\mathbb{C}^{2m\times m}$–valued solution of the equation

$$-iJ\frac{\mathrm{d}}{\mathrm{d}\,t}X(t,\lambda) - V(t)X(t,\lambda) = \lambda X(t,\lambda), \tag{1.3}$$

subject to the boundary conditions

$$(I_m \ \ -I_m)X(0,\lambda) = 0, \qquad (I_m \ 0)X(t,\lambda) = e^{-i\lambda t}I_m + o(1) \ \ (t \to \infty).$$

Such a solution exists and is unique (see [20], [11]). It has the further property that there exists a $\mathbb{C}^{m\times m}$ matrix $S(\lambda)$ such that

$$(0 \ I_m)X(t,\lambda) = S(\lambda)e^{i\lambda t} + o(1) \ \ (t \to \infty).$$

The function $\lambda \mapsto S(\lambda)$ is called the scattering matrix function and it belongs to the Wiener algebra $\mathcal{W}^{m\times m}$. Recall that this algebra $\mathcal{W}^{m\times m}$ consists of the matrix–valued functions of the form

$$Z(\lambda) = D - \int_{-\infty}^{\infty} z(t)e^{i\lambda t}dt \tag{1.4}$$

where $D \in \mathbb{C}^{m\times m}$ and $z \in L_1^{m\times m}(\mathbb{R})$. Note that $D = \lim_{\lambda\to\pm\infty} Z(\lambda)$; we will use the notation $D = Z(\infty)$.

The scattering function S has the following properties: it takes unitary values, belongs to $\mathcal{W}^{m\times m}$, $S(\infty) = I_m$ and it admits a Wiener–Hopf factorization:

$$S(\lambda) = S_-(\lambda)S_+(\lambda), \tag{1.5}$$

where S_- and its inverse are in $\mathcal{W}_-^{m\times m}$, and S_+ and its inverse are in $\mathcal{W}_+^{m\times m}$. Here the subalgebra $\mathcal{W}_-^{m\times m}$ consists of the elements of the form (1.4) for which the support of $z(t)$ is in $\mathbb{R}_-$, and $\mathcal{W}_+^{m\times m}$ consists of the elements of the form (1.4) for which the support of $z(t)$ is in $\mathbb{R}_+$.

The inverse scattering problem consists in recovering the function k (and hence the potential) from the scattering function S. There is a rich literature about this problem. We follow the approach suggested in [18], [19], [20].

We now turn to the spectral function. The operator H defined by (1.1) is selfadjoint when restricted to the space D_H of $\mathbb{C}^{2m}$–valued functions f which are absolutely continuous and which satisfy the initial value $(I_m - I_m)f(0) = 0$. Let W be a $\mathbb{C}^{m\times m}$–valued function which is continuous on the real line and for which $W(\lambda) > 0$ for all real λ. It is called a spectral function for the operator H if there is a unitary mapping $U : L_2^{2m}(0,\infty) \to L_2^m(W)$ such that $(UHf)(\lambda) = \lambda(Uf)(\lambda)$ for $f \in D_H$, where $L_2^m(W)$ is the Hilbert space of $\mathbb{C}^m$–valued measurable functions g such that $\int_{-\infty}^{\infty} g(t)^*W(t)g(t)dt < \infty$. If S given by (1.5) is the scattering function of the operator (1.1), then the function

$$W(\lambda) = S_-(\lambda)^{-1}S_-(\lambda)^{-*} \tag{1.6}$$

is a spectral function of H, and the map U is given in terms of the continuous orthogonal polynomials of M.G. Krein (see [17], [11]). The definitions of these functions and of the map U are given in the next section. We will call this function the spectral function of the operator H; it is uniquely determined from the scattering function S and the condition $W(\infty) = I_m$. Let $W \in \mathcal{W}^{m\times m}$, with $W(\infty) = I_m$. The function W admits Wiener–Hopf factorizations $W = W_+W_+^* = W_-W_-^*$, where W_- and its inverse are in $\mathcal{W}_+^{m\times m}$ and W_- and its inverse are in $\mathcal{W}_-^{m\times m}$. The function W is the spectral function of the differential operator (1.1) with scattering matrix–function $S = W_-^{-1}W_+$. The inverse spectral problem consists of recovering the function k from the spectral function W.

We will also consider the case where the reflection coefficient matrix function $R(\lambda)$ is known and rational. We recall that $R(\lambda) = X_{21}\ (0,\lambda)X_{11}\ (0,\lambda)^{-1}$ where $X = (X_{ij})$ is the (unique) $\mathbb{C}^{2m\times 2m}$ solution of equation (1.3) subject to the asymptotic property

$$X(t,\lambda) = \begin{pmatrix} e^{-i\lambda t}I_m & 0 \\ 0 & -e^{i\lambda t}I_m \end{pmatrix} + o(1) \quad (t \to \infty).$$

In this paper we present explicit formulas for k when the spectral matrix function (or equivalently the scattering matrix function or the reflection coefficient function) is rational, and give applications of these formulas to the equivalence between Kreĭn's and Marchenko's approach to inverse problems in the rational case. This leads us to a new relationship between the coefficient matrix functions of the Carathéodory–Toeplitz and Nehari extension problems. We also discuss a solution of the direct scattering problem in the rational case which turns out to be related to a problem of partial realization considered in [14]. In general, the results of this paper are obtained by a method which is based on the state space method from system theory. The main results with complete proofs can be found in the papers [2], [1], [3], [5], [4] and [6]. A topic not discussed here is the discrete case (see [1]).

Some words on notation: we denote by $\mathbb{C}^{m\times n}$ the space of m–rows and n–columns matrices with complex entries, and $\mathbb{C}^m$ is short for $\mathbb{C}^{m\times 1}$; the identity matrix of $\mathbb{C}^{m\times m}$ is denoted by I_m, or simply by I. The adjoint of a matrix A is denoted by A^*.

2 The Approaches of Kreĭn and Marchenko

The approach of M.G. Kreĭn's to the inverse spectral problem is as follows: let

$$W(\lambda) = I_m - \int_{-\infty}^{\infty} h(u)e^{i\lambda u}du$$

with $h \in L_1^{m\times m}(\mathbb{R})$ be the spectral function. Since $W(\lambda) > 0$ for all real λ, the integral equation

$$\gamma_\tau(t,s) - \int_0^\tau h(y-u)\gamma_\tau(u,s)du = h(t-s), \qquad t,s \in [0,\tau] \tag{2.1}$$

has a unique solution $\gamma_\tau(t,s)$ for every $\tau > 0$. Then, the potential $k(t)$ is given by the formula

$$k(t) = -2i\gamma_{2t}(0,2t). \tag{2.2}$$

Let

$$P(t,\lambda) = e^{i\lambda t}\left(I_m + \int_0^{2t} \gamma_{2t}(u,0)e^{-i\lambda u}du\right) \tag{2.3}$$

and

$$R(t,\lambda) = e^{i\lambda t}\left(I_m + \int_0^{2t} \gamma_{2t}(2t-u,2t)e^{-i\lambda u}du\right). \tag{2.4}$$

The unitary map U between the spaces $L_2^{2m}[0,\infty)$ and $L_2^m(W)$ alluded to in the introduction is given by

$$(Uf)(\lambda) = \sqrt{2\pi}\int_0^\infty (P(t,-\lambda) \quad R(t,\lambda))f(t)dt. \tag{2.5}$$

Marchenko's approach is concerned with inverse scattering. Let

$$S(\lambda) = I_n - \int_{-\infty}^{\infty} \sigma(u)e^{-i\lambda u}du \quad \lambda \in \mathbb{R}$$

be the scattering matrix function where $\sigma \in L_1^{m\times m}(\mathbb{R})$, and set

$$\xi(u) = \begin{pmatrix} 0 & \sigma(u)^* \\ \sigma(u) & 0 \end{pmatrix}. \tag{2.6}$$

Marchenko's approach consists in solving the equation

$$M(t,s) - \xi(t+s) - \int_t^\infty M(t,u)\xi(u+s)du = 0 \tag{2.7}$$

for $0 \le t \le s < \infty$ (see [20, equation (1.10)]) with the unknown matrix $M(t,s) = (m_{ij}(t,s))_{i,j=1,2}$ (where the block m_{ij} are $\mathbb{C}^{n\times n}$–valued).

The potential is then given by $k(t) = -2im_{21}(t,t)$.

3 Review of the State Space Technique

We recall a number of facts from the theory of realization of matrix–valued rational functions. Any $\mathbb{C}^{m\times m}$–valued rational function W, analytic on the real line and at infinity with $W(\infty) = I_m$, can be written as

$$W(\lambda) = I_m + C(\lambda I_n - A)^{-1}B, \tag{3.1}$$

where $A \in \mathbb{C}^{n\times n}$, $B \in \mathbb{C}^{n\times m}$ and $C \in \mathbb{C}^{m\times n}$.

Such an expression (3.1) is called a realization of W. The realization is called minimal if the number n in (3.1) is as small as possible and the minimal such n is called the McMillan degree of W. Two minimal realizations of W are similar: namely, if $W(\lambda) = I_m + C_i(\lambda I_n - A_i)^{-1}B_i$, $i = 1,2$ are two minimal realizations of W, there exists a (uniquely defined and invertible) matrix $S \in \mathbb{C}^{n\times n}$ such that

$$A_2 = SA_1S^{-1} \quad B_2 = SB_1 \quad C_2 = C_1S^{-1}. \tag{3.2}$$

For these facts and more information on the theory of realization of matrix–valued functions, we refer to [8] and [21].

If W is a $\mathbb{C}^{m\times m}$–valued function analytic on the real line and at infinity with $Z(\infty) = I_m$, with minimal realization (3.1), it is of the form (1.4) with

$$z(u) = \begin{cases} iCe^{-iuA}(I_m - P)B & u > 0 \\ \\ -iCe^{-iuA}PB & u < 0 \end{cases} \tag{3.3}$$

where P is the Riesz projection corresponding to the eigenvalues of A in $\mathbb{C}_+$. The function z has absolutely summable entries and thus Z is in the Wiener algebra and it is therefore meaningful to consider the case of rational scattering and spectral functions.

A factorization $R = R_-R_+$ of R into two $\mathbb{C}^{n\times n}$–valued functions analytic at infinity is called a (right) canonical (Wiener-Hopf or spectral) factorization if R_- and its inverse are analytic in the closed lower half plane and R_+ and its inverse are analytic in the closed upper half plane. Similarly, the factorization $R = R_+R_-$ with R_+ and R_- as above is called left a spectral factorization. Wiener Hopf factorizations need not exist, and we refer to [8] for a complete discussion.

Theorem 3.1 [*8, Section* 4.5], [*12, Section* 13.6]. *Let R be a $\mathbb{C}^{n\times n}$-valued rational function analytic at infinity with $R(\infty) = I_n$ and let $R(\lambda) = I_n + C(\lambda I_m - A)^{-1}B$ be a minimal realization of R. Assume that A has no real eigenvalues. Then R admits a right canonical Wiener–Hopf factorization relative to the real line if and only if the following two conditions are fulfilled:*

(*i*) *$A^\times = A - BC$ has no real eigenvalues.*

(*ii*) *$\mathbb{C}^m = M \oplus M^\times$*

where M (resp. $M^\times$) is the space spanned by the eigenvectors and generalized eigenvectors corresponding to the eigenvalues of A (resp. $A^\times$) in the upper (resp. lower) half plane. Furthermore, in that case, R admits a canonical factorization $R(\lambda) = R_-(\lambda)R_+(\lambda)$ with $R_-(\lambda) = I_n + C(\lambda I_m - A)^{-1}(I - \pi)B$ and $R_+(\lambda) = I_n + C\pi(\lambda I_m - A)^{-1}B$. Then

$$\begin{aligned} R_-(\lambda)^{-1} &= I_n - C(I-\pi)(\lambda I_m - A^\times)^{-1}B \\ R_+(\lambda)^{-1} &= I_n - C(\lambda I_m - A^\times)^{-1}\pi B \end{aligned}$$

where π is the projection of $\mathbb{C}^m$ along M onto $M^\times$.

See [16, p. 2] for this theorem and further discussion. The selection of papers [16] also contains a paper of Ball and Ran where the following problem is considered: how to compute a right spectral factorization when a left spectral factorization is given. The result is:

Theorem 3.2 [*7, Theorem* 2.1 *p.* 13]. *Suppose that the rational $\mathbb{C}^{n\times n}$-valued function W admits a left canonical factorization $W(\lambda) = Y_+(\lambda)Y_-(\lambda)$ where*

$$Y_+(\lambda) = I_n + C_+(\lambda I_{m_+} - A_+)^{-1}B_+,$$

and

$$Y_-(\lambda) = I_n + C_-(\lambda I_{m_-} - A_-)^{-1}B_-.$$

We assume that both A_- and $A_-^\times = A_- - B_-C_-$ have their spectra in the open upper half-plane and A_+ and $A_+^\times = A_+ - B_+C_+$ have their spectra in the open lower half-plane. Let P and Q denote the unique solutions of the Lyapunov equations

$$A_-^\times P - PA_+^\times = B_-C_-$$

$$A_+Q - QA_- = -B_+C_+.$$

Then, W admits a right canonical factorization if and only if $I - QP$ is invertible. When this is the case the factors W_- and W_+ for the right factorization $W(\lambda) = W_-(\lambda)W_+(\lambda)$ are given by

$$W_-(\lambda) = I_n + (C_+Q + C_-)(\lambda I_{m_-} - A_-)^{-1}(I_{m_-} - PQ)^{-1}(-PB_+ + B_-)$$

and

$$W_+(\lambda) = I_n + (C_+ + C_-P)(I_{m_+} - QP)^{-1}(\lambda I_{m_+} - A_+)^{-1}(B_+ - QB_-).$$

Using these two theorems we obtain:

Theorem 3.3 *(see [3]). Let S be the scattering function of a differential operator* (1.1) *and let $S = S_-S_+$ be its right Wiener-Hopf factorization. Let $S_-(\lambda) = I_n + c(\lambda I_p - a)^{-1}b$ be a minimal realization of the spectral factor S_-. Then*

$$S_+(\lambda) = I_n - (ic\Omega - b^*)(\lambda - a^{\times *})^{-1}(I_p + Y\Omega)^{-1}(c^* + iYb) \tag{3.4}$$

is a minimal realization of the spectral factor S_+. In this expression, Ω and Y are the solutions of the Lyapunov equations

$$i(\Omega a^{\times *} - a^{\times}\Omega) = bb^* \tag{3.5}$$

and

$$i(Ya - a^*Y) = -c^*c. \tag{3.6}$$

A minimal realization of S is given by $S(\lambda) = I_n + C(\lambda I_{2p} - A)^{-1}B$, where

$$A = \begin{pmatrix} a & -b(ic\Omega - b^*) \\ 0 & a^{\times *} \end{pmatrix}, \quad B = \begin{pmatrix} b \\ (I_p + Y\Omega)^{-1}(c^* + iYb) \end{pmatrix} \tag{3.7}$$

$$C = (c \quad ic\Omega - b^*). \tag{3.8}$$

The associated hermitian matrix is equal to

$$H = \begin{pmatrix} -\Omega & iI_p \\ -iI_p & -Y(I_p + \Omega Y)^{-1} \end{pmatrix}. \tag{3.9}$$

A rational matrix valued function is the scattering function of a differential operator of the form (1.1) *if and only if it has a minimal realization similar to a realization of the form* (3.7)–(3.8).

4 The Main Theorems

In this section we present the main results, namely explicit formulas for the function $k(t)$ in terms of various realizations of the spectral matrix function $S(\lambda)$, of the matrix spectral function $W(\lambda)$ or of the reflection coefficient matrix function $R(\lambda)$. We also present here the main ideas of the proofs.

The first result is the formula for the potential when the spectral function is given in realized form. It appears in [2].

Theorem 4.1 *(see [6]) Let W be a $\mathbb{C}^{n\times n}$–valued rational function analytic at infinity and on the real line and assume that $W(\infty) = I_n$ and $W(\lambda) > 0$ for all $\lambda \in \mathbb{R}$. Then W is the spectral function of a differential operator of the form (1.1). Let $W(\lambda) = I_n + C(\lambda I_m - A)^{-1}B$ be a minimal realization of W. The function k is given by the formula*

$$k(t) = 2C(Pe^{-2itA^{\times}}|_{\text{Im } P})^{-1}PB. \tag{4.1}$$

In this expression, $A^{\times} = A - BC$ and P is the Riesz projection corresponding to the eigenvalues of A in the open upper half–plane.

The proof is based on the work [9], where an explicit formula for the solution of equation (2.1) is given when W is rational.

The second result is a formula for the potential when the scattering function is given in realized form. It is proved in [3].

Theorem 4.2 *Let S be a $\mathbb{C}^{n\times n}$–valued rational function analytic at infinity with $S(\infty) = I_n$. Assume that S takes unitary values on the real line and admits a Wiener–Hopf factorization. Then S is the scattering function of a differential operator of the form (1.1). Let $S(\lambda) = I_n + C(\lambda I_m - A)^{-1}B$ be a minimal realization of S. Then the potential is equal to*

$$V(t) = \begin{pmatrix} 0 & k(t) \\ k(t)^* & 0 \end{pmatrix}$$

where

$$k(t) = -2CP\left(\left(Pe^{-2itA}P - PX_2P^*e^{-2itA^*}P^*X_1P\right)|_{\text{Im } P}\right)^{-1}PB. \tag{4.2}$$

In this expression, P denotes the Riesz projection corresponding to the eigenvalues of A in the open upper half plane $\mathbb{C}_+$, and X_1 and X_2 are such that

$$i((P^*X_1P)(AP) - (AP)^*(P^*X_1P)) = -P^*C^*CP \tag{4.3}$$

and

$$i((AP)(PX_2P^*) - (PX_2P^*)(AP))^* = -PBB^*P^*. \tag{4.4}$$

Furthermore, the asymptotic equality holds

$$k(t) = -2CPe^{-2itA}PB(I + o(e^{-(\alpha+\epsilon)t})) \quad t \to +\infty \tag{4.5}$$

where ε is any strictly positive number and

$$\alpha = 4\inf\{\text{Im } \lambda; \quad \lambda \in \sigma(A) \cap \mathbb{C}_+\}.$$

The idea of the proof is as follows: when S is rational, the function $\xi(u)$ in equation (2.7) can be written as

$$\xi(u) = \begin{pmatrix} 0 & iB^*P^*e^{-iuA^*}P^*C^* \\ -iCe^{iuA}PB & 0 \end{pmatrix} = Fe^{uT}G$$

where

$$F = i\begin{pmatrix} 0 & B^*P^* \\ -CP & 0 \end{pmatrix}, \tag{4.6}$$

$$T = \begin{pmatrix} iAP & 0 \\ 0 & -iA^*P^* \end{pmatrix}, \tag{4.7}$$

and

$$G = \begin{pmatrix} PB & 0 \\ 0 & P^*C^* \end{pmatrix}. \tag{4.8}$$

Thus the kernel $\xi(u+s) = (Fe^{uT})(e^{sT}G)$ is separable, and this fact allows us to obtain the explicit solution

$$M(t,s) = Fe^{tT}(I - e^{tT}Ze^{tT})^{-1}e^{sT}G. \tag{4.9}$$

where Z is the solution of

$$-GF = TZ + ZT. \tag{4.10}$$

For the next formula, we start from the factor S_- of a right spectral factorization $S = S_-S_+$ of the scattering function and take $S_-(\lambda) = I_n + c(\lambda I_p - a)^{-1}b$ a minimal realization of S_-. Then:

Theorem 4.3 ([2], [3]) *Let S be a scattering rational function and let*

$$S_-(\lambda) = I_n + c(\lambda I_p - a)^{-1}b$$

be a minimal realization of its left factor S_- in the Wiener Hopf factorization $S = S_-S_+$. Then

$$k(t) = -2ce^{ita}(I + \Omega(Y - e^{-2ita^*}Ye^{2ita}))^{-1}(b + i\Omega c^*). \tag{4.11}$$

In this expression, Ω and Y are the solutions of the Lyapunov equations (3.5) *and* (3.6).

We prove formula (4.11) in two different ways; first from the minimal realization $S_-(\lambda) = I_n + c(\lambda I_p - a)^{-1}b$ of S_-, we obtain the minimal realization $W(\lambda) = I + C(\lambda I - A)^{-1}B$ of $W = S_-S_-^{-*}$ with

$$A = \begin{pmatrix} a - bc & -bb^* \\ 0 & (a-bc)^* \end{pmatrix}, \quad B = \begin{pmatrix} b \\ c^* \end{pmatrix}, \quad C = (-c \;\; -b^*), \tag{4.12}$$

and we apply formula (4.1). The second method is as follows: starting from a minimal realization of S_- we obtain a minimal realization of S by Theorem 3.3. Finally, we use this realization in formula (4.5).

We get to the same formula (4.11) using both methods, thus showing directly, in the rational case, the equivalence between Kreĭn's and Marchenko's approaches to inverse problems.

In the next theorem, we compute the potential (now called the local reflexivity coefficient function) when the reflection coefficient function is rational.

Theorem 4.4 *Let R be the reflection coefficient function of the differential expression (1.2) and assume that R is rational. Let $R(\lambda) = -C(\lambda I_n - A)^{-1}B$ be a minimal realization of R. Then*

$$W(\lambda) = I_m + \mathbf{C}(\lambda I - \mathbf{A})^{-1}\mathbf{B}$$

is a realization of the spectral function W, and the local reflexivity coefficient function is given by

$$k(t) = 2\mathbf{C}(\mathbf{P}e^{-2it\mathbf{A}^\times}|\mathrm{Im}\ \mathbf{P})^{-1}\mathbf{PB}.$$

In these expressions

$$\mathbf{A} = \begin{pmatrix} A-BC & -BC & 0 & -BB^* \\ 0 & A & BB^* & 0 \\ 0 & 0 & A^* & -C^*B^* \\ 0 & 0 & 0 & (A-BC)^* \end{pmatrix}, \qquad \mathbf{B} = \begin{pmatrix} B \\ 0 \\ C^* \\ C^* \end{pmatrix}$$

and

$$\mathbf{C} = -(C\ \ C\ \ 0\ \ B^*).$$

The operators $\mathbf{A}$ and $\mathbf{A}^\times$ have no real spectrum and $\mathbf{P}$ denotes the Riesz projection corresponding to the eigenvalues of $\mathbf{A}$ in $\mathbb{C}_+$.

5 Relationship between the Carathéodory–Toeplitz and Nehari Extension Problems

Theorem 4.3 leads us to the following connection between two different extension problems, namely the Carathéodory–Toeplitz extension problem and the Nehari extension problem. We first recall the relevant definitions.

The Carathéodory–Toeplitz extension problem. *Given $\tau > 0$ and given $h \in L_1^{m\times m}[-2\tau, 2\tau]$ satisfying $h(t) = h(-t)^*$ for $|t| \leq 2\tau$, find all extensions of h in $L_1^{m\times m}(\mathbb{R})$ such that for all $\lambda \in \mathbb{R}$*

$$I_m - \hat{h}(\lambda) > 0. \tag{5.1}$$

For this problem the following is known (see [13]).

Theorem 5.1 *Given $\tau > 0$ and given $h \in L_1^{m\times m}[-2\tau, 2\tau]$ satisfying $h(t) = h(-t)^*$. The Carathéodory–Toeplitz problem has a solution if and only if the operator $I - H_\tau$ from $L_2^m[0, 2\tau]$ into itself which to f associates the function*

$$f(t) - \int_0^{2\tau} h(t-s)f(s)ds \tag{5.2}$$

is strictly positive. Let this condition hold. Then, h is a solution of the Carathéodory–Toeplitz interpolation problem if and only if $Z(\lambda) = I_m - 2\int_0^\infty h(s)e^{is\lambda}ds$ has the form

$$Z = (E_{2\tau}^\circ - F_{2\tau}^\circ R)(E_{2\tau} + F_{2\tau}R)^{-1} \tag{5.3}$$

where R is any matrix function from $\mathcal{W}_+^{m\times m}$ with $R(\infty) = 0$ which takes strictly contractive values on the real line and the various coefficients of the linear fractional transformation are computed as follows. Let x and y denote the unique solutions of the integral equations

$$x(t) - \int_0^{2\tau} h(t-s)x(s)ds = h(t), \qquad 0 \le t \le 2\tau \tag{5.4}$$

$$y(t) - \int_{-2\tau}^{0} h(t-s)y(s)ds = h(t), \qquad -2\tau \le t \le 0. \tag{5.5}$$

Let $c \in \mathcal{W}_+^{m\times m}$ be such that

$$c(\lambda) + c(\lambda)^* = (I_m + \hat{x}(\lambda))^{-*}(I_m + \hat{x}(\lambda))^{-1} \tag{5.6}$$

and $2c(\infty) = I_m$. Then,

$$E_{2\tau}(\lambda) = I_m + \hat{x}(\lambda), \qquad F_{2\tau}(\lambda) = e^{2i\lambda\tau}(I_m + \hat{y}(\lambda)) \tag{5.7}$$

and

$$E_{2\tau}^\circ(\lambda) = c(\lambda)(I_m + \hat{x}(\lambda)) \qquad F_{2\tau}^\circ(\lambda) = e^{2i\lambda\tau}c^\sharp(\lambda)(I_m + \hat{y}(\lambda)) \tag{5.8}$$

(where $c^\sharp(\lambda) = c(\lambda^*)^*$).

The matrix–valued function

$$\Theta(\tau, \lambda) = \frac{1}{\sqrt{2}}\begin{pmatrix} F_\tau & E_\tau \\ F_\tau^\circ & -E_\tau^\circ \end{pmatrix} \tag{5.9}$$

will be called the coefficient matrix function of the Carathéodory–Toeplitz extension problem.

The Nehari extension problem. *Given $\tau \ge 0$ and $\sigma \in L_1^{m\times m}[2\tau, \infty)$, find all summable extensions of σ to $\mathbb{R}$ and all $M \in \mathbb{C}^{m\times m}$ such that*

$$S(\lambda) = M - \int_{\mathbb{R}} e^{-i\lambda u}\sigma(u)du \tag{5.10}$$

is a contraction for all $\lambda \in \mathbb{R}$.

Theorem 5.2 *(see [15]) Let $\tau \geq 0$ and $\sigma_\tau \in L_1^{m\times m}[2\tau,\infty)$. In order that there exists an extension of σ to $\mathbb{R}$ such that*

$$\| \int_{\mathbb{R}} e^{-i\lambda u}\sigma(u)du\| < 1$$

for all $\lambda \in \mathbb{R}$, it is necessary and sufficient that the Hankel operator from $L_2[\tau,\infty)$ into itself defined by

$$\mathbf{H}f(t) = \int_\tau^\infty \sigma(u+s)f(s)ds \tag{5.11}$$

has norm strictly less than 1. The solutions of the Nehari problem are then described by the formula

$$S = (H_{21} + H_{22}R)(H_{11} + H_{12}R)^{-1} \tag{5.12}$$

where R is in $\mathcal{W}_+^{m\times m}$ and takes strictly contractive values on the real line and $H = (H_{ij})$ is the decomposition into four $\mathbb{C}^{m\times m}$–valued block of the function H

$$H(\tau,\lambda) = \begin{pmatrix} e^{i\lambda\tau} & 0 \\ 0 & e^{-i\lambda\tau} \end{pmatrix} + \int_\tau^\infty \begin{pmatrix} e^{i\lambda u} & 0 \\ 0 & e^{-i\lambda u} \end{pmatrix} K(\tau,u)du. \tag{5.13}$$

Here K is the unique solution of the equation

$$K(\tau,s) - \int_\tau^\infty \xi(u+s)K(\tau,u)du = \xi(\tau+s), \qquad 0 \leq \tau \leq s < \infty \tag{5.14}$$

with ξ defined by (2.6).

We will call H the coefficient matrix associated to the Nehari problem. The main result of this section is the next theorem.

Theorem 5.3 *(see [5]) Let $W = I_m - \hat{h}$ be in the Wiener algebra $\mathcal{W}^{m\times m}$ and suppose that $W(\lambda) > 0$ for all $\lambda \in \mathbb{R}$. Let $W = S_-^{-1}S_-^{-*} = S_+S_+^*$ be its right and left Wiener–Hopf factorization (with $s_-(\infty) = s_+(\infty) = I_m$) and let*

$$S(\lambda) = S_-(\lambda)S_+(\lambda) = I_m - \int_{\mathbb{R}} e^{-iu\lambda}\sigma(u)du, \tag{5.15}$$

where $\sigma \in L_1^{m\times m}(\mathbb{R})$. Then, the Hankel operator defined by (5.11) *has norm strictly less than 1. For $\tau \geq 0$ let $\Theta(\tau,\lambda)$ and $H(\tau,\lambda)$ be the coefficient matrix functions associated respectively to the Carathéodory–Toeplitz problem associated to h and to the Nehari problem associated to σ. Then,*

$$H(\tau,\lambda) = H(0,\lambda)JM\Theta(2\tau,\lambda)Je^{-i\lambda\tau}. \tag{5.16}$$

where

$$M = \frac{1}{\sqrt{2}}\begin{pmatrix} I_m & I_m \\ I_m & -I_m \end{pmatrix}.$$

We prove this theorem first for the case where W is rational using the explicit expressions for the solutions of the equations (2.1) and (5.14) and the equivalence between Kreĭn's and Marchenko's approaches. We show that both the functions H and Θ are solutions of the same canonical differential equation of the form (1.3). The general case is done by a limit argument.

6 The Direct Scattering Problem

Formula (4.1) leads to the following method of computing the rational matrix spectral function if the potential is known. To present the method it is convenient to express some formulas in terms of the function $\kappa(t)$ introduced by

$$\kappa(t) = C\left(Pe^{tA^{\times}}|_{\mathrm{Im}\ P}\right)^{-1} PB. \tag{6.1}$$

Then, $k(t) = 2\kappa(-2it)$ and for $\ell \geq 0$,

$$k^{(\ell)}(0) = 2(-2i)^{\ell}\kappa^{(\ell)}(0). \tag{6.2}$$

Let $X(t) = Pe^{tA^{\times}}|_{\mathrm{Im}\ P}$. Then, $X'(t) = PA^{\times}e^{tA^{\times}}|_{\mathrm{Im}\ P}$. Furthermore,

$$(X^{-1})'(t) = -X^{-1}(t)X'(t)X^{-1}(t).$$

Using these formulas it can be computed that

$$\begin{aligned}
\kappa(0) &= CPB, \\
\kappa^{(1)}(0) &= -CPAB + (CPB)^2, \\
\kappa^{(2)}(0) &= CPA^2B - (CPAB)(CPB) - (CBP)(CPAB) + \\
&\quad +2(CPB)^3 - (CPB)(CB)(CPB), \\
\kappa^{(3)}(0) &= -CPA^3B + (CPA^2B)(CBP) + (CPB)(CPA^2B) \\
&\quad -(CPB)(CAB)(CPB) + 6(CPB)^4 + (CPAB)^2 \\
&\quad -3(CPAB)(CPB)^2 - 3(CPB)^2(CPAB) \\
&\quad +2(CPB)(CB)(CPAB) + 2(CPAB)(CB)(CPB) \\
&\quad +(CPB)(CB)^2(CPB).
\end{aligned}$$

The realization (4.12) (and in fact any minimal realization of W) satisfies the condition

$$CA^{\ell}B = CA^{\ell}PB + (CA^{\ell}PB)^*. \tag{6.3}$$

Using (6.3) we can invert the previous equations and obtain the following equalities (in fact, we use (6.3) only starting with the equality for CA^2PB):

$$CPB = \kappa(0),$$

$$\begin{aligned}
CAPB &= -\kappa'(0) + (CPB)^2 \\
&= -\kappa'(0) + \kappa(0)^2, \\
CA^2PB &= \kappa''(0) + (CPAB)(CPB) + (CPB)(CPAB) - 2(CPB)^3 \\
&\quad -(CPB)(CB)(CPB) \\
&= \kappa''(0) + \kappa(0)^3 - \kappa(0)\kappa^{(1)}(0) \\
&\quad -\kappa^{(1)}(0)\kappa(0) + \kappa(0)\kappa(0)^*\kappa(0),
\end{aligned}$$

and similarly for CA^3PB. In fact, these computations can be extended and for every $\ell \geq 1$, the matrix $CA^\ell PB$ (and hence using (6.3) the matrix $CA^\ell B$) is a noncommutative polynomial function of the matrices

$$\kappa(0), \ldots, \kappa^{(\ell)}(0)$$

and of their conjugates. This remark is used in order to solve the direct spectral problem and to calculate the spectral function from the values of the reflection coefficient function $k(t)$ and of a number of its derivatives at the origin, in the case when it is known that the spectral function is rational. This is based on the results of partial realization from system theory, which allows us to reconstruct C, A and B and hence the function $I_m + C(\lambda I - A)^{-1}B$ from a finite number of matrices $CA^\ell B$ and especially on the results from [14].

Theorem 6.1 *Let $W(\lambda) = I_m + C(\lambda I_n - A)^{-1}B$ be a minimal realization of the spectral function W. Let k be given by (4.1). Then, $k(0) = 2CPB$ and for every $\ell \geq 0$, there exists a noncommutative polynomial $\mathbf{p}_\ell$ in the 2ℓ variables $k(0), k(0)^*, \ldots, k^{(\ell-1)}(0), k^{(\ell-1)}(0)^*$ (with coefficients independent of A, B, C) such that*

$$CA^\ell PB = (-1)^\ell c_\ell k^{(\ell)}(0) + \mathbf{p}_\ell(k(0), k(0)^*, \ldots, k^{(\ell-1)}(0), k^{(\ell-1)}(0)^*) \quad (6.4)$$

where $c_\ell = -(2(-2i)^\ell)^{-1}$.

The equations (6.4) can be inverted, and for every $\ell \geq 1$, there exists a noncommutative polynomial $\mathbf{q}_\ell$ in the 2ℓ variables CA^jPB, CA^jB, $j = 0, \ldots, \ell - 1$ such that

$$k^{(\ell)}(0) = (-1)^\ell c_\ell{}^{-1} CA^\ell PB + \mathbf{q}_\ell(CB, (CPB), \cdots, CA^{\ell-1}B, CPA^{\ell-1}B).$$

In the next theorem we show how the potential can be constructed from the first coefficients of the Laurent expansion of the spectral function at infinity.

Theorem 6.2 *Let W be the spectral function of a differential operator of the form (1.1), with $k(t) \in L_1^{m\times m}(0, \infty)$. Assume that W is rational and analytic at infinity, with $W(\infty) = I_m$, and that the McMillan degree of W is n. Then, the potential can be expressed from the $2n$ matrices $k(0), \ldots, k^{(2n-1)}(0)$ and their adjoints as follows:*

1. *Set* $M_0 = \frac{k(0)}{2}$ *and compute the* $2n-1$ *matrices*

$$M_j = c_j k^{(j)}(0) + \mathbf{p}_j(k(0), k(0)^*, \ldots) \quad j = 1, \ldots, 2n-1, \tag{6.5}$$

where the constants c_j *and the polynomials* $\mathbf{p}_j$ *are defined in Theorem* 6.1, *and set* $N_j = M_j + M_j^*$. *The matrix* N_j *is the* $(j+1)$*–th coefficient of the Laurent expansion of the spectral function* W *at infinity.*

2. *Using the reduction procedure of [14] compute a minimal triple*

$$(A, B, C) \in \mathbb{C}^{n\times n} \times \mathbb{C}^{n\times m} \times \mathbb{C}^{m\times n}$$

such that:

$$N_j = CA^jB, \qquad j = 0, \ldots, 2n-1. \tag{6.6}$$

3. *Compute the Riesz projection corresponding to the spectrum of* A *in* $\mathbb{C}_+$, *and use the obtained formulas for* A, B, C, P *in formula* (4.1).

Functions arising as potentials of differential operators of the form (1.1) with rational spectral functions are characterized in the next theorem:

Theorem 6.3 *Let* $k(t)$ *be a* $\mathbb{C}^{m\times m}$*–valued function analytic in a neighborhood of the origin and define matrices* M_j, $j = 0, 1, \ldots$ *by* $M_0 = \frac{1}{2}k(0)$ *and by* (6.5) *for* $j \geq 1$. *Then, a necessary condition for* k *to be the potential associated to a differential equation of the form* (1.1) *with a rational weight is that there exists an integer* ℓ_0 *such that, for* $\ell \geq \ell_0$,

$$\text{Rank } (M_{i+j})_{i,j=0}^{\ell} = p. \tag{6.7}$$

When condition (6.7) *is in force, the function*

$$Z(\lambda) = \frac{I_m}{2} + \sum_0^{\infty} \frac{M_j}{\lambda^{j+1}} \tag{6.8}$$

is rational and analytic at infinity, of McMillan degree p. *Suppose that* Z *is analytic in* $\mathbb{C}_+ \cup \mathbb{R}$ *and that*

$$W(\lambda) = 2\text{Re } Z(\lambda) > 0 \tag{6.9}$$

for all real λ. *Then,* k *is the potential of the differential equation with spectral function* W. *The function* W *can be computed from the procedure described in Theorem* 6.2.

References

[1] D. Alpay and I. Gohberg. Inverse spectral problems for difference operators with rational scattering matrix function. *Integral Equations and Operator Theory* **20** (1994), 125-170.

[2] D. Alpay and I. Gohberg. Inverse spectral problem for differential operators with rational scattering matrix functions. *Journal of Differential Equations* **118** (1995), 1-19.

[3] D. Alpay and I. Gohberg. Inverse scattering problem for differential operators with rational scattering matrix functions. To appear *Operator Theory: Advances and Applications.*

[4] D. Alpay and I. Gohberg. Potentials associated to rational weights. To appear *Operator Theory: Advances and Applications.*

[5] D. Alpay and I. Gohberg. A relationship between the Nehari and the Carathéodory-toeplitz extension problem. To appear *Integral Equations and Operator Theory.*

[6] D. Alpay, I. Gohberg, and L. Sakhnovich. Inverse scattering for continuous transmission lines with rational reflection coefficient function. Vol. 87. *Operator Theory: Advances and Applications.* Boston: Birkhäuser, 1996. 1-16.

[7] J. Ball and A. Ran. Left versus right canonical Wiener-Hopf factorization. Vol. 21. *Operator Theory: Advances and Applications.* Boston: Birkhäuser, 1986. 9-38.

[8] H. Bart, I. Gohberg, and M. Kaashoek. Minimal factorization of matrix and operator functions. Vol. 1. *Operator Theory: Advances and Applications.* Boston: Birkhäuser, 1979.

[9] H. Bart, I. Gohberg, and M. Kaashoek. Convolution equations and linear systems. *Integral Equations and Operator Theory* **5** (1982), 283-340.

[10] A. Bruckstein, B. Levy, and T. Kailath. Differential methods in inverse scattering. *SIAM Journal of Applied Mathematics* **45** (1985), 312-335.

[11] H. Dym and A. Iacob. *Positive Definite Extensions, Canonical Equations and Inverse Problems.* Vol. 12. *Operator Theory: Advances and Applications.* Boston: Birkhäuser, 1984. 141-240.

[12] I. Gohberg, S. Goldberg, and M. Kaashoek. *Classes of Linear Operators. Vol. I.* Vol. 49. *Operator Theory: Advances and Applications.* Boston: Birkhäuser, 1990.

[13] I. Gohberg, S. Goldberg, and M. Kaashoek. *Classes of Linear Operators. Vol. II.* Vol. 63. *Operator Theory: Advances and Applications.* Boston: Birkhäuser, 1993.

[14] I. Gohberg, M. Kaashoek, and L. Lerer. On minimality in the partial realization problem. *Systems and Control Letters* **9** (1987), 97-104.

[15] I. Gohberg, M. Kaashoek, and H. Woerderman. The band method for positive and contractive extension problems. *Journal of Operator Theory* **22** (1989), 109-155.

[16] I. Gohberg and M.A Kaashoek, Editors. *Constructive Methods of Wiener-Hopf Factorizations.* Vol. 21. *Operator theory: Advances and Applications.* Boston: Birkhäuser, 1986.

[17] M.G. Kreĭn. Continuous analogues of propositions for polynomials orthogonal on the unit circle. *Dokl. Akad. Nauk. SSSR* **105** (1955), 637-640.

[18] M.G. Kreĭn and F.E. Melik-Adamyan. On the theory of S-matrices of canonical equations with summable potentials. *Dokl. Akad. Nauk. SSSR* **16** (1968), 150-159.

[19] F.E. Melik-Adamyan. Canonical differential operators in Hilbert space. *Izvestya Akademii Nauk. Armyanskoi SSR Matematica* **12** (1977), 10-31.

[20] F.E. Melik-Adamyan. On a class of canonical differential operators. *Izvestya Akademii Nauk. Armyanskoi SSR Matematica* **24** (1989), 570-592. In English: *Soviet Journal of Contemporary Mathematics* **24** (1989), 48-69.

[21] M.W. Wonham. *Linear Multivariable Control: Geometric Approach.* New York: Springer-Verlag, 1979.

Department of Mathematics, Ben-Gurion University of the Negev, PO box 653, 84105 Beer-Sheva, Israel

School of Mathematical Sciences, The Raymond and Beverly Slacker Faculty of Exact Sciences, Tel-Aviv University, Tel-Aiv, Ramat-Avaiv 69989, Israel

New Developments in the Theory of Positive Systems

B.D.O. Anderson[1]

1 Introduction

This paper deals with some special finite-dimensional linear systems problems, broadly speaking ones where the underlying matrices in state-variable descriptions of the systems considered contain nonnegative or positive entries.

The problems tend to be difficult, for a number of reasons. These include the fact that one of the common tools of linear system theory, that of replacing a triple $\{A, b, c\}$ realizing a transfer function matrix $H(z) = c^T(zI - A)^{-1}b$ by $\{TAT^{-1}, Tb, (T^{-1})^T c\}$ for an arbitrary nonsingular T, is in general not available, since such a transformation in general will destroy the nonnegativity.

The paper actually focuses on three problems: the so-called positive linear system realization problem, the problem of exponential forgetting of initial conditions and an associated smoothing issue in hidden Markov models, and the problem of realizing a hidden Markov model, given the collection of probabilities of output strings. The latter problem very much draws together ideas from the first two problems.

We begin however with two motivational sections: one sets out some examples of positive systems and the other records some broad questions associated with such systems. We then briefly review the Perron-Frobenius theory on the eigenstructure of nonnegative matrices before tackling the three problems above. For the latter two problems, a sort of time-varying generalization of the Perron-Frobenius theory is needed; we use the treatment of [1] as a base.

In the final section, we record some open problems.

The following individuals helped greatly in the development of the ideas of this paper: Manfred Deistler, Subhrakanti Dey, Lorenzo Farina, Hajime Maeda, John Moore and Louis Shue.

[1]The author wishes to acknowledge the funding of the activities of the Cooperative Research Centre for Robust and Adaptive Systems by the Australian Commonwealth Government under the Cooperative Research Centres Program.

2 Some Examples of Positive Systems

For convenience, we shall restrict attention to discrete-time systems. In nearly all instances, to a discrete-time result there corresponds a continuous-time result.

2.1 Deterministic Systems

Consider the equation

$$x(k+1) = Ax(k) + bu(k). \tag{2.1}$$

This equation is a discrete-line Leslie population model [2], [3] when each entry of $x(k)$ corresponds to members of a population in a given age cohort at time instant k. In a closed population, $b = 0$ and entries of A are nonnegative (most are in fact zero). If immigration is allowed, then b has nonnegative entries and $u(\cdot)$ is nonnegative. Obviously all entries of $x(\cdot)$ are nonnegative, and the total population is given by

$$y(k) = cx(k), \tag{2.2}$$

where $c = [1 \dots 1]$. Variants of this model can be used to distinguish subsets of the population, e.g. males and females.

A second class of positive deterministic systems is exemplified by compartmental models, [4, 5] where the entries of x corresponds to the quantities of different entities (chemicals, water, heat, telephone calls, etc.)

As a final example, we cite *charge routing networks*, which are a special form of realization of a digital filter using MOS technology, [6]. The entries of A, b, c, x, y and u are necessarily nonnegative.

2.2 Markov Chains and Hidden Markov Models

If $X_k, k = 0, 1, \dots$ is a state moving at discrete time instants between one of a finite number of states $1, 2, \dots, N$ in a Markov manner, it is termed a finite state Markov chain, [7]. If $Pr\,[X_{k+1} = i \mid X_k = j] = a_{ij}$ is independent of k, it is stationary. Let

$$\pi(k) = \text{ vector with i-th entry } Pr[X_k = i]. \tag{2.3}$$

Then

$$\pi(k+1) = A\pi(k). \tag{2.4}$$

Here, entries of A and π are nonnegative.

In a hidden Markov model [8, 9] in addition to the state process, there is an observation or measurement process Y_k, assuming values in the set

$\{1, 2, \ldots, M\}$, with Y_k depending probabilistically on X_k, via an $M \times N$ matrix C with $c_{ij} = Pr[Y_k = i \mid X_k = j]$. Let

$$\sigma(k) = \text{ vector with j-th entry } Pr[Y_k = j]. \tag{2.5}$$

Then

$$\sigma(k) = C\pi(k). \tag{2.6}$$

Now (2.4) and (2.5) together define a positive system.

2.3 Multidimensional Positive Systems

Equations of the form

$$\begin{aligned} x(h+1,k+1) &= Ax(h,k+1) + Bx(h+1,k) \\ &\quad + Cu(h,k+1) + Du(h+1,k) \\ y(h,k) &= Hx(h,k) + Ju(h,k), \end{aligned}$$

in which all matrices and vectors are nonnegative, have been used in the modeling of river pollution, gas absorption, and the diffusion and advection of biological material,[10]. Of course, h, k refer to two spatial variables.

Equations of the form

$$x(h+1,k+1) = Ax(h,k+1) + Bx(h+1,k),$$

where x is a vector of probabilities, and $A = \alpha P, B = (1-\alpha)Q$ with $0 \leq \alpha \leq 1$ and P^T, Q^T stochastic matrices have been used to model two-dimensional Markov chains, [11]. Such chains can be used as signal models in designing image processing algorithms.

3 Some Broad Questions

Based on standard ideas of linear systems, deterministic and stochastic, there are a number of ideas which present themselves in relation to positive systems. Some of these are as follows:

(i) Is there a realization theory, i.e. a way of passing from an external (input/output or transfer function) description to an internal (state variable) description of a deterministic positive system?

(ii) Is there a realization theory for hidden Markov models, i.e. a way of passing from the collection of joint probabilities associated with the output process to a HMM description?

(iii) How may one approximate a high order positive system by a low order positive system?

(iv) Given one internal description of a positive system, can all other internal descriptions be readily found from it? Is it minimal?

(v) Can one build a finite state filter for a hidden Markov model, such that the filter forgets old measurements, and allows expansion to become a fixed-lag smoother?

This paper will summarize answers to some of these questions, principally (i), (ii) and (v).

4 Review of Perron-Frobenius Theory

This section recalls key eigenvalue/eigenvector properties of nonnegative matrices, [12]. Let $A = (a_{ij}) \epsilon R^{N \times N}$ have $a_{ij} \geq 0$ for all i, j. Let

$$\rho(A) = \overset{max}{i} \ |\lambda_i(A)| \,. \tag{4.1}$$

If A cannot be brought by simultaneous row and column permutation to the form

$$A = \begin{bmatrix} A_1 & A_2 \\ 0 & A_3 \end{bmatrix}, \tag{4.2}$$

it is termed indecomposable or irreducible.

If A is irreducible:

(i) $\rho(A)$ is a simple eigenvalue of A.

(ii) There exists a positive x with $Ax = \rho(A)x$.

(iii) For some positive integer k, perhaps 1, the entire spectrum of A is invariant under rotation around the origin of $2\pi/k$ radians.

If A^m is positive for some m, A (which is necessarily irreducible) is termed primitive, and

(iv) $\rho(A)$ is the only eigenvalue of modulus $max\,|\lambda_i(A)|$ (and there is no rotational invariance of the spectrum).

A general nonnegative A has an eigenvalue set which is the union of sets associated with irreducible matrices.

5 Deterministic Positive Realization Problem

In this section, we shall clarify the nature of the positive realization problem, review some earlier results relating its solvability to a question regarding existence of cones, and then review the results of [13] which effectively provide a solution.

Let $A \epsilon R^{N \times N}, b \epsilon R^N, c \epsilon R^N$ with nonnegative entries. Then $H(z) = c^T(zI - A)^{-1}b$ and $h(k) = c^T A^{k-1} b, k = 1, 2, \ldots$ is a nonnegative sequence. This raises the converse question. Suppose a prescribed $H(z)$ is rational and has nonnegative impulse response.

(i) Is there a positive/nonnegative realization A, b, c of finite dimension N?

(ii) If so, how may it be found?

(iii) What is the minimal value for N over all realizations?

(iv) Is there a set of realizations, and how are members of the set related, especially those of minimal dimension?

Example. (The possible need for nonminimality of the conventional type). Consider

$$H(z) = \frac{1}{z + 0.8} + \frac{1}{z - 0.8} + \frac{1}{z + 0.5}, \tag{5.1}$$

for which $h(k) = 2[(0.8)^{2k-1}] - (0.5)^{2k-1}, h = 1, 2, \ldots$ and $h(k) = (0.5)^{2k}$ for $k = 1, 2, \ldots$ There obviously exists a 3rd order realization for $H(z)$ without a nonnegativity constraint. Suppose there is a nonnegative realization. Could it have dimension 3? The eigenvalues of A would have to be $+0.8, -0.8, -0.5$. Since $\rho(A) = 0.8$ and there are two eigenvalues of A of maximum modulus, viz ± 0.8, the spectrum of A must be invariant through rotation by π radians. Since $+0.5$ is not an eigenvalue of A, this is impossible. So *there is no* 3×3 *nonnegative* A which could be used in a realization. There is however a realization with a 4×4 A:

$$A = \begin{bmatrix} 0 & 0.25 & 0 & 0 \\ 1 & 0 & 0 & 0 \\ 0 & 0.39 & 0 & 0.8 \\ 0 & 0 & 0.8 & 0 \end{bmatrix} \qquad b = \begin{bmatrix} 1 \\ 0 \\ 0 \\ 0 \end{bmatrix} \qquad c = \begin{bmatrix} 1 \\ 1.1 \\ 0 \\ 2 \end{bmatrix}. \tag{5.2}$$

Normally, nonminimal realizations of a rational transfer can have an A matrix with an eigenvalue set that is arbitrary, provided only that it includes the poles of the transfer function. It is remarkable that if A is primitive, this is not the case:

Lemma 5.1 Let $H(z)$ be a nonzero rational transfer function with nonnegative realization $c^T(zI - A)^{-1}b$ in which A is primitive. Then $\rho(A)$ is necessarily a pole of $H(z)$.

Proof. By primitivity, there exist positive v and w for which $Av = \rho(A)v, w^T A = w^T \rho(A)$; since $\rho(A)$ is the only eigenvalue of A of this magnitude, it is easily checked that $\rho(A)^{-k} A^k \to vw^T$ as $k \to \infty$. Hence

$$\rho(A)^{-k} c^T A^k b \to c^T v w^T b.$$

Since $H(z)$ is nonzero, c and b are not identically zero. Hence $c^T v > 0$, $w^T b > 0$ and $c^T v w^T b > 0$. Thus $\rho(A)^k$ shows up in $h(k)$, i.e. $\rho(A)$ is a pole of $H(z)$.

What if we drop the assumption that A is primitive? Then a simple argument shows that either $\rho(A)$ is a pole of $H(z)$, or with no loss of nonnegativity, uncontrollable and/or unobservable states can be removed to construct a smaller dimension nonnegative realization. We give the unobservable case in detail.

Lemma 5.2. Let $H(z)$ be a nonzero rational transfer function with nonnegative realization $c^T(zI - A)^{-1}b$. Then either $\rho(A)$ is observed or unobservable states can be removed without losing nonnegativity.

Proof. Suppose $Av = \rho(A)v$, and suppose states are ordered so that $c^T = (c_1^T \quad 0 \quad 0)$ and $v^T = (0 \quad 0 \quad v_3^T)$ with c_1, v_3 positive. Then the nonnegativity/
positivity constraints and the equation

$$\begin{pmatrix} A_{11} & A_{12} & A_{13} \\ A_{21} & A_{22} & A_{23} \\ A_{31} & A_{32} & A_{33} \end{pmatrix} \begin{pmatrix} 0 \\ 0 \\ v_3 \end{pmatrix} = \rho(A) \begin{pmatrix} 0 \\ 0 \\ v_3 \end{pmatrix}$$

imply $A_{13} = 0, A_{23} = 0$, and a lower dimension realization of $H(z)$ is provided by

$$\begin{pmatrix} A_{11} & A_{12} \\ A_{21} & A_{22} \end{pmatrix}, \begin{pmatrix} b_1 \\ b_2 \end{pmatrix}, \begin{pmatrix} c_1 \\ 0 \end{pmatrix}.$$

We have an immediate consequence.

Theorem 5.1 If a rational $H(z)$ with nonnegative impulse response has a nonnegative realization, the poles of $H(z)$ of maximum modulus must be a subset of those which are allowed eigenvalues of maximum modulus of a nonnegative matrix, and include a positive real pole.

Example. (Rational $H(z)$, nonnegative $h(k)$ with no finite dimensional nonnegative realization). Consider $h(k) = (\frac{1}{2})^k sin^2 k$ for which

$$H(z) = \frac{1}{2}\left[\frac{\frac{1}{2}}{z-\frac{1}{2}} - \frac{z(\frac{1}{2}cos2) - \frac{1}{4}}{z^2 - zcos2 + \frac{1}{4}}\right].$$

The maximum modulus poles of $H(z)$ are $\frac{1}{2}, \frac{1}{2}exp(\pm 2j)$, and these are not of the form $\frac{1}{2}\omega_i$, where ω_i is a k-th root of unity for some integer k. So no finite dimensional nonnegative realization exists.

A major advance was obtained by reformulating the realization problem using cones. Let $X = \{x_i, i = 1,2\ldots\}, x_i \epsilon R^n$. Then $\mathcal{X} =$ cone $X = \{\sum \alpha_i x_i, \alpha_i \geq 0, x_i > 0$ for finitely many $i\}$. The dual of $\mathcal{X}$, is $\mathcal{X}^* = \{y \mid y^T x \geq 0 \,\forall\, x \epsilon \mathcal{X}\}$. For a matrix $P, \mathcal{P} =$ cone P is the cone generated by the columns of P.

Ohta, Maeda and Kodama [14] proved

Theorem 5.2 Let an n-th degree $H(z)$ with nonnegative impulse response have a minimal realization $h^T(zI-F)^{-1}g$. Let $\mathcal{R} =$ cone $[g, Fg, F^2g, \ldots]$. Then $H(z)$ has a nonnegative realization of dimension N if and only if there exists an $n \times N$ P and $\mathcal{P} =$ cone P with

$$\mathcal{R} \subset \mathcal{P} \qquad F\mathcal{P} \subset \mathcal{P} \qquad h \epsilon \mathcal{P}^*$$

Note that if P is known, construction of nonnegative A, b and c is easy: $\mathcal{R} \subset \mathcal{P} \Rightarrow g \epsilon \mathcal{P} \Rightarrow g = Pb$ for some $b \geq 0$; $F\mathcal{P} \subset \mathcal{P} \Rightarrow FP = PA$ for some $A \geq 0$; $h \epsilon \mathcal{P}^* \Rightarrow h^T P = c^T$ for $c \geq 0$. Then $h^T F^k g = h^T F^k P b = h^T P A^k b = c^T A^k b$.

When does P exist? How may it be found? Answers are to be found in [13].

Theorem 5.3 Suppose an n-th degree $H(z)$ with nonnegative impulse response and minimal realization $h^T(zI-F)^{-1}g$ has just one pole of maximum modulus, and this pole is simple and positive real. Then a P as in Theorem 5.2 can be found.

The idea behind constructing P is as follows:

(i) Normalize $g, Fg, F^2g \ldots \epsilon R^n$ to unit length (the normalized vectors span the same cone)

(ii) Let $Fv = \rho(F)v, v > 0$. Then $v/\|v\|$ is the limit of the normalized vectors in (i).

(iii) Let $\bar{B}_\epsilon$ be a box of side length 2ϵ symmetrically positioned about $v/\|v\|$. Let B_ϵ be the cone formed by vectors from the origin to the corners of $\bar{B}_\epsilon$. Then for certain finite M_1, M_2 one can take

$$\mathcal{P} = \text{cone}\,[g \quad Fg \ldots F^{M1}g \quad B_\epsilon \quad FB_\epsilon \ldots F^{M2}B_\epsilon].$$

The above result assumes there is just a single and simple maximum modulus pole. Extending to the case of a single multiple pole is easy, [13]. With more work, the case of more than one pole of maximum modulus can be treated. Partial results are in [13] and the complete results are in unpublished work of Farina, Maeda and Kitano, [15, 16]. The key to the complete results is the following theorem.

Theorem 5.4 Suppose an n-th degree $H(z)$ has a nonnegative impulse response $h(k)$. For any integer K define K impulse responses obtained by sampling $h(k)$ at K time units apart:

$$\begin{aligned} h_1(k) &= h(kK) \\ h_2(k) &= h(kK+1) \\ &\vdots \\ h_K(k) &= h(hK+\overline{K-1}). \end{aligned}$$

Then $h(k)$ has a nonnegative realization if and only if $h_1(k), \ldots, h_K(k)$ all have nonnegative realizations.

Outline of Proof. If $H(z) = c^T(zI - A)^{-1}b$ with nonnegative A, b, c then $H_i(z) = c^T(zI - A^K)^{-1}A^{i-1}b$. Also, if $H_i(z) = c_i^T(zI - A_i)^{-1}b_i$ for nonnegative A_i, b_i, c_i with impulse response $h_i(k)$, define $\bar{H}_i(z)$ to have impulse response $\{0, \ldots, 0, h_i(1), 0, 0, \ldots h_i(2), \ldots\}$ with $h_i(1)$ the response at the i-th time, $h_i(2)$ the response at the $(K+i)$-th time etc. Then $\bar{H}_i(\bar{z}) = \bar{c}_i^T\ (zI - \bar{A}_i)^{-1}\bar{b}_i$ where

$$\bar{A}_i = \begin{bmatrix} 0 & I & 0 & \ldots & 0 \\ 0 & 0 & I & \ldots & 0 \\ \vdots & \vdots & \vdots & \ddots & \\ & & & & I \\ A_i & 0 & 0 & \ldots & 0 \end{bmatrix}, \qquad \bar{b}_i = \begin{bmatrix} 0 \\ 0 \\ \vdots \\ 0 \\ b_i \end{bmatrix},$$

$$\bar{c}_i^T = [0 \ldots c_i^T 0 \ldots 0],$$

with c_i^T occurring in block $K - i + 1$. Finally $h(k) = \sum \bar{h}_i(k), H(z) = \sum \bar{H}_i(z)$ and a nonnegative realization is immediate.

How is this applied? To test for and obtain a nonnegative realization first refer to Theorem 5.1. Assume the necessary conditions of Theorem 5.1 are fulfilled.

If $H(z)$ has a single maximum modulus pole, Theorem 5.3 gives realizability just when the pole is positive real. If $H(z)$ has more than one maximum modulus pole, a necessary condition for realizability (by Theorem 5.1) is that for some integer K, λ^K for any maximum modulus pole λ

is positive real. Now each transfer function associated with $h_1(k), \ldots h_K(k)$ either has a single maximum modulus pole at λ^K, in which case Theorem 5.3 gives a positive realization, or there is one or more maximum modulus poles at a smaller value than λ^K and *at the same time* a smaller degree than $H(z)$. (This second possibility will arise if the maximum modulus poles of $H(z)$ are not observed in a particular set of K-spaced impulse response samples).

If the transfer function resulting in this second case has more than one maximum modulus pole, realizability is either ruled out on account of lack of rotational symmetry of the maximum modulus poles of the spectrum (Theorem 5.1), or subsampling is used again. The whole process must terminate in view of the degree reduction which occurs if a subsampled transfer function has more than one pole of maximum modulus.

It is not straightforward to define the minimal dimension of nonnegative realizations, nor to characterize the set of (possibly minimal) nonnegative realizations in terms of one, nor to approximate (with a nice error formula) one nonnegative realization by a lower dimension one. It is in principle possible to answer the question: "does there exist a realization of a particular dimension $N \geq n$?" using Tarski-Seidenberg Theory [17]. For the question can be restated as: do there exist "$a_{ij}, b_i, c_j, i = 1, 2, \ldots, N, j = 1, 2, \ldots N$ such that $a_{ij} \geq 0, \quad b_i \geq 0, \quad c_j \geq 0, \quad h_k = c^T A^{k-1} b, \quad k = 1, 2, \ldots, 2N$"? This is an existence question involving real solutions of polynomial equalities and inequalities, which is the subject of the Tarski-Seidenberg theory, [17].

6 Initial Condition Forgetting and Smoothing in Hidden Markov Models

In Section 2.2, we defined a hidden Markov model. *Filtering* for a hidden Markov model is the process of computing recursively the filtered probability vector $\pi_{k/k}$ with i-th entry $Pr[X_k = i \mid Y_k = y_k, Y_{k-1} = y_{k-1}, \ldots]$. It is also useful to work with $\pi_{k+1/k}$, the one-step-ahead prediction probability vector, with i-th entry $Pr[X_{k+1} = i \mid Y_k = y_k, Y_{k-1} = y_{k-1} \ldots]$.

6.1 Initial Condition Forgetting

The first question we want to consider is: as $k \to \infty$, does $\pi_{k/k}$ become independent of $\pi_{0/0}$ (initial condition) or equivalently, old measurements? Why is this an important question? First, in practical applications of an HMM problem, there will be many situations in which initial condition data is very poor; one would like to know that the filter can "recover" from such inadequacy. Second, if there is not some form of exponential forgetting, there is a risk that round-off and quantization errors present in the numerical implementation of a filter may accumulate to the point where for large time, computational accuracy is lost and the filter is of no use.

To treat the exponential forgetting question, we shall explain how $\pi_{k/k}$ evolves [8]. Let

$$C_m = diag[c_{m1}, c_{m2}, \ldots, c_{mN}],$$

and let C_{y_k} denote that matrix in the set $C_1, C_2, \ldots, C_M$ resulting when Y_k assumes the value y_k. Bayes' rule leads to

$$\pi_{k+1/k} = A\pi_{k/k}(\text{time update}) \tag{6.1}$$

$$\pi_{k+1/k+1} = \frac{C_{y_{k+1}}\pi_{k+1/k}}{[1 \quad 1\ldots 1]C_{y_{k+1}}\pi_{k+1/k}}(\text{ measurement update}). \tag{6.2}$$

If we let $\tilde{\pi}_{k+1/k}$ and $\tilde{\pi}_{k+1/k+1}$ denote positively scaled versions of $\pi_{k+1/k}$, $\pi_{k+1/k+1}$ (the latter have entries summing to unity), we have unnormalized update equations:

$$\tilde{\pi}_{k+1/k} = A\tilde{\pi}_{k/k} \qquad \tilde{\pi}_{k+1/k} = C_{y_{k+1}}\tilde{\pi}_{k+1/k}$$

or

$$\tilde{\pi}_{k+1/k+1} = (C_{y_{k+1}}A)\tilde{\pi}_{k/k}. \tag{6.3}$$

Evidently,

$$\tilde{\pi}_{k/k} = E_k E_{k-1}\ldots E_1\tilde{\pi}_{0/0}, \tag{6.4}$$

where each E_i is drawn from the set of matrices $\{C_1A, C_2A, \ldots, C_MA\}$.

Products of nonnegative and positive matrices exhibit some properties like those of powers of nonnegative or positive matrices. An extensive analysis has been presented by Seneta [1] using a tool called the Birkhoff contraction coefficient. The key result is that if A, C are both positive, then the columns of $D(k) = E_k E_{k-1}\ldots E_1$ tend to proportionality exponentially fast as $k \to \infty$, i.e.

$$D(k) \quad \to \quad \begin{bmatrix} d_{11}(k) & v_2 d_{11}(k) & \ldots & v_N d_{11}(k) \\ \vdots & \vdots & & \vdots \\ d_{N1}(k) & v_2 d_{N1}(k) & \ldots & v_N d_{N1}(k) \end{bmatrix}$$

$$= \begin{bmatrix} d_{11}(k) \\ \vdots \\ d_{N1}(k) \end{bmatrix} [1 \quad v_2 \ldots v_N].$$

(The convergence is also provable if A is primitive with C positive, or if A is positive and C is nonnegative, see [18]). It follows that

$$\tilde{\pi}_{k/k} \to \begin{bmatrix} d_{11}(k) \\ \vdots \\ d_{N1}(k) \end{bmatrix} [1 \quad v_2 \ldots v_N]\tilde{\pi}_{0/0},$$

or that

$$\pi_{k/k} \rightarrow \frac{1}{\sum_j d_{j1}(k)} \begin{bmatrix} d_{11}(k) \\ \vdots \\ d_{N1}(k) \end{bmatrix},$$

and so there is exponential forgetting of $\tilde{\pi}_{0/0}$.

A related result can be found in [19], obtained by a different but similar tool, perhaps more suited to considering products of stochastic matrices, rather than general nonnegative matrices. Full details of the above argument are in [18], which also indicates how convergence rate bounds are computable, and demonstrates that forgetting occurs at least as fast as the underlying Markov state process forgets its initial probability vector.

As noted above, the convergence results allow A to be only nonnegative as long as it is primitive, or C to be nonnegative, but apparently not both relaxations simultaneously. Indeed, one can construct examples of primitive A and nonnegative C where in the worst case, there is no forgetting of an initial condition, although on average there is a forgetting.

These results are similar to those obtained in Kalman filtering problems, where exponential forgetting of initial conditions is a consequence of detectability and stabilizability assumptions [20]; without exponential forgetting, the numerical behaviour of both an HMM filter and a Kalman filter is likely to be unreliable.

6.2 Fixed-Lag Smoothing of Hidden Markov Models

The vector $\pi_{k/k}$ sums up what the measurements up till time k tell us about X_k. For some fixed Δ (and variable k), the vector $\pi_{k/k+\Delta}$ with i-th entry $Pr[X_k = i \mid Y_j = y_j, j \leq k+\Delta]$ (termed a fixed lag smoothed probability vector) sums up what the measurements up till k *and* then on till $k+\Delta$ tell us about X_k. The vector cannot be computed until time $k+\Delta$, but it should tell us more about X_k than $\pi_{k/k}$. Provided the delay in availability is acceptable, $\pi_{k/k+\Delta}$ (which is not difficult to compute) is a more useful vector than $\pi_{k/k}$, because it will be more informative about X_k. A key question is: how does $\pi_{k/k+\Delta}$ behave with Δ. Work in [18] shows that for a fixed k, $\pi_{k/k+\Delta}$ approaches a limit as $\Delta \rightarrow \infty$, at an exponentially fast rate that is the same as that associated with the forgetting of initial conditions (and is certainly uniform in k). Thus when Δ is taken as several times the time constant of this exponential rate, virtually all the improvement which smoothing offers over filtering is extracted.

These results verify conjectures (which were bolstered with simulation data made more than two decades ago, [21]), and parallel results in Kalman filter theory, [22].

7 The HMM Realization Problem

For convenience, we shall work with a slightly modified definition of an HMM in this section. As before, X_k is a stationary Markov state process which can assume N discrete values, $1, 2, \ldots, N$. We shall assume that the output process Y_k can assume M output levels, and that outputs are associated with states according to a set of stationary conditional probabilities:

$$a_{ij}(y) = Pr(X_{k+1} = j, Y_{k+1} = y \mid X_k = i). \tag{7.1}$$

It follows that

$$A = A(1) + \ldots + A(M) \tag{7.2}$$

is a stochastic matrix, with $\pi^T A = \pi^T$ for some nonnegative vector π and $Ae = e$ where $e = [1 \quad 1 \ldots 1]^T$. It is reasonable to assume that A is irreducible, so that π is positive.

Let $y_{\alpha 1} y_{\alpha 2} \ldots y_{\alpha q}$ be an output string (starting with $y_{\alpha 1}$ and ending with $y_{\alpha q}$). Then it is not hard to verify that $Pr(Y_t = y_{\alpha 1},\ Y_{t+1} = y_{\alpha 2}, \ldots Y_{t+q-1} = y_{\alpha q})$, written $Pr(y_{\alpha 1} y_{\alpha 2} \ldots y_{\alpha q})$ is given by

$$Pr(y_{\alpha 1} y_{\alpha 2} \ldots y_{\alpha q}) = \pi^T A(y_{\alpha 1}) A(y_{\alpha 2}) \ldots A(y_{\alpha q}) e. \tag{7.3}$$

The realization problem is: given the set of quantities $Pr(y_{\alpha 1} y_{\alpha 2} \ldots y_{\alpha q})$, find $\pi > 0$, $A(1), \ldots, A(M)$ with $A = \sum_i A(i)$ stochastic, and $\pi^T A = \pi^T$ such that the formula (7.3) holds. (Embedded within the problem is of course the question of existence: what conditions on the collection of $Pr(y_{\alpha 1}\ y_{\alpha 2} \ldots y_{\alpha q})$ allow solvability of the problem? Obviously, if the quantities $Pr(y_{\alpha 1} y_{\alpha 2} \ldots y_{\alpha q})$ are known to result from some *unknown* HMM, then the existence question is bypassed).

The HMM problem is a complication of the normal linear system realization problem in several respects:

(i) It is a multivariable problem. In the usual linear system realization problem, one is given a sequence $c^T b, c^T Ab, c^T A^2 b, \ldots$ corresponding to the terms in a Laurent series expansion of $c^T(zI - A)^{-1} b$ and one has to find c, A, b. Here, in case there are just two output levels, one is given $c^T b$, $c^T A_1 b$, $c^T A_2 b$, $c^T A_1^2 b$, $c^T A_1 A_2 b$, $c^T A_2 A_1 b$, $c^T A_2^2 b$ etc., these terms corresponding to coefficients in a Laurent series expansion of $c^T(I - z_1^{-1} A_1 - z_2^{-1} A_2)^{-1} b$. More generally, one has coefficients in the expansion of $c^T \left[I - \sum_{i=1}^M z_i^{-1} A(i)\right]^{-1} b$, and one has to find the $A(i)$, b and c

(ii) It is a *nonnegative* realization problem, just like the problem of Section 5.

(iii) There are special constraints: $A = \sum_{i=1}^{M} A(i)$ must be stochastic, $Ae = e$ and $\pi^T A = \pi^T$.

The problem has been studied for almost three decades, see e.g. [23, 30]. The most comprehensive results are to be found in [31]. We shall give only an outline of the results here.

We shall first describe how the multivariable nature of the problem is handled.

7.1 The Generalized Hankel Matrix

Let Y^* denote the set of finite strings of Y, let Y_t^+ denote $\{Y_{t+1}, Y_{t+2}, \ldots\}$ and let Y_t denote $\{\ldots, Y_{t-1}, Y_t\}$.

Let us enumerate the strings u_i of Y^* lexicographically, ordering entries of a string from left to right, such that the length $|u_i|$ increases monotonically with i. Include the empty sequence as the first element of the enumeration. Thus the ordering is $\{\phi, 0, 1, 00, 10, 01, 11, 000, 100, \ldots\}$. Further, position u_i along the time axis so that the right most (or last occurring) symbol occurs at time t. Thus $u_i \epsilon Y_t^-$.

Consider a second enumeration $\{v_j\}$, identical save that string entries are ordered from right to left; thus the ordering is $\phi, 0, 1, 00, 01, 10, 11, 000, 100, \ldots$ Further, v_j is so positioned that the left most symbol (the first occurring in time) occurs at time $t+1$. Thus $v_j \epsilon Y_t^+$.

The generalized Hankel matrix H is defined to have $i-j$ element $p(u_i v_j)$, where u_i, v_j are the $i-th$ and $j-th$ elements of the two enumerations.

Thus with two output symbols the top left corner of H will appear as

$$
\begin{array}{c}
\\ \phi \\ 0 \\ 1 \\ 00 \\ 10 \\ 01 \\ 11 \\ \\
\end{array}
\begin{array}{c|cc|ccccc}
\phi & 0 & 1 & 00 & 01 & 10 & 11 & \\
\hline
1 & p(0) & p(1) & p(00) & p(01) & p(10) & p(11) & \cdot \\
\hline
p(0) & p(00) & p(01) & p(000) & p(001) & p(010) & p(011) & \cdot \\
p(1) & p(10) & p(11) & p(100) & p(101) & p(110) & p(111) & \cdot \\
\hline
p(00) & p(000) & p(001) & p(0000) & p(0001) & p(0010) & p(0011) & \cdot \\
p(10) & p(100) & p(101) & p(1000) & p(1001) & p(1010) & p(1011) & \cdot \\
p(01) & p(010) & p(011) & p(0100) & p(0101) & p(0110) & p(0111) & \cdot \\
p(11) & p(110) & p(111) & p(1100) & p(1101) & p(1110) & p(1111) & \cdot \\
\cdot & \cdot & \cdot & \cdot & \cdot & \cdot & \cdot &
\end{array}.
$$

We shall let H_{KL} denote the block matrix of H given by

$$[p(u_i v_j) / \, |u_i| = K, |v_j| = L].$$

The top left corner of H above is partitioned as

$$
H = \begin{bmatrix}
H_{00} & H_{01} & H_{02} & \cdot \\
H_{10} & H_{11} & H_{12} & \cdot \\
H_{20} & H_{21} & H_{22} & \cdot \\
\cdot & \cdot & \cdot & \cdot
\end{bmatrix}.
$$

Conventional Hankel matrices have a structure associated with blocks parallel to the anti-diagonal. This is true also of H. Consider

$$H_{K,L+1} = [H_{K,L}(y_1) \quad H_{K,L}(y_2) \ldots H_{K,L}(y_m)],$$

with the definition

$$H_{K,L}(y_1) = p(u_i y_1 v_j), \text{with } |u_i| = K, |v_j| = L.$$

(The enumeration scheme for the columns is relevant here).

Because of the row enumeration scheme, we can also evidently write

$$H_{K+1,L} = \begin{bmatrix} H_{K,L}(y_1) \\ \vdots \\ H_{K,L}(y_n) \end{bmatrix}.$$

Thus although $H_{K+1,L} \neq H_{K,L+1}$ (as is normal for Hankel matrices), the identity is true provided same rearrangement of entries is permitted.

It is then possible to show

Theorem 7.1 Let H be the infinite generalized Hankel matrix associated with a hidden Markov model with N states. Then rank $H \leq N$.

Suppose now we were simply presented with an infinite generalized Hankel matrix H with finite rank N, say. We can proceed a modest distance towards obtaining an HMM, by taking care of the multivariable finite-dimensional realization problem without taking account of nonnegativity. This is done in the following theorem, which states how vectors x_ϕ, y_ϕ and matrices F_i can be defined satisfying an analog of (7.3), but without a nonnegativity constraint on the entries.

Theorem 7.2 Let H be the infinite generalized Hankel matrix of output string probabilities $Pr(u_i v_j)$ associated with an unknown hidden Markov model with an unknown but finite number of states. Let rank $H = N$, and factor H as

$$H = XY = \begin{bmatrix} X_\phi \\ X_1 \\ X_2 \\ \vdots \end{bmatrix} [Y_\phi \quad Y_1 \quad Y_2 \ldots], \tag{7.4}$$

where X_K and Y_L correspond to u_i and v_j of length K and L respectively, and have N columns and rows respectively. Let $\bar{Y}_N$ denote the submatrix of $[Y_\phi \quad Y_1 \quad Y_2 \ldots]$ containing the first N linearly independent columns, indexed by strings $\bar{v}_1, \bar{v}_2, \ldots \bar{v}_N$. Define $\bar{Y}_{iN}$ to be those columns of Y indexed by $y_i\bar{v}_1, y_i\bar{v}_2, \ldots, y_i\bar{v}_N$ and set

$$F_1 = \bar{Y}_{1N}\bar{Y}_N^{-1} \ldots F_M = \bar{Y}_{MN}\bar{Y}_N^{-1}. \tag{7.5}$$

Let x_ϕ^T, y_ϕ denote the first row and first column of X and Y.

Then

$$p(u_i v_j) = x_\phi^T F_1^{\alpha_0} \ldots F_M^{\xi_0} F_1^{\alpha_1} \ldots y_\phi, \tag{7.6}$$

where $u_i v_j$ consists of α_0 ones, followed by β_0 twos, etc. (Any of the α_i, β_i etc. may be zero).

It is obvious that x_ϕ, y_ϕ and the F_i may not be nonnegative, let alone satisfy further special constraints; e.g. $\sum F_i$ is stochastic with x_ϕ^T and y_ϕ left and right eigenvectors. Also, if H comes from a HMM, the number of states in the HMM may exceed N - in fact (just as in the positive systems realization problem) there may be no HMM realizing H with precisely N states.

7.2 Introducing Nonnegativity via Cones

The result of Ohta et al allowing reformulation of the positive system realization problem has its parallel in the HMM problem, in that it gives the key for introducing nonnegative quantities in place of x_ϕ, y_ϕ and the F_i in (7.6) above, [3, 8, 14, 29].

Theorem 7.3 Adopt the hypotheses of Theorem 6. Then a necessary and sufficient condition for the existence of nonnegative c, b and $A_i, i = 1, \ldots m$ for which

$$p(u_i v_j) = c^T A_1^{\alpha_0} \ldots A_M^{\xi_0} A_1^{\alpha_1} \ldots b, \tag{7.7}$$

is that there exists a matrix R with associated $\mathcal{R} =$ cone R with

$$F_i \mathcal{R} \subset \mathcal{R} \tag{7.8}$$

$$y_\phi \epsilon \mathcal{R} \tag{7.9}$$

$$x_\phi \epsilon \mathcal{R}^* \tag{7.10}$$

7.3 Cone Existence and Construction

Theorem 7 reformulates, rather than solves, a multivariable nonnegative realization problem. In this subsection, we indicate sufficient (but very broad) conditions for the existence (and constructability) of a cone, [31].

Suppose first that H of finite rank N comes from an (unknown) HMM in which all the probabilities $Pr(X_{k+1} = j, Y_{k+1} = y \mid X_k = i)$ are positive.

- Consider the submatrix of the generalized Hankel matrix comprising the first N linearly independent rows. It has an infinite number of columns.
- Consider a factorization of this N rowed matrix as $\bar{X}_N Y$, where

$$Y = [Y_0 \quad Y_1 \quad Y_2 \ldots] = [y_\phi \vdots F_1 y_\phi \ldots F_M y_\phi \vdots F_1^2 y_\phi \quad F_1 F_2 y_\phi \ldots]$$

  and $\bar{X}_N$ is a certain set of N linearly independent rows of X with

$$X^T = [x_\phi \vdots F_1^T x_\phi \ldots F_M^T x_\phi \vdots (F_1^T)^2 x_\phi \ldots].$$

  Suppose that $x_\phi, F_i, i = 1, \ldots, M$ and y_ϕ have been identified, as in Theorem 7.2.
- Change the coordinate basis so that in the new basis $X_N^T = I$. Then $x_\phi^T = [1 \quad 0 \ldots 0]$ and Y is simply an N-rowed submatrix of the Hankel matrix.
- Evidently, the infinite set of columns of Y span a cone containing y_ϕ, x_ϕ is in the dual cone, and the cone is F_i−invariant for $i = 1, 2, \ldots, M$. If the cone is polyhedral, we would be done. However, given that the cone has an infinite set of generators, it may not be polyhedral.
- It turns out that a large but finite number of columns of Y, together with perturbations around the columns, do define a polyhedral cone containing all the columns of Y that is also F_i-invariant for $i = 1, 2, \ldots, M$.
- The reason is that columns of the generalized Hankel matrix corresponding to very long strings obey (in the limit) some alignment properties. The same will be true of subvectors of these columns containing N entries. The subvectors will then be in a cone defined using early subvectors of the infinite sequence, together with perturbations thereof.

The alignment property just referred to is a consequence of assuming positivity of $Pr(X_{k+1} = j, Y_{k+1} = y \mid X_k = i)$ i.e. of the matrices $A(i), i = 1, 2, \ldots, M$ and actually holds under weaker conditions; as we know, long products of positive and some nonnegative matrices approach a rank one matrix as the number of terms in the product approaches infinity. This idea was used in Section 6 to establish an exponential forgetting property of an HMM filter. Here, it can be used to establish the forgetting property

$$\begin{bmatrix} p(\bar{u}_1 \mid v\tilde{v}) \\ \vdots \\ p(\bar{u}_N \mid v\tilde{v}) \end{bmatrix} \rightarrow \begin{bmatrix} p(\bar{u}_1 \mid v) \\ \vdots \\ p(\bar{u}_N \mid v) \end{bmatrix},$$

when $|v| \to \infty$, and $|\tilde{v}| = 1$. The convergence is exponential with $|v|$, and uniform in $v, \tilde{v}$. By Bayes' rule, $p(\bar{u}_i \mid v\tilde{v}) = p(\bar{u}_i v\tilde{v})/p(v\tilde{v})$ and $p(\bar{u}_i \mid v) = p(\bar{u}_i v)/p(v)$. Hence the two conditional probability vectors above are scaled versions of two columns of the N-rowed submatrix of H, corresponding to the columns indexed by v and $v\tilde{v}$.

7.4 Satisfying the Special Constraints

At this stage, let us suppose there has been constructed nonnegative c, b and $A_i, i = 1, 2, \ldots, M$ so that (7.7) holds. We now consider how the special constraints (stochasticity and eigenvector properties) can be achieved. By way of a preliminary calculation, let Ψ^k denote a don't care sequence of length k. If there is an underlying (but unknown) hidden Markov model with positive transition probabilities, it is straightforward to show that

$$Pr(u\Psi^k v) \to Pr(u)Pr(v) \text{ as } k \to \infty. \tag{7.11}$$

(This result is hardly unexpected). Appealing to the formula (7.6), one can show that

$$(\sum_{i=1}^{M} F_i)^k \to y_\phi x_\phi^T. \tag{7.12}$$

Additionally, we have

$$\sum_{j=1}^{M} p(\bar{u}_i y_j) = p(\bar{u}_i) \qquad i = 1, 2, \ldots, N,$$

or

$$\bar{X}_N[F_1 y_\phi + \ldots + F_M y_\phi] = \bar{X}_N y_\phi,$$

whence

$$(\sum_{i=1}^{M} F_i) y_\phi = y_\phi. \tag{7.13}$$

Similarly,

$$x_\phi^T (\sum_{i=1}^{M} F_i) = x_\phi^T. \tag{7.14}$$

These facts establish that x_ϕ^T, y_ϕ are the only left and right eigenvectors of $\sum_{i=1}^{M} F_i$ corresponding to eigenvalue 1, and all other eigenvalues are of lesser magnitude.

These facts can be combined with the cone formula $F_i\mathcal{R} \subset \mathcal{R}$ etc. (from which $F_i R = RA_i$) to conclude that c^T, b are unique left and right eigenvectors of $\sum_{i=1}^{M} A_i$ corresponding to eigenvalue 1. A positive diagonal basis transformation taking A_i to $\Lambda A_i \Lambda^{-1} = A(i), b$ to $\Lambda b = e$ and c^T to $c^T \Lambda^{-1}$ ensures that $\sum_{i=1}^{M} A_i$ is stochastic. Simply take $\Lambda^{-1} = \text{diag}\ (b)$.

8 Open Problems

There are a considerable number of open problems, some already alluded to. We list some:

(i) For the nonnegative realization problem, how can one simply figure the minimum dimension of nonnegative realizations?

(ii) How are all nonnegatively minimal realizations of rational $H(z)$ with $h_k \geq 0$ connected?

(iii) How can one approximate (systematically and preferably with an easily interpreted error bound) a high state dimension nonnegative realization by a lower dimension nonnegative realization ?

(iv) One can state variations on the above problems for HMMs.

References

[1] E. Seneta. *Nonnegative Matrices and Markov Chains.* New York: Springer Verlag, 1981.

[2] S. Rinaldi and L. Farina. *I Sistemi Lineari Positivi.* Milan: Citta Studi Edizione, 1995.

[3] P. H. Leslie. On the use of matrices in certain population mathematics. *Biometrika* **35** (1945), 183-212.

[4] H. Maeda, S. Kodama and F. Kajiya. Compartmental system analysis: Realization of a class of linear systems with physical constraints. *IEEE Transactions on Circuits and Systems* **CAS-24**(1) (1977), 8-14.

[5] J.A. Jacquez. *Compartmental Analysis in Biology and Medicine.* New York: Elsevier, 1972.

[6] A. Gersho and B B Gopinath. Charge-routing networks. *IEEE Trans Circuits and Systems* **CAS-26** (1979), 81-92.

[7] M. Rosenblatt. *Markov Processes; Structure and Asymptotic Behaviour.* Berlin: Springer Verlag, 1971. 81-92.

[8] L.R. Rabiner. A tutorial on hidden Markov models and selected applications in speech recognition. *Proceedings of the IEEE* **77**(2) (1989), 257-285.

[9] R.J. Elliott, L. Aggoun and J.B. Moore. *Hidden Markov Model: Estimation and Control.* New York: Springer Verlag, 1994.

[10] E. Fornasini and M.E. Valcher. On the spectral and combinatorial structure of 2D positive systems. To appear *Linear Algebra and Applications.*

[11] E. Fornasini. 2D Markov chains. *Linear Algebra and its Applications* **140** (1990), 101-127 .

[12] R.A. Horn and C.R. Johnson. *Matrix Analysis.* Cambridge: Cambridge University Press, 1985.

[13] B.D.O. Anderson, M. Deistler, L. Farina and L. Benvenutti. Nonnegative realization of a linear system with nonnegative impulse response. *IEEE Transactions on Circuits and Systems* **4** (1996), 134-142.

[14] Y. Ohta, H. Maeda and S. Kodama. Realizability, observability and realizability of continuous-time positive systems. *SIAM Journal of Control and Optimization* **22** (1984), 171-180.

[15] T. Kitano and H. Maeda. Positive realization by downsampling the impulse response. Preprint, 1996.

[16] L. Farina. On the existence of a positive realization. Preprint, 1996.

[17] N. Jacobson. *Basic Algebra I.* San Francisco: Freeman and Co., 1974.

[18] L. Shue, B.D.O. Anderson and S. Dey. Exponential stability of filters and smoothers for hidden Markov models. Preprint, 1996.

[19] A. Arapostathis and S.I. Marcus. Analysis of an identification algorithm arising in the adaptive estimation of Markov chains. *Mathematics of Control, Signals and Systems* **3** (1990), 1-29.

[20] B.D.O. Anderson and J.B. Moore. Detectability and stabilizability of time-varying discrete-time linear systems. *SIAM Journal of Control and Optimization* **18** (1981), 20-32.

[21] D. Clements and B.D.O. Anderson. A nonlinear fixed-lag smoother for finite-state Markov processes. *IEEE Trans. Information Theory* **2** (1975), 446-452.

[22] B.D.O. Anderson and J.B. Moore. *Optimal Filtering.* Englewood Cliff, N.J.: Prentice Hall, 1981.

[23] D. Blackwell and L. Koopmans. On the identifiability problem for functions of finite Markov chains. *Annals of Mathematical Statitics* **28** (1957), 1011-1015.

[24] E.J. Gilbert. On the identifiability problem for functions of finite Markov chains. *Annals of Mathematical Statistics* **30** (1959), 688-697.

[25] S.W. Dharmadhikari. Functions of finite Markov chains. *Annals of Mathematical Statistics* **34** (1963), 1022-1032.

[26] S.W. Dharmadhikari. Sufficient conditions for a stationary process to be a function of a finite Markov chain. *Annals of Mathematical Statistics* **34** (1963), 1033-1041.

[27] A. Heller. On stochastic processes derived from Markov chains. *Annals of Mathematical Statistics* **36** (1965), 1286-1291.

[28] M. Fliess. Series rationnelles positives et processus stochastiques. *Ann. Inst. Henri Poincare* **11** (1975), 1-21.

[29] G. Picci. On the internal structure of finite state stochastic processes. *Recent Developments in Variable Structure Systems, Economics and Biology.* (R.R. Mohler and A. Rubert, Eds.). Vol. 162 in *Lecture Notes in Economics and Math. Systems.* Berlin: Springer Verlag, 288-304.

[30] H. Ito, S-I. Amani and K. Kobayashi. Identifiability of hidden Markov information sources and their minimum degrees of freedom. *IEEE Transactions on Information Theory* **38** (1992), 324-333.

[31] B.D.O. Anderson. The realization problem for hidden Markov models. Preprint, 1996.

Department of Systems Engineering and Cooperative Research, Centre for Robust and Adaptive Systems Research, School of Information Sciences and Engineering, Australian National University, Canberra, ACT 0200, Australia

Modeling Methodology for Elastomer Dynamics

H.T. Banks and Nancy Lybeck

Abstract: As engineering applications of elastomers increase in complexity, knowledge of the behavior of these materials, and the ability to predict these behaviors, becomes increasingly valuable. Elastomers exhibit a complex variety of mechanical properties, including nonlinear constitutive laws, strong damping and hysteresis (loss of kinetic and potential energy, respectively), and the dependence of strain on its history. Most current models for rubber-like materials assume a form of the strain energy function (SEF), such as a cubic Mooney-Rivlin form or an Ogden form. While these methods can produce good results, they are only applicable to static behavior, and they ignore hysteresis and damping. We discuss a dynamic partial differential equation (PDE) formulation based on large deformation theory elasticity as an alternative approach to the SEF formulation. Models using the PDE formulation are presented for both simple extension and generalized simple shear.

1 Introduction

In recent years the use of rubber in engineering applications has expanded well beyond traditional products such as tires and belts. Today rubber (and, more generally, elastomers) is employed in a diverse set of applications including sealing, vibration damping, and load bearing (see [5, 9]). The applications of rubber are becoming increasingly sophisticated, as exemplified by the use of rubber and rubber–like polymers in building supports which protect the structure during an earthquake (see [7]).

Traditionally, elastomers are filled with carbon black or silica particles, inactive substances chosen to change the physical properties of the material to match the needs of the given application. A controllable elastomer, resulting from the addition of active fillers such as conductive, magnetic, or piezoelectric particles, could be used in products such as active vibration suppression devices. As these new composite materials are developed, the applications of rubber will become more complex, and design will play an even more prominent role in the development of components. As a consequence, the capability to predict the dynamic mechanical response of the components undergoing a variety of deformations will become most important.

Many complications arise in the process of formulating models for elastomers, some of which can be attributed to the desirable characteristics of rubber as a design component, including the ability to undergo large elastic deformations, good damping properties, and near incompressibility. For example, rubber components often undergo large deformations, hence infinitesimal based strain theory is not appropriate. Damping is highly significant, and the nonlinear constitutive laws cannot be modeled by Hooke's law. The mechanical response is affected in a nontrivial manner by environmental temperature, amount and type of filler, rate of loading, and strain history. Moreover, many elastomers, especially those with a synthetic rubber base, exhibit strong hysteresis characteristics similar to those found in shape memory alloys and piezoceramic materials.

Researchers have made substantial progress in developing elastomer models (see [6, 13, 15] for basic texts), the majority of which are phenomenological, based on finite strain (FS) and strain energy function (SEF) theories. SEF theories are typically used for *static* finite element analysis (see [4]), as the strain energy functions contain information about the elastic properties of elastomers, but do not describe either damping or hysteresis. The SEF material models are based on the principal extension ratios λ_i (a misnomer widely found in the literature – "principal stretches" is a more appropriate terminology) which represent the deformed length of unit vectors parallel to the principal axes (the axes of zero shear stress). Rivlin proposed ([12]) that the SEF should depend only on the strain invariants $I_1 = \lambda_1^2 + \lambda_2^2 + \lambda_3^2$, $I_2 = \lambda_2^2\lambda_3^2 + \lambda_1^2\lambda_3^2 + \lambda_1^2\lambda_2^2$ and $I_3 = \lambda_1^2\lambda_2^2\lambda_3^2$. Many SEF models are founded on this assumption, including the Mooney SEF $U = C_1(I_1 - 3) + C_2(I_2 - 3)$, or more generally, the modified expression $U = C_1(I_1 - 3) + f(I_2 - 3)$, where f has certain qualitative properties. This class of models is most appropriate for components where the rubber is not tightly confined and where the assumption of absolute incompressibility (implying $\lambda_1\lambda_2\lambda_3 = 1$ or $I_3 = 1$) is a reasonable approximation. The more general Rivlin SEF $U = \sum_{i+j\geq 1}^{N} C_{ij}(I_1 - 3)^i(I_2 - 3)^j$ and its generalization for near incompressibility (see [4]) permit higher order dependence of the SEF on the invariants. An important departure form Rivlin's proposal is found in the work of Ogden, as well as that of Valanis and Landel, where the models use strain energy functions that depend only on the extension ratios [10, 14].

The finite strain elastic theory of Rivlin [12, 15] is developed with a generalized Hooke's law in an analogy to infinitesimal strain elasticity but makes no "small deformation" assumption and includes higher order exact terms in its formulation. In addition, finite stresses are defined relative to the *deformed* body and hence are the "true stresses" as contrasted to the "nominal" or "engineering" stresses (relative to the *undeformed* body) one encounters in the usual infinitesimal linear elasticity employed with metals. This Eulerian measure of strain (relative to a coordinate system convected with the deformations) - as opposed to the usual Lagrangian measure (relative to a fixed coordinate system for the undeformed body) - is an important feature of any development of models for use in analytical/computation/experimental investigations of rubber-like material bodies. The finite strain elasticity of Rivlin can be directly related to the strain energy function formulations through equations relating the finite strains $\tilde{e}_{x_1x_1}, \tilde{e}_{x_2x_2}, \tilde{e}_{x_3x_3}$ to the extension ratios $\lambda_1, \lambda_2, \lambda_3$ used in the SEF. However, the finite strain approach can be formulated in a somewhat more general framework in the context of classical modeling of elastic solids and fluids. Unfortunately, elastomers and filled rubbers are not exactly in either category. In the next section, we present a general procedure for developing models using large deformation elasticity theory. This will be followed by examples on simple extension and generalized simple shear illustrating the use of this methodology in developing models.

2 Modeling in large deformation elasticity

In this section we outline approaches to modeling the dynamics of elastic bodies where large deformations are of primary interest. Rather detailed discussions can be found in the texts by Ogden [11] and Marsden-Hughes [8]. For our discussions we consider the body, depicted in Figure 1, with reference configuration Ω_0 in the fixed principal $\vec{X}$ coordinate system. Translations, rotations, and deformations are all possible results of applied

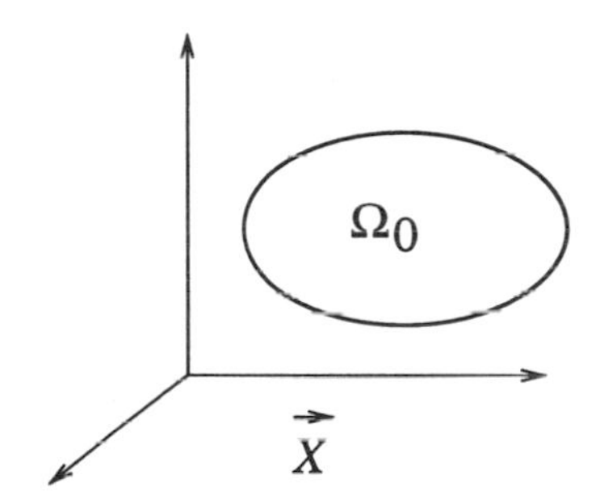

Figure 1: *Reference* configuration

body forces. Our primary interest is in *deformations* of this body. Let Ω be the current configuration of the (deformed) body in the principal $\vec{x}$ coordinate system, as depicted in Figure 2.

The *Lagrangian* or fixed coordinate system, $\vec{X}$, is appropriate for small displacements from which the body fully recovers. The *Eulerian* or moving coordinate system, $\vec{x}$, is appropriate for large deformations from which the body does not fully recover. Elastomers typically undergo large deformations from which they fully (or almost fully) recover. Since it is not always clear which coordinate system is best to use in elastomers, the ability to translate quantities between the Lagrangian coordinate system and those quantities in the Eulerian coordinate system is highly desirable, indeed essential.

Using a summation convention, one can write $\vec{x} = x_i\vec{e}_i$ and $\vec{X} = X_i\vec{E}_i$,

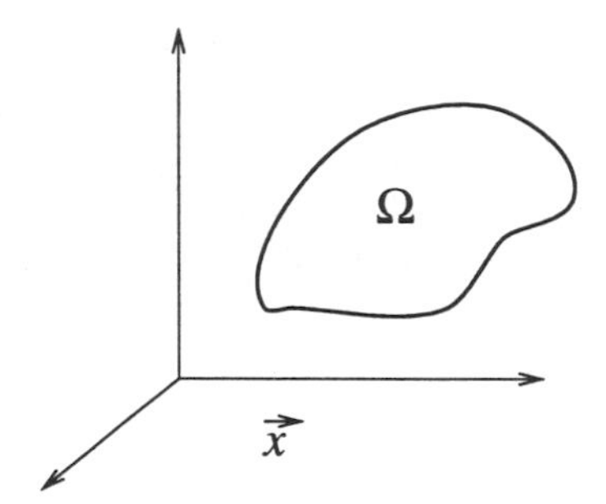

Figure 2: *Deformed* or *current* configuration

where $\vec{e}_i$ and $\vec{E}_i$ are unit vectors parallel to the x_i and X_i axes, respectively. The X_i are called "referential," or Lagrangian coordinates while the x_i are called "current" or Eulerian coordinates of a point. We may define a "configuration," or "position," map $\vec{x} = \phi(\vec{X})$. A "motion" or "trajectory" for a particle initially at $\vec{X}$ is given by $\vec{x}(t) = \phi(t, \vec{X})$. The configuration map is sometimes also called the "deformation map," but this is a misnomer, as the map does not give the deformation

$$\begin{aligned} u(t,\vec{X}) &= \phi(t,\vec{X}) - \phi(0,\vec{X}) \\ &= \phi(t,\vec{X}) - \vec{X}\ , \end{aligned}$$

as usually defined in elasticity (where the deformation is zero for an undeformed body). The *Lagrangian* description of a physical phenomenon associated with a deformation of a body involves vector and scalar fields defined over the reference configuration Ω_0 while the *Eulerian* description involves quantities defined over Ω.

In studying deformations, it is desirable to transform physical quantities defined over a region in the Lagrangian coordinate system to quantities over a region in the Eulerian coordinate system (and conversely). To this end we consider the "configuration gradient"

$$A = \left(\frac{\partial x_i}{\partial X_j}\right) = \frac{\partial \vec{x}}{\partial \vec{X}} = \frac{\partial \phi(\vec{X})}{\partial \vec{X}}$$

which is the nonsingular, nonsymmetric gradient of the "configuration" map. Here $dx_i = \frac{\partial x_i}{\partial X_j} dX_j$, so $d\vec{x} = A d\vec{X}$ and A is literally the "change of configuration" gradient. Let $J = \det A$, $B = (A^{-1})^T$. Then a standard elementary calculus result yields

$$\int_{\Omega} d\omega = \int_{\Omega_0} J d\omega_0$$

since volume elements in the Eulerian system are related to volume elements in the Lagrangian system via $d\omega = J d\omega_0$. For a volume-preserving (*isochoric*) material (a common approximation used in rubber based elastomers), $J = 1$.

If the usual deformation $u(\vec{X}) = \phi(\vec{X}) - \vec{X}$ is considered, then $\frac{\partial u}{\partial \vec{X}} = A - I$ is the actual deformation gradient. The matrix A is sometimes termed the "deformation gradient," but this is misleading since our undeformed body thus has the identity as its "deformation gradient" if this terminology is adopted. The right Cauchy-Green tensor, defined by $A^T A = \left(\frac{\partial \vec{x}}{\partial \vec{X}}\right)^T \frac{\partial \vec{x}}{\partial \vec{X}} = \left(\sum_k \frac{\partial x_k}{\partial X_i} \frac{\partial x_k}{\partial X_j}\right)$, and the left Cauchy-Green tensor, $AA^T = \left(\sum_k \frac{\partial x_i}{\partial X_k} \frac{\partial x_j}{\partial X_k}\right)$, are both symmetric, and play a fundamental role in strain analysis.

Given the above formulations there are several distinct ways to define a strain tensor. To begin with, we consider the change in squared lengths from the reference configuration to the current configuration

$$|d\vec{x}|^2 - |d\vec{X}|^2 = d\vec{X} \cdot (A^T A - I) d\vec{X} \ ,$$

and $A^T A - I$ is a measure of the strain (sometimes called the Lagrangian or Green strain). Thus, the body is unstrained if and only if $A^T A = I$.

One could then define the strain tensor

$$E \equiv \frac{1}{2}(A^T A - I) \ ,$$

so that (using a summation convention)

$$\begin{aligned} E_{\alpha\beta} &= \frac{1}{2}(A_{i\alpha} A_{i\beta} - \delta_{\alpha\beta}) \\ &= \frac{1}{2}\left(\frac{\partial x_i}{\partial X_\alpha}\frac{\partial x_i}{\partial X_\beta} - \delta_{\alpha\beta}\right) \ . \end{aligned}$$

In elasticity one usually finds more terms in the strain tensor since it most often is defined in terms of the deformations instead of configurations. If one considers $u(\vec{X}) = \phi(\vec{X}) - \vec{X}$, then the true "deformation gradient" is given by

$$D \equiv \frac{\partial u}{\partial \vec{X}} = A - I,$$

and hence

$$A^T A - I = D^T D + D + D^T \ .$$

Substituting into the definition of strain, we find that

$$E_{\alpha\beta} = \frac{1}{2}\left(\frac{\partial u_i}{\partial X_\alpha}\frac{\partial u_i}{\partial X_\beta} + \frac{\partial u_\alpha}{\partial X_\beta} + \frac{\partial u_\beta}{\partial X_\alpha}\right) \ .$$

This is the usual Green strain found in most elasticity books.

The above definitions and formulations involve static concepts. We are interested in dynamics of deformable bodies where the motion is described by $\vec{x}(t) = \phi(t, \vec{X})$ with $\vec{x}(0) = \vec{X}$. The Lagrangian velocity is given by

$$\vec{V}(t, \vec{X}) = \frac{\partial \vec{x}}{\partial t}(t, \vec{X}) = \frac{\partial \phi}{\partial t}(t, \vec{X}) ,$$

while the Eulerian velocity is defined by

$$\vec{v}(t, \vec{x}) = \vec{V}(t, \phi^{-1}(\vec{x})) .$$

The material time derivative is given by $\frac{\partial}{\partial t}\big|_{\vec{X}} = \frac{\partial}{\partial t}\big|_{\vec{x}} + \vec{v}\frac{\partial}{\partial \vec{x}}$, so that for any function $\psi(t, \vec{x}(t))$,

$$\begin{aligned} \frac{d}{dt}\psi(t, \vec{x}(t)) &= \frac{\partial \psi}{\partial \vec{x}}\frac{\partial \vec{x}}{\partial t} + \frac{\partial \psi}{\partial t} \\ &= \text{grad}\psi \cdot \vec{v} + \frac{\partial \psi}{\partial t} \end{aligned}$$

Balance laws may be formulated in either the Lagrangian or Eulerian system. The linear momentum is defined by

$$\int_\Omega \rho(t, \vec{x})\vec{v}(t, \vec{x})d\omega \qquad \text{(Eulerian form of momentum)}$$

or

$$\int_{\Omega_0} \rho_0(t, \vec{X})\vec{v}(t, \phi(t, \vec{X}))d\omega_0 \qquad \text{(Lagrangian form of momentum)}$$

and the balance of linear momentum (one of Euler's laws of motion) is given (in the current configuration) by

$$\begin{aligned} \frac{d}{dt}\int_\Omega \rho(t, \vec{x})\vec{v}(t, \vec{x})d\omega &\equiv \int_\Omega \rho(t, \vec{x})\dot{\vec{v}}(t, \vec{x})d\omega \\ &= \int_\Omega \rho(t, \vec{x})\vec{f}(t, \vec{x})d\omega + \int_{\partial\Omega} \vec{t}(\partial\Omega, \vec{x})da , \end{aligned}$$

where $\vec{f}$ is the body-force density due to the applied force and $\vec{t}$ is the contact-force density (i.e., the contact stress or traction).

Cauchy's theorem yields that a stress vector $\vec{t}$ at the point x on $\partial\Omega$, which depends on the (unit) normal $\hat{n}$ to the surface at $\vec{x}$, is related to the (symmetric) *Cauchy* or *true* stress tensor $\mathbf{T}(\vec{x})$ by

$$\vec{t}(\hat{n}, \vec{x}) = \mathbf{T}(\vec{x})\hat{n} . \tag{1}$$

Using this notation, we may rewrite the linear momentum as

$$\int_\Omega \rho(t, \vec{x})\vec{f}(t, \vec{x})d\omega + \int_{\partial\Omega} \mathbf{T}(t, \vec{x})\hat{n}da = \int_\Omega \rho\dot{\vec{v}}d\omega .$$

Applying the divergence theorem to the surface integral we obtain

$$\int_{\partial\Omega} \mathbf{T}(t,\vec{x})\hat{n}da = \int_{\Omega} \text{div}\mathbf{T}^T d\omega \ ,$$

where div means divergence with respect to the Eulerian or current coordinate system. Hence

$$0 = \int_{\Omega} \left(\rho(t,\vec{x})\vec{f}(t,\vec{x}) + \text{div}\mathbf{T}^T - \rho\dot{\vec{v}}\right) d\omega \ ,$$

which leads to Cauchy's first law of motion

$$\rho\dot{\vec{v}} = \rho\vec{f} + \text{div}\mathbf{T}^T \ .$$

This is usually coupled to $\mathbf{T}^T = \mathbf{T}$, and the equation of continuity: $\dot{\rho} + \rho\text{div}\vec{v} = 0$ (conservation of mass). These are called the Eulerian field equations or the Eulerian equations of motion.

To write these equations of motion in the Lagrangian or reference coordinate systems, we must make a change of variables and the configuration gradient becomes important. Recall $B = (A^{-1})^T$, $J = \det A$. The resultant contact force on the boundary $\partial\Omega$ of the current configuration Ω may be rewritten as

$$\int_{\partial\Omega} \mathbf{T}\hat{N}da = \int_{\partial\Omega_0} J\mathbf{T}\vec{B}\hat{N}da_0 \ ,$$

where $\hat{N}$ is the unit outward normal to the boundary $\partial\Omega_0$. We then define

$$\mathbf{S}^T = J\mathbf{T}\vec{B} = J\mathbf{T}(A^{-1})^T \ , \tag{2}$$

where $\mathbf{S}^T$ is the First Piola–Kirchhoff stress tensor (it is *not* symmetric). Then $\mathbf{S} = JB^T\mathbf{T} = JA^{-1}\mathbf{T}$ is the nominal (engineering) stress tensor.

Physically, the First Piola–Kirchhoff stress tensor can be used to represent the loading in the reference configuration:

$$\begin{aligned} d\vec{l} &= \mathbf{T}da \\ &= \mathbf{S}^T da_0 \ . \end{aligned}$$

Using these notions the linear momentum balance equation may be rewritten in terms of integrals over Ω_0 and $\partial\Omega_0$ (i.e., with respect to the reference or Lagrangian configuration)

$$\int_{\Omega_0} \rho_0(t,\vec{X})\vec{f_0}(t,\vec{X})d\omega_0 + \int_{\partial\Omega_0} \mathbf{S}^T(t,\vec{X})\hat{N}da_0 = \int_{\Omega_0} \rho_0(t,\vec{X})\ddot{\vec{\phi}}(t,\vec{X})d\omega_0 \ .$$

Thus we have the Lagrangian equation of motion

$$\text{Div}\mathbf{S} + \rho_0\vec{f_0} = \rho_0\ddot{\vec{\phi}} = \rho_0\ddot{\vec{u}} \tag{3}$$

where Div means divergence with respect to the reference coordinate system. We note that $\ddot{x} = \ddot{u}$ but $x \neq u$! Moreover, conservation of mass implies $\det A = \frac{\rho_0}{\rho}$.

The concepts summarized above can be readily related to the usual SEF formulations in terms of principal stretches or "extension ratios." To do this, let $\vec{M}$ be a unit vector along $d\vec{X}$ and let $\vec{m}$ be a unit vector along $d\vec{x}$. Then $\vec{m}|d\vec{x}| = A\vec{M}|d\vec{X}|$ and $|d\vec{x}|^2 = \vec{M} \cdot A^T A\vec{M}|d\vec{X}|^2$. The ratio of the deformed length to the original length of these vectors is given by

$$\frac{|d\vec{x}|}{|d\vec{X}|} = |A\vec{M}| = \{\vec{M} \cdot A^T A\vec{M}\}^{(1/2)} \equiv \lambda(\vec{M}) ,$$

and is called the *stretch* in the direction $\vec{M}$ at $\vec{X}$. In the literature the λ's are often, in a misnomer, referred to as extension ratios. However, $\lambda(\vec{M})-1$ are the true extension ratios involving the ratio of the extension $|d\vec{x}| - |d\vec{X}|$ to the original length $|d\vec{X}|$. If $\vec{M}$ is a principal direction for $A^T A$, then $\lambda^2(\vec{M})$ is an eigenvalue for the symmetric matrix $A^T A$ and $\lambda(\vec{M})$ is called a *principal stretch.*

For a body undergoing homogeneous pure strain in the principal axis system we have that the principal stretches are eigenvalues of A and (no summation)

$$\begin{aligned} x_i &= \lambda_i X_i \\ \lambda_i &= \frac{\partial x_i}{\partial X_i} = \frac{\partial u_i}{\partial X_i} + 1 . \end{aligned}$$

3 Examples

In this section we present examples to illustrate use of the methodologies presented in Section 2 to formulate models of simple extension and generalized simple shear. In each case we will begin by formulating the model for a neo–Hookean material (SEF $U = C(I_1 - 3)$). In the simple extension example we will extend this to describe more general materials.

3.1 Simple Extension

We consider a rod under uniform extension with lateral contraction where we may begin with a choice of the SEF or with Rivlin's finite strain formulation, and use this along with the above deformation theory to derive dynamic models. We use a simple example to illustrate this: an isotropic, isochoric rubber–like rod, with a tip mass, with a finite applied stress in the direction of the principal axis $x_1 = x$, as depicted in Figure 3. (Here, following standard convention, we use lower case letters to denote the Lagrangian coordinates.)

For incompressible *neo–Hookean* materials (SEF $U = \mu(I_1 - 3)$) the Cauchy (true) stress tensor is given by

$$\mathbf{T} = \begin{bmatrix} \mu\lambda_1^2 - p & & 0 \\ 0 & \mu\lambda_2^2 - p & 0 \\ 0 & 0 & \mu\lambda_3^2 - p \end{bmatrix},$$

where p is an arbitrary hydrostatic stress (necessary due to the incompressibility assumption). The configuration gradient A is given by

$$A = \begin{bmatrix} \lambda_1 & & 0 \\ 0 & \lambda_2 & 0 \\ 0 & 0 & \lambda_3 \end{bmatrix},$$

and thus the engineering stress tensor

$$\mathbf{S} = \begin{bmatrix} \mu\lambda_1 - \frac{p}{\lambda_1} & & 0 \\ 0 & \mu\lambda_2 - \frac{p}{\lambda_2} & 0 \\ 0 & 0 & \mu\lambda_3 - \frac{p}{\lambda_3} \end{bmatrix}.$$

It is useful to note that, for any homogeneous pure strain deformation, the relationship between the principal stretches and the normal components of finite strain is given by

$$\lambda_i^2 = 1 + 2\tilde{e}_{x_i x_i} = \left(1 + \frac{\partial u_i}{\partial x_i}\right)^2 .$$

Thus the incompressibility condition $J = 1$ reduces to $\lambda_1^{-1/2} = \lambda_2 = \lambda_3$ or

$$\left(1 + \frac{\partial u_1}{\partial x_1}\right)^{-\frac{1}{2}} = \left(1 + \frac{\partial u_2}{\partial x_2}\right) = \left(1 + \frac{\partial u_3}{\partial x_3}\right) . \tag{4}$$

In order to solve the full three dimensional problem, one would solve the three partial differential equations (with four unknowns u_1, u_2, u_3, and p) which arise from substituting $\mathbf{S}$ into the Lagrangian equation of motion (3) subject to the constraint (4).

An easier approach is to reduce the problem to a one dimensional motion by considering the deformation along the axis of the rod (which is also one of the principal axes for the deformation). Once the motion $u = u_1(t, x_1)$ along the axis of the rod has been calculated, the other motions u_2 and u_3 can be calculated using (4). Along the x_1 axis there is no deformation in the other directions in the one-dimensional formulation (i.e., $u_2(t) = 0$, $u_3(t) = 0$) and this yields the same equation as arising from setting $\mathbf{T}_{22} = \mathbf{T}_{33} = 0$ (which implies that $p = \frac{\mu}{\lambda_1}$). Thus the the stress tensors $\mathbf{T}$, $\mathbf{S}$ of (1), (2) reduce to one nontrivial component

$$T = \mathbf{T}_{11} = \frac{E}{3}(\lambda_1^2 - \frac{1}{\lambda_1}) ,$$

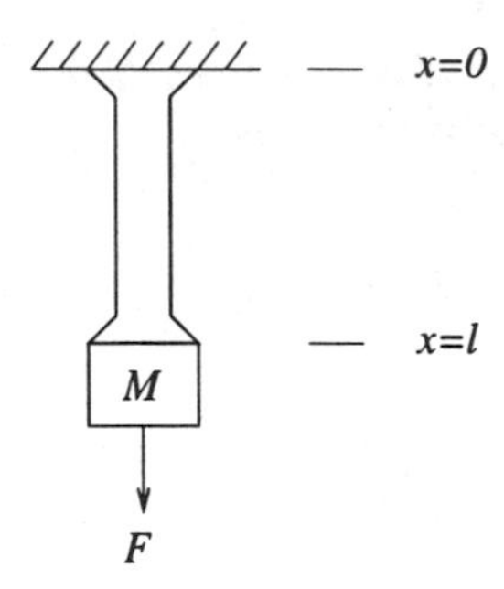

Figure 3: Rod with tip mass under tension

and

$$S = \mathbf{S}_{11} = \frac{T}{\lambda_1} = \frac{E}{3}(\lambda_1 - \frac{1}{\lambda_1^2}) \; .$$

Here $E = 3\mu$ is a generalized modulus of elasticity and we note these formulations are restricted to $\lambda_1 > 0$.

This can be used in the Lagrange formulation for longitudinal vibrations of a rubber rod with a tip mass to obtain

$$\begin{aligned} \rho A_c \frac{\partial^2 u}{\partial t^2} + \gamma \frac{\partial u}{\partial t} - \frac{\partial S_R}{\partial x} &= 0 \quad 0 < x < l \\ M \frac{\partial^2 u}{\partial t^2}(t,l) = -S_R\big|_{x=l} &+ F(t) + Mg \; , \end{aligned} \tag{5}$$

where the mass density $\rho = \rho_o$ since the body is isochoric, $F(t)$ is the applied external force, A_c is the cross sectional area, M is the tip mass, g is the gravitational constant, γ is the air damping coefficient, and S_R, the internal (engineering) stress resultant, is given by

$$\begin{aligned} S_R &= \frac{A_c E}{3}(\lambda_1 - \frac{1}{\lambda_1^2}) + C_D A_c \frac{\partial \lambda_1}{\partial t} \\ &= \frac{A_c E}{3} \tilde{g}\left(\frac{\partial u}{\partial x}\right) + A_c C_D \frac{\partial^2 u}{\partial t \partial x} \end{aligned}$$

with $\tilde{g}(\xi) = 1 + \xi - (1+\xi)^{-2}$. Here we have included a Kelvin-Voigt damping term (C_D is the Kelvin-Voigt damping coefficient) in the stress resultant S_R as a first approximation to internal damping. This leads to the nonlinear partial differential equation IBVP

$$\begin{aligned} \rho A_c \frac{\partial^2 u}{\partial t^2} + \gamma \frac{\partial u}{\partial t} - \frac{\partial}{\partial x}\left(\frac{E A_c}{3} \tilde{g}\left(\frac{\partial u}{\partial x}\right) + A_c C_D \frac{\partial^2 u}{\partial t \partial x}\right) &= 0, \\ &\quad 0 < x < l \\ M \frac{\partial^2 u}{\partial t^2}(t,l) = -\left(\frac{A_c E}{3} \tilde{g}\left(\frac{\partial u}{\partial x}\right) + C_D A_c \frac{\partial^2 u}{\partial t \partial x}\right)\Bigg|_{x=l} &+ F(t) + Mg \end{aligned} \tag{6}$$

$$u(t,0)=0, \quad u(0,x)=\Delta(x), \quad \dot{u}(0,x)=0$$

for dynamic longitudinal displacements of a neo-Hookean material rod in extension. In the case of small displacements, this does reduce to the usual longitudinal deformation equation for Hookean materials, which can be confirmed by considering the series expansion $\tilde{g}(\xi)=3\xi-3\xi^2+4\xi^3-\ldots$. This model can also be used with a more general $\tilde{g}$ to encompass other constitutive laws (which may arise from a different SEF or from estimations made using experimental results). Identification results using experimental data with more general nonlinear constitutive laws can be found for static problems in [2] and for dynamic problems in [3]. An abstract well–posedness theoretical framework that includes systems such as (6) as well as the simple shear example of the next section is presented in [1].

3.2 Generalized Simple Shear

We consider a body in generalized simple shear as depicted in Figure 4. True simple shear (which is characterized by a constant angle γ) is rarely achieved in laboratory situations, since for most bodies the angle γ depends on y. A generalization of simple shear for a neo–Hookean material provides an enlightening example when represented in terms of our continuum model. As in the simple extension example above, we use x, y, and z for the Lagrangian coordinates.

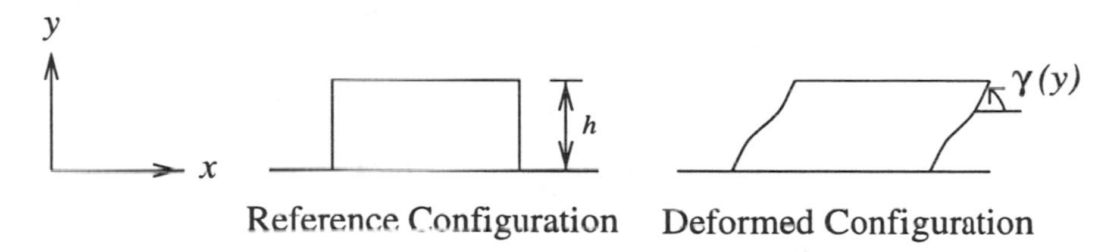

Figure 4: Simple Shear Configuration

For an incompressible, isotropic neo–Hookean material undergoing simple shear one finds [11] the Cauchy stress tensor

$$\mathbf{T}=\begin{bmatrix} \mu\gamma^2+\mu-p & \mu\gamma & 0 \\ \mu\gamma & \mu-p & 0 \\ 0 & 0 & \mu-p \end{bmatrix},$$

where μ is a material-dependent constant, p is an arbitrary hydrostatic stress (as in simple extension, p is necessary due to assumed incompressibility). Recall that, in general, the deformation $u(\vec{x})=\phi(\vec{X})-\vec{X}$, and

(with Lagrangian coordinates x, y, z,)

$$\begin{aligned} A &= \left(\frac{\partial \phi}{\partial \vec{x}}\right) \\ &= \left(\frac{\partial u}{\partial \vec{x}}\right) + I \\ &= \begin{bmatrix} 1 + \frac{\partial u}{\partial x} & \frac{\partial u}{\partial y} & \frac{\partial u}{\partial z} \\ \frac{\partial v}{\partial x} & 1 + \frac{\partial v}{\partial y} & \frac{\partial v}{\partial z} \\ \frac{\partial w}{\partial x} & \frac{\partial w}{\partial y} & 1 + \frac{\partial w}{\partial z} \end{bmatrix} . \end{aligned}$$

For simple shear

$$A = \begin{bmatrix} 1 & \gamma & 0 \\ 0 & 1 & 0 \\ 0 & 0 & 1 \end{bmatrix}$$

and hence $\gamma = \dfrac{\partial u}{\partial y}$. The engineering stress tensor is then given by

$$\mathbf{S} = \begin{bmatrix} \mu - p & p & 0 \\ \mu\gamma & \mu - p & 0 \\ 0 & 0 & \mu - p \end{bmatrix} .$$

The Lagrangian equations of motion (3) reduce to the linear wave equation

$$\rho \frac{\partial^2 u}{\partial t^2} = \mu \frac{\partial^2 u}{\partial y^2}, \qquad 0 < y < h \tag{7}$$

with boundary conditions

$$u(t,0) = 0 \ , \ \mu \frac{\partial u}{\partial y}(t,h) = F(t)$$

and initial conditions

$$u(0,y) = 0 \ , \ \dot{u}(0,y) = 0 \ .$$

This interesting result (a linear wave equation for a neo–Hookean material) is consistent with the fact that the neo–Hookean SEF ($U = C_1(I_1 - 3)$) is based upon a Hookean assumption in shear (see [13], [15]). The simple shear model as presented in (7) can be generalized to a nonlinear equation by including damping terms and assuming that the Cauchy stress tensor includes appropriate terms such as $\mathbf{T}_{12} = \mu\bar{g}(\gamma)$ for some nonlinear function $\bar{g}$.

For a neo–Hookean material in simple shear, the dynamics are simple linear shear waves moving up and down the thickness of the sample. Fitting neo–Hookean models to the data is equivalent to fitting Hookean models

– the material will be characterized by only *one parameter* μ which is essentially a *linear* modulus of elasticity. (This will be true whether one uses a SEF approach or a dynamic approach.) If one wants to understand any of the nonlinear material properties, one should not expect to achieve this with simple shear tests using neo–Hookean material constitutive laws, or, indeed, any constitutive laws arising from a SEF of the Mooney form.

It is also interesting to note that, for simple shear, the principal stretch ratios satisfy $\lambda_2 = \frac{1}{\lambda_1}$ and $\lambda_3 = 1$. Thus $I_1 = \lambda_1^2 + \lambda_2^2 + \lambda_3^2 = \lambda_1^2 + \lambda_1^{-2} + 1$ and $I_2 = \lambda_1^2\lambda_2^2 + \lambda_2^2\lambda_3^2 + \lambda_1^2\lambda_3^2 = 1 + \lambda_1^{-2} + \lambda_1^2$. Hence under simple shear the relationship $I_1 = I_2$ holds. Thus simple shear is not an appropriate test for identifying coefficients for Rivlin type SEFs.

4 Summary

In this note we have presented one approach to the modeling of dynamic behavior of composite elastomer structures. Such structures typically involve concepts familiar to both elasticity and fluid mechanics, although elastomers do not behave like either classical elastic structures or viscous fluids. We present a methodology for model development which is based on classical large deformation formulations for deformable bodies. This methodology permits the combining of computationally tractable Lagrangian descriptions of the motion with Euler variables, such as "true stress," which arise naturally in large deformations.

To illustrate the use of this methodology we have discussed two simple but very practical examples – simple uniform extension and generalized simple shear. The methodology presented in these examples has been the basis of our mathematical, computational, and experimental efforts which have been reported elsewhere [1, 2, 3].

Acknowledgments. The authors are grateful to B. Muñoz, L. Yanyo, and M. Gaitens of the Thomas Lord Research Center, Lord Corporation, and to Y. Zhang of North Carolina State University, for many helpful discussions and collaborations on the models presented here. The authors also gratefully acknowledge that this research was carried out with support in part by the U. S. Air Force Office of Scientific Research under grant AFOSR F49620-95-1-0236, and the National Science Foundation under grant NSF DMS-9508617 (with matching funds for N.J.L. from the Lord Corporation).

References

[1] H. T. Banks, D. S. Gilliam, and V. I. Shubov, *Global solvability for damped abstract nonlinear hyperbolic systems*, Tech. Rep. CRSC-TR95-25, NCSU, 1995; Differential and Integral Equations, to appear.

[2] H. T. Banks, N. J. Lybeck, M. J. Gaitens, B. C. Muñoz, and L. C. Yanyo, *Computational methods for estimation in the modeling of nonlinear elastomers*, Tech. Rep. CRSC-TR95-40, NCSU, 1995; Kybernetika, to appear.

[3] ——, *Modeling the dynamic mechanical behavior of elastomers*, Tech. Rep. CRSC-TR96-26, NCSU, 1996; Rubber Chemistry and Technology, submitted.

[4] D. J. Charlton, J. Yang, and K. K. Teh, *A review of methods to characterize rubber elastic behavior for use in finite element analysis*, Rubber Chemistry & Technology, 67 (1994), pp. 481–503.

[5] R. J. Crawford, *Plastics and Rubbers: Engineering Design and Applications*, Mechanical Engineering Publications, Ltd., London, 1985.

[6] J. D. Ferry, *Viscoelastic Properties of Polymers*, John Wiley & Sons, New York, 1980.

[7] A. N. Gent, *Engineering with Rubber: How to Design Rubber Components*, Hanser Publishers, New York, 1992.

[8] J. E. Marsden and T. J. R. Hughes, *Mathematical Foundations of Elasticity*, Prentice–Hall, Englewood Cliffs, NJ, 1983.

[9] K. Nagdi, *Rubber as an Engineering Material: Guideline for Users*, Hanser Publishers, New York, 1993.

[10] R. W. Ogden, *Large deformation isotropic elasticity – on the correlation of theory and experiment for incompressible rubberlike solids*, Proc. R. Soc. Lond. A, 326 (1972), pp. 565 – 584.

[11] ——, *Non–Linear Elastic Deformations*, Ellis Horwood Limited, Chichester, 1984.

[12] R. S. Rivlin, *Large elastic deformations of isotropic materials I, II, III*, Phil. Trans. Roy. Soc. A, 240 (1948), pp. 459–490, 491–508, 509–525.

[13] L. R. G. Treloar, *The Physics of Rubber Elasticity*, Clarendon Press, Oxford, 3rd ed., 1975.

[14] K. C. Valanis and R. F. Landel, *The strain–energy function of a hyperelastic material in terms of the extension ratios*, Journal of Applied Physics, 38 (1967), pp. 2997–3002.

[15] I. M. Ward, *Mechanical Properties of Solid Polymers*, John Wiley & Sons, New York, 2nd ed., 1983.

Center for Research in Scientific Computation, North Carolina State University, Raleigh, NC 27695-8205

Numerical Methods for Linear Control Systems

D. Boley[1] and B.N. Datta[2]

1 General Introduction

The design and analysis of linear control systems:

$$\dot{x}(t) = Ax(t) + Bu(t), \qquad y = Cx(t) \tag{1.1}$$

and

$$x_{k+1} = Ax_k + Bu_k, \qquad y_k = Cx_k \tag{1.2}$$

give rise to many interesting linear algebra problems. Some of the important ones are: *controllability and observability problems, the problem of computing the exponential matrix* $\mathbf{e}^{\mathbf{A}t}$, the matrix equations problems: *(Lyapunov equations, Sylvester equations, the algebraic Riccati equations), the pole-placement problems, stability problems, and frequency response problems.* These problems have been very widely studied in the literature. There exists a voluminous work both on theory and computations. Theory is very rich. Unfortunately, the same can not be remarked about computations. Many of the earlier methods were developed before the computer era, and are not based on numerically sound techniques. Fortunately, in the last twenty years or so, numerically effective techniques have been developed for most of these problems, and numerical analysis aspects of these methods (e.g. study of stability by round-off error analysis) and of the problems themselves (e.g. study of sensitivity) have been studied.

The purpose of this paper is: (a) first, to point out the difficulties in computational setting associated with many important theoretical methods; (b) second, to outline some of the best ones from numerical view points; and (c) third, to indicate the sensitivity issues of some of the problems that are available in the literature. Because of the lack of space, this is done only for a very few selected problems and mostly just for continuous-time models. The *survey is far from exhaustive.* Most of the methods for the continuous-time system (1.1) have discrete counterparts.

Some material of the paper has been taken from the book: *Numerical Methods for Linear Control Systems Design and Analysis*, currently being completed by one of the authors, Biswa Datta. The book will also have a MATLAB based software implementing most of the algorithms described in the book, including algorithms for second-order control systems, not discussed in this paper.

---

[1]Research supported by NSF under grant CCR-9405380.
[2]Research supported by NSF under grant DMS-9212629.

Part I: Numerical Methods for Dense Problems by B.N. Datta

2 Introduction

This part of the paper is concerned with the computational methods for some selected problems of moderate size (about a few hundred or so) with no exploitable structure.

The selected problems include some of the important ones mentioned in the introduction. It is assumed that the readers are familiar with the control theory background. Most linear systems theory books such as Chen (1984), Kailath (1980) have this material. As far as the numerical linear algebra background is concerned, as much as the author wanted to discuss it in details, was unable to do so due to space limitations. Only some crucial concepts such as *stability and conditioning*, and the methods for obtaining important canonical forms such as *the Hessenberg, Real Schur* and *Controller-Hessenberg* forms are briefly mentioned. For other linear algebra topics, essential for understanding material of this paper, such as ***LU*** and ***QR*** *factorizations* and their applications to *solutions of linear systems and least square problems*, the *QR* and *QZ* algorithms for eigenvalue and generalized eigenvalue problems, and the *singular value decomposition*, the readers are referred to the recent numerical linear algebra text of the author: *Numerical Linear Algebra and Applications* [Datta (1995) and the standard reference book in the subject: *Matrix Computations* (Golub and Van Loan (1989)]. The recent reprint book *Numerical Linear Algebra Techniques for Systems and Control*, edited by R.V. Patel, Alan J. Laub, and Paul M. Van Dooren (1993) contains most of the relevant papers.

3 Numerical Stability, Conditioning, Accuracy, and Efficiency

The numerical stability is a property of a computational algorithm, and the conditioning is a property of the problem, but both affect the accuracy of the solution.

The numerical stability of an algorithm is established by means of round-off error analysis. There are two types of stability: *forward stability* and *backward stability*. The backward stability is widely used in practice in numerical linear algebra. *In this paper, by numerical stability, we will mean backward stability.*

An algorithm is said to be *backward stable* if the computed solution obtained by the algorithm is the exact solution of a nearby problem. For example, an algorithm for solving $Ax = b$ is backward stable, if the computed solution $\hat{x}$ satisfies $(A + E)\hat{x} = b + \delta b$, where the matrix E and the vector δb are small in some measure.

A problem (with respect to a given set of data) is said to be *ill-conditioned* if a small relative error in data of the problem causes a large relative error in the solution, regardless of the method used to solve the problem. Otherwise it is *well-conditioned.* Usually numerical analysts associate a number, called the *condition number*, with the problem that gives an indication of the conditioning, e.g. the condition number of the linear systems problem $Ax = b$ is $\text{Cond}(A) = \|A\| \, \|A^{-1}\|$. *An important fact to note is that the backward stability of the algorithm does not guarantee an accurate solution.* However, when a backward stable algorithm is applied to a well-conditioned problem, the solution should be accurate. *The condition numbers of only a few control problems have been identified. Some of them will be mentioned in this paper.*

Another important aspect of a numerical algorithm is its *efficiency*. A matrix algorithm involving computations with $n \times n$ matrices is said to be *efficient* if it does not require more than $O(n^3)$ floating point operations (*Flops*). *Note that an algorithm may be efficient but unstable.* The Gaussian elimination algorithm without pivoting is efficient, but unstable in general.

4 Canonical Forms

4.1 *Ill-Conditioned Similarity Transformations and Potentially Numerically Dangerous Canonical Forms*

A common strategy for solving control problems is:

1. Transform the problem by reducing the matrices A, B, and C to some convenient "*condensed*" forms using transformations that preserve the desirable properties of the problem in hand.

2. Solve the transformed problem by exploiting the structure of the condensed forms of the matrices A, B, and C obtained in step 1.

3. Recover the solution of the original problem from the solution of the transformed problem.

Extreme caution should be taken in implementing step 1 in floating point computations. Of course, steps 2 and 3 must be implemented also in a numerically stable way, but things might go wrong in the first place.

Let $\text{fl}(s)$ denote the floating point computation of a quantity s, and let the matrix A be transformed by similarity to some condensed form. Then it can be shown [see Golub and Van Loan (1989), page 339-340], that: $\text{fl}(X^{-1}AX) = X^{-1}AX + E$, where $\|E\|_2 \approx \mu \text{Cond}_2(X)\|A\|_2$, ($\mu$ is the machine-precision). $\text{Cond}_2(X)$ is the condition number with respect to the 2-norm.

Since the error matrix E clearly depends upon $\text{Cond}_2(X)$, it follows from the above theorem that *we should avoid ill-conditioned transforming matrices in transforming* $\boldsymbol{A}$ *to a condensed form.*

Two condensed forms that have been used often in the past in control literature are: the *Jordan Canonical Form* (JCF) and the *Frobenius* (or *Block Companion*) Form (a variation of this is known as the *Luenberger Canonical Form*).

Exploitation of rich structures of these forms often makes it much easier to solve a problem.

Unfortunately, determination of both these forms might require very ill-conditioned similarity transformations. Thus, in these cases, the solutions obtained might be highly sensitive and therefore, the subsequent computations involving those solutions might be inaccurate.

It is difficult to determine exactly the Jordan structure of a matrix in case the matrix is defective or nearly defective. The eigenvector matrix X in this case is very ill-conditioned and this ill-conditioning contaminates the accurate determination of the JCF. The process of finding the companion form of a matrix A (or *equivalently finding the characteristic polynomial*) might be highly unstable. There are two stages: 1) reduction of A to a Hessenberg matrix H and 2) further reduction of H to a companion matrix. *The first stage can be achieved in a numerically stable* way using Householder's or Givens' method [see Datta (1995)] with an orthogonal transformation (note that an orthogonal matrix is well-conditioned); however, *the second stage can be highly unstable.* The transforming matrix will be severely ill-conditioned if the product of the entries on the subdiagonal of the upper Hessenberg matrix H is small.

4.2 Examples of Inaccurate Computations with the JCF and Companion Matrices

We give below a few examples to illustrate how the use of the JCF and companion matrices may give inaccurate solutions in certain cases.

Example. (Pole-Placement using the Characteristic Polynomial Approach)

> $A = 12 \times 12$ randomly generated matrix with entries between 0 and 1. $b = (0.999, 0, 0, \cdots, 0)^T$. $s = (1, 2, \cdots, 12)$: The eigenvalues to be assigned.

The feedback vector k is obtained by using the well-known Ackerman formula:

$$k = e_n^T C_M^{-1} \phi(A),$$

where C_M is the controllability matrix, and $\phi(x)$ is the desired closed-loop characteristic polynomial, The computed eigenvalues of $(A - bk)$ are : $1, 2, 2.9999, 4.0049, 4.9552, 6.0032, 6.8524 \pm 0.8937i, 9.1970 \pm 2.1155i, 12.4686 \pm 1.4509i$. (*Note some of them are even complex*).

Example. (Solving the Lyapunov Equation using the Jordan Canonical Form)

$$A = \begin{pmatrix} 1 & 0 \\ 2 & 0.9999 \end{pmatrix}, \; C = \begin{pmatrix} 6 & 3.9999 \\ 3.9999 & 1.9998 \end{pmatrix}.$$

The Lyapunov equation $XA + A^T X = C$ was solved by reducing A to a diagonal matrix first.

The computed solution $\hat{X} = \begin{pmatrix} 1.0009 & 0.9995 \\ 0.9995 & 1.0009 \end{pmatrix}$, whereas the exact solution $X = \begin{pmatrix} 1 & 1 \\ 1 & 1 \end{pmatrix}$.

Residual. $\|XA + A^T X - C\| = 0(10^{-2})$.

Suggestions. *Avoid the use of the JCF and Companion canonical forms in numerical computations, and use only canonical forms that can be obtained using well-conditioned transforming matrices, such as orthogonal transformations. The Hessenberg form, the Hessenberg-Controller form, the Real Schur forms, etc., are examples of such canonical forms.*

4.3 Reduction to Hessenberg Form

An arbitrary $n \times n$ matrix H can be transformed to an upper Hessenberg matrix H_U by orthogonality similarity:

$$PAP^T = H_U.$$

There are at least two numerically viable methods: *Householder's method* and *Given's method.* Householder's method is more efficient than Given's method, but both are numerically stable.

A matrix H is called *Householder matrix* if $H = I - \frac{2uu^T}{u^T u}$, where u is a vector. ***H** is symmetric and orthogonal.*

In Householder Hessenberg reduction, the matrix P is constructed as the product of $(n-2)$ Householder matrices P_1 through P_{n-2}. P_1 is constructed to create zeros in the first column of A below the entry $(2,1)$, P_2 is determined to create zeros below the entry $(3,2)$ of the second column of $P_1 A P_1^T$, and so on. At the end of $(n-2)$ the step, one, therefore, has $H_U = PAP^T$, where $P = P_{n-2}P_{n-3}\cdots P_1$. *Neither the matrices $\boldsymbol{P_i}$ nor the products $\boldsymbol{P_i A P_i^T}$ need to be computed explicitly.* The method requires $\frac{5n^3}{3}$

flops to compute H_U and another $\frac{2}{3}n^3$ flops to compute P. *The method is numerically stable.* For details, see Datta (1995, 146-152).

4.4 QR Factorization and the Reduction to Real Schur Form

A real matrix is said to be in *Real Schur Form* (RSF) if it is quasi-triangular - a triangular matrix with $1x1$ or 2×2 block matrices on the diagonal. The 2×2 blocks are referred to as *bumps. Every matrix A can be transformed to a RSF by orthogonal similarity:* $PAP^T = S$. This is routinely done using *QR iteration algorithm.* The QR iteration algorithm is based on *QR Factorization* of a matrix. A factorization of a matrix A in the form $A = QR$, where Q is orthogonal and R is upper triangular, is called QR factorization of A.

There are several ways to achieve this factorization: *Householder's method, Givens' method, the Classical Gram-Schmidt Process.*

Modified Gram-Schmidt Process. Again Householder's method is preferred in practice. To obtain a QR factorization of A, the Householder matrices H_1 through H_{n-1} are computed successively to create zeros below the diagonal in the first column of A, in the second column of H_1A, and so on. Thus, at the end of $(n-1)th$ step, we have $H_{n-1}H_{n-2}\cdots H_2H_1A = R$.

Taking $Q = H_1H_2\cdots H_{n-1}$, we have $A = QR$. Note that Q is orthogonal.

Again, neither the Householder matrices $\mathbf{H_i}$ *nor the products* $\mathbf{H_iA}$ *need to be computed explicitly.* The process takes about $\frac{2}{3}n^3$ flops to compute R and another $\frac{2}{3}n^3$ flops to compute Q. *It is numerically stable.*

To obtain RSF, the process is used in an iterative fashion.

A sequence of matrices $\{A_k\}$, starting with $A = A_0$ is constructed as: $A_k = R_{k-1}Q_{k-1} = Q_kR_k$, where Q_kR_k is the $Q - R$ factorization of A_k.

Each member of the sequence $\{\mathbf{A_k}\}$ *is orthogonally similar to the previous one and therefore, to the original matrix* $\mathbf{A}$. Under some mild conditions, the sequence $\{A_k\}$ can be shown to converge to a Real Schur matrix S.

This is the *basic QR iteration method.* It is not practical, because to carry out n iterations, $O(n^4)$ flops will be needed. To make it practical, the matrix A is first transformed to an upper Hessenberg matrix H_U once and for all, and then the above iterative process is applied to H_U. The convergence, even when A is first transformed to H_U, can be painfully slow in the presence of two nearly multiple eigenvalues. To speed up the convergence, the process is used with some suitable shifts which are approximate eigenvalues. In practice, double shifts are used at each iteration, and the matrices with shifts are constructed implicitly. The resulting process is known as the *implicit QR iteration algorithm with double-shifts.* For details, see Datta (1995, 435-450). The method requires *about* $13n^3$ flops to obtain both P and S. *It is stable.*

4.5 Reduction to Controller-Hessenberg Form

Given the pair of matrices (A, B), there always exists an orthogonal matrix P such that

$$PB = R = \begin{bmatrix} \bar{B} \\ 0 \end{bmatrix}, PAP^T = H,$$

where H is in block Hessenberg form with $n_1 = \rho(B) \equiv \text{rank}(B)$. This is called the *controller-Hessenberg* form for the pair (A, B).

Algorithm 41 The Controller-Hessenberg Decomposition

Input: $A \in R^{n \times n}$ and $B \in R^{n \times m}$
Output: The controller-Hessenberg form $(H, R) = (PAP^T, PB)$
of the pair (A, B).
Step 1 Compute the QR decomposition $R = Q_0 B$, set $n_1 = \rho B \equiv \text{rank } B$.
Step 2 Compute $A_1 = Q_0 A Q_0^T$.
Step 3 Compute the Block Hessenberg Decomposition $H = Q A_1 Q^T$:
For $i = 1, 2, \cdots$
Compute the QR decomposition $R_i = Q_i B_i$, where

$$A_i = \begin{bmatrix} * & * \\ B_i & \tilde{A}_i \end{bmatrix} \text{ and } B_i \text{ is } \left(n + 1 - \sum_{j=1}^{i-1} n_j \right) \times n_i.$$

Set $n_{i+1} = \rho(B_i)$.
Compute $A_{i+1} = Q_i \tilde{A}_i Q_i^T$
Compute $Q = \begin{bmatrix} I & 0 \\ 0 & Q_i \end{bmatrix} Q$
If $n_{i+1} = 0$ or $\sum_{j=1}^{i+1} n_j = n$ then Quit
Step 4 Compute $P = QQ_0$
See Paige (1981) for more details.

Flop-count and Stability. In case the system is controllable, the algorithm requires about $\frac{5}{3}n^3 + n^2m + \frac{7}{2}n^2$ flops. *The algorithm is backward stable.*

5 Solving the State Equations: Computation of e^{At}

It is well-known that the solution $x(t)$ of the control system (1.1) can be written as

$$x(t) = e^{At}x_0 + \int_0^t e^{(t-s)A}Bu(s)ds.$$

The major problem here is, therefore, computing the exponential matrix e^{At}. There is a wide variety of methods to compute the exponential matrix. Several of these methods have been carefully analyzed, with respect to efficiency and numerical stability, in an authoritative paper on the subject by Moler and Van Loan (1978). The methods can be broadly classified as: *Series Methods, Ordinary Differential Equation Methods, and the Matrix Decomposition Methods.*

Out of these methods, the method based on Padé approximation with scaling and squaring and the one based on the reduction of A to a Real Schur form are numerically attractive and widely used. ODE Methods are useful for large and sparse problems.

The drawbacks to the Taylor series method, which is a natural choice for the problem, is that a large number of terms is needed for convergence, and even when convergence occurs, the answer can be totally wrong.

Consider the following example from Moler and Van Loan (1978).

$$A = \begin{pmatrix} -49 & 24 \\ -64 & 31 \end{pmatrix}.$$

A total of $k = 59$ terms was required for convergence and the computed output was

$$e^A \approx \begin{pmatrix} -22.25880 & -1.432766 \\ -61.49931 & -3.474280 \end{pmatrix},$$

which is nowhere close to the true answer (to 6 decimal places)

$$e^A \approx \begin{pmatrix} -0.735759 & 0.551819 \\ -1.471518 & 1.103638 \end{pmatrix}.$$

The reason is that catastrophic cancellation took place in the evaluation of $\frac{A^{16}}{16!} + \frac{A^{17}}{17!}$. These two terms have almost the same magnitude but are of opposite signs.

Among the matrix-decomposition methods, the ones based on diagonalization of A, and the reduction of A to a companion matrix, should be avoided, due to reasons mentioned earlier. The *Schur-method* based on the reduction of A to a triangular matrix can certainly be used. If $P^TAP = R$ is upper triangular, then the matrix e^R is easily computed by exploiting the triangular structure of R. Unfortunately, *large round-off errors might occur if there are multiple or near multiple eigenvalues.*

Sensitivity of the Matrix Exponential Problem. The matrix exponential problem can be highly ill-conditioned unless $\boldsymbol{A}$ is a normal matrix. It can be shown that there exists a perturbation matrix E for which the relative error can be as large as

$$\kappa(A,t)\frac{\|E\|_2}{\|A\|_2}, \text{ where } \kappa(A,t) = \max_{\|E\|_2=1} \| \int_0^t e^{A(t-s)}Ee^{As}ds\|_2 \; \frac{\|A\|_2}{\|e^{At}\|_2}.$$

6 Controllability, Observability and Distance to Uncontrollability

The well-known mathematical criteria of controllability and observability such as the *Kalman rank criterion, the Hautus spectrum criteria, the disjoint spectrum criteria*, and criteria based on computing the *controllability and observability Grammians.* [see Kailath (1980), Chen (1984), etc], do not yield numerically effective tests for controllability and observability. For example, if $A = \text{diag}\left(1, 2^{-1}, \cdots, 2^{1-n}\right)$, and $B = (1, 1, \cdots, 1)^T$, then the pair (A, B) is obviously controllable. However, with $n = 10$, the controllability matrix $(B, AB, \cdots, A^9B)$ will have three *small singular values*: $0.613 \times 10^{-12}, 0.364 \times 10^{-9}$, 0.712×10^{-7}. (Note that the most *numerically viable way to find the rank of a matrix is to compute its singular values*) Thus, using the Kalman criterion on a computer whose machine precision is no smaller than 10^{-12}, one will erroneously conclude that the pair is uncontrollable. *Similar examples can be given with the other mathematical criteria.* [see Datta (1997)].

The most numerically effective way to determine the controllability of the pair $(\boldsymbol{A}, \boldsymbol{B})$ *is to transform it to the controller-Hessenberg form, described previously.*

Thus, (A, B) is controllable if in the pair $(H, \bar{B}), H$ is an *unreduced block* Hessenberg matrix, meaning that the subdiagonal blocks of H have full rank. Similarly, (A, C) is observable if in the pair $(H, \bar{C})$, H is a block unreduced upper Hessenberg matrix, where $PAP^T = H,\ CP^T = (0, \bar{C})$).

Distance from an Uncontrollable System. The concepts of controllability and observability are generic ones. Since determining if a system is controllable depends upon whether or not a certain matrix (or matrices) has full rank, it is immediately obvious that a controllable system may be very close to an uncontrollable system. For example, [Eising (1984)]: if A is an upper Hessenberg matrix with unit subdiagonal and $B = e_1$, then the pair (A, b) is obviously controllable. However, it is easily verified that if we add $(-2^{1-n}, -2^{1-n}, \cdots, -2^{1-n})$ to the last row of $[A, b]$, we obtain an uncontrollable system. Clearly, when n is large, the perturbation 2^{1-n} is small, implying that the original controllable system (A, b) is close to an uncontrollable system. *Thus, what is important in practice is knowledge of how close a controllable system is to an uncontrollable one rather than determining if a system is controllable or not.* To this end, we introduce, following Paige (1981), *a measure of controllability:*

$$\mu(A,B) \triangleq \text{ minimum } \|\Delta A, \Delta B\|_2 \text{ such that the system defined by } (A+\Delta A, B+\Delta B) \text{ is uncontrollable.}$$

Here ΔA and ΔB are complex. *If $\mu(A, B)$ is small, then the original controllable system is close to an uncontrollable system. If this distance is large, then the system is far from an uncontrollable system.*

Theorem 6.1 *(Eising (1984)) $\mu(A,B) = \min \sigma_n(sI-A,\ B)$, where $\sigma_n(sI-A,\ B)$ is the smallest singular value of $[sI-A, B]$ and s runs over all complex numbers. Based on the above theorem, several algorithms (see Patel, Laub, and Van Dooren (1993)), have been developed in the last few years to compute $\mu(A, B)$. Another recent algorithm for computing $\mu(A, B)$ is due to Elsner and He (1991). The algorithm is a Newton-type algorithm and is based on finding zeros of a certain function.*

7 Numerical Methods for Matrix Equations

In this section, we consider the well-known matrix equations in control theory:

Lyapunov Equation: $XA + A^TX = C$,
Sylvester Equation: $AX + XB = C$,
Sylvester Observer Equation: $XA - FX = DC$,
Algebraic Riccati Equation: $XA + A^TX - XBR^{-1}B^TX + Q = 0$.

The Sylvester-observer equation is a variation of the Sylvester equation, and the Lyapunov equation is a special case of the algebraic Riccati equation.

There exists a wide variety of methods for the Lyapunov and the Sylvester equations: finite and infinite series solutions methods, and methods based on reduction of the matrices A and B to convenient forms such as companion or Schwarz, Hessenberg Forms, and methods based on computing the *explicit expressions* for solutions of these equations. *Out of these, the Schur method for the Lyapunov equation*, based on the reduction of A to RSF, devised by Bartels and Stewart (1972), and the *Hessenberg-Schur method* for the *Sylvester equation*, based on the reductions of A to Hessenberg and B^T to RSF (assuming that B is of smaller dimension than A), devised by Golub, Nash, and Van-Loan (1979), are *the best from the numerical point of view and are widely used.* There also exists an important variation of the Schur method by Hammarling (1982) for constructing the Cholesky factor of the symmetric positive definite solution of the Lyapunov matrix equation, without computing the solution itself.

7.1 The Hessenberg-Schur Method for the Sylvester Equation

Let A be $m \times m$ and B be $n \times n$; $n < m$. The matrix equation $AX + XB = C$ is first reduced to the equation $HY + YS^T = C'$, where $H = U^T AU$, $V^T B^T V = S$, and $C' = U^T CV$. H is upper Hessenberg, S is in RSF, and U and V are orthogonal. The reduced Hessenberg-Schur problem is then solved for Y by finding one or two columns at a time. Finally, the solution X is recovered from $X = UYV^T$.

The Schur-Method for the Lyapunov Equation. In the case of Lyapunov equation we have $B = A^T$. Thus we obtain the Schur method for the Lyapunov equation.

Flop-Count. The Hessenberg-Schur method for the Sylvester equation requires about $\frac{5}{3}m^3 + 13n^3 + 5m^2n + \frac{5}{2}mn^2$ flops.

The Schur method for the Lyapunov equation requires about $13n^3 + \frac{7}{2}n^3$ flops.

Stability of the Schur and Hessenberg-Schur Algorithms. The round-off error analysis performed by Golub, Nash, and Van Loan (1979), and more recently in detail, by Higham (1996), shows that *these algorithms are at best conditionally backward stable*, which might be satisfactory for all practical purposes.

Hessenberg methods for the Sylvester and Lyapunov Equations

Since the reduction of a matrix to Real Schur form is rather expensive, and on the other hand, as an initial step to the Schur-reduction, one needs to transform the matrix to a Hessenberg matrix, it is natural to wonder if methods can be devised that are based solely on Hessenberg reductions. Such a method has been proposed by Datta and Datta (1987) for certain Lyapunov equations. *This algorithm followed by some iterative refinement procedure might produce acceptable results*, but the algorithms itself is not guaranteed to be stable.

7.2 The Sylvester-observer Matrix Equation

The matrix equation $XA - FX = DC$, where A and C are given such that (A, C) is observable, and X, F, and D have to be found such that (i)F has an arbitrary spectrum and (ii) (F, D) is controllable, is called the *Sylvester-observer equation.* The equation is so-called, because, it is a variation of the classical Sylvester equation considered in the last section, and this equation arises in the design of Luenberger observer [see Datta (1994) for more details on this equation].

The numerical methods for solving this equation include a method by Van Dooren (1984), a parallel/high-performance algorithm by Bischof, Datta and Purkayastha (1996), and Arnoldi-type methods for constructing *orthogonal solutions* to the equation by Datta and Saad (1991), and Datta and Hetti (1996).

The Van Dooren method is based on orthogonal reduction of the observable pair (A, C) to the observer-Hessenberg form. It has been shown that the reduced Hessenberg problem $YH - FY = D\tilde{C}$ can be solved by choosing Y as the unit-upper triangular matrix and F as lower triangular matrix with arbitrarily given diagonal entries, the remaining entries of Y and D can be computed simultaneously, by finding one row of each matrix at a time.

Neither the Van Dooren method nor the Hessenberg-Schur method for the Sylvester equation constructs, in general, an orthogonal solution. On the other hand, since the solution to a Sylvester-observer equation is used in subsequent computations in the design process of a control system, it is highly desirable, from numerical and practical view points, to have a well-conditioned solution. The Aroldi type methods of Datta and Saad (1991) and Datta and Hetti (1996) construct such solutions. The Datta-Saad method is a projection method; on the other hand the Datta-Hetti method is based on a new generalization of the classical Arnoldi-recursion.

The Bischof-Datta-Purkayastha (1996) algorithm constructs the matrix Y with a block of rows at a time by choosing the matrix F as a block bidiagonal matrix with arbitrarily chosen diagonal blocks.

7.3 The Algebraic Riccati Equation

The continuous-time algebraic Riccati equation (CARE)

$$XA + A^T X - XSX + Q = 0,$$

where $S = BR^{-1}B^T$, arises in *linear quadratic regular problem, H-infinity control problems, filter theory,* etc.

The solution of most practical interests in these applications is the *symmetric positive semidefinite stabilizing solution.*

The numerical methods for solving the CARE include (i) *The Eigenvector Methods,* (ii) *The Schur-methods* [Laub (1979)], and Byers (1986), (iii) *The Matrix Sign Function Method* [Roberts (1980)], the *generalized eigenvalue-eigenvector methods* [Van Dooren (1981), Arnold and Laub (1984), etc.].

The eigenvector and Schur methods are based on constructing a basis for an appropriate invariant subspace of the associated Hamiltonian matrix

$$H = \begin{pmatrix} A & -S \\ -Q & -A^T \end{pmatrix}.$$

*The eigenvector method can be highly unstable if there are multiple or near multiple eigenvalues of **H**.* This difficulty in the eigenvector method can be overcome by using the Schur-vectors (the vectors of the orthogonal transforming matrix that transforms H to Real Schur form), resulting in the Schur method for the problem.

Thus if $U = \begin{pmatrix} U_{11} & U_{12} \\ U_{21} & U_{22} \end{pmatrix}$ is an orthogonal matrix such that $UHU^T = \begin{pmatrix} T_{11} & T_{12} \\ 0 & T_{22} \end{pmatrix}$, where the n eigenvalues with negative real parts of H are contained in T_{11}, then the matrix $X = U_{21}U_{11}^{-1}$ is the required solution of the CARE.

The ordinary Schur method, however, fails to exploit the special Hamiltonian structure of H. The Hamiltonian-Schur method due to Byers (1986) is based on the transformation of H to *Hamiltonian-Schur form*, which preserves the special rich structure of H. The method is, however, restricted to the single-input case only.

The matrix sign function method is based on constructing the matrix sign function of the Hamiltonian matrix H, and is convenient for use on parallel machines. *It is not stable as such. But, when used in conjunction with an iterative refinement technique, produces satisfactory results.*

The generalized eigenvector and Schur methods are based on finding the eigenvectors and Schur vectors of the pencil $M - \lambda N$, where

$$M = \begin{pmatrix} A & 0 & B \\ -Q & -A^T & 0 \\ 0 & B^T & R \end{pmatrix}, N = \begin{pmatrix} I & 0 & 0 \\ 0 & I & 0 \\ 0 & 0 & 0 \end{pmatrix}.$$

The methods are, therefore, suitable when R is singular or ill-conditioned.

The pencil $M - \lambda N$ can be further compressed into an $n \times n$ pencil using an orthogonal reduction of the matrix $\begin{pmatrix} B \\ R \end{pmatrix}$. *Again, for reasons of numerical instability, the generalized Schur method is preferred.*

Newton's method is naturally an iterative method. It is, therefore, suitable for refining an approximate solution obtained by another method.

In fact, *a common practice to solve the CARE is to use the Schur method (or the generalized Schur method) followed by Newton's method.*

Newton's method is based upon iterative solutions of the Lyapunov equations: $A_i^T X_i + X_i A_i A_i + Q_i = 0$, where $A_i = A - BK_i$, $K_i = R^{-1}B^T X_i$, and $Q_i = Q + K_i^T R K_i$, $i = 0, 1, \ldots$. The matrix K_0 is so chosen that A_0 is stable.

Sensitivity of the Riccati and Lyapunov Equations. A perturbation analysis[(Byers (1985), Petkov et al (1991)] shows that the CARE is ill-conditioned if the norm of the solution matrix $\boldsymbol{X}$ is very small or very large and/or $\boldsymbol{sep}\left((\boldsymbol{A}-\boldsymbol{SX})^T, -(\boldsymbol{A}-\boldsymbol{SX})\right)$ is small, where $sep(A,-B) = \sigma_{min}(I \otimes A + B^T \otimes I)$, σ_{min} stands for the minimum singular value and $\otimes$ denotes the Kronecker product. The number $\frac{\|A\|_F + \|B\|_F}{sep(A,-B)}$ is considered as a *condition number* for the *Sylvester equation.* With $B = A$, we obtain the *condition number* for the *Lyapunov equation.* There is a *LINPACK-style estimator* for $sep(A, B)$ by Byers (1984) that avoids computing the Kronecker product and the singular values.

8 Pole-placement Problems

Given an $n \times n$ matrix A, and $n \times m$ matrix $B(m \leq n)$, and a set $S = \{s_1, \cdots, s_n\}$ of numbers, closed under complex conjugation, the problem of finding a matrix K such that $A - BK$ has the spectrum S, is known as the *pole-placement* or the *eigenvalue assignment* problem.

The problem is central to the design of a control system. There exists a wide variety of methods for the problem. Many of the earlier developed methods are based on computing the characteristic polynomial of A or equivalently on the reduction of A to companion or Luenberger canonical form are discarded for reasons of numerical instability. *The most recent numerically attractive algorithms are based on the orthogonal reduction of* $(\boldsymbol{A}, \boldsymbol{B})$ *to the controller-Hessenberg form* $(\boldsymbol{H}, \bar{\boldsymbol{B}})$.

The different algorithms differ in the way the reduced Hessenberg problem is solved. Such existing algorithms include recursive algorithms of Datta (1987), and Arnold and Datta (1990); QR algorithms of Miminis and Paige (1988), Patel and Misra (1984), Petkov et al (1991), and Arnold and Datta (1995); the Schur-methods of Varga (1981); and the matrix equation method of Bhattacharyya and DeSouza (1982).

The QR and Schur methods are believed to be numerically stable. In fact, stability of some of these algorithms have been proven by detailed round-off error analyses [Miminis and Paige (1988), (Cox and Moss (1989), Arnold and Datta (1995)].

The efficiency of all these algorithms, except the recursive ones, are comparable. *The recursive algorithm of Datta (1987) and that of Arnold and Datta (1990) are, the fastest algorithms developed so far, respectively, for the single-input and multi-input problems.*

Furthermore, it has been shown in Arnold (1993) and Arnold and Datta (1995) that all the single-input QR algorithms are related to each other - this relationship has been exposed via RQ (rather than QR) factorizations of the deflated matrices at each step. The RQ version of the single-input recursive algorithms has provided such a unified framework.

8.1 Recursive Algorithms for Pole-Placement

The single-input algorithm [Datta (1987)] is based on solution of the Sylvester-observer equation $HL - L\Lambda = (0, f)$, where Λ is a bidiagonal matrix whose diagonal entries are the eigenvalues to be assigned and f is the required feedback vector. The matrix L can be computed column by column recursively starting with $l_1 = e_n^T$. Once L is found, f is easily computed. The algorithm requires $\frac{n^3}{6}$ flops. It is not always stable, but is reliable. *The stability can be monitored or controlled by monitoring or controlling the condition number of L which can be easily computed.* Its RQ version is stable (see Arnold and Datta (1995)). The multi-input version is a straightforward generalization of the single-input algorithm [Arnold and Datta (1990)].

8.2 Robust Pole Assignment

A robust pole assignment for a *multi-input problem* is concerned with assigning not only eigenvalues, but also a set of linearly independent eigenvectors associated with the required eigenvalues such that the matrix of eigenvectors is as well-conditioned as possible. The idea is to find a feedback matrix F such that the assigned poles are as insensitive to perturbations as possible. Mathematically, given (A, B) and the set $s = \{\lambda_1, \cdots, \lambda_n\}$, find a matrix F and a non-singular matrix X satisfying

$$(A + BF)X = X\Lambda,$$

where $\Lambda = \text{diag}(\lambda_1, \cdots, \lambda_n)$, such that the eigenproblem is as well-conditioned as possible. Kautsky, Nichols, and Van Dooren (1984) have given *several iterative schemes* for choosing such an X, and an explicit expression for F, when it exists: $F = Z^{-1}U_0^T \left(X\Lambda X^{-1} - A\right)$, where $B = [U_0, U_1] \begin{bmatrix} Z \\ 0 \end{bmatrix}$.

8.3 A New Arnoldi-type Algorithm

In a recent thesis, Hetti (1995) has given an algorithm for constructing a feedback matrix F for the multi-input problem using an *orthogonal solution* to a certain Sylvester-observer matrix equation. The matrix equation is solved using a *generalization* of the *classical Arnoldi scheme*.

8.4 Conditioning of the Pole-placement and the Feedback Problems

The pole-placement is intrinsically an ill-conditioned problem. For even moderate order systems (say $n \geq 15$) the eigenvector matrix of the closed-loop matrix is usually highly ill-conditioned [He, Laub, and Mehrmann (1995)]. The condition number for the single-input *feedback problem* has recently been identified by Arnold (1993), who has also given several cheap estimators for the condition number.

9 Control Problems for Second-order Systems

There have been some fine developments in numerical solutions for control problems associated with the second-order control system: $M\ddot{x}+D\dot{x}+K = Bu$.

The traditional engineering approach is *Modal Space Control Approach*, which requires complete knowledge of eigenvalues and eigenvectors of the associated quadratic pencil $P(\lambda) = \lambda^2 M + \lambda D + K$, and is not practical for large and sparse systems. Nonmodal and partial modal approaches for the feedback stabilization, eigenvalue assignment, and partial eigenvalue assignment have been developed, which are suitable for large and sparse problems (e.g. those arising in the design of large flexible structures). The readers are referred to the papers: Datta and Rincon (1993), Datta, Elhay, and Ram (1995, 1996) and Chu and Datta (1995).

Part II: Krylov Space Methods for Control Problem by D.L. Boley

10 Krylov Space Methods

We give an overview of various Lanczos/Krylov space methods and how they are being used for solving certain problems in Control Systems Theory based on state-space models. The matrix methods used are based on Krylov sequences and are closely related to modern iterative methods for standard matrix problems such as sets of linear equations and eigenvalue calculations. We show how these methods can be applied to problems in Control Theory such as controllability, observability and model reduction. All the methods are based on the use of state-space models, which may be very sparse and of high dimensionality.

A block Krylov sequence is generated by an $n \times n$ matrix A and an $n \times p$ matrix B as follows

$$K(A, B, k) \equiv (B, AB, A^2B, \cdots, A^{k-1}B),$$

and the corresponding column space is called the *k-th Krylov space* and is denoted by $\mathcal{K}(A, B, k)$. When working with problems with very large dimensionality, it is necessary and useful to project the large problem onto a smaller subspace, and it turns out that the Krylov space is an excellent choice for such a smaller subpace. There are two main reasons for this: (a) the resulting projections tend to inherit most of the useful properties from the original operator, such as the extreme eigenvalues or the leading Markov moments, and (b) there are very efficient incremental algorithms for generating good bases for this space. We first sketch some of the principal methods for generating these spaces, then we sketch some examples where these methods have found great usefulness. Space does not permit a detailed description, but many of the details can be found in [Boley (1994].

10.1 Methods

The Arnoldi method [Arnoldi (1951), Boley and Golub (1991), and Wilkinson (1965)] is used to generate an orthogonal basis $\mathbf{X}$ for the Krylov Space, starting with an $n \times n$ matrix A and a set of p starting vectors B. The algorithm starts by orthogonalizing the starting vectors by means of a QR decomposition: $X_1R = B$. The algorithm then proceeds by recursively filling in the first few columns in the relation

$$A\mathbf{X} = \mathbf{X}\mathbf{H} \text{ subject to } \mathbf{X}^{\mathrm{T}}\mathbf{X} = I,$$

where $\mathbf{X} = (X_1, X_2, \ldots)$ is an $n \times r_{\max}$ matrix of orthonormal columns spanning the Krylov space of maximal rank $r_{\max}$ and $\mathbf{H}$ is an $r_{\max} \times r_{\max}$ block upper Hessenberg matrix. At each step j, the algorithm expands the Krylov space by forming the product AX_j, then orthogonalizes the product against all the blocks $(X_1, \ldots, X_j)$ computed so far, and finally orthonormalizes the result using the QR decomposition (with pivoting if necessary), yielding the new block of vectors X_{j+1}. The coefficients from the orthogonalizing steps are collected into the block upper Hessenberg matrix $\mathbf{H}$. When we start with only a single starting vector, some of these steps are simplified: each X_i consists of a single vector, and $\mathbf{H}$ is "scalar" upper Hessenberg.

Unless the starting vectors are deficient in certain eigendirections, $r_{\max} = n$. However, the Arnoldi algorithm is never carried out this far, but is stopped after $r \leq r_{\max}$ steps, in which case we have the orthonormal basis $\mathbf{X}_r = (X_1, \ldots, X_r)$ for the partial Krylov space $\mathcal{K}(A, B, r)$. This algorithm is the basic method behind the popular GMRES method [Saad and Schultz (1986)] for solving large sparse system of linear equations.

As a special case, if the matrix A is symmetric, then so will be the block upper Hessenberg matrix $\mathbf{H}$; hence it will be block tridiagonal. For the algorithm, this implies that at each step j, the product AX_j need be orthogonalized against only X_j and X_{j-1}. In this way it is possible to drop all the previous history, saving only the most recent blocks of vectors, at least in exact arithmetic. The resulting algorithm is called the symmetric Lanczos algorithm Lanczos (1950). It is possible to extend this savings to the nonsymmetric case, but only by sacrificing another desirable property, namely orthogonality. The resulting algorithm is usually called the non-symmetric Lanczos algorithm, and involves two Krylov spaces: the right space $\mathcal{K}(A, B, r_{\mathrm{r}})$, and the left space $\mathcal{K}(A^{\mathrm{T}}, C, r_{\mathrm{l}})$. When starting with multiple starting vectors, it is called the the block nonsymmetric Lanczos algorithm.

In the block nonsymmetric Lanczos algorithm, we generate two sets of vectors $\mathbf{X}, \mathbf{Y}$, and two block upper Hessenberg matrices of orthogonalization coefficients $\mathbf{H}, \mathbf{G}$ satisfying:

$$\text{(a) } AX = XH, \text{ (b) } A^T Y = YG, \text{ (c) } Y_i^T X_j = 0 \text{ if } i \neq j$$
$$\text{(d) } Y_i^T X_i = D_i \text{ is nonsingular, for } i \leq l$$

and $(\mathbf{x}_1, \ldots, \mathbf{x}_k)$, $(\mathbf{y}_1, \ldots, \mathbf{y}_k)$ span the first k independent columns of $\mathcal{K}(A, B, \infty)$, $\mathcal{K}(A^{\mathrm{T}}, C, \infty)$, respectively, for all $k = 1, 2, 3, \ldots$.

In the above, we have grouped the columns of $\mathbf{X}, \mathbf{G}$ into l clusters: X_1, $X_2, \ldots, X_l$, $Y_1, Y_2, \ldots, Y_l$, but these clusters do not correspond to those obtained in the block Arnoldi algorithm (101). Instead these clusters are normally just single vectors, even when starting with multiple starting vectors. Normally the *bi-orthogonality condition* (102c) is enforced between individual vectors, but if $\mathbf{D} = \mathbf{G}^{\mathrm{T}}\mathbf{H}$ is singular it is impossible to do that, so we group the vectors into *look-ahead clusters* such that the bi-orthogonality condition is enforced only between different clusters. From (102c), we see that $\mathbf{D} = \mathbf{G}^{\mathrm{T}}\mathbf{H}$ is block diagonal, and from that it follows that $\mathbf{H}, \mathbf{G}$ must both be block tridiagonal. The look-ahead clusters are defined by enforcing the condition (102d) for all the clusters, except for the last one which may be incomplete due to termination of the algorithm.

A brief sketch of the block non-symmetric Lanczos process is as follows, in which the vectors sequences $\mathbf{X}, \mathbf{Y}$ are computed one column at a time (even in the block case). We refer the reader to [Aliaga et al (1996)] for the precise details. The process is initialized by applying the two-sided Gram-Schmidt process [Parlett (1992)] to the starting vectors B, C. Suppose we have k right vectors and k left vectors:

$$\mathbf{X}_{r_{\mathrm{r}}} = (X_1, \ldots, X_l) = (\mathbf{x}_1, \ldots, \mathbf{x}_k) \text{ and } \mathbf{Y}_{r_{\mathrm{l}}} = (Y_1, \ldots, Y_l) = (\mathbf{y}_1, \ldots, \mathbf{y}_k)$$

partitioned into l clusters as indicated, where the last cluster is either empty or incomplete (meaning that the l-th diagonal block of $\mathbf{D}$, $D_l = Y_l^{\mathrm{T}} X_l$, is singular), but $\mathbf{D}_{l-1} = (Y_1, \ldots, Y_{l-1})^{\mathrm{T}}(X_1, \ldots, X_{l-1})$ is nonsingular.

Having initialized the process, the Lanczos method proceeds by (1) applying the operator A to the first vector in $\mathbf{X}$ not yet multiplied by A in this fashion, (2) biorthogonalizing the result against the more recent previous *complete* clusters $Y_{l-1}, Y_{l-2}, \ldots$ [filling in (102a)], and (3) orthogonalizing the result against the other vectors (if any) in the current *incomplete* cluster X_l. If the result from these steps is zero (meaning the new vector is already a linear combination of all the previous right Krylov vectors), then it is discarded and and a new vector is obtained by repeating steps (1), (2), (3). The nonzero result from these steps, after suitable scaling, is the next right Krylov vector, and appended to the current cluster X_l. All of this is repeated in analogous fashion for the left Krylov vector. Finally, if the resulting look-ahead cluster X_l, Y_l are such that $D_l = Y_l^{\mathrm{T}} X_l$ is nonsingular, then the current cluster is marked *complete* and closed, and we open a new empty look-ahead cluster X_{l+1}, Y_{l+1}.

10.2 Relation to Dynamical Systems

We mention some of the principal ways the Krylov space methods have been used in control and dynamical systems. For reasons of space, we limit this discussion to the case of single starting vectors ("single input single output") though most of these are easily extendible to the multiple vector case.

10.2.1 Moment Matching

A popular use of Krylov space methods is in model reduction. Suppose we have a dynamical system represented by $(A, \mathbf{b}, \mathbf{c})$ whose transfer function is

$$f(z) = \mathbf{c}^{\mathrm{T}}(zI - A)^{-1}\mathbf{b} = \sum_{i=0}^{\infty} \frac{\mathbf{c}^{\mathrm{T}} A^i \mathbf{b}}{z^{i+1}} = \sum_{i=0}^{\infty} \frac{m_i}{z^{i+1}},$$

where m_i is the i-th high frequency moment (or impulse response if in discrete time). Suppose also that we have used the scalar Lanczos to generate k vectors: $A\mathbf{X}_k = \mathbf{X}_k \mathbf{H}_k + \mathbf{x}_k \mathbf{e}_k^{\mathrm{T}} \beta_k$, $A^{\mathrm{T}} \mathbf{Y}_k = \mathbf{Y}_k \mathbf{G}_k + \mathbf{y}_k \mathbf{e}_k^{\mathrm{T}} \gamma_k$, where we have included the remainder term after the k-th step. Then one can show that the first $2k$ moments of the original system $f(z)$ are preserved in the reduced k-th order system represented by the matrices $(\mathbf{H}_k, \mathbf{e}_1, D_1^{\mathrm{T}} \mathbf{e}_1)$, where the quantities $D_1, \mathbf{H}_K$ are exactly those arising from the Lanczos algorithm, and $\mathbf{e}_1$ is the first coordinate unit vector. This idea is the basis for the model reduction techniques for space structures in [Kim and Craig (1988), (1990)] and in the Padé-Lanczos waveform analysis of circuits in

[Feldmann and Freund (1995)]. In these applications, it is also necessary to find the low frequency moments $\mathbf{c}^{\mathrm{T}}A^{-k}\mathbf{b}$, $k = 1, 2, \ldots$, which involves an implicit inversion of the system matrix A.

We note that if the Lanczos process is carried to completion so that $\mathbf{X}_k$ is a basis for the entire Krylov space $\mathcal{K}(A, \mathbf{b}, \infty)$, then the "reduced" order system $(\mathbf{H}_k, \mathbf{e}_1, D_1^{\mathrm{T}}\mathbf{e}_1)$ is the *minimal realization* of the original system, and $\mathbf{Y}_k$ is a basis for the orthogonal complement to the unobservable space. Furthermore the look-ahead clustering from the Lanczos also yields directly bases for the controllable-observable, controllable-unobservable, and uncontrollable-observable spaces, namely, $(X_1, \ldots, X_{l-1})$, X_l, and Y_l, respectively, as fully discussed in [Boley and Golub (1991), Boley (1994)].

10.2.2 Balanced Realization, Lyapunov and Riccati Equations

Another popular approach for model reduction is via the balanced realization. It is well known that in order to obtain the balanced realization of a system represented by the matrices (A, B, C), it is necessary to obtain the reachability and observability grammians: $\mathbf{W}_{\mathrm{c}}, \mathbf{W}_{\mathrm{o}}$. In the time-invariant infinite horizon case, these are obtained by solving the respective Lyapunov equations $A\mathbf{W}_{\mathrm{c}} + \mathbf{W}_{\mathrm{c}}A^{\mathrm{T}} = -BB^{\mathrm{T}}$ and $A^{\mathrm{T}}\mathbf{W}_{\mathrm{o}} + \mathbf{W}_{\mathrm{o}}A = -C^{\mathrm{T}}C$, in continuous time, and $A\mathbf{W}_{\mathrm{c}}A^{\mathrm{T}} - \mathbf{W}_{\mathrm{c}} = -BB^{\mathrm{T}}$ and $A^{\mathrm{T}}\mathbf{W}_{\mathrm{o}}A - \mathbf{W}_{\mathrm{o}} = -C^{\mathrm{T}}C$, in discrete time. The Arnoldi method has been proposed as a way of projecting the original system to a lower order system in which the corresponding projected Lyapunov equations can be easily solved [Saad (1990), Jaimoukha and Kasenally (1994)] for continuous time; [Boley (1994)] for discrete time. For example, in discrete time the reachability Lyapunov equation is projected to $\mathbf{H}_k G \mathbf{H}_K^{\mathrm{T}} + \begin{pmatrix} R \\ 0 \end{pmatrix} \begin{pmatrix} R^{\mathrm{T}} & 0 \end{pmatrix} = G$. Then the approximate solution is $\widetilde{\mathbf{W}}_{\mathrm{c}} = \mathbf{X}_k G \mathbf{X}_k^{\mathrm{T}}$, where $R, \mathbf{X}_k, \mathbf{H}_k$ are the matrices obtained from the block Arnoldi algorithm. It has been shown that this solution satisfies an algebraic Galerkin condition and also satisfies an error bound which converges to zero as the dimension of the Arnoldi approximation increases [Boley (1994)].

Similar techniques have been proposed for the more general Sylvester equation $AX - XB = C$ in [Hu and Reichel (1992)], in which the approximate solutions can be recursively generated as the Arnoldi process advances. These ideas have even been proposed as a method for solving large order Riccati equations [Hodel and Poolla (1988)], but the theory still needs further development.

References

[1] J. Aliaga, D. Boley, R. Freund, and V. Hernández. A Lanczos-type

method for multiple starting vectors. Preprint, 1996.

[2] M. Arnold. *Algorithms and Conditioning for Eigenvalue Assignment.* Ph.D. Dissertation. Northern Illinois University. DeKalb, Illinois. 1993.

[3] M. Arnold and B.N. Datta. An algorithm for the multi-input eigenvalue assignment problem. *IEEE Trans. Auto. Contr.* **35**(10) (1990), 1149-1152.

[4] M. Arnold M. and B.N. Datta. The single-input eigenvalue assignment algorithm: a close-look. To appear *SIAM J. Matrix Anal. Appl.*

[5] W.E. Arnoldi. The principle of minimized iterations in the solution of the matrix eigenvalue problem. *Quart. Appl. Math.* **9** (1951), 17-29.

[6] W.F. Arnold and A.J. Laub. Generalized eigenproblem algorithms and software for algebraic Riccati equations. *Proc. IEEE* **72** (1984), 1746-54.

[7] S.P. Bhattacharyya and E. DeSouza. Pole assignment via Sylvester's equation. *Syst. and Contr.* **1**(4), 261-263.

[8] C.H. Bischof, B.N. Datta and A. Purkayastha. A parallel algorithm for the Sylvester observer equation. *SIAM J. Sci. Comp.* **17** (1996), 686-698.

[9] D.L. Boley and G.H. Golub. The Lanczos algorithm and controllability. *Systems & Control letters* **4** (1984), 317-324.

[10] D.L. Boley. Krylov space methods on state-space control models. *Circ. Syst. & Signal Proc.* **13** (1994), 733-758.

[11] R. Byers. A LINPACK-style condition estimator for the equation $AX - XB^T = C$. *IEEE Trans. Autom. Control* **AC-29** (1984), 926-8.

[12] R. Byers. Numerical condition of the algebraic Riccati equation. *Contemporary Math.* **47** (1985), 35-49.

[13] R. Byers. A Hamiltonian QR algorithm. *SIAM J. Sci. Stat. Comput.* **7** (1986), 212-29.

[14] C.T. Chen. *Linear System Theory and Design.* New York: CBS College Publishing, 1984.

[15] E. Chu and B.N. Datta. Numerical robust pole placement for second order systems. To appear *Int. J. Control.*

[16] C.L. Cox and W.F. Moss. Backward error analysis for a pole assignment algorithm. *SIAM J. Matrix Anal. Appl.* **10** (1989), 446-456.

[17] B.N. Datta. Numerical methods for linear control systems design and analysis. Preprint, 1995.

[18] B.N. Datta. An algorithm to assign eigenvalues in a Hessenberg matrix: single-input case. *IEEE Trans. Auto. Cont.* **AC-32** (1987), 414-417.

[19] B.N. Datta. Linear and numerical linear algebra in control theory: some research problems. *Lin. Alg. Appl.* **197** (1994), 755-790.

[20] B.N. Datta. *Numerical Linear Algebra and Applications.* New York: Brooks/Cole Publishing Company, 1995.

[21] B.N. Datta and C. Hetti. An Arnoldi-type method for the Sylvester-observer equation. To appear *Proc. Eur. Contr. Conf.*

[22] B.N. Datta and F. Rincon. Feedback stabilization of the second-order model: a nonmodal approach. *Lin Alg. Appl.* **188** (1993), 138-161.

[23] B.N. Datta and Y. Saad. Arnoldi methods for large Sylvester-like observer matrix equations, and an associated algorithm for partial spectrum assignment. *Lin. Alg.Appl.* **154-156** (1991), 225-244.

[24] B.N. Datta and K. Datta. A Hessenberg method for the positive semidefinite Lyapunov equation. *Proc. 26th IEEE Conf. Dec. Contr. 1987.*

[25] B.N. Datta, S. Elhay and Y. Ram. An algorithm for the multi-input pole placement of second-order system. To appear *Proc. IEEE Conf. Dec. Contr.*

[26] B.N. Datta, S. Elhay, and Y. Ram. Orthogonality and partial pole placement for a quadratic pencil. To appear *Lin. Alg. Appl.*

[27] R. Eising. Between controllable and uncontrollable. *Systems & Control Lett.* **4** (1984), 263-4.

[28] L. Elsner and C. He. An algorithm for computing the distance to uncontrollability. *Systems & Control Letters* **17** (1991), 453-464.

[29] P. Feldmann and R. Freund. Efficient linear circuit analysis by Padé approximation via the Lanczos process. *IEEE Trans. CAD Integr. Circ. & Syst.* **14** (1995), 639-649.

[30] G. Golub, S. Nash and C. Van Loan. A Hessenberg Schur method for the problem $AX + XB = C$. *IEEE Trans. Auto. Contr.* **24**(6) (1979), 209-213.

[31] G. Golub and C. Van Loan. *Matrix Computations.* Baltimore: The Johns Hopkins University Press, 1989.

[32] S.J. Hammarling. Numerical solution of the stable, nonnegative definite Lyapunov equation. *IMA J. Numer. Anal.* **2** (1982), 303-23.

[33] C. He, A. Laub, and V. Mehrmann. Placing plenty of poles is pretty preposterous. To appear *IEEE Trans. Automatic Control.*

[34] C. Hetti, *On Numerical Solutions of the Sylvester-observer Equation and the Multi-input Eigenvalue Assignment Problem.* Ph.D. Dissertation. Northern Illinois University. DeKalb, Illinois. 1996.

[35] N.J. Higham N.J. *Accuracy and Stability of Numerical Algorithms.* Philadelphia: SIAM, 1996.

[36] A. Hodel and K. Poolla. Heuristic approaches to the solution of very large sparse Lyapunov and algebraic Riccati equations. *27th IEEE Conf. Dec. Contr. 1988*, 2217-2222.

[37] D.Y. Hu and L. Reichel. Krylov Subspace Methods for the Sylvester Equation. *Lin. Alg. Appl. and Appl.* **172** (1992), 283-313.

[38] I.M. Jaimoukha and E. M. Kasenally. Krylov subspace methods for solving large Lyapunov equations. *SIAM J. Num. Anal.* **31** (1994), 227-251.

[39] T. Kailath. **Linear Systems**. Englewood Cliffs, N.J.: Prentice Hall, 1980.

[40] J. Kautsky, N.K. Nichols and P. Van Dooren. Robust pole assignment in linear state feedback. *Int. J. Control* **41**(5) (1985), 1129-1155.

[41] H.M. Kim and R. R. Craig, Jr. Structural dynamics analysis using an unsymmetric block Lanczos algorithm. *International Journal for Numerical Methods in Engineering* **26** (1988), 2305-2318.

[42] H.M. Kim and R.R. Craig Jr. Computational enhancement of an Unsymmetric block Lanczos algorithm. *Int. J. Numerical Methods in Engineering* **30** (1990), 1083-1089.

[43] C. Lanczos. An iteration method for the solution of the eigenvalue problem linear differential and integral operators. *J. Res. Natl. Bur. Stand.* **45** (1950), 255-282.

[44] A.J. Laub. A Schur method for solving algebraic Riccati equations. *IEEE Trans. Autom. Control* **AC-24** (1979), 913-21.

[45] G.S. Miminis and C.C. Paige. A direct algorithm for pole assignment of time-invariant multi-input linear systems using state feedback. *Automatica* **24** (1988), 343-56.

[46] C. Moler and C. Van Loan (1978). Nineteen dubious ways to compute the exponential of a matrix. *SIAM Review* **20** (1978), 801-836.

[47] C.C. Paige. Properties of numerical algorithms related to computing controllability. *IEEE Trans. Aut. Contr.* **AC-26** (1981), 130-138.

[48] B.N. Parlett. Reduction to tridiagonal form and minimal realizations. *SIAM J. Matr. Anal.* **13** (1992), 567-593.

[49] R.V. Patel and P. Misra. Numerical algorithms for eigenvalue assignment by state feedback. *Proc. IEEE* **72** (1984), 1755-1764.

[50] R.V. Patel, A. Laub, and P. Van Dooren. *Numerical Linear Algebra Techniques for Systems and Control.* Piscataway, N.J.: IEEE Press, 1993.

[51] P.Hr. Petkov, N.D. Christov, and M.M Konstantinov. *Computational Methods for Linear Control Systems.* London: Prentice Hall Int'l, 1991.

[52] J.D. Roberts. Linear model reduction and solution of the algebraic Riccati equation by use of the sign function. *Int. J. Contr.* **32** (1980), 677-87.

[53] Y. Saad. Numerical solutions of large Lyapunov equations. *Signal Processing, Scattering, Operator Theory, and Numerical Methods.* (M.A. Kaashoek, J. H. Van Schuppen and A. C. Ran, Eds.). Boston: Birkäuser, 1990. 503-511.

[54] Y. Saad and M. H. Schultz. GMRES: A generalized minimal residual algorithm for solving unsymmetric linear systems. *SIAM J. Sci. Stat. Comput.* **7** (1986), 856-869.

[55] P. Van Dooren. A generalized eigenvalue approach for solving Riccati equations. *SIAM J. Sci. Stat. Comput* (1981) **2**, 121-135.

[56] A. Varga. A Schur method for pole assignment. *IEEE Trans. Autom. Control* **AC-26** (1981), 517-519.

[57] H.F. Walker. Implementation of the GMRES method using Householder transformations. *SIAM J. Sci. Stat. Comput.* **9** (1988), 152-163.

[58] J.H. Wilkinson. *The Algebraic Eigenvalue Problem.* Oxford: Oxford University Press, 1988.

Dept. of Computer Science, University of Minnesota, Minneapolis, Minnesota 55455
Dept. of Mathematical Sciences, Northern Illinois University, DeKalb, Illinois 60115

Notes on Stochastic Processes on Manifolds

Roger Brockett[1]

1 Introduction

Vision, as a sensing modality, differs from sensing a position of a shaft or the voltage from a thermocouple in that the data comes in the form of a two dimensional array coded in such a way that the location of objects, typically the information to be used in defining the feedback signal, must be extracted from the array through some auxiliary process involving image segmentation, computation of centroids, etc. It is both a blessing and a curse that sensing via light is a noncontact, "remote" modality so that the meaning of what is being sensed depends in a significant way, on the positioning of the vision sensor relative to the objects being sensed.

For these reasons, and others having to do with variations in lighting and the dependence of reflectance on geometry, closing a feedback loop using visual sensing often requires many more algorithmic steps than are required in traditional servo problems. Moreover, because the elemental sensors used to measure the amount of light that is imaged on an array by a lens system are imperfect and subject to a degree of randomness, stochastic analysis is required. Different parts of visible spectrum are transformed by the lens system in different ways, further complicating matters.

The literature on computer vision is large and growing. The books of Koenderink [1] and Faugeras [2] are among the more mathematically oriented. In them one can find a wide variety of interesting problem formulations.

2 Probability Distributions on the Circle

Before beginning a more general discussion, we consider aspects of a specific example illustrating how one can extend the theory of linear stochastic systems. The example leads to the construction of a two parameter family of probability distributions on the circle, considered as the interval $[-\pi, \pi]$, with the end points identified. In a vision context such a situation might arise if one is steering to a beacon and wants to model the error between the true heading defined by the beacon and its measured value.

---

[1]This work was supported in part by the National Science Foundation under the Engineering Research Center Program, NSF EEC-94-02384, by the U.S. Army Research Office under grant DAAL03-92-G0115 (Center for Intelligent Control Systems), and by the Office of Naval Research under Grants N00014–1887 and N00014-89-J-1023.

On the real line, the gaussian densities are characterized by a mean and a variance. Recall that the solution of the heat equation

$$\frac{\partial \rho}{\partial t} = \frac{1}{2}\frac{\partial^2 \rho}{\partial x^2}$$

corresponding to the initial condition $\rho(0,x) = \delta(x-m)$ is

$$\rho(t,x) = \frac{1}{\sqrt{2\pi t}} e^{-(x-m)^2/2t} \ ; \ t \geq 0.$$

Of course the integral of ρ is one for all time. We may say that the solution of the heat equation at time t, given that the initial condition was a delta function at $x = m$ is a gaussian of mean m and variance t. We can also consider the heat equation on the circle. The evolution equation is as above. However, because the domain is different, the solution is different. It is easily shown that either of the two expressions considered on the interval $(0 < t < \infty)$

$$\rho(t,\theta) = \frac{1}{\sqrt{2\pi t}}\left(\dots e^{-(\theta-2\pi)^2/2t} + e^{-\theta^2/2t} + e^{-(\theta+2\pi)^2/2t} + \dots\right)$$

$$= \sum_{n=-\infty}^{\infty} \frac{1}{\sqrt{2\pi t}} e^{-(\theta-2\pi n)^2/2t}$$

or

$$\rho(t,\theta) = \frac{1}{2\pi} + e^{-\frac{1}{2}t}\cos\theta + e^{-\frac{4}{2}t}\cos 2\theta + \dots = \frac{1}{2\pi} + \sum_{n=1}^{\infty} e^{-n^2t/2}\cos 2\pi n\theta$$

satisfies the diffusion equation on the circle and, when interpreted in the usual sense, approach the delta function as t approaches zero through positive values. We may think about these expressions in the following way. Consider the real line as an additive group and consider the multiples of 2π as a subgroup, isomorphic to the additive group of the integers. The quotient $\mathbb{R}/2\pi\mathbb{Z}$ is just the real numbers, modulo 2π. This can be identified with the interval $[-\pi,\pi]$ provided we regard $-\pi$ and π as being the same. Matters being so, a point x on the real line is identified with the point θ on the circle if $x - 2\pi n = \theta$ for some integer n. The first sum can be thought of as "reducing a gaussian modulo 2π." The second sum, on the other hand, can be thought of as having come from an eigenfunction expansion based on

$$\frac{1}{2}\frac{\partial^2 \cos n\theta}{\partial \theta^2} = -\frac{1}{2}n^2 \cos n\theta$$

which implies that any sum of the form

$$\rho(t,\theta) = \sum c_n e^{-n^2t/2}\cos n\theta$$

is a solution to the heat equation on the circle. These matters are discussed, for example, in Bellman [3] and Mumford [4].

This procedure works in the same way if we want to construct probability distributions on the product of circles, say the two dimensional torus, but they must be modified considerably if, for example, we want to construct a two parameter family of distributions on the sphere. The distinction is that the two dimensional sphere, S^2, cannot be expressed $\mathbb{R}^2/\Gamma$, in any suitable way. We must go to noncommutative groups, e.g. the noncommutative orthogonal groups and work with formulas such as $S^2 = \mathbb{S}o(3)/\mathbb{S}o(2)$.

In view of the ease with which one can wrap the gaussian around the circle, the reader may expect that it is possible to recast recursive estimation theory in this setting with the conditional densities being theta functions. However, a substantial problem arises. If $\theta_{m,\sigma}(x)$ denotes the theta function of mean m and variance σ, it does not happen that the product $\theta_{m,\sigma}(x)\theta_{p,\tau}(x)$ can be normalized to give a theta function. This fact frustrates the straight-forward application of Bayes' rule, preventing the direct extrapolation of Kalman filtering to this context. (For a detailed treatment of the estimation problem see Lo and Willsky [5].)

3 Some Mathematical Preliminaries

We now introduce some notation and preliminary ideas. Reference [6] covers these ideas in much more detail and is written with a view towards applications.

By a $C^{(k)}$ *differentiable manifold* we understand a triple $(\mathcal{X}, \Phi, \tau)$ such that $(\mathcal{X}, \tau)$ is a locally euclidean topological space, say of dimension n, and Φ is a collection of invertible mappings

$$\phi_\alpha : U_\alpha \to V_\alpha \subset \mathbb{R}^n$$

with the U's being open subsets covering $\mathcal{X}$ such that if two domains, say U_1 and U_2 overlap, then the map

$$\phi_2\phi_1^{-1} : V_{12} \to \hat{V_{12}}$$

is k times differentiable. The collection of maps, including the specification of the domains, is called an *atlas*. To avoid making trivial distinctions between objects that are the same for purposes of analysis, it is standard to ask that the set of maps in Φ be maximal in the sense that any map that is compatible with the given ones should be in Φ.

A *Lie algebra* is a triple, $(\mathcal{L}, [\ ,\ ], \mathcal{F})$ such that $\mathcal{F}$ is a field, $[\ ,\ ] : \mathcal{L} \times \mathcal{L} \to \mathcal{L}$ is a bilinear mapping satisfying the three properties, $[x,x] = 0$, $[x,y] = -[y,x]$, $[x,[y,z]] + [y,[z,x]] + [z,[x,y]] = 0$. The last of these conditions is called the *Jacobi Identity*.

Example 3.1 Consider the set of all real n by n skew-symmetric matrices with the bilinear operation being $[A, B] = AB - BA$. Verification of the properties requires only straightforward manipulation. Likewise, the set of all real n by n matrices whose trace is zero with the same bilinear operation form a Lie algebra. However, the set of n by n symmetric matrices does not form a Lie algebra under this definition of the bracket because the bracket of two symmetric matrices is skew-symmetric.

A *Group* is a pair $(\mathcal{G},*)$ with $* : \mathcal{G} \times \mathcal{G} \to \mathcal{G}$ being associative, having an identity, and such that for each $g \in \mathcal{G}$, there exists $g^{-1} \in \mathcal{G}$ such that $gg^{-1} =$ identity. A *Lie group* is a group whose underlying set of elements admits the structure of a differentiable manifold in such a way that the group operations of inversion and multiplication are both continuous.

Example 3.2 The set of all real n by n invertible matrices admits the structure of a Lie group. It is usually denoted as $\mathbb{GL}(n, \mathbb{R})$. The set of all real orthogonal matrices having positive determinants admits the structure of a Lie Group, often denoted as $\mathbb{SO}(n, \mathbb{R})$. It admits the structure of an $(n(n-1)/2)$-dimensional manifold consisting of those n by n matrices satisfying

$$\mathbb{S}o(n) = \{\Theta \mid \Theta^T\Theta = I, \det\Theta = 1\}.$$

It has an associated Lie algebra consisting of the skew-symmetric matrices

$$so(n) = \{\Omega \mid \Omega + \Omega^T = 0\}.$$

Given an arbitrary differentiable manifold there is, typically, no obvious or natural choice of coordinates. When it comes to Lie groups, however, the situation is better. If we are given an ordered basis for the Lie algebra, $L_1, L_2, ..., L_k$ then the product of the exponential functions

$$\Theta(\theta_1, \theta_2, ..., \theta_k) = e^{L_1\theta_1}e^{L_2\theta_2} \dots e^{L_k\theta_k}$$

can be used to specify points in the neighborhood of the identity matrix.

An Exception. The geometry of a matrix Lie group is usually such that it is impossible to choose a single global coordinate system. Any attempt will lead to the appearance of singularities. However, one does have some control over where these singularities appear. For example, a standard way of defining Euler angles for describing points on the orthogonal group $\mathbb{S}o(3)$ is to write

$$\Theta = e^{\Omega_1\phi}e^{\Omega_2\theta}e^{\Omega_1\psi}$$

with Ω_1 and Ω_2 being defined as

$$\Omega_1 = \begin{bmatrix} 0 & 1 & 0 \\ -1 & 0 & 0 \\ 0 & 0 & 0 \end{bmatrix} \quad \Omega_2 = \begin{bmatrix} 0 & 0 & 1 \\ 0 & 0 & 0 \\ -1 & 0 & 0 \end{bmatrix}.$$

This "reuse" of Ω_1 , rather than introducing

$$\Omega_3 = \begin{bmatrix} 0 & 0 & 0 \\ 0 & 0 & 1 \\ 0 & -1 & 0 \end{bmatrix}$$

has the effect of making the identity matrix a singular point.

4 Riemannian Manifolds

A differentiable manifold is said to be a *riemannian manifold* if there is given, at each point in the manifold, a positive definite quadratic form, G depending continuously on the choice of the point. This quadratic form is interpreted as defining a line element

$$(ds)^2 = \sum_{i=1}^{n} \sum_{j=1}^{n} g_{ij} dx_i dx_j .$$

It also defines a volume measure on the manifold

$$dv = \sqrt{\det G} dx_1 dx_2 ... dx_k .$$

One can think of the definition of length as being analogous to the pythagorean formula and the definition of volume as being analogous to the formula for the volume of a parallelepiped in n-dimensional euclidean space.

Example 4.1 Consider the spherical shell in three dimensional space. Suppose its diameter is two and its center is at the origin of an orthonormal coordinate system in $\mathbb{E}^3$. We introduce spherical coordinates, (ϕ, θ) with ϕ (latitude) measuring the angle down from the "north pole" and θ (longitude) measuring the angle in the $x-y$-plane. Note that the distance between a point on the sphere with coordinates (ϕ_0, θ_0) and $(\phi_0 + \epsilon, \theta_0 + \delta)$ is, for small values of ϵ and δ,

$$d \approx \sqrt{\epsilon^2 + \sin^2 \phi \, \delta^2}.$$

Thus in this case the matrix defining the quadratic form is

$$G(\phi, \theta) = \begin{bmatrix} 1 & 0 \\ 0 & \sin^2 \phi \end{bmatrix}.$$

One also sees this written in terms of the "line element"

$$(ds)^2 = (d\phi)^2 + (\sin \phi d\theta)^2 .$$

Given a manifold there may be no obvious choice of distance measure. But, as W. Killing observed more than 100 years ago, the situation in the case of Lie groups is better. In general, one can use the definition of the Lie bracket, $[x, y]$ to define a bilinear form on a Lie algebra. This works in the following way. For any element z of the Lie algebra, the mapping of $\mathcal{L}$ into $\mathcal{L}$ sending an element z into the element w given by $w = [x, [y, z]]$ is a linear map of the Lie algebra into itself. Any linear map of a vector space into itself has eigenvalues and a trace equal to the sum of the eigenvalues. Thought of in this way, we see that $\text{tr}[x, [y, \cdot]]$ is well defined and is clearly linear in both x and y. This bilinear form is called the *Killing form*. The nondegeneracy and sign definiteness of the Killing form play a central role in the study of Lie algebras. For our present purposes it is important to know that the Killing form is negative definite when evaluated on the space of real n by n skew-symmetric matrices, provided that $n \geq 3$. In fact, in this case the map $<,>: so(n) \times so(n) \to \mathbb{R}$ is a multiple of the Frobenius norm

$$\begin{aligned} <\Omega, \Omega> &= \text{tr}[\Omega, [\Omega, \cdot]] \\ &= \text{sum of eigenvalues of } [\Omega, [\Omega, \cdot]] \\ &= -(n-1)\|\Omega\|^2 \end{aligned}$$

The minus sign in the last equation should not come as a surprise. After all, in those situations in which the operator $[\Omega, \cdot]$ is skew-symmetric it has purely imaginary eigenvalues; its square will have negative eigenvalues. A similar calculation applies to $Su(n)$ (defined below) and the other real Lie algebras associated with simple compact Lie groups.

Example 4.2 Consider $\mathbb{S}o(n)$ and let $\Omega_{ij} = E_{ij} - E_{ji}$. We introduce a coordinate system in the neighborhood of the origin using the product of exponentials as described above. If we differentiate this product with respect to the coordinates the differential takes the form

$$d\Theta = (\Omega_{12} d\theta_{12} + e^{\Omega_{12}\theta_{12}} \Omega_{13} d\theta_{13} e^{-\Omega_{12}\theta_{12}} + \ldots)\Theta.$$

We define the riemannian metric by

$$\sum\sum g_{ij} dx_i dx_j = \|\Omega_{12} d\theta_{12} + e^{\Omega_{12}\theta_{12}} \Omega_{13} d\theta_{13} e^{-\Omega_{12}\theta_{12}} + \ldots\|^2.$$

This representation also leads to an explicit formula for the volume

$$dv = \sqrt{\det G}\, dx_1 dx_2 \ldots dx_n$$

Example 4.3 We can use this to get an expression for the riemannian metric on $\mathbb{S}o(3)$ in terms of the Euler angles. Starting from an evaluation of the exponential representation

$$\Theta = \begin{bmatrix} \cos\phi & \sin\phi & 0 \\ -\sin\phi & \cos\phi & 0 \\ 0 & 0 & 1 \end{bmatrix} \begin{bmatrix} \cos\theta & 0 & \sin\theta \\ 0 & 1 & 0 \\ -\sin\theta & 0 & \cos\theta \end{bmatrix} \begin{bmatrix} \cos\psi & \sin\psi & 0 \\ -\sin\psi & \cos\psi & 0 \\ 0 & 0 & 1 \end{bmatrix}.$$

Differentiating this with respect to the angles gives a formula for the riemannian metric

$$(ds)^2 = \begin{bmatrix} d\phi & d\theta & d\psi \end{bmatrix} \begin{bmatrix} 1 & 0 & \cos\theta \\ 0 & 1 & 0 \\ \cos\theta & 0 & 1 \end{bmatrix} \begin{bmatrix} d\phi \\ d\theta \\ d\psi \end{bmatrix}.$$

In the case of the orthogonal group $\mathbb{S}o(3)$ the volume measure here is

$$\sqrt{\det G} = \sqrt{(1-\cos^2\theta)} = \sin\theta.$$

In riemannian geometry the *gradient* of a function ϕ is a vector field

$$\nabla\phi = \sum_{i=1}^{n}\sum_{j=1}^{n} g^{ij}\frac{\partial\phi}{\partial x_i}\frac{\partial}{\partial x_j}$$

where g^{ij} denotes the ij^{th} entry in the matrix G^{-1}. Defined in this way, the gradient is an intrinsic quantity whose meaning is dependent on the choice of riemannian metric but independent of the choice of coordinates. In this same vein, the *divergence* of a vector field is given by

$$\operatorname{div} f = \frac{1}{\sqrt{\det G}}\sum_{i=1}^{n}\frac{\partial}{\partial x_i}\sqrt{\det G} f_i.$$

If ϕ and ψ are scalar functions then it is easy to verify the identities

$$\text{grad } \phi\psi = \psi \text{grad } \phi + \phi \text{grad } \psi.$$

Likewise, if ψ is a function and v is a vector, then

$$\text{div } \psi v = \psi \text{div } v + \langle \text{grad}\psi, v\rangle_G$$

In euclidean coordinates Laplace's operator, $\nabla^2\phi = \text{div grad}\phi$ takes the form

$$\nabla^2\phi = \left(\frac{\partial^2}{\partial x_1^2} + \frac{\partial^2}{\partial x_2^2} + \ldots + \frac{\partial^2}{\partial x_n^2}\right)\phi.$$

The *Laplace-Beltrami Operator*, which is defined on any riemannian manifold, generalizes this definition. Using the expressions for div and grad given above,

$$\nabla^2 = \sum_{i=1}^{n}\sum_{j=1}^{n}\frac{1}{\sqrt{\det G}}\frac{\partial}{\partial x_i}\sqrt{\det G}(g^{ij})\frac{\partial}{\partial x_j}.$$

Example 4.4 (Euler Again) We continue our discussion of the Lie Group $\mathbb{S}o(3)$. Using the coordinate system defined by the Euler angles, the inverse of the metric is

$$G^{-1} = \begin{bmatrix} \sin^{-2}\theta & 0 & -\cos\theta\sin^{-2}\theta \\ 0 & 1 & 0 \\ -\cos\theta\sin^{-2}\theta & 0 & \sin^{-2}\theta \end{bmatrix}.$$

Thus the Laplace-Beltrami operator is

$$\nabla^2 = \frac{1}{\sin^2\theta}\frac{\partial^2}{\partial\phi^2} + \frac{1}{\sin\theta}\frac{\partial}{\partial\theta}\sin\theta\frac{\partial}{\partial\theta} + 2\frac{\cot\theta}{\sin\theta}\frac{\partial}{\partial\phi}\frac{\partial}{\partial\psi} + \frac{1}{\sin^2\theta}\frac{\partial^2}{\partial\psi^2}.$$

5 Diffusion Process

Closely related to the Laplace-Beltrami operator is the heat equation on a riemannian manifold,

$$\frac{\partial\rho}{\partial t} = \frac{1}{2}\nabla^2\rho.$$

There is a close analogy between the diffusion of heat and the spreading of probability mass. We exploit this analogy in what follows. Let x be an n-vector and let w be an m-vector of standard white noises with components w_i, assumed to be independent. Our notation for stochastic equations of the Itô type will be

$$dx = f(x)dt + B(x)dw \;\; ; \;\; B = (b_{ij}).$$

We assume, without further explicit mention, that the differential equations are such that solutions exist for all positive time. Some sufficient conditions for this to be true are given in Gihman and Skorohod [7]. They include the hypothesis that $||b||$ and $||f||$ should grow no faster than linearly in $||x||$. The reader may also consult McKean [8] for an account of stochastic equations of this type. If it is useful to display the columns of B, we use a Greek letter index and write

$$dx = f(x)dt + \sum_{\alpha=1}^{m} b_\alpha(x)dw_\alpha.$$

Associated with the given differential equation is the differentiation rule for real valued functions

$$d\psi(x) = \sum_{i=1}^{n}\frac{\partial\psi}{\partial x_i}(f_i dt + \sum_{j=1}^{n} b_{ij}dw_j) + \frac{1}{2}\sum_{\alpha=1}^{m} b_\alpha^T\left(\frac{\partial^2\psi}{\partial x_i\partial x_j}\right)b_\alpha dt$$

and the formula for the derivative of the expectation of ψ,

$$\frac{d\mathcal{E}\psi(x)}{dt} = \sum_{k=1}^{n}\mathcal{E}\frac{\partial\psi}{\partial x_k}f_k + \frac{1}{2}\sum_{\alpha=1}^{m}\mathcal{E}b_\alpha^T\left(\frac{\partial^2\psi}{\partial x_i\partial x_j}\right)b_\alpha.$$

It is useful to abbreviate this latter formula as

$$\frac{d\mathcal{E}\psi(x)}{dt} = \mathcal{E}L^*\psi.$$

The Fokker-Plank operator, appearing on the right-hand side of the evolution equation for the probability density, is the adjoint of the operator appearing in differentiation formula. Letting $\rho(t,x)$ denote the probability density with respect to the volume measure $dx_1 dx_2 ... dx_n$, an integration-by-parts argument yields

$$\frac{\partial \rho(t,x)}{\partial t} = -\sum_{i=1}^{n} \frac{\partial}{\partial x_i} f_i \rho + \frac{1}{2} \sum_{i=1,j=1}^{n,n} \sum_{k=1}^{n} \frac{\partial^2}{\partial x_i \partial x_j} b_{ik} b_{jk} \rho.$$

If β is an everywhere positive function we can consider a different description of the density as suggested by

$$\hat{\rho}(x)\beta(x) dx_1 dx_2 ... dx_n = \frac{\rho(x)}{\beta(x)} \beta(x) dx_1 dx_2 ... dx_n.$$

The evolution equation for $\hat{\rho}$, obtained by an obvious change of variables, is

$$\frac{\partial \hat{\rho}}{\partial t} = \frac{1}{\beta} L \beta \hat{\rho} = \hat{L} \hat{\rho}$$

where L is the Fokker-Plank operator for ρ.

We now consider a riemannian space analog of the gaussian process in euclidean space. Let (X, G) be a riemannian metric and let B be such that $BB^T = G^{-1}$. (This factorization of G^{-1} is not unique but for our present purposes the nonuniqueness does not matter.)

Proposition 5.1 Suppose that $x \in X$ satisfies the Itô equation

$$dx = f(x)dt + B(x)dw.$$

Assume that BB^T is nonsingular and introduce the notation $\beta = \sqrt{\det BB^T}$, $G = (BB^T)^{-1}$. Then if

$$f_i(x) = -\frac{1}{2} \sum_{j=1}^{n} \left(\frac{1}{\beta} g^{ij} \frac{\partial \beta}{\partial x_j} - \frac{\partial g^{ij}}{\partial x_j} \right)$$

or, equivalently,

$$f = -\frac{1}{2\beta} \nabla \beta + \frac{1}{2} \sum_{\alpha=1}^{n} \left((\beta^{-1} \text{div } \beta b_\alpha) b_\alpha + \frac{\partial b_\alpha}{\partial x} b_\alpha \right)$$

the probability density with respect to the volume measure

$$d\mu = \sqrt{\det G} dx_1 dx_2 ... dx_n = \frac{1}{\beta} dx_1 dx_2 ... dx_n$$

which we write as ρ_R, satisfies

$$\frac{\partial \rho_R}{\partial t} = \frac{1}{2} \nabla^2 \rho_R$$

with ∇^2 being the Laplace-Beltrami operator corresponding to G.

Supporting Calculations. A detailed study of a more general (and less specific!) version of this problem appears in reference [9]. We give the relevant computations here. For an arbitrary f the Fokker-Planck equation for this system is

$$\frac{\partial \rho}{\partial t} = -\sum_{i=1}^{n} \frac{\partial}{\partial x_i} f_i \rho + \frac{1}{2} \sum_{i=1}^{n} \sum_{j=1}^{n} \frac{\partial^2}{\partial x_i x_j} g^{ij} \rho.$$

Thus $\rho_R = \rho_\beta$ satisfies

$$\frac{\partial \rho_R}{\partial t} = -\mathrm{div} f \rho_R + \frac{\beta}{2} \sum_{i=1}^{n} \sum_{j=1}^{n} \frac{\partial^2}{\partial x_i x_j} g^{ij} \frac{1}{\beta} \rho_R.$$

We expand the second term, moving the final $1/\beta$ inside the partial differentiation with respect to x_j, in accordance with

$$\frac{\beta}{2} \sum_{i=1}^{n} \sum_{j=1}^{n} \frac{\partial^2}{\partial x_i x_j} g^{ij} \frac{1}{\beta} = -\frac{\beta}{2} \sum_{i=1}^{n} \frac{\partial}{\partial x_i} \frac{1}{\beta^2} \sum_{j=1}^{n} g^{ij} \frac{\partial \beta}{\partial x_j} + \frac{\beta}{2} \sum_{i=1}^{n} \frac{\partial}{\partial x_i} \frac{1}{\beta} \sum_{j=1}^{n} \frac{\partial}{\partial x_j} g^{ij}.$$

The second of these terms can, in turn, be expanded as

$$\frac{\beta}{2} \sum_{i=1}^{n} \frac{\partial}{\partial x_i} \frac{1}{\beta} \sum_{j=1}^{n} \frac{\partial}{\partial x_j} g^{ij} = \frac{\beta}{2} \sum_{i=1}^{n} \frac{\partial}{\partial x_i} \frac{1}{\beta} \sum_{j=1}^{n} \frac{\partial g^{ij}}{\partial x_j} + \frac{\beta}{2} \sum_{i=1}^{n} \sum_{j=1}^{n} \frac{\partial}{\partial x_i} \frac{1}{\beta} g^{ij} \frac{\partial}{\partial x_j}.$$

Putting these steps together gives

$$\frac{\beta}{2} \sum_{i=1}^{n} \sum_{j=1}^{n} \frac{\partial^2}{\partial x_i x_j} g^{ij} \frac{1}{\beta} = -\frac{1}{2} \mathrm{div} \frac{1}{\beta} \sum_{j=1}^{n} g^{ij} \frac{\partial \beta}{\partial x_j} + \frac{1}{2} \mathrm{div} \sum_{j=1}^{n} \frac{\partial g^{ij}}{\partial x_j} + \frac{1}{2} \nabla^2.$$

Thus

$$\frac{\partial \rho_R}{\partial t} = -\mathrm{div} \left(f - \frac{1}{2\beta} \nabla \beta - \frac{1}{2} \sum_{j=1}^{n} \frac{\partial g^{ij}}{\partial x_j} \right) \rho_R + \frac{1}{2} \nabla^2 \rho_R$$

The above argument shows that, for the given choice of f, the Fokker-Plank operator and the Laplace-Beltrami operator are related by the change of measure formula $\sqrt{\det G}^{-1} \mathrm{L} \sqrt{\det G} = \frac{1}{2} \nabla^2$. We can express f more explicitly in terms of B using

$$(g^{ij}) = (\sum_{\alpha=1}^{n} b_{i\alpha} b_{j\alpha}).$$

In fact,

$$\sum_{j=1}^{n}\frac{\partial g^{ij}}{\partial x_j}=\sum_{j=1}^{n}\sum_{\alpha=1}^{n}\frac{\partial b_{i\alpha}b_{j\alpha}}{\partial x_j}=\sum_{j=1}^{n}\sum_{\alpha=1}^{n}\left(\frac{\partial b_{j\alpha}}{\partial x_j}\right)b_{i\alpha}+\sum_{j=1}^{n}\sum_{\alpha=1}^{n}\left(\frac{\partial b_{i\alpha}}{\partial x_j}\right)b_{j\alpha},$$

That is,

$$\sum_{j=1}^{n}\frac{\partial g^{ij}}{\partial x_j}=\sum_{\alpha=1}^{m}\operatorname{tr}\left(\frac{\partial b_\alpha}{\partial x}\right)b_\alpha+\sum_{\alpha=1}^{m}\left(\frac{\partial b_\alpha}{\partial x}\right)b_\alpha$$

which implies

$$f=-\frac{1}{2\beta}\nabla\beta+\frac{1}{2}\sum_{\alpha=1}^{n}\left(\operatorname{tr}\left(\frac{\partial b_\alpha}{\partial x}\right)b_\alpha+\frac{\partial b_\alpha}{\partial x}b_\alpha\right).$$

This is easily seen to be equivalent to the expression in terms of the divergence given in the statement of the proposition.

Example 5.1 Consider the scalar system

$$dx=f(x)dt+b(x)dw.$$

The associated metric is $(ds)^2=(dx)^2/b^2$ and the volume measure is dx/b. the Fokker-Planck equation is

$$\frac{\partial\rho}{\partial x}=-\frac{\partial}{\partial x}f\rho+\frac{1}{2}\frac{\partial^2}{\partial x^2}b^2\rho.$$

The effect of the change of measure is

$$b\left(-\frac{\partial}{\partial x}f\rho+\frac{1}{2}\frac{\partial^2}{\partial x^2}b^2\right)b^{-1}=-b\frac{\partial}{\partial x}fb^{-1}+\frac{b}{2}\frac{\partial^2}{\partial x^2}b.$$

Recasting the last term slightly gives

$$-b\frac{\partial}{\partial x}fb^{-1}+\frac{b}{2}\frac{\partial^2}{\partial x^2}b=-b\frac{\partial}{\partial x}fb^{-1}+\frac{b}{2}\frac{\partial}{\partial x}b_x+\frac{b}{2}\frac{\partial}{\partial x}b\frac{\partial}{\partial x}.$$

If $f=b_xb/2$ this reduces to the Laplace-Beltrami operator and so

$$\frac{\partial\rho_R}{\partial x}=-\frac{1}{2}\nabla^2\rho_R.$$

Remark. If we were using the Stratonovich calculus instead of the Itô calculus the final term in the formula for f in the proposition would not be present. Instead, we would have

$$f_S=-\frac{1}{2\beta}\nabla\beta+\frac{1}{2}\sum_{\alpha=1}^{n}(\beta^{-1}\operatorname{div}\beta b_\alpha)b_\alpha.$$

The difference between f_S and f is sometimes referred to as an "Itô correction term." We may think of the formula for f as being the sum of a gradient term, a divergence term, and an "Itô correction term". In the previous example each of these three involve $b_x b$. In the next example the gradient term is nonzero whereas both the divergence term and Itô term are zero.

Example 5.2 Consider the sphere S^2 with the standard polar coordinates (ϕ, θ) with $0 \leq \phi \leq \pi$ and $0 \leq \theta \leq 2\pi$. The riemannian metric is given by

$$(ds)^2 = (d\phi)^2 + (\sin\phi)^2 (d\theta)^2$$

Thus $\sqrt{\det G} = \sin\phi$ and

$$G^{-1} = \begin{bmatrix} 1 & 0 \\ 0 & \sin^{-2}\phi \end{bmatrix}.$$

The Laplace-Beltrami operator is the familiar laplacian on the sphere

$$\begin{aligned} \nabla^2 &= \frac{1}{\sin\phi} \begin{bmatrix} \frac{\partial}{\partial\phi} & \frac{\partial}{\partial\theta} \end{bmatrix} \begin{bmatrix} \sin\phi & 0 \\ 0 & \sin^{-1}\phi \end{bmatrix} \begin{bmatrix} \frac{\partial}{\partial\phi} \\ \frac{\partial}{\partial\theta} \end{bmatrix} \\ &= \frac{1}{\sin\phi}\frac{\partial}{\partial\phi}\sin\phi\frac{\partial}{\partial\phi} + \frac{1}{\sin^2\phi}\frac{\partial^2}{\partial\theta^2}. \end{aligned}$$

We can express G^{-1} as

$$G^{-1} = b_1 b_1^T + b_2 b_2^T = \begin{bmatrix} 1 \\ 0 \end{bmatrix} \begin{bmatrix} 1 & 0 \end{bmatrix} + \begin{bmatrix} 0 \\ \sin^{-1}\phi \end{bmatrix} \begin{bmatrix} 0 & \sin^{-1}\phi \end{bmatrix}.$$

The divergence of βb_1 and the divergence of βb_2 are zero. The Itô correction term also vanishes in both cases, but

$$\nabla\beta = -\begin{bmatrix} \frac{\cos\phi}{\sin^2\phi} \\ 0 \end{bmatrix}.$$

According to Proposition 5.1, the Itô equations

$$d\phi = \frac{1}{2}\cot\phi\, dt + dw$$

$$d\theta = \frac{1}{\sin\phi} d\nu$$

define a process on S^2 whose density, relative to the riemannian volume, evolves as dictated by the heat equation. We verify this as follows. Using the formula for the Fokker-Planck equation given above, we see that the density with respect to the measure $d\phi d\theta$ satisfies

$$\frac{\partial\rho}{dt} = -\frac{1}{2}\frac{\partial}{\partial\phi}\cot\phi + \frac{1}{2}\left(\frac{\partial^2}{\partial\phi^2} + \frac{1}{\sin^2\phi}\frac{\partial^2}{\partial\theta^2}\right)\rho.$$

Thus the probability density ρ_R, defined with respect to the riemannian volume measure $\sin\phi d\phi d\theta$ satisfies the equation

$$\frac{\partial\rho_R}{dt} = \frac{1}{2}\frac{1}{\sin\phi}\left(-\frac{\partial}{\partial\phi}\cot\phi + \frac{\partial^2}{\partial\phi^2} + \frac{1}{\sin^2\phi}\frac{\partial^2}{\partial\theta^2}\right)\sin\phi\ \rho_R$$

but

$$\frac{1}{2}\frac{1}{\sin\phi}\left(-\frac{\partial}{\partial\phi}\cot\phi + \frac{\partial^2}{\partial\phi^2}\right)\sin\phi = \frac{1}{2}\frac{1}{\sin\phi}\frac{\partial}{\partial\phi}\sin\phi\frac{\partial}{\partial\phi}$$

and so ρ_R satisfies the heat equation on S^2.

6 Processes Defined on Submanifolds

A subset of $\mathbb{R}^n$ of the form

$$X = \{x|c_1(x) = 0; c_2(x) = 0; ...c_r(x) = 0\}$$

admits the structure of a k-dimensional differentiable manifold provided that the rank of

$$C = \begin{bmatrix} \frac{\partial c_1}{\partial x_1} & \frac{\partial c_1}{\partial x_2} & \cdots & \frac{\partial c_1}{\partial x_n} \\ \frac{\partial c_2}{\partial x_1} & \frac{\partial c_2}{\partial x_2} & \cdots & \frac{\partial c_2}{\partial x_n} \\ \cdots & \cdots & \cdots & \cdots \\ \frac{\partial c_r}{\partial x_1} & \frac{\partial c_r}{\partial x_2} & \cdots & \frac{\partial c_r}{\partial x_n} \end{bmatrix}$$

equals $n-k$ in a neighborhood of X. Such submanifolds can be given a system of local coordinates that can be extended to a full set of local coordinates on $\mathbb{E}^n$; they inherit a riemannian structure from $\mathbb{E}^n$, etc. In particular, if B is a rank k matrix such that $CB = 0$, then the matrix BB^T can, by a possibly x-dependent, orthogonal transformation, be put in block form

$$\Theta^T BB^T\Theta = \begin{bmatrix} \tilde{G}_{11}^{-1} & 0 \\ 0 & 0 \end{bmatrix}$$

with $\tilde{G}_{11}$ invertible. In other words, $\tilde{G}_{11}^{-1}$ is the restriction of BB^T to the tangent space of X. It is necessarily positive definite and hence $\tilde{G}_{11}$ itself defines a riemannian metric on X. More specifically, letting $\tilde{g}_{ij}$ denote the ij^{th} component of $\tilde{G}_{11}$, we can choose local coordinates $y_1, y_2, ..., y_k$ for an open subset of X such that on X

$$(ds)^2 = \sum_{i=1}^{k}\sum_{j=1}^{k}\tilde{g}_{ij}dy_idy_j$$

is a metric.

Proposition 6.1 Suppose that $X \subset \mathbb{E}^n$ is given by

$$X = \{x|c_1(x) = 0; c_2(x) = 0; ...c_r(x) = 0\}$$

with C being of rank $n-k$ in a neighborhood of X. Suppose that $x \in X$ satisfies the Itô equation

$$dx = f(x)dt + B(x)dw$$

with $CB = 0$, BB^T of rank k, and

$$\sum_{j=1}^{n} \frac{\partial c_i}{\partial x_j} f_j(x)dt + \frac{1}{2} b_\alpha^T \frac{\partial^2 c_i}{\partial x^2} b_\alpha = 0 \; ; \; i = 1, 2, ...r.$$

Define β to be the square root of the product of the k nonzero eigenvalues of BB^T and let g^{ij} denote the ij^{th} entry of BB^T. Then if

$$f_i(x) = -\frac{1}{2} \sum_{j=1}^{n} \left(\frac{1}{\beta} g^{ij} \frac{\partial \beta}{\partial x_j} - \frac{\partial g^{ij}}{\partial x_j} \right)$$

or, equivalently,

$$f = -\frac{1}{2\beta} \nabla \beta + \frac{1}{2} \sum_{\alpha=1}^{n} \left(\mathrm{tr} \left(\frac{\partial \beta b_\alpha}{\partial x} \right) b_\alpha + \frac{\partial b_\alpha}{\partial x} b_\alpha \right),$$

the probability density with respect to the volume measure

$$d\mu = \frac{1}{\beta(x)} dy_1 dy_2 ... dy_n$$

satisfies

$$\dot{\rho_R} = \frac{1}{2} \nabla^2 \rho_R$$

with ∇^2 being the Laplace-Beltrami operator on X corresponding to the metric $\tilde{G}$.

Supporting Calculations. The first thing to observe is that the stochastic differential equation actually evolves on X. To see this, notice that an application of the Itô differentiation rule shows that

$$dc_i = \sum_{j=1}^{n} \frac{\partial c_i}{\partial x_j} f_j(x)dt + \sum_{\alpha=1}^{m} b_\alpha^T \left(\frac{\partial^2 c_k}{x_i x_j} \right) b_\alpha dt + \sum_{\alpha=1}^{m} \frac{\partial c_i}{\partial x} b_\alpha dw_\alpha = 0.$$

The coefficient of dw_α vanishes by virtue of $CB = 0$ and the coefficient of dt vanishes by explicit hypothesis. Thus the c_i are constant. Matters being so, we can pick local coordinates for X, say $y_1, y_2, ..., y_k$, and rewrite the stochastic evolution in terms of y

$$dy = \tilde{f}(y)dt + \tilde{B}dw.$$

But now we are in the situation covered by Proposition 5.1.

Example 6.1 Let $n = 3$ and consider the single constraint $c(x) = x_1^2 + x_2^2 + x_3^2 - 1 = 0$. As an application of the differentiation rule shows, the Itô equation

$$\begin{bmatrix} dx_1 \\ dx_2 \\ dx_3 \end{bmatrix} = \begin{bmatrix} -x_1 \\ -x_2 \\ -x_3 \end{bmatrix} dt + \begin{bmatrix} x_2 \\ -x_1 \\ 0 \end{bmatrix} dw_1 + \begin{bmatrix} x_3 \\ 0 \\ -x_1 \end{bmatrix} dw_2 + \begin{bmatrix} 0 \\ x_3 \\ -x_2 \end{bmatrix} dw_3$$

define a flow on X. In this case

$$B(x)B(x)^T = \begin{bmatrix} x_2^2 + x_3^2 & -x_1x_2 & -x_1x_3 \\ -x_1x_2 & x_1^2 + x_3^2 & -x_2x_3 \\ -x_1x_3 & -x_2x_3 & x_1^2 + x_2^2 \end{bmatrix}.$$

It is easy to see that x spans the null space of BB^T and that the two nonzero eigenvalues of BB^T are one. The formula for f given in the proposition statement simplifies because β is constant. We easily verify that

$$(f_i(x)) = -\frac{1}{2}\left(\sum_{j=1}^{n} \frac{1}{\beta} g^{ij} \frac{\partial \beta}{\partial x_j} - \frac{\partial g^{ij}}{\partial x_j}\right) = \begin{bmatrix} -x_1 \\ -x_2 \\ -x_3 \end{bmatrix}.$$

Thus the given equation provides an alternative for the description of the stochastic process on S^2 given in Example 5.2.

Example 6.2 The set of n by n matrices is a vector space of dimension n^2. The sum of the squares of the entries defines a positive definite quadratic form and hence a euclidean structure. The n by n orthogonal matrices, i.e. the subset of this space defined by

$$X = \{\Theta | \Theta^T \Theta = I\},$$

can be given the structure of an $n(n-1)/2$ dimensional manifold and also the structure of a Lie group. The associated Killing form is negative definite. Its negative defines a riemannian metric on the orthogonal group and, by implication, a Laplace-Beltrami operator. Let Θ take on values in X. Consider the stochastic equation

$$d\Theta = -\frac{n-1}{2}\Theta dt + \sum (E_{ij} - E_{ji})\Theta dw_{ij}$$

with w_{ij} being independent Wiener processes. An application of the Itô calculus shows that $\Theta^T(t)\Theta(t) = I$ for all time, assuming that $\Theta^T(0)\Theta(0) = I$. The restriction of BB^T to the tangent space can be shown to be the identity. More details are given in [11].

7 Degenerate Diffusions

There are situations in which the number of independent Wiener processes entering the right-hand side of the differential equations noise terms are fewer than the dimension of the manifold and thus BB^T is not invertible, but that the c's of Proposition 6.1 fail to exist because the subspace of the tangent space defined by range B is not integrable. Even so, and this is amply confirmed by standard results in linear theory, it can happen that smooth densities exist. We consider the treatment of some situations of this type in the proposition below.

Observe that in the expression for the Laplace-Beltrami operator, it is only in the divergence that G itself appears, and then only through its determinant. Otherwise, the operator depends only on $g^{ij} = \Sigma b_{ik} b_{jk}$, which exists even in a degenerate situation. Moreover, in the divergence

$$\text{div} \equiv \frac{1}{\sqrt{\det G}} \sum \frac{\partial}{\partial x_i} \sqrt{\det G}$$

the determinant and its reciprocal "almost" cancel and do, in fact, when $\det G$ does not depend on x. This suggests that under suitable assumptions it should be possible to generalize the previous analysis.

Suppose that (X, G) is a riemannian manifold with $G = G(\epsilon)$. Assume that $G(\epsilon)$ is such that $G^{-1}(\epsilon)$ depends continuously on ϵ in a neighborhood of $\epsilon = 0$ and that $\det G^{-1}(\epsilon) = \psi(\epsilon)\beta^{-2}(x)$. (It would, of course, defeat the purpose to assume that $G(\epsilon)$ has a limit as ϵ goes to zero.) We can, under these circumstances, associate with $G(\epsilon)$ an *Extended Laplace-Beltrami Operator*

$$\nabla_e^2 = \sum_{i=1}^{n} \sum_{j=1}^{n} \frac{1}{\beta(x)} \frac{\partial}{\partial x_i} \beta(x) g^{ij} \frac{\partial}{\partial x_j}$$

which is simply a suitable interpretation of the limit

$$\nabla_e^2 = \lim_{\epsilon \to 0} \sum_{i=1}^{n} \sum_{j=1}^{n} \frac{1}{\sqrt{\det G}} \frac{\partial}{\partial x_i} \sqrt{\det G} g^{ij} \frac{\partial}{\partial x_j}.$$

Thus we see that under the given hypothesis it is possible to allow the rank of G^{-1} to drop and still have a meaningful expression.

If we extend B by adding additional columns multiplied by ϵ, we may hope to put ourselves in the situation described above with

$$\tilde{G}(\epsilon) = (B, \epsilon\hat{B})(B, \epsilon\hat{B})^T.$$

In doing so, it can happen that the choice of $\hat{B}$ influences the normalization $\beta = \sqrt{\det G(\epsilon)}$. For this reason, a specific choice of $\hat{B}$ must be made. With this in mind, we introduce a type of mixing hypothesis, or controllability condition, and use the additional structure that it provides to specify $\hat{B}$. Let $[b_\alpha, b_\beta]$ denote the Lie bracket

$$[b_\alpha, b_\beta] = \frac{\partial b_\beta}{\partial x} b_\alpha - \frac{\partial b_\alpha}{\partial x} b_\beta.$$

We describe a canonical choice of G in the special case where the vectors b_α together with vectors of the form $[b_\alpha, b_\beta]$ span the space. Let

$$\hat{B} = ([b_1, b_2], [b_1, b_3], ...[b_{n-1}, b_n])$$

take $G(\epsilon)$ to be

$$G(\epsilon) = (B, \epsilon([b_1, b_2], [b_1, b_3], ...[b_{n-1}, b_n]))(B, \epsilon([b_1, b_2], [b_1, b_3], ...[b_{n-1}, b_n]))^T$$

and assume that $G(\epsilon)$ is nonsingular when ϵ is nonzero. (Compare with the construction of a metric in reference [10], page 26.)

Proposition 7.1 Suppose that $x \in X$ satisfies the Itô equation

$$dx = f(x)dt + B(x)dw$$

with B such that $\tilde{B} = (b_1, b_2, ...b_m, \epsilon[b_1, b_2], \epsilon[b_1, b_3], ..., \epsilon[b_{m-1}, b_n])$ is of rank n for $\epsilon \neq 0$. Let $\tilde{G}^{-1}(\epsilon) = \tilde{B}\tilde{B}^T$ and introduce the notation $\tilde{\beta} = \sqrt{\det \tilde{B}\tilde{B}^T}$, Suppose, further, that $\tilde{\beta} = \epsilon^k \beta(x)$. Then if

$$f_i(x) = -\frac{1}{2} \sum_{j=1}^{n} \left(\frac{1}{\beta} g^{ij} \frac{\partial \beta}{\partial x_j} - \frac{\partial g^{ij}}{\partial x_j} \right)$$

or, equivalently,

$$f = -\frac{1}{2\beta} \nabla \beta + \frac{1}{2} \sum_{\alpha=1}^{n} \left(\text{tr} \left(\frac{\partial \beta b_\alpha}{\partial x} \right) b_\alpha + \frac{\partial b_\alpha}{\partial x} b_\alpha \right)$$

the probability density with respect to the volume measure

$$d\mu = \frac{1}{\beta(x)} dx_1 dx_2 ... dx_n$$

satisfies

$$\frac{\partial \rho_\beta}{\partial t} = \frac{1}{2} \nabla_e^2 \rho_\beta$$

with ∇_e^2 being the extended Laplace-Beltrami operator corresponding to $\tilde{G}$.

Example 7.1 We may apply this to the Itô equations

$$dx = dw$$

$$dy = d\nu$$
$$dz = ydw - xd\nu.$$

In this case B is given by

$$B(x,y) = \begin{bmatrix} 1 & 0 \\ 0 & 1 \\ y & -x \end{bmatrix}$$

and

$$[b_1, b_2] = \begin{bmatrix} 0 \\ 0 \\ -2 \end{bmatrix}.$$

Make the definition

$$\tilde{B}(x,y) = \begin{bmatrix} 1 & 0 & 0 \\ 0 & 1 & 0 \\ y & -x & 2\epsilon \end{bmatrix}.$$

Clearly $\det \tilde{B} = 2\epsilon$ and the corresponding extended Laplace-Beltrami operator is the limit as ϵ goes to zero of the nondegenerate form associated with the metric

$$\tilde{G} = (BB^T)^{-1} = \frac{1}{4\epsilon^2} \begin{bmatrix} 4\epsilon^2 + y^2 & -xy & -y \\ -xy & 4\epsilon^2 + x^2 & x \\ -y & x & 1 \end{bmatrix}.$$

The associated measure is $dxdydz$ and

$$\frac{\partial \rho_\beta}{\partial t} = \frac{1}{2}\left(\left(\frac{\partial}{\partial x} + y\frac{\partial}{\partial z}\right)^2 \left(\frac{\partial}{\partial y} - x\frac{\partial}{\partial z}\right)^2\right)\rho.$$

8 Gradient Systems

If the manifold X is connected and if it has finite volume then the probability density associated to the differential equation of Proposition 5.1 approaches the constant $1/\mathrm{vol}(X)$ as t approaches infinity; if the volume is infinite there is no invariant measure. However, the insertion of the gradient of a potential function into the differential equation opens up new possibilities. If ϕ is a function that takes on a minimum at x_0 and is such that its sub-level sets

$$S_a = \{x | \phi(x) \leq a\}$$

are strictly convex, then the gradient flow associated with ϕ gives rise to a restoring force that causes the solutions of

$$\dot{x} = -\frac{1}{2}\nabla\phi(x)$$

to flow to x_0. In a stochastic setting a white noise term will tend to cause trajectories to wander away from zero. The combination of the two effects may be expected to give rise to a dynamic equilibrium characterized by a steady state invariant measure. We now describe some situations in which this happens with the invariant density taking the form

$$\rho_R(\infty, x) = \frac{1}{N} e^{-\phi(x)}.$$

The results we give can be thought of as generalizations of the fact that for

$$dx = -\frac{1}{2}\frac{\partial \phi}{\partial x} + dw.$$

There is an invariant measure of the form

$$\rho_R(x) = \frac{1}{N} e^{-\phi(x)}$$

This is readily verified from the Fokker-Planck equation and the identities

$$\frac{\partial^2}{\partial x^2} e^{-\phi(x)} = (-\phi_{xx} + \phi_x^2) e^{-\phi(x)}$$

and

$$\frac{\partial}{\partial x} \phi_x e^{-\phi(x)} = (\phi_{xx} - \phi_x^2) e^{-\phi(x)}.$$

Proposition 8.1 Suppose that $x \in X$ satisfies the Itô equation

$$dx = f(x)dt + B(x)dw.$$

Assume that BB^T is nonsingular and introduce the notation $\beta = \sqrt{\det BB^T}$, $G = (BB^T)^{-1}$. Then if

$$f_i(x) = -\frac{1}{2}\sum_{j=1}^{n} g^{ij}\frac{\partial \phi}{\partial x_j} - \frac{1}{2}\sum_{j=1}^{n}\left(\frac{1}{\beta} g^{ij}\frac{\partial \beta}{\partial x_j} - \frac{\partial g^{ij}}{\partial x_j}\right)$$

or, equivalently,

$$f = -\frac{1}{2}\nabla\phi - \frac{1}{2\beta}\nabla\beta + \frac{1}{2}\sum_{\alpha=1}^{n}\left(\operatorname{tr}\left(\frac{\partial b_\alpha}{\partial x}\right) b_\alpha + \frac{\partial b_\alpha}{\partial x} b_\alpha\right)$$

the probability density with respect to the riemannian volume measure

$$d\mu = \frac{1}{\beta} dx_1 dx_2 \ldots dx_n$$

satisfies

$$\dot{\rho}_R = \frac{1}{2}\operatorname{div}(\nabla\phi)\rho_R + \frac{1}{2}\nabla^2 \rho_R$$

with ∇^2 being the Laplace-Beltrami operator corresponding to G. Moreover, if ϕ is such that the integral

$$N = \int_X e^{-\phi(x)} dx_1 dx_2 ... dx_n$$

exists, there exists a steady state density in the form

$$\frac{\partial \rho_\beta}{\partial t} = \frac{1}{N} e^{-\phi(x)}.$$

Supporting Calculations. An obvious modification of the calculation offered in support of Proposition 5.1 shows that in the present situation the equation for the density ρ_R can be expressed as

$$\frac{\partial \rho_R}{\partial t} = \frac{1}{2}\beta \sum_{i=1}^{n} \frac{\partial}{\partial x_i} \sum_{j=1}^{n} g^{ij} \left(\frac{\partial \phi}{\partial x_j}\right) \frac{1}{\beta}\rho_R + \frac{1}{2}\nabla^2 \rho_R$$

which is equivalent to

$$\frac{\partial \rho_R}{\partial t} = \frac{1}{2}\text{div }(\text{grad}\phi)\rho_R + \frac{1}{2}\nabla^2 \rho_R.$$

Using the fact that

$$\nabla^2 e^{-\phi(x)} = -\text{div}(\nabla\phi) e^{-\phi(x)}$$

we see that any multiple of $e^{-\phi(x)}$ is a steady state solution.

Example 8.1 Consider again a process on the sphere S^2 with coordinates (ϕ, θ) and riemannian metric as in Example 5.2. We alter the Itô equations given there by the addition of a gradient term corresponding to the potential $-\cos\phi$

$$d\phi = -\frac{1}{2}\sin\phi - \frac{1}{2}\cot\phi \, dt + dw$$

$$d\theta = \frac{1}{\sin\phi} d\nu.$$

The invariant density is

$$\rho_R(\infty, \phi, \theta) = \frac{1}{N} e^{\cos\phi}.$$

Example 8.2 Let $\mathbb{E}^n$ denote cartesian n-space considered as a riemannian manifold with the usual distance function. Let $\phi : \mathbb{E}^n \to \mathbb{R}$ be a differentiable function and consider the Itô equation in $\mathbb{E}^n$

$$dx = -\frac{1}{2}\nabla\phi(x) dt + dw.$$

The corresponding Fokker-Planck equation is just

$$\frac{\partial\rho(t,x)}{\partial t}=\frac{1}{2}\sum_{i=1}^{n}\frac{\partial}{\partial x_i}\left(\frac{\partial\phi}{\partial x_i}\right)\rho(t,x)+\frac{1}{2}\left(\frac{\partial^2}{\partial x_1^2}+...+\frac{\partial^2}{\partial x_n^2}\right)\rho(t,x).$$

If ϕ grows sufficiently rapidly with $||x||$ so that the integral

$$N=\int_{\mathbb{E}^n}e^{-\phi(x)}dx_1dx_2...dx_n$$

exists, then it is straightforward to verify that the time independent density

$$\rho(\infty,x)=\frac{1}{N}e^{-\phi(x)}$$

is a solution. We now extend this.

Proposition 8.2 Suppose that $X\subset\mathbb{E}^n$ is given by

$$X=\{x|c_1(x)=0;c_2(x)=0;...c_r(x)=0\}$$

with C being of rank $n-k$ in a neighborhood of X. Suppose that $x\in X$ satisfies the Itô equation

$$dx=f(x)dt+B(x)dw$$

with $CB=0$, BB^T of rank k, and

$$\sum_{j=1}^{n}\frac{\partial c_i}{\partial x_j}f_j(x)dt+\frac{1}{2}b_\alpha^T\left(\frac{\partial^2c_i}{\partial x^2}\right)b_\alpha=0\ ;\ i=1,2,...r.$$

Let G^{-1} denote the restriction of BB^T to the tangent space of X. Then if

$$f_i(x)=-\frac{1}{2}\sum_{j=1}^{n}g^{ij}\frac{\partial\phi}{\partial x_j}-\frac{1}{2}\sum_{j=1}^{n}\left(\frac{1}{\beta}g^{ij}\frac{\partial\beta}{\partial x_j}-\frac{\partial g^{ij}}{\partial x_j}\right)$$

or, equivalently,

$$f=-\frac{1}{2}\nabla\phi-\frac{1}{2\beta}\nabla\beta+\frac{1}{2}\sum_{\alpha=1}^{n}\left(\mathrm{tr}\left(\frac{\partial b_\alpha}{\partial x}\right)b_\alpha+\frac{\partial b_\alpha}{\partial x}b_\alpha\right)$$

the probability density with respect to the volume measure

$$d\mu=\frac{1}{\beta}(x)dx_1dx_2...dx_n$$

satisfies

$$\frac{\partial\rho_\beta}{\partial t}=\frac{1}{2}(\mathrm{div}\nabla\phi)\rho_\beta+\frac{1}{2}\nabla^2\rho_\beta$$

with ∇^2 being the Laplace-Beltrami operator on X corresponding to the metric G. Moreover, if ϕ is such that the integral

$$N = \int_X e^{-\phi(x)} dx_1 dx_2 ... dx_n$$

exists, there exists a steady state density in the form

$$\rho_R(x) = \frac{1}{N} e^{-\phi(x)}.$$

Example 8.3 This is an extension of Example 6.2. Consider the function $\mathrm{tr}(\Theta^T Q \Theta N)$ on $\mathbb{S}o(n)$ and the stochastic equation

$$d\Theta = [\Theta^T Q \Theta, N]\Theta dt - \frac{n-1}{2}\Theta dt + \sum_{i=1}^{n}\sum_{j=1}^{n}(E_{ij} - E_{ji})\Theta dw_{ij}.$$

The invariant density with respect to the Killing metric is

$$\rho_R(\Theta) = \frac{1}{N} e^{-tr\Theta^T Q \Theta N}.$$

The following proposition extends the ideas of Proposition 7.1.

Proposition 8.2 Suppose that $x \in X$ satisfies the Itô equation

$$dx = f(x)dt + B(x)dw$$

with B such that $\tilde{B} = (b_1, b_2, ... b_m, \epsilon[b_1, b_2], \epsilon[b_1, b_3], ..., \epsilon[b_{m-1}, b_n])$ is of rank n for $\epsilon \neq 0$. Let $G^{-1}(\epsilon) = \tilde{B}\tilde{B}^T$ and introduce the notation $\tilde{\beta} = \sqrt{\det \tilde{B}\tilde{B}^T}$, Suppose, further, that $\tilde{\beta} = \epsilon^k \beta(x)$. Then if

$$f_i(x) = -\frac{1}{2} g^{ij} \frac{\partial \phi}{\partial x_j} - \frac{1}{2}\sum_{j=1}^{n} \frac{1}{\beta(x)} g^{ij} \frac{\partial}{\partial x_j}\beta(x) - \frac{\partial g^{ij}}{\partial x_j}$$

or, equivalently,

$$f = -\frac{1}{2}\nabla\phi - \frac{1}{2\beta}\nabla\beta + \frac{1}{2}\sum_{\alpha=1}^{n}\left(\mathrm{tr}\left(\frac{\partial \beta b_\alpha}{\partial x}\right) b_\alpha + \frac{\partial b_\alpha}{\partial x} b_\alpha\right),$$

the probability density with respect to the volume measure

$$d\mu = \frac{1}{\beta(x)} dx_1 dx_2 ... dx_n$$

satisfies

$$\frac{\partial \rho_R}{\partial t} = \mathrm{div}(\nabla\phi)\rho_R + \frac{1}{2}\nabla^2 \rho_R$$

with ∇_e^2 being the extended Laplace-Beltrami operator corresponding to $\tilde{G}$. Moreover, if ϕ is such that the integral

$$N = \int_X e^{-\phi(x)} dx_1 dx_2 ... dx_n$$

exists, there exists a steady state density in the form

$$\rho_R(x) = \frac{1}{N} e^{-\phi(x)}.$$

Supporting Calculations. This is, again, just a calculation. A key point involves a modification of the Laplace-Beltrami operator described in Proposition 6.1.

Example 8.4 We extend Example 7.1 by introducing a gradient term. Consider $\phi(x, y, z) = (x^2 + y^2 + z^2)/2$. The term

$$f = \sum_{j=1}^{n} g^{ij} \frac{\partial \phi}{\partial x_j}$$

is just

$$f = \begin{bmatrix} 1 & 0 & y \\ 0 & 1 & -x \\ y & -x & x^2 + y^2 \end{bmatrix} \begin{bmatrix} x \\ y \\ z \end{bmatrix}.$$

If we insert this in the Itô equations given in Example 7.1 we get

$$dx = -(x + yz)dt + dw$$

$$dy = -(y - xz)dt + d\nu$$

$$dz = -(x^2 + y^2)zdt + ydw - xd\nu.$$

As indicated in Example 7.1, there is an extension of B which has a constant determinant. Relative to the volume measure $dxdydz$ we have the invariant density for this set of equations

$$\rho_\beta(\infty, x, y, z) = \frac{1}{\sqrt{(2\pi)^3}} e^{-(x^2+y^2+z^2)/2}.$$

Qualitatively speaking, it is the strong growth of the restoring force on z that overcomes the growth of the noise terms, allowing a gaussian density to exist in steady state. However, the reader is warned that the standard existence results of reference [7] do not apply because of the quadratic growth in the drift terms.

9 Skew Terms

If we have an invariant measure and if we alter the drift vector field in such a way as to add a term that is everywhere tangential to the equiprobable surfaces, then the invariant measure will still exist.

Proposition 9.1 Suppose that $x \in X$ satisfies the Itô equation

$$dx = f(x)dt + B(x)dw$$

with B such that $\tilde{B} = (b_1, b_2, ...b_m, \epsilon[b_1, b_2], \epsilon[b_1, b_3], ..., \epsilon[b_{m-1}, b_n])$ is of rank n for $\epsilon \neq 0$. Let $G^{-1}(\epsilon) = \tilde{B}\tilde{B}^T$ and introduce the notation $\tilde{\beta} = \sqrt{\det \tilde{B}\tilde{B}^T}$, Suppose, further, that $\tilde{\beta} = \epsilon^k \beta(x)$. Then if

$$f_i(x) = s(x) - \frac{1}{2} g^{ij} \frac{\partial \phi}{\partial x_j} - \frac{1}{2} \sum_{j=1}^{n} \frac{1}{\beta(x)} g^{ij} \frac{\partial}{\partial x_j} \beta(x) - \frac{\partial g^{ij}}{\partial x_j}$$

or, equivalently,

$$f = s(x) - \frac{1}{2} \nabla \phi(x) - \frac{1}{2\beta} \nabla \beta + \frac{1}{2} \sum_{\alpha=1}^{n} \left(\text{tr} \left(\frac{\partial \beta b_\alpha}{\partial x} \right) b_\alpha + \frac{\partial b_\alpha}{\partial x} b_\alpha \right),$$

the probability density with respect to the volume measure

$$d\mu = \frac{1}{\beta(x)} dx_1 dx_2 ... dx_n$$

satisfies the equation

$$\frac{\partial \rho_R}{\partial t} = -\text{div } (s\rho_R) + \text{div } ((\nabla \phi)\rho_R) + \frac{1}{2} \nabla^2 \rho_R.$$

Moreover, if ϕ is such that the integral

$$N = \int_X e^{-\phi(x)} dx_1 dx_2 ... dx_n$$

exists, and if

$$\text{div} s = \langle s, \nabla \phi \rangle_{G^{-1}},$$

there exists a steady state density in the form

$$\rho_R(x) = \frac{1}{N} e^{-\phi(x)}.$$

Supporting Calculations. The basic idea is to find conditions on s so that the invariant measure obtained without s persists when s is present.

Without s we have an invariant measure which is a multiple of $e^{-\phi(x)}$. We want s to satisfy

$$\beta \sum_{i=1}^{n} \frac{\partial}{\partial x_i} \frac{s_i}{\beta} e^{-\phi(x)} = 0$$

and the condition in the proposition is necessary and sufficient.

Example 9.1 Let B be constant and suppose

$$dx = Axdt + Bdt.$$

Consider the potential term $x^T Qx$ with Q positive definite. Thus the equation

$$dx = -2BB^T Qx + Bdw$$

has an invariant measure

$$\rho(\infty, x) = \frac{1}{N} e^{-x^T Qx}.$$

However, unless we assume that B is of full rank there are many other normalized invariant solutions to the steady state equation. If $\Omega Q + Q\Omega^T = 0$, then we can add a term Ωx to the drift without changing the invariant measure. Thus the system

$$dx = -BB^T Q + \Omega xdt + Bdw \ ; \ \ \Omega Q + Q\Omega^T = 0$$

has the same invariant measure. With Ω present it is possible to maintain uniqueness even under a relaxation of the rank condition on B; the condition for the invariant measure to be unique is simply that the (Ω, B) should be a controllable pair.

Example 9.2 Our final example extends Examples 7.1 and 8.4, introducing a skew term. Consider

$$dx = -(x + yz)dt - zydt + dw = -xdt + dw$$

$$dy = -(y - xz)dt + zxdt + d\nu = -ydt + d\nu$$

$$dz = -(x^2 + y^2)zdt + ydw - xd\nu.$$

The insertion of the skew term

$$s = \begin{bmatrix} -zy \\ +zx \\ 0 \end{bmatrix}$$

does not affect the status of the invariant measure of Example 8.4.

References

[1] J. Koonderink. *Solid Shape.* Cambridge, MA: MIT Press, 1990.

[2] O. Faugeras. *Three-dimensional Computer Vision: A Geometric Viewpoint.* Cambridge, MA: MIT Press, 1993.

[3] R. Bellman. *A Brief Introduction to the Theta Function.* New York: Holt, 1968.

[4] D. Mumford. *Tata Lectures on Theta.* Boston: Birkhäuser, 1981.

[5] J. Lo and A. Willsky. Estimation for rotational processes with one degree of freedom. *IEEE Trans. AC* **AC-20** (1975), 10-20.

[6] Y. Choquet-Bruhat, et al. *Analysis, Manifolds, and Physics.* Amsterdam: North Holland, 1982.

[7] I.I. Gihman and A.V. Skorohod. *Stochastic Differential Equations.* Berlin: Springer-Verlag, 1972.

[8] H.P. McKean. *Stochastic Integrals.* New York: Academic Press, 1969.

[9] R. Gangolli. On the construction of certain diffusions on a differentiable manifold. *Z. Wahrscheinlichkeitstheorie* **2** (1964), 406-419.

[10] R.W. Brockett. *Control Theory and Singular Riemannian Geometry.* (Peter Hilton and Gail Young, Eds.). *New Directions in Applied Mathematics.* New York: Springer-Verlag, 1982.

[11] R.W. Brockett. Lie algebras and lie groups in control theory. *Geometric Control Theory.* (David Mayne and R. W. Brockett, Eds.). Dordecht, the Netherlands: Reidel, 1973.

[12] R.J. Muirhead. *Aspects of Multivariate Statistics.* New York: J. Wiley, 1982.

Division of Engineering and Appied Sciences, Harvard University, Cambridge, Massachusetts

On Duality between Filtering and Interpolation

C.I. Byrnes and A. Lindquist[1]

1 Introduction

This paper is a survey on the rational covariance extension problem, a problem with historical roots in the beginning of the century going back to work by Carathéodory, Toeplitz and Schur on interpolation [20, 21, 60, 59]. Carathéodory's interest was in classifying all bounded harmonic functions with prescribed first n derivatives at a given point, such as ∞ in our analysis. This problem was also studied by Toeplitz [60] and Schur [59], who was able to develop a complete parameterization of the class of such interpolants defining meromorphic functions $v(z)$ which are strictly positive real. From the perspective of classical analysis, however, the question of which meromorphic interpolants were rational would not have played a major role.

Rationality is a requirement added by systems theoretical considerations, important applications being speech synthesis [23], spectral estimation [36, 54], stochastic systems theory [38], and systems identification [51]. Since these application areas focus principally on mathematical models for devices, such as circuits, which can be physically realized with a finite number of active elements, the covariance extension problem in these contexts insist that the solution to the Carathéodory extension problem be rational of a degree no larger than the number of given correlation coefficients, as well as being positive real. This makes the problem considerably more challenging. Only recently has it been proved that there is a complete parameterization of such extensions in terms of the zeros of the corresponding minimum-phase spectral factor [15], thereby extending a result by Georgiou [28, 29] and proving a longstanding conjecture by him.

The need to construct stochastic models from a finite window of correlation coefficients has led to the study of several problems involving the description of classes of stationary linear stochastic systems having outputs which match a given partial covariance sequence. One of these is the *partial stochastic realization problem*, which consists of describing all such stochastic systems having the smallest possible degree, which we refer to as the *positive degree* of the partial covariance sequence. Kalman motivated the study of the partial stochastic realization problem by describing minimal realizations as being the simplest class of models capable of describing the given data. A well-known and simply computable solution is the *maximum entropy filter*, which may be interpreted as maximizing some measure of

[1]The research of both authors was supported in part by grants from AFOSR, NSF, TFR, the Göran Gustafsson Foundation, and Southwestern Bell.

the "entropy" of the covariance window and, in this way, assumes as little as possible about the completion of the correlation sequence. However, the output of a maximum entropy filter has a spectral density without zeros, which makes it less desirable in many applications. For example, in speech synthesis it produces a "flat" speech, and hence more general solutions with spectral zeros would be preferable.

The body of the paper is outlined as follows. In Section 2 the rational covariance extension problem is formulated, Schur's classical parameterization of the not necessarily rational solutions is outlined, and Georgiou's result is stated. The partial stochastic realization problem is introduced in Section 3, and modeling filters and applications to speech processing are discussed. We present the maximum entropy filter and the Georgiou-Kimura parameterization, the latter of which serves as a device to reparameterize the problem. Section 4 presents a fast Kalman filtering algorithm as a device for spectral factorization and as a preamble for Section 5, in which we formalize an observation in [15] that filtering and interpolation induce dual, or complementary, decompositions of the space of positive real rational functions of degree less than or equal to n. From this basic result, in Section 6 we provide a complete parameterization of all positive rational extensions of a given partial covariance sequence and give a new proof, based on a global inverse function theorem, that the problem is well-posed. We begin Section 7 with an alternative complete parameterization of all rational extensions in terms of the unique positive semidefinite solutions of a nonstandard Riccati-type matrix equation. The rank of the unique semidefinite solution is related to the positive degree of the covariance sequence, an invariant which is central to the minimal partial stochastic realization problem. In this context, we also note that, in sharp contrast to the minimal partial deterministic realization problem, the positive degree does not assume any generic value. We conclude the paper in Section 8 with some simulations.

2 The rational covariance extension problem

The following interpolation problem, apparently first studied in this form by Kalman [38], has been a fundamental open problem in systems theory. Given a finite sequence of real numbers

$$c_0, c_1, c_2, \ldots, c_n \tag{1}$$

which is positive in the sense that the Toeplitz matrix

$$T_n = \begin{bmatrix} c_0 & c_1 & c_2 & \cdots & c_n \\ c_1 & c_0 & c_1 & \cdots & c_{n-1} \\ \vdots & \vdots & \vdots & \ddots & \vdots \\ c_n & c_{n-1} & c_{n-2} & \cdots & c_n \end{bmatrix} \tag{2}$$

is positive definite, find a complete parameterization of the class of all infinite extensions

$$c_{n+1}, c_{n+2}, c_{n+3}, \ldots \tag{3}$$

of (1) with the properties that the function $v(z)$ defined by

$$v(z) = \frac{1}{2}c_0 + c_1 z^{-1} + c_2 z^{-2} + \ldots \tag{4}$$

in the neighborhood of infinity is

(i) *rational* of at most degree n

(ii) *strictly positive real*; i.e., it is analytic for $|z| \geq 1$ and satisfies

$$v(z) + v(z^{-1}) > 0 \tag{5}$$

at each point of the unit circle.

It can be shown that any such extension has the property that the infinite Toeplitz matrix

$$T_\infty = \begin{bmatrix} c_0 & c_1 & c_2 & \cdots \\ c_1 & c_0 & c_1 & \cdots \\ c_2 & c_1 & c_0 & \cdots \\ \vdots & \vdots & \vdots & \ddots \end{bmatrix} \tag{6}$$

is positive definite. This problem is called the *rational covariance extension problem*, since the positivity of the Toeplitz matrix (2) or (6) is the condition required for the corresponding sequence to be a *bona fide* covariance sequence. As we must have $c_0 > 0$, it is no restriction to normalize the problem by setting $c_0 := 1$. This will be done in the rest of the paper.

This parameterization problem has historical roots going back to important work by Carathéodory and Schur in potential theory [20, 21, 59]. In fact, if the rationality condition (i) is removed, the problem is called the *Carathéodory extension problem* and a complete parameterization of all extensions was given by Schur [59] in 1918. This solution will be discussed next.

However, our interest in the problem is motivated by its connection to speech synthesis [23], spectral estimation [36, 54], stochastic systems theory [38], and systems identification [51], application areas which are mainly concerned with mathematical models for devices, such as circuits, which can be physically realized with a finite number of active elements. In these contexts, therefore, it is required that the solution to the Carathéodory extension problem be rational as well as being positive real. Indeed, rational, positive real functions also arise in circuit theory as the mathematical models for the impedance, or transfer function, of an RLC network, where the degree of the rational function is precisely the sum of the number of capacitors and inductors and where the positivity reflects the fact that the network resistors are positive. For these reasons, systems-theoretic formulations of the Carathéodory extension problem insist on rationality as well, and then rationality of at most a prescribed degree.

The rational covariance extension problem should not be confused with another rational interpolation problem arising in linear systems theory, *the deterministic partial realization problem* [39, 40, 58, 31], as has often been the case. In this problem, one insists on rational interpolants which are not necessarily positive real, so hence condition (ii) is suppressed. This partial realization problem is considerably simpler. On the other hand, the Schur parameterization gives a solution to the problem if one suppresses rationality. The combination of these two design requirements has made this problem more elusive, despite its importance in stochastic system theory, spectral analysis and speech synthesis.

For the moment, let us disregard the rationality condition (i) required in systems theory and consider the classical Carathéodory extension problem. Using what are now known as Schur parameters, Schur introduced a complete parameterization of the class of extensions defining meromorphic functions $v(z)$, analytic for $z \geq 1$ and satisfying $\Re v(z) \geq 0$ there. Such functions are called *Carathéodory functions.* Clearly all $v(z)$ satisfying (i) and (ii) are Carathéodory functions.

More precisely, recall that the Szegö polynomials

$$\varphi_t(z) = z^t + \varphi_{t1} z^{t-1} + \cdots + \varphi_{tt} \tag{7}$$

are monic polynomials orthogonal on the unit circle [1, 32], which can be determined recursively [1] via the *Szegö-Levinson equations*

$$\begin{aligned} \varphi_{t+1}(z) &= z\varphi_t(z) - \gamma_t \varphi_t^*(z) \quad \varphi_0(z) = 1 &\quad (8a)\\ \varphi_{t+1}^*(z) &= \varphi_t^*(z) - \gamma_t z \varphi_t(z) \quad \varphi_0^*(z) = 1, &\quad (8b) \end{aligned}$$

where $\gamma_0, \gamma_1, \gamma_2, \ldots$ are the Schur parameters

$$\gamma_t = \frac{1}{r_t} \sum_{k=0}^{t} \varphi_{t,t-k} c_{k+1}, \tag{9}$$

and where $(r_0, r_1, r_2, \ldots)$ are generated by

$$r_{t+1} = (1 - \gamma_t^2) r_t \quad r_0 = 1. \tag{10}$$

Similarly, the Szegö polynomials

$$\psi_t(z) = z^t + \psi_{t1} z^{t-1} + \cdots + \psi_{tt} \tag{11}$$

of the second kind are obtained from (8) by merely exchanging γ_t for $-\gamma_t$ everywhere.

For each t, the Schur parameters $\gamma_0, \gamma_1, \dots, \gamma_{t-1}$ are uniquely determined by the covariance parameters $c_1, c_2, \dots, c_t$ via (8), (9) and (10). Conversely, it can be shown that $c_1, c_2, \dots, c_t$ are uniquely determined by $\gamma_0, \gamma_1, \dots, \gamma_{t-1}$ so that there is a bijective correspondence between partial covariance and Schur sequences of the same length [59]. Moreover the function $v(z)$ having the Laurent expansion

$$v(z) = \frac{1}{2} + c_1 z^{-1} + c_2 z^{-2} + c_3 z^{-3} + \dots \tag{12}$$

for $|z| > 1$ is a Carathéodory function if and only if

$$|\gamma_t| < 1 \quad \text{for } t = 0, 1, 2, \dots, \tag{13}$$

and, as was shown by Schur [59], (12) and (13) provide us with complete parameterization of all meromorphic Carathéodory functions.

Now returning to the covariance extension problem, $c_1, c_2, \dots, c_n$ are fixed, and hence $\gamma_0, \gamma_1, \dots, \gamma_{n-1}$ are also fixed. The assumption that the Toeplitz matrix T_n is positive definite is equivalent to the condition that $|\gamma_t| < 1$ for $t = 0, 1, \dots, n-1$. Covariance extension then amounts to selecting the remaining Schur parameters

$$\gamma_n, \gamma_{n+1}, \gamma_{n+2}, \dots \tag{14}$$

arbitrarily subject to the positivity constraint (13). An important special case, the *maximum entropy solution*, is obtained by setting all Schur parameters (14) equal to zero, a choice that certainly satisfies (13). This yields the rational Carathéodory function

$$v(z) = \frac{1}{2} \frac{\psi_n(z)}{\varphi_n(z)}, \tag{15}$$

where $\varphi_n(z)$ and $\psi_n(z)$ are the degree n Szegö polynomials of first and second kind respectively.

The maximum entropy solution also happens to satisfy the rationality condition (i). In general, however, an arbitrary extension (14) satisfying the Schur condition (13) can only be guaranteed to be meromorphic, not rational of degree at most n as required in our case, and, as pointed out in [15], there is no way to characterize the rational solutions by a finite number of inequalities. Indeed, adding rationality changes the character of the problem considerably.

If $v(z)$ is rational of at most degree n, then, in view of (4), $v(z)$ can be written

$$v(z) = \frac{1}{2} \frac{b(z)}{a(z)}, \tag{16}$$

where $a(z)$ and $b(z)$ are monic polynomials of degree n. Moreover, $v(z)$ is strictly positive real if and only if the pseudo-polynomial

$$d(z) := \frac{1}{2}[a(z)b(z^{-1}) + a(z^{-1})b(z)] > 0 \tag{17}$$

on the unit circle and the denominator polynomial $a(z)$ is a *Schur polynomial,* i.e., has all its roots on the open unit disc. Since the function $1/v(z)$ is strictly positive real if and only if $v(z)$ is, we may replace the last condition with the numerator polynomial $b(z)$ being a Schur condition. In fact, both $a(z)$ and $b(z)$ need to be Schur polynomials for $v(z)$ to be positive real.

Using a very innovative application of topological degree theory Georgiou [28, 29] proved the following theorem.

Theorem 2.1 (Georgiou) *Given a finite sequence of real numbers* (1) *with* $c_0 = 1$ *which is positive in the sense that the Toeplitz matrix* (2) *is positive definite and an arbitrary pseudo-polynomial*

$$d(z) = d_0 + d_1(z + z^{-1}) + \cdots + d_n(z^n + z^{-n}) \tag{18}$$

of at most degree n *which is positive on the unit circle, there exists two Schur polynomials*

$$a(z) = z^n + a_1 z^{n-1} + \cdots + a_n \tag{19}$$

and

$$b(z) = z^n + b_1 z^{n-1} + \cdots + b_n \tag{20}$$

such that

$$\frac{1}{2}[a(z)b(z^{-1}) + a(z^{-1})b(z)] = d(z) \tag{21}$$

and the interpolation condition

$$\frac{1}{2}\frac{b(z)}{a(z)} = \frac{1}{2} + \hat{c}_1 z^{-1} + \hat{c}_2 z^{-2} + \ldots \qquad \hat{c}_i = c_i \quad \text{for} \quad i = 1, 2, \ldots, n \tag{22}$$

is fulfilled.

As we shall see in Section 3, this is an important result, but it does not provide a complete parameterization of the rational covariance extension problem. For this we also need the solution to be unique. Georgiou conjectured uniqueness in [29] and left the question of completeness of the parameterization open. In addition, for computational and other reasons, a set-theoretical bijection is not sufficient, but the solution needs to be continuous in the given data so that the problem is well-posed. As we shall see in Section 6, a strong version of such a result was presented in [15].

3 Modeling filters and speech synthesis

In signal processing and speech processing [29, 42, 23, 19, 53, 41], a signal is often modeled as a stationary random sequence $\{y(t)\}_{t\in\mathbb{Z}}$ which is the output of a linear stochastic system

$$\begin{cases} x(t+1) = Ax(t) + Bu(t) \\ \quad y(t) = Cx(t) + Du(t) \end{cases} \tag{23}$$

obtained by passing (normalized) white noise $\{u(t)\}_{t\in\mathbb{Z}}$ through a filter

$$\text{white noise} \xrightarrow{u} \boxed{w(z)} \xrightarrow{y}$$

with a stable transfer function

$$w(z) = C(zI - A)^{-1}B + D \tag{24}$$

and letting the system come to a statistical steady state. Here stability amounts to the matrix A having all its eigenvalues strictly inside the unit circle.

Consequently, the stationary stochastic process $\{y(t)\}_{t\in\mathbb{Z}}$ is given by the convolution

$$y(t) = \sum_{k=-\infty}^{t} w_{t-k}u(k) \quad t = 0, 1, 2, \ldots, \tag{25}$$

where $w_0 = D$ and $w_k = CA^{k-1}B$ for $k = 1, 2, 3, \ldots$, and where

$$w(z) = w_0 + w_1 z^{-1} + w_2 z^{-2} + w_3 z^{-3} + \ldots. \tag{26}$$

The process $\{y(t)\}_{t\in\mathbb{Z}}$ has a rational spectral density

$$\Phi(z) = w(z)w(z^{-1}), \tag{27}$$

which we assume to be positive on the unit circle. In other words, $w(z)$ is a stable spectral factor of Φ which we shall take to be *minimum-phase*; i.e., the rational function $w(z)$ has all its poles and zeros in the open unit disc and $w_0 = w(\infty) \neq 0$. In systems-theoretical language we say that y is the output of a *shaping filter* driven by a white noise input, with the transfer function w.

It is well-known that the spectral density Φ has the Fourier representation

$$\Phi(z) = c_0 + \sum_{k=1}^{\infty} c_k(z^k + z^{-k}), \tag{28}$$

where

$$c_0, c_1, c_2, c_3, \dots \tag{29}$$

is the covariance sequence defined as

$$c_k = \mathrm{E}\{y(t+k)y(t)\} \qquad k = 0, 1, 2, 3, \dots. \tag{30}$$

Such a covariance sequence has the property that the infinite Toeplitz matrix (6) is positive definite.

The corresponding stochastic realization problem is the inverse problem of determining the stochastic system (23) given the *infinite* covariance sequence (29). The condition that $\Phi(z)$ be rational introduces a finiteness condition on the covariance sequence (29). In fact, the positive real part

$$v(z) = \frac{c_0}{2} + \sum_{i=1}^{\infty} c_i z^{-i} \tag{31}$$

of

$$\Phi(z) = v(z) + v(z^{-1}) \tag{32}$$

is rational and may be written

$$v(z) = \frac{1}{2}\frac{b(z)}{a(z)}, \tag{33}$$

where $a(z)$ and $b(z)$ are monic polynomials (19) and (20) of degree n. The property that $v(z)$ be strictly positive real is equivalent to $a(z)$ and $b(z)$ being Schur polynomials and satisfying

$$a(z)b(z^{-1}) + a(z^{-1})b(z) > 0 \tag{34}$$

on the unit circle. Therefore, once $a(z)$ and $b(z)$ has been determined, say, by identifying coefficient of like powers of z in $2a(z)v(z) = b(z)$, the unique stable minimum-phase spectral factor of Φ, i.e., the solution

$$w(z) = \rho\frac{\sigma(z)}{a(z)}, \tag{35}$$

of (27) such that $\rho \in \mathbb{R}_+$ and $\sigma(z)$ is a monic Schur polynomial

$$\sigma(z) = z^n + \sigma_1 z^{n-1} + \cdots + \sigma_n, \tag{36}$$

may be determined via the polynomial spectral factorization problem

$$\frac{1}{2}[a(z)b(z^{-1}) + a(z^{-1})b(z)] = \rho^2\sigma(z)\sigma(z^{-1}). \tag{37}$$

Next let us consider the problem in systems identification to determine the system (23) from an observed string of output data. Let us first consider the idealized situation that we have an infinite string of output data

$$y_0, y_1, y_2, y_3, \dots \tag{38}$$

satisfying the appropriate ergodicity property. Then the covariance sequence $(c_0, c_1, c_2 \dots)$ can be determined as

$$c_k = \lim_{T\to\infty} \frac{1}{T}\sum_{t=0}^{T} y_{t+k}y_t, \tag{39}$$

which defines a unique spectral density and hence a unique (minimum phase) shaping filter $w(z)$.

However, in practice only a finite string of observed data

$$y_0, y_1, y_2, \dots, y_N \tag{40}$$

is typically available. If N is sufficient large, there is a $T < N$ such that

$$\frac{1}{T}\sum_{t=0}^{T} y_{t+k}y_t \tag{41}$$

is a good approximation of c_k, but now only a finite covariance sequence

$$c_0, c_1, c_2, \dots, c_n, \tag{42}$$

where $n << N$, can be produced. Consequently, we have a rational covariance extension problem, and we have one solution for each rational extension (3) and hence one shaping filter $w(z)$. We shall call each such shaping filter $w(z)$ a *modeling filter* of the partial covariance sequence (42).

As an illustration from speech synthesis, recall that artificial speech is produced from a synthesis of two classes of phonemes, one kind for voiced sounds (such as vowels) and one kind for unvoiced sounds (for consonants such as "s" or "t"). A speech sample is typically broken into 20 ms segments which are then regarded as the synthesis of a combination of voiced and unvoiced segments, each of which is viewed as the output of the same transfer function, driven by periodic "pulse trains" and by white noise, respectively. For each 20 ms segment a new transfer function is computed from the speech sample. In particular, on a sufficiently small interval of time the unvoiced speech pattern, which is produced by passing white noise through a shaping filter, can be regarded as a realization of a stationary random sequence y with covariances

$$c_k := E\{y(t+k)y(t)\}$$

and with a spectral density (28). For speech synthesis the covariance sequence $c_0, c_1, c_2, \ldots$ on each stationary interval of time needs to be determined from output data via an ergodic limit (39) or some equivalent procedure. Of course, in practice only a finite string of observed data (40) is typically available, in which case only a finite covariance sequence (42) can be produced.

The usual method for determining modeling filters for each of the partial covariance sequences is the maximum entropy procedure, basically because this is the only solution to the rational covariance extension problem for which there has been computational procedures. As mentioned in Section 2, the maximum entropy filter is obtained by setting

$$\gamma_i = 0 \quad \text{for } i = n, n+1, n+2, \ldots,$$

yielding the Carathéodory function

$$v(z) = \frac{1}{2}\frac{\psi_n(z)}{\varphi_n(z)}, \tag{43}$$

where $\{\varphi_t(z)\}$ and $\{\psi_t(z)\}$ are the Szegö polynomials of the first and second kind respectively. In fact, it can be shown that

$$\frac{1}{2}\frac{\psi_n(z)}{\varphi_n(z)} = \frac{1}{2} + c_1 z^{-1} + c_2 z^{-2} + \cdots + c_n z^{-n} + \ldots \tag{44}$$

and that

$$\varphi_n(z)\psi_n(z^{-1}) + \varphi_n(z^{-1})\psi_n(z) = r_n > 0. \tag{45}$$

Consequently, since $\varphi_n(z)$ and $\psi_n(z)$ are Schur polynomials, v is strictly positive real and

$$v(z) + v(z^{-1}) = \frac{r_n}{\varphi_n(z)\varphi_n(z^{-1})}, \tag{46}$$

yielding the modeling filter

$$w(z) = \frac{\sqrt{r_n} z^n}{\varphi_n(z)}, \tag{47}$$

the *maximum entropy filter*.

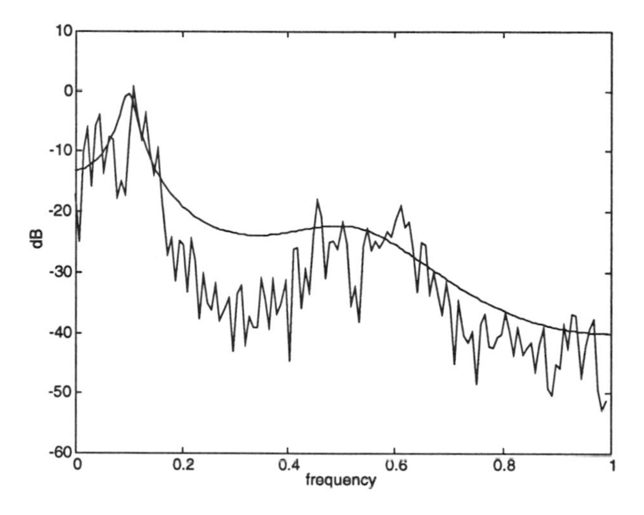

Figure 3.1

Since the maximum entropy solution has the property that the corresponding spectral density (46) lacks finite zeros, the speech becomes rather "flat." This is illustrated in Figure 3.1 where a true spectrum has been approximated by that of a 6th order maximum entropy filter. In Figure 3.2 we depict another solution, determined by our methods, of the same order but with appropriately chosen zeros.

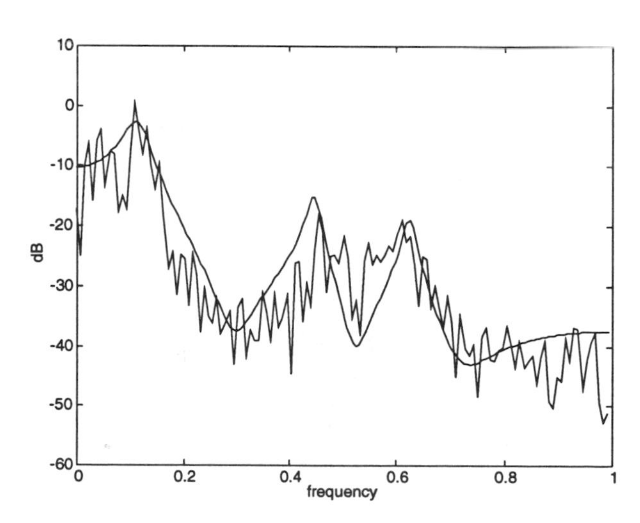

Figure 3.2

As these examples illustrate, in many speech processing applications zeros are desired, and the question arises whether it is possible to assign zeros arbitrarily, while still satisfying the interpolation condition. In view of (37) and (21), Georgiou's result (Theorem 2.1) answers this question in the affirmative, even though the question of uniqueness had been left open. However, a computational procedure is needed. To this end, Georgiou [29] and Kimura [42] independently observed that the formula (44) could be generalized to

$$v(z) = \frac{1}{2}\frac{\psi_n(z) + \alpha_1\psi_{n-1}(z) + \cdots + \alpha_n\psi_0(z)}{\varphi_n(z) + \alpha_1\varphi_{n-1}(z) + \cdots + \alpha_n\varphi_0(z)} \tag{48}$$

$$= \frac{1}{2} + c_1 z^{-1} + c_2 z^{-2} + \cdots + c_n z^{-n} + \ldots, \tag{49}$$

thus expressing $v(z)$ in terms of the $2n$ parameters (α, γ), where $\alpha = (\alpha_1, \alpha_2, \ldots, \alpha_n)' \in \mathbb{R}^n$ and $\gamma = (\gamma_0, \gamma_1, \ldots, \gamma_{n-1})' \in \mathbb{R}^n$. In fact, it was shown in [29] and later also in [17] that the interpolation condition in (48) holds for all α, and consequently the *Georgiou-Kimura parameterization* (48) characterizes rationality but not positivity. We denote by $\mathcal{P}_n$ the subset of $\mathbb{R}^{2n}$ for which $v(z)$ is strictly positive real, and let

$$\mathcal{P}_n(\gamma) = \{(\alpha, \gamma) \in \mathcal{P}_n \mid \gamma \text{ fixed}\} \subset \mathbb{R}^n$$

be the positive real region for fixed covariance data.

Of course, given the partial covariance data γ, the choice $\alpha = 0$ is the maximum entropy solution, but in general it is very complicated to characterize those other α for which $v(z)$ is positive real, i.e., to characterize the sets $\mathcal{P}_n(\gamma)$. For $n = 1$ the representation (48) takes the form

$$v(z) = \frac{1}{2}\frac{z + \gamma_0 + \alpha_1}{z - \gamma_0 + \alpha_1}. \tag{50}$$

The strictly positive real region is the diamond depicted in Figure 3.3 below, and fixing the partial covariance data γ_0, the admissible α are the ones on the open interval in the figure.

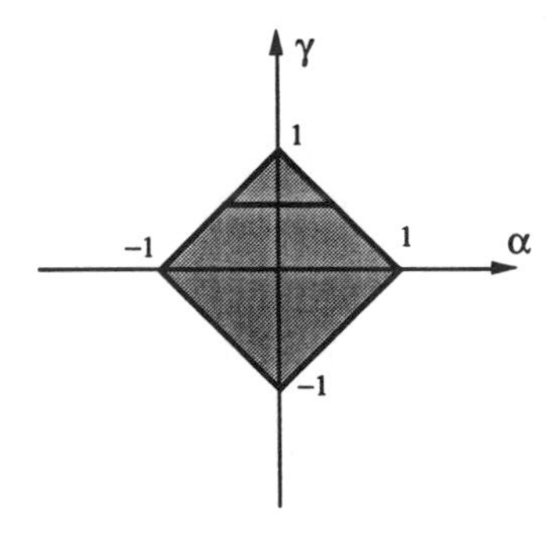

Figure 3.3

Next, let us consider the case $n = 2$. Fixing the covariance data at $\gamma_0 = \frac{1}{2}$ and $\gamma_1 = \frac{1}{3}$, we obtain

$$v(z) = \frac{1}{2}\frac{z^2 - \frac{2}{3}z + \frac{1}{3} + \alpha_1(z + \frac{1}{2}) + \alpha_2}{z^2 + \frac{1}{3}z + \frac{1}{3} + \alpha_1(z - \frac{1}{2}) + \alpha_2}, \tag{51}$$

and the region of positive real $\alpha = (\alpha_1, \alpha_2)$ is as depicted in Figure 3.4.

The higher-dimensional cases become much more complicated. While it is true that $\mathcal{P}_n(\gamma)$ is always diffeomorphic to Euclidean space [11], any good solution to the rational covariance extension problem would give such a parameterization in terms of familiar systems theoretic objects. In this direction, the possibility of parameterizing those filters which are positive real by arbitrarily prescribing the zeros of a modeling filter was suggested by Georgiou in Theorem 2.1 and his conjecture. Recently we proved an amplification of Georgiou's conjecture that, for any desired choice of spectral density zero structure, there is one and only one positive extension, i.e., one and only one modeling filter. This result was obtained by viewing a certain fast filtering algorithm as a nonlinear dynamical system defined on the space of positive real rational functions of degree less than or equal to n. It is then observed that filtering and interpolation induce complementary, or "dual" decompositions (or foliations) of this space. From this assertion about the geometry of positive real functions follows the first complete parameterization of all positive rational extensions [15]. This will be the topic of Sections 4–6.

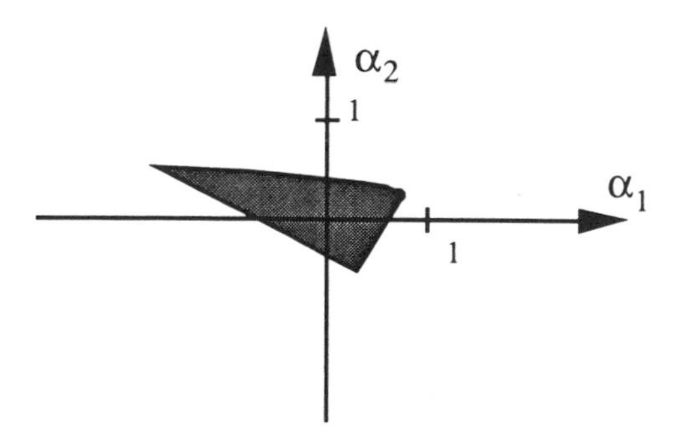

Figure 3.4

4 The fast filtering algorithm: a dynamical system which computes spectral factors

The connection between the Carathéodory function $v(z)$ of a rational covariance extension and the corresponding modeling filter $w(z)$ is through spectral factorization. More precisely, given a strictly positive real $v(z)$, the modeling filter $w(z)$ is the minimum phase solution of

$$w(z)w(z^{-1}) = v(z) + v(z^{-1}). \tag{52}$$

If

$$v(z) := \frac{1}{2}\frac{b(z)}{a(z)}, \tag{53}$$

where $a(z)$ and $b(z)$ are monic Schur polynomials (19) and (20), we saw in Section 3 that

$$w(z) = \rho\frac{\sigma(z)}{a(z)}, \tag{54}$$

where $\sigma(z)$ is the Schur polynomial solution (36) of the polynomial spectral factorization problem

$$\rho^2\sigma(z)\sigma(z^{-1}) = d(z) := \frac{1}{2}[a(z)b(z^{-1}) + a(z^{-1})b(z)]. \tag{55}$$

The spectral factorization problem (52) then amounts to determining $\sigma(z)$ from $a(z)$ and $b(z)$.

There is a well-known connection between spectral factorization and Kalman filtering that we shall exploit next. In fact, solving the spectral factorization problem by iterating the Riccati equation of Kalman filtering to steady state is a common procedure. Here we shall use the same idea but instead applied to a certain fast algorithm for Kalman filtering.

We shall formulate the Kalman filtering problem in terms of covariance data or, equivalently, in terms of the Carathéodory function $v(z)$. Defining

$$g(z) := \frac{1}{2}[b(z) - a(z)] = g_1z^{n-1} + g_2z^{n-2} + \cdots + g_n, \tag{56}$$

we may write

$$v(z) = \frac{1}{2} + \frac{g(z)}{a(z)}, \tag{57}$$

or, alternatively,

$$v(z) = \frac{1}{2} + h'(zI - F)^{-1}g, \tag{58}$$

where, without lack of generality, we have taken (F, g, h) in the observer canonical form

$$F = \begin{bmatrix} -a_1 & 1 & 0 & \cdots & 0 \\ -a_2 & 0 & 1 & \cdots & 0 \\ \vdots & \vdots & \vdots & \ddots & \vdots \\ -a_{n-1} & 0 & 0 & \cdots & 1 \\ -a_n & 0 & 0 & \cdots & 0 \end{bmatrix} \quad g = \begin{bmatrix} g_1 \\ g_2 \\ \vdots \\ g_n \end{bmatrix} \quad h = \begin{bmatrix} 1 \\ 0 \\ \vdots \\ 0 \end{bmatrix} \tag{59}$$

and where prime ($'$) denotes transpose. Sometimes it is convenient to write

$$F = J - ah' \tag{60}$$

where a is the column vector $(a_1, a_2, \ldots, a_n)'$ and J is the obvious shift matrix.

Now, let $w(z)$ be the corresponding modeling filter, and consider the stationary random sequence $\{y(t)\}_{t\in\mathbb{Z}}$ obtained by passing white noise through the shaping filter

$$\text{white noise} \xrightarrow{u} \boxed{w(z)} \xrightarrow{y}$$

with transfer function $w(z)$ or some other spectral factor of $v(z) + v(z^{-1})$. Then the one-step predictor, i.e., the linear least squares estimate $\hat{y}(t)$ of $y(t)$ given $y(0), y(1), \ldots, y(t-1)$, is generated by the Kalman filter

$$\begin{cases} \hat{x}(t+1) = F\hat{x}(t) + k(t)[y(t) - h'\hat{x}(t)] \\ \hat{y}(t) = h'\hat{x}(t) \end{cases} \tag{61}$$

where the gain $k(t)$ can be determined via a matrix Riccati equation. Apparently less well-known is that the gain can also be determined via the fast algorithm

$$\begin{cases} a(t+1) = \frac{1}{1-g_1(t)}[a(t) + (I-J)g(t)] & a(0) = a \\ g(t+1) = \frac{1}{1-g_1(t)^2}[-g_1(t)a(t) + (J - g_1(t)I)g(t)] & g(0) = g \end{cases} \tag{62}$$

consisting of $2n$ nonlinear first-order difference equations, in terms of which

$$k(t) = a(t) + g(t) - a. \tag{63}$$

We say that it is "fast" since, for $n > 1$, $2n$ is less than the number $\frac{1}{2}n(n+1)$ of scalar equations in the corresponding matrix Riccati equation.

This algorithm is a version, appearing in [47], of the fast Kalman filtering algorithm introduced in [46]. (Also see [16] where these matters are reviewed.) It is also shown in [47] that the equality

$$\frac{1}{2}r_t[a_t(z)b_t(z^{-1}) + a_t(z^{-1})b_t(z)] = d(z) \tag{64}$$

is preserved along the trajectory of (62), where $r_t := \prod_{k=0}^{t-1}[1-g_1(t)^2]$ and the monic polynomials $a_t(z)$ and $b_t(z) := a_t(z) + 2g_t(z)$ are formed from $a(t)$ and $b(t) := a(t) + 2g(t)$ as above, and that $a_t(z)$ and $b_t(z)$ have all their zeros in the unit disc $|z| < 1$.

An important property of the fast algorithm (62) is that the $2n$ parameters (a, g) of the problem appear only in the initial conditions and not in the dynamical system, which is invariant under parameter changes. In fact, the algorithm updates the parameters and can therefore be regarded as an iteration in parameter space. Moreover, the algorithm makes sense also for initial data (a, g) which does not correspond to a positive real $v(z)$.

We now express this dynamical system in the Georgiou-Kimura coordinates. In fact, the parameterization (48) can be given a nice geometric interpretation as a birational diffeomorphic change of coordinates in the strictly positive real region

$$\mathcal{P} = \{(a,g) \in \mathbb{R}^{2n} \mid v(z) \text{ is strictly positive real}\}$$

[18]. In fact, the Georgiou-Kimura parameterization yields

$$\begin{cases} a(z) = \varphi_n(z) + \alpha_1\varphi_{n-1}(z) + \cdots + \alpha_n\varphi_0(z) \\ b(z) = a(z) + 2g(z) = \psi_n(z) + \alpha_1\psi_{n-1}(z) + \cdots + \alpha_n\psi_0(z) \end{cases} \tag{65}$$

where the Szegö polynomials $\{\varphi_n(z)\}_0^n$ and $\{\psi_n(z)\}_0^n$ are determined from $\gamma := (\gamma_0, \gamma_1, \ldots, \gamma_{n-1})'$ via the Szegö recursions (8) and (11).

The fast filtering algorithm (62) can now be reformulated as a nonlinear dynamical system in (α, γ)-space [18]. More precisely, if $(\alpha, \gamma) \in \mathcal{P}_n$ and the maps $A, G : \mathbb{R}^n \to \mathbb{R}^{n\times n}$ are defined as

$$A(\gamma) = \begin{bmatrix} \frac{1}{1-\gamma_{n-1}^2} & \frac{\gamma_{n-1}\gamma_{n-2}}{(1-\gamma_{n-1}^2)(1-\gamma_{n-2}^2)} & \cdots & \frac{\gamma_{n-1}\gamma_0}{(1-\gamma_{n-1}^2)\cdots(1-\gamma_0^2)} \\ 0 & \frac{1}{1-\gamma_{n-2}^2} & \cdots & \frac{\gamma_{n-2}\gamma_0}{(1-\gamma_{n-2}^2)\cdots(1-\gamma_0^2)} \\ \vdots & \vdots & \ddots & \vdots \\ 0 & 0 & \cdots & \frac{1}{1-\gamma_0^2} \end{bmatrix} \tag{66}$$

and

$$G(\alpha) = \begin{bmatrix} 0 & 1 & 0 & \cdots & 0 \\ 0 & 0 & 1 & \cdots & 0 \\ \vdots & \vdots & \vdots & \ddots & \vdots \\ 0 & 0 & 0 & \cdots & 1 \\ -\alpha_n & -\alpha_{n-1} & -\alpha_{n-2} & \cdots & -\alpha_1 \end{bmatrix}, \tag{67}$$

then the dynamical system

$$\begin{cases} \alpha(t+1) = A(\gamma(t))\alpha(t), & \alpha(0) = \alpha \\ \gamma(t+1) = G(\alpha(t+1))\gamma(t), & \gamma(0) = \gamma \end{cases} \tag{68}$$

initiated at (α, γ) evolves on the invariant manifold defined by (64). In fact, the preserved pseudo-polynomial (64) yields, after change of coordinates and identification of coefficients of like powers in z, $n+1$ equations in the $2n+1$ variables $\{\alpha(t), \gamma(t), r_t\}$ which, upon elimination of r_t by dividing the last n equations by the first, in turn yields n integrals

$$f_i(\alpha(t), \gamma(t)) = \kappa_i \quad i = 1, 2, \ldots, n \tag{69}$$

for the dynamical system (68), where $\kappa_1, \kappa_2, \ldots, \kappa_n$ are constants, which can be determined from the initial conditions (α, γ). Moreover, $\gamma(t)$ is updated by shifting so that

$$\gamma_k(t) = \gamma_{t+k}. \tag{70}$$

Theorem 4.1 ([18]) *The strictly positive real region $\mathcal{P}_n$ is invariant under the dynamical system* (68), *which is globally convergent on $\mathcal{P}_n$. More specifically, $(\alpha(t), \gamma(t))$ tends to $(\alpha_\infty, 0)$, where $\alpha_\infty \in \mathcal{P}_n(0) \subset \mathbb{R}^n$ so that*

$$\alpha_\infty(z) = z^n + \alpha_{\infty 1} z^{n-1} + \cdots + \alpha_{\infty n} \tag{71}$$

is a monic Schur polynomials of degree n and $\mathcal{P}_n(0)$ is the equilibrium set. The global stable manifold $\mathcal{W}^s(\alpha_\infty, 0)$ of the equilibrium $(\alpha_\infty, 0)$ is defined by (69) *with $\kappa_i = f_i(\alpha_\infty, 0)$ for $i = 1, 2, \ldots, n$.*

A consequence of this theorem and (70) is that

$$\gamma_t \to 0 \quad \text{as } t \to \infty. \tag{72}$$

Therefore, since the Szegö polynomials $\varphi_k(z)$ and $\psi_k(z)$ become z^k when the Schur parameters are zero, this implies in view of (65) that

$$a_t(z) \to \alpha_\infty(z) \quad \text{and} \quad b_t(z) \to \alpha_\infty(z) \quad \text{as } t \to \infty \tag{73}$$

in (64). Likewise,

$$r_t \to r_\infty \quad \text{as } t \to \infty, \tag{74}$$

and consequently the pseudo-polynomial $d(z)$ corresponding to the stable manifold $\mathcal{W}^s(\alpha_\infty, 0)$ becomes

$$d(z) = r_\infty \alpha_\infty(z) \alpha_\infty(z^{-1}). \tag{75}$$

Therefore, evaluating (64) for $t = 0$ and noting that $r_0 = 1$ we have

$$\frac{1}{2}[a(z)b(z^{-1}) + a(z^{-1})b(z)] = r_\infty \alpha_\infty(z) \alpha_\infty(z^{-1}), \tag{76}$$

which should be compared with the spectral factorization equation (55). Consequently, the required spectral factor is

$$w(z) = \sqrt{r_\infty} \frac{\alpha_\infty(z)}{a(z)}, \tag{77}$$

where $\alpha_\infty(z)$ is obtained from the limit of the dynamical system and r_∞ is easiest determined from (76) by identifying coefficient of like power in z.

5 The complementary foliations

In this section we shall investigate the geometry of the strictly positive real region, i.e. the open subset $\mathcal{P} \subset \mathbb{R}^{2n}$ of (α, γ) such that

$$v(z) = \frac{1}{2}\frac{\psi_n(z) + \alpha_1\psi_{n-1}(z) + \cdots + \alpha_n\psi_0(z)}{\varphi_n(z) + \alpha_1\varphi_{n-1}(z) + \cdots + \alpha_n\varphi_0(z)} \tag{78}$$

is strictly positive real, where $\{\varphi_k(z)\}_0^n$ and $\{\psi_k(z)\}_0^n$ are the Szegö of the first and second type respectively corresponding to the covariance data γ. Similarly, as before let

$$\mathcal{P}_n(\gamma) = \{(\alpha, \gamma) \in \mathcal{P}_n \mid \gamma \text{ fixed}\} \subset \mathbb{R}^n \tag{79}$$

be the strictly positive real region for fixed covariance data.

Geometrically, the decomposition

$$\mathcal{F}_1 : \quad \mathcal{P}_n = \bigcup_{\gamma} \mathcal{P}_n(\gamma) \tag{80}$$

is an important example of what is known as a foliation of the open manifold $\mathcal{P}_n$. Intuitively, a foliation is a decomposition of a manifold into disjoint connected submanifolds, called leaves, with the additional property that in the neighborhood of any point the leaves vary in a sufficiently smooth way. More precisely, a *foliation* $\mathcal{F}$ of dimension m on a smooth manifold M of dimension n is a partition of M into a family of disjoint, connected m-dimensional submanifolds L_β, called the *leaves* of the foliation, such that (i) $M = \cup_\beta L_\beta$, (ii) each point $x \in M$ has a Euclidean neighborhood U and coordinates $(x_1, ..., x_n)$ for which the equations

$$x_1 = 0, \quad x_2 = 0, \quad \ldots, \quad x_{n-m} = 0$$

define the connected components of the nonempty intersections $U \cap L_\beta$. In fact, it was shown in [11] that $\mathcal{P}_n(\gamma)$ is diffeomorphic to Euclidean space and is therefore connected. With this in mind, we note (see [7]) that the Kimura-Georgiou parameterization shows that this decomposition is sufficiently regular to define a foliation $\mathcal{F}_1$ of $\mathcal{P}_n$ into the leaves $\mathcal{P}_n(\gamma)$ (see [15].

In Section 4 we showed that for any $(\alpha, \gamma) \in \mathcal{P}_n$ the dynamical system (68) converges to a limit $(\alpha_\infty, 0) \in \mathcal{P}_n$ along the stable manifold $\mathcal{W}^s(\alpha_\infty, 0)$. As proved in [15], the decomposition of $\mathcal{P}_n$ as a union of the global stable manifolds $\mathcal{W}^s(\alpha_\infty, 0)$ defines a second foliation

$$\mathcal{F}_2 : \quad \mathcal{P}_n = \bigcup_{\alpha_\infty \in \mathcal{P}(0)} \mathcal{W}^s(\alpha_\infty, 0) \tag{81}$$

of $\mathcal{P}_n$. In the special case $n = 1$, Figure 5.1 depicts the global stable manifold $\mathcal{W}^s(\alpha_\infty, 0)$ at $(\frac{1}{3}, 0)$ as a subset of $\mathcal{P}_1$. Also depicted is $\mathcal{P}_1(\frac{1}{2})$.

As Figure 5.1 suggests, the leaves of the foliations $\mathcal{F}_1$ and $\mathcal{F}_2$ are transverse; i.e. at a point of intersection of the leaves of these two foliations the corresponding tangent spaces are complementary subspaces. Indeed, using the characterization of the tangent spaces to the stable manifold $\mathcal{W}^s(\alpha_\infty, 0)$ developed in [18], in [15] we proved that the intersection of $\mathcal{W}^s(\alpha_\infty, 0)$ with $\mathcal{P}_n(\gamma)$ is in fact always transverse. In such a case, one says that two foliations are complementary.

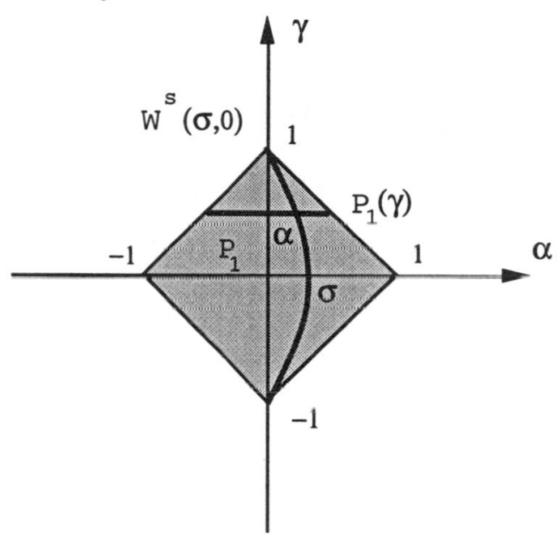

Figure 5.1

Theorem 5.1 *The positive real region $\mathcal{P}_n$ is connected and is foliated by the stable manifolds $\mathcal{W}^s(\alpha_\infty, 0)$ of the equilibrium set $\mathcal{P}_n(0)$. The set $\mathcal{P}_n$ is also foliated into leaves given by the submanifolds $\mathcal{P}_n(\gamma)$. Moreover, these foliations are complementary.*

Consequently, there are two complementary foliations of $\mathcal{P}_n$, one indexed by the partial covariance data and one by the zero polynomial of the modeling filter. Theorem 5.1 suggests that, given a partial covariance sequence and a stable zero polynomial, there is a unique solution of the rational covariance problem represented by the intersection between the corresponding leaves of the foliations $\mathcal{F}_1$ and $\mathcal{F}_2$. The fact that these foliations are complementary says that this uniqueness does occur to first order, in the following sense.

Corollary 5.2 *For each $(\lambda, \alpha) \in \mathbb{R}_+ \times \mathcal{P}_n(\gamma)$, the Jacobian matrix Jac f_γ of f_γ is nonsingular.*

In fact, in [15] Corollary 5.2 formed the basis for a degree theoretic argument which demonstrated that to any point σ in $\mathcal{P}_n(0)$, there is one and only one α such that $(\alpha, \gamma) \in \mathcal{P}_n(\gamma)$. This α defines a modeling filter $w(z)$ having the zeros of $\sigma(z)$.

It is interesting to note that, conversely, any α such that $(\alpha, \gamma) \in \mathcal{P}_n(\gamma)$ determines a Schur polynomial $\sigma(z)$, which can be computed via the convergence of the dynamical system (68) with initial condition determined by (α, γ). In fact, the dynamical system defines a map

$$g : \mathcal{P}_n(\gamma) \to \mathcal{P}_n(0) \tag{82}$$

sending a strictly positive real choice of the parameters α in the Georgiou-Kimura parameterization (78) to the corresponding choice of stable zeros

$$g(\alpha) = \sigma. \tag{83}$$

As an illustration in the case $n = 2$, Figure 5.2 depicts the connected open submanifolds $\mathcal{P}_2(\gamma)$ and $\mathcal{P}_2(0)$, the latter corresponding to the monic Schur polynomials in $\mathcal{S}_2$, for $\gamma = (1/2, 1/3)$. These sets form the domain and codomain of the function g sending α to σ .

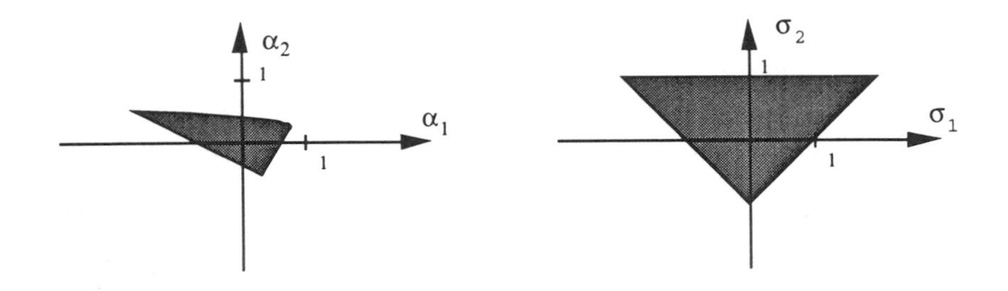

Figure 5.2

In the next section we shall sketch a new proof that the function g is a diffeomorphic bijection, using a theorem of Hadamard to show that the rational covariance extension problem is well-posed.

6 A complete parameterization of all rational covariance extensions

As pointed out above, the maximum entropy solution gives rise to a rational spectral density with no finite zeros, and hence a modeling filter with all zeros in the origin. In many applications, it turns out to be important to be able to design filters with prescribed zeros and which shape processes with observed correlation coefficients. The important question as to which zeros can be prescribed, and in which manner, has been a limiting factor in filter design.

In order to address this issue, Georgiou [29] launched an investigation of which zeros could be prescribed using degree theory a tool for studying the existence of solutions to nonlinear equations. In 1983, he proved that any Schur polynomial is possible as the numerator of a modeling filter which interpolated the given covariance data (Theorem 2.1) and conjectured that there is a unique zero polynomial to each rational covariance extension of degree at most n.

In practice, however, we would require more, e.g., that the solutions should depend continuously in the problem data, so that small variations in problem data would give rise to small variations in the solution. In [15] we proved Georgiou's conjecture in this stronger form as a corollary of Theorem 5.1, which, among other things, provides a complete (and analytic) parameterization of all rational covariance extensions of degree at most n.

Theorem 6.1 ([15]) *Suppose one is given a finite sequence of real numbers*

$$c_0, c_1, c_2, \dots, c_n \tag{84}$$

which is positive in the sense that the Toeplitz matrix

$$T_n = \begin{bmatrix} c_0 & c_1 & c_2 & \cdots & c_n \\ c_1 & c_0 & c_1 & \cdots & c_{n-1} \\ \vdots & \vdots & \vdots & \ddots & \vdots \\ c_n & c_{n-1} & c_{n-2} & \cdots & c_n \end{bmatrix} \tag{85}$$

is positive definite. Then, to each Schur polynomial

$$\sigma(z) = z^n + \sigma_1 z^{n-1} + \cdots + \sigma_n, \tag{86}$$

there corresponds a unique Schur polynomial

$$a(z) = z^n + a_1 z^{n-1} + \cdots + a_n \tag{87}$$

such that, for some suitable uniquely defined positive number ρ,

$$w(z) = \rho \frac{\sigma(z)}{a(z)} \tag{88}$$

satisfies the interpolation condition

$$w(z)w(z^{-1}) = 1 + \sum_{i=k}^{\infty} \hat{c}_i (z^k + z^{-k}); \qquad \hat{c}_i = c_i \quad \text{for} \quad i = 1, 2, \dots, n. \tag{89}$$

Moreover, this one-one correspondence is an analytic diffeomorphism.

We begin by noting that Theorem 6.1 would follow if we could prove that the function $f_\gamma : \mathbb{R}_+ \times \mathcal{P}_n(\gamma) \to \mathcal{D}_n$, given by

$$f_\gamma(\lambda, \alpha) = \frac{1}{2}\lambda[a(z)b(z^{-1}) + a(z^{-1})b(z)] \tag{90}$$

is a diffeomorphism. Here $a(z)$ and $b(z)$ depend on $\alpha \in \mathbb{R}^n$ via (65), and $\mathcal{D}_n$ is the space of pseudo-polynomials

$$d(z) = d_0 + d_1(z + z^{-1}) + \cdots + d_n(z^n + z^{-n}), \tag{91}$$

of degree at most n which are positive on the unit circle. In fact, if f_γ is a diffeomorphism for all γ satisfying the Schur condition (13), then it is in particular a diffeomorphism for $\gamma = 0$ so that the map $f_0 : \mathbb{R}_+ \times \mathcal{P}_n(\gamma) \to \mathcal{D}_n$ defined via

$$f_0(\mu, \sigma) = \mu\sigma(z)\sigma(z^{-1})$$

is a diffeomorphic bijection. Then the commutative diagram

$$\begin{array}{ccc} \mathbb{R}_+ \times \mathcal{P}_n(0) & \xrightarrow{h} & \mathbb{R}_+ \times \mathcal{P}_n(\gamma) \\ f_0 \searrow & & \nearrow f_\gamma^{-1} \\ & \mathcal{D}_n & \end{array}$$

defines a homeomorphic bijection h under which

$$\frac{1}{2}\lambda[a(z)b(z^{-1}) + a(z^{-1})b(z)] = \mu\sigma(z)\sigma(z^{-1}).$$

Setting $\rho^2 := \mu/\lambda$ is equivalent to

$$\frac{1}{2}\frac{b(z)}{a(z)} + \frac{1}{2}\frac{b(z^{-1})}{a(z^{-1})} = \rho^2 \frac{\sigma(z)}{a(z)}\frac{\sigma(z^{-1})}{a(z^{-1})},$$

where

$$\frac{1}{2}\frac{b(z)}{a(z)} = \frac{1}{2} + c_1 z + \cdots + c_n z^{-n} + \ldots$$

interpolates the given partial covariance sequence so that

$$w(z) = \rho\frac{\sigma(z)}{a(z)}$$

is a modeling filter. Therefore, Theorem 6.1 would follow.

A new proof of the fact that f_γ is a diffeomorphic bijection can in fact be based on a theorem by Hadamard [33, 34, 35], as we shall now see. Hadamard formalized the concept of well-posedness of problems described by maps between Euclidean n-spaces was as follows. For such an f, the problem of finding solutions to $f(x) = y$ is said to be *well-posed* provided

f is (i) surjective, (ii) injective, and (iii) has a continuous inverse. The criterion for well-posedness given by Hadamard reposes upon a property of maps which reflects the existence of a priori bounds on the size of solutions, given bounds on the size of the problem data. Topologically, this can be expressed in terms of properness. Recall that a function $f : \mathbb{R}_n \to \mathbb{R}_n$ is said to be *proper* if, and only if, $f^{-1}(K)$ is compact for every compact K. In these terms, there are several related criteria for well-posedness, the earliest such result being in essence a global inverse function theorem.

Theorem 6.2 (Hadamard's Theorem) *Suppose $f : \mathbb{R}_n \to \mathbb{R}_n$ is a C^1 map. If f is proper and satisfies* det *Jac $f \neq 0$ for every $x \in \mathbb{R}^n$, then f is a diffeomorphism onto $\mathbb{R}^n$.*

Hadamard's Theorem can be proven by either degree theory or by the theory of covering spaces [5]. For these reasons, there are also extensions of this theorem to classes of spaces and maps to which either of these theories apply. However, one should also expect the topology of more general spaces to complicate the conclusion of the analogues of this basic theorem. As an example, consider the map f defined on the unit circle S^1 in the complex plane, with the unit circle as its range, defined via $f(z) = z^2$. The map f satisfies all the hypotheses in Hadamard's Theorem, except that of having Euclidean spaces for it domain and range. However, f is not a diffeomorphism but rather exhibits its domain as a double covering of its range.

In our application of Hadamard's Theorem it is of course essential that the domain and range be Euclidean spaces. To illustrate this point further, we shall consider as a preliminary example the question of whether f_γ is a diffeomorphism in the case, $\gamma = 0$, which is the problem of spectral factorization.

More precisely, given a symmetric pseudo-polynomial

$$d(z) = d_0 + d_1(z + z^{-1}) + \cdots + d_n(z^n + z^{-n}), \tag{92}$$

of degree at most n, which is positive on the unit circle, find a stable polynomial $a(z)$ of degree n, i.e., a polynomial

$$a(z) = a_0 z^n + a_1 z^{n-1} + \cdots + a_n, \quad a_0 > 0 \tag{93}$$

having all its roots strictly inside the unit circle, such that

$$a(z)a(z^{-1}) = d(z). \tag{94}$$

Such spectral factors $a(z)$ are of course unique if they exist. In fact, $a(z)$ has n roots each of which is either zero, and hence canceling with $a(z^{-1})$, or a root of $d(z)$. Conversely, a root of $d(z)$ located in the open unit disc is nonzero and a root of $a(z)$. Consequently, all polynomials satisfying (94) have the same roots. Then the a_0 must also be the same, and hence the polynomials are identical. Although simple to see in a more direct way, to illustrate our point of view we shall give a topological proof for existence of a solution to (94).

Lemma 6.3 *The space $\mathcal{Z}_n$ of all stable polynomials of degree n is diffeomorphic with $\mathbb{R}^{n+1}$.*

Proof. Denote by $\mathcal{S}_n$ the space of Schur polynomials, that is the space of monic polynomials

$$b(z) = z^n + b_1 z^{n-1} + \cdots + b_n, \tag{95}$$

which have all its roots strictly inside the unit circle. The space $\mathcal{Z}_n$ of all stable polynomials of degree n is diffeomorphic to the product $\mathbb{R}_+ \times \mathcal{S}_n$ via the mapping $\varphi : a(z) \to (a_0, a(z)/a_0)$. By identifying a monic polynomial with its roots, the space $\mathcal{S}_n$ may be identified with the space of real divisors of order n in the open unit disc, where by a real divisor of order n we mean a self-conjugate, unordered sets of n points. Therefore, by identifying the open unit disc with the complex plane via a diffeomorphism which preserves conjugation, one can also identify the space of all real divisors of order n in the open unit disc with the space of all real divisors of order n in the complex plane, i.e., with the space of all real monic polynomials of degree n. In this way, $\mathcal{S}_n$ is diffeomorphic to $\mathbb{R}^n$. Therefore, $\mathcal{Z}_n$ is diffeomorphic with a product of Euclidean spaces and hence diffeomorphic with $\mathbb{R}^{n+1}$.

For any $a \in \mathcal{Z}_n$, define the operator $S(a) : \mathcal{Z}_n \to \mathcal{W}_n$ from the vector space $\mathcal{Z}_n$ into the $n+1$-dimensional vector space $\mathcal{W}_n$ of symmetric pseudo-polynomials of degree at most n via

$$S(a)b = \frac{1}{2}[a(z)b(z^{-1}) + a(z^{-1})b(z)]. \tag{96}$$

In view of the unit circle version of Orlando's formula [27], $S(a)$ is nonsingular for all $a \in \mathcal{Z}_n$. (Also see, e.g., [24] where a determinantal expression is given.) Let $\mathcal{D}_n \subset \mathcal{W}_n$ be the space of pseudo-polynomials (92) which are positive on the unit circle. Then for any $d \in \mathcal{D}_n$, $S(a)b = d$ uniquely defines a strictly positive real function $v(z) = \frac{1}{2}\frac{b(z)}{a(z)}$ and hence $b = S(a)^{-1}d \in \mathcal{Z}_n$. The operator $G : \mathcal{Z}_n \to \mathcal{Z}_n$ defined by

$$Ga = S(a)^{-1}d \tag{97}$$

is an involution, i.e., $G^2 a = a$. In fact, in view of (96), $Ga = b$ and $Gb = a$. By Smith's Theorem [6], an involution which maps a Euclidean space into itself has a fixed point, and therefore there exists a solution $a \in \mathcal{Z}_n$ to (94) for all $d \in \mathcal{D}_n$.

Remark 6.4 *The existence of spectral factors can of course be proven in a much more straight-forward manner. On the other hand, polynomial spectral factorization is the simplest form of the rational covariance extension problem and also illustrates one of the key points to which we shall now turn. That is, our proof relies heavily on the fact that $\mathcal{Z}_n$ is homeomorphic to $\mathbb{R}^{n+1}$ since on many non-Euclidean spaces there exist involutions without fixed points.*

We now proceed with the proof of Theorem 6.1 using the global inverse function theorem, Theorem 6.2. We note that f_γ is easily shown to be proper, as in [15, Lemma A.2]. Next, the duality between filtering and interpolation implies that for each $(\lambda, \alpha) \in \mathbb{R}_+ \times \mathcal{P}_n(\gamma)$ the matrix Jac f_γ is in fact nonsingular, as we noted in Corollary 5.2.

Finally, it remains to check that the domain and range of f_γ are diffeomorphic to Euclidean space. By spectral factorization, the open manifold $\mathcal{D}_n$ is diffeomorphic to $\mathcal{Z}_n$, which we know is Euclidean by Lemma 6.3. Thus, it remains to prove that the domain of f_γ is Euclidean.

The fact that the open manifold $\mathbb{R}_+ \times \mathcal{P}_n(\gamma)$ is diffeomorphic to Euclidean space follows of course from the same assertion about $\mathcal{P}_n(\gamma)$. That $\mathcal{P}_n(\gamma)$ is diffeomorphic to Euclidean space was shown in [11] using the Brown-Stallings criterion [55], which asserts that an n-manifold is diffeomorphic to Euclidean n-space if and only if every compact subset has a Euclidean neighborhood. Very briefly, the proof uses two facts. The first is that to say $v(z)$ is positive real is to say that for any $\mu \in \mathbb{C}_+$ the polynomial $b(z) + \mu a(z)$ has all its roots in the unit disc. This allows one to pass to a problem about compact sets of divisors in the unit disc. The second tool is a general method for recognizing Euclidean spaces of polynomials from their divisors: If U is a self conjugate open subset of $\mathbb{C}$ with a simple, closed, rectifiable, orientable curve as boundary, then the space of all monic real polynomials with all of their roots lying in U is diffeomorphic to $\mathbb{R}^n$. The proof, of course, follows the proof of Lemma 6.3, *mutatis mutandis*.

This concludes the proof of Theorem 6.1.

Caveat. This solution to the rational covariance extension problem expresses the choice of free parameters in familiar systems theoretic terms, viz. the numerator of the resulting modeling filter. One might imagine that a similar situation holds if one were to first choose the pole polynomial. However, although the zeros of the modeling filter can be chosen arbitrarily, this is not the case for the poles, as can be seen from the following simple counter example.

Counter Example. Consider the partial covariance sequence $(1, c_1) = (1, \frac{1}{2})$. Then $n = 1$ and $\gamma_0 = \frac{1}{2}$. Now suppose we would like to chose the stable pole polynomial $a(z) = z + a_1$ with $a_1 = \frac{3}{4}$. Then it follows from (50) that $\alpha_1 = a_1 + \gamma_0 = \frac{5}{4}$. However, as can be seen from Figure 3.3, the point $(\alpha, \gamma) = (\frac{5}{4}, \frac{1}{2})$ does not belong to the strictly positive real region $\mathcal{P}_1$. This can also be seen from the fact that $b(z) = z + \frac{7}{4}$, which is not a Schur polynomial.

7 Minimality

An important and partially open question in partial stochastic realization theory is to find the rational covariance extension of minimum McMillan degree. More precisely, given a partial covariance sequence

$$1, c_1, c_2, \dots, c_n, \tag{98}$$

find an infinite extension $c_{n+1}, c_{n+2}, \dots$ with the property that the corresponding positive real function

$$v(z) = \frac{1}{2} + c_1 z^{-1} + c_2 z^{-2} + c_3 z^{-3} + \dots$$

has minimum degree. This degree is called the ***positive degree*** of the partial covariance sequence (98). It is easy to see, and follows readily from classical stochastic realization theory, that the degree of the corresponding modeling filter $w(z)$ is the same as that of $v(z)$. If $p < n$, the polynomials $a(z)$, $b(z)$ and $\sigma(z)$ thus must have common factors.

This is equivalent to finding the minimum triplet (F, g, h) of matrices such that

(i) $h'F^{k-1}g = c_k$ for $k = 1, 2, \dots, n$ and

(ii) $v(z)$ is strictly positive real,

i.e. the matrices F, g and h having the dimensions $p \times p$, $p \times 1$ and $p \times 1$ respectively with the smallest possible p such that (i) and (ii) are satisfied. This p is precisely the positive degree of (98). If the positivity requirement (ii) is removed, the problem reduces to the much simpler deterministic partial realization problem [39, 40, 31], The corresponding degree, the *algebraic degree* of (98), is of course smaller or equal to the positive degree p. Nevertheless, they are often confused in the literature.

As a starting point for studying the positive degree of a partial covariance sequence, it would be helpful to be able to compute the degree of the positive real rational function which we know corresponds to any fixed choice of zero polynomial $\sigma(z)$. To this end, we shall next introduce a Riccati-type equation, called the *Covariance Extension Equation*, which is formulated in terms of the partial covariance data and a choice of desired modeling-filter zeros [14, 12, 13]. This is a nonstandard Riccati equation, the positive semidefinite solutions of which parameterize the solution set of the rational covariance extension problem in terms of the partial covariance sequence and the zeros of the desired modeling filter. While it is interesting in its own right, in the present setting of partial covariance data, the

Covariance Extension Equation (CEE) replaces the usual algebraic Riccati equation of stochastic realization theory required when the covariance data is complete.

It is convenient to represent the given covariance data $1, c_1, c_2, \ldots, c_n$ in terms of the first n coefficients in the expansion

$$\frac{z^n}{z^n + c_1 z^{n-1} + \cdots + c_n} = 1 - u_1 z^{-1} - u_2 z^{-2} - u_3 z^{-3} - \ldots \tag{99}$$

about infinity and to define

$$u = \begin{bmatrix} u_1 \\ u_2 \\ \vdots \\ u_n \end{bmatrix} \qquad U = \begin{bmatrix} 0 & & & & \\ u_1 & 0 & & & \\ u_2 & u_1 & & & \\ \vdots & \vdots & \ddots & & \\ u_{n-1} & u_{n-2} & \cdots & u_1 & 0 \end{bmatrix}. \tag{100}$$

Likewise, we may collect the coefficients of the desired zero polynomial

$$\sigma(z) = z^n + \sigma_1 z^{n-1} + \cdots + \sigma_n, \tag{101}$$

in the matrices

$$\sigma = \begin{bmatrix} \sigma_1 \\ \sigma_2 \\ \vdots \\ \sigma_n \end{bmatrix}, \quad \Gamma = \begin{bmatrix} -\sigma_1 & 1 & 0 & \cdots & 0 \\ -\sigma_2 & 0 & 1 & \cdots & 0 \\ \vdots & \vdots & \vdots & \ddots & \vdots \\ -\sigma_{n-1} & 0 & 0 & \cdots & 1 \\ -\sigma_n & 0 & 0 & \cdots & 0 \end{bmatrix} \quad \text{and} \quad h = \begin{bmatrix} 1 \\ 0 \\ \vdots \\ 0 \end{bmatrix}. \tag{102}$$

The Covariance Extension Equation is given in terms of these parameters as

$$P = \Gamma(P - Phh'P)\Gamma' + (u + U\sigma + U\Gamma Ph)(u + U\sigma + U\Gamma Ph)' \tag{103}$$

where the $n \times n$ matrix P is the unknown.

Our principal result concerning the CEE concerns existence and uniqueness of the positive semi-definite solution, similar in spirit to the classical existence and uniqueness theorems for the Riccati equations arising in filtering and control, and the connection of this solution to the corresponding modeling filter (104). This, of course, is of considerable independent interest in partial stochastic realization theory. The following theorem was first presented in [12, 13] and proved in [14].

Theorem 7.1 *Let $(1, c_1, \cdots, c_n)$ be a given positive partial covariance sequence. For every Schur polynomial $\sigma(z)$, there exists a unique positive* *semidefinite solution P of the Covariance Extension Equation satisfying $h'Ph < 1$, to which in turn there corresponds a unique modeling filter*

$$w(z) = \rho \frac{\sigma(z)}{a(z)}, \tag{104}$$

for which the denominator polynomial

$$a(z) = z^n + a_1 z^{n-1} + \cdots + a_n, \tag{105}$$

is given by

$$a = (I - U)(\Gamma P h + \sigma) - u \tag{106}$$

and $\rho \in (0, 1]$ is a real number given by

$$\rho = (1 - h'Ph)^{1/2}. \tag{107}$$

All modeling filters are obtained in this way. Moreover, the degree of $w(z)$, and hence that of $v(z)$, equals the rank of P.

From the last statement of Theorem 7.1 we see that the degree of any rational covariance extension, and hence the positive degree of a partial covariance sequence, is connected to the corresponding solution of the Covariance Extension Equation. In fact, one can derive the following corollary of Theorem 7.1.

Corollary 7.2 *The positive degree of the partial covariance sequence* (98) *is given by*

$$p = \min_{\sigma \in \mathcal{S}_n} \operatorname{rank} P(\sigma), \tag{108}$$

where $P(\sigma)$ is the unique solution of the Covariance Extension Equation corresponding to (98) *and the zero polynomial σ, and the minimization is over the space $\mathcal{S}_n$ of all Schur polynomials of degree n. The modeling filter corresponding to a minimizing σ is a minimal partial stochastic realization of* (98).

This corollary does not of course provide us with an applicable algorithm for determining the positive degree or a minimal rational covariance extension, but it does explain the connection to the CEE.

Since positive degree has often been confused with algebraic degree, and since the algebraic degree of sequences (98) has the generic value $\left[\frac{n+1}{2}\right]$, it is important to investigate whether there is a generic positive degree and, if so, what it is. To this end, represent the covariance data as a vector

$$c := \begin{bmatrix} c_1 \\ c_2 \\ \vdots \\ c_n \end{bmatrix} \in \mathcal{C}_n \subset \mathbb{R}^n,$$

where $\mathcal{C}_n$ is the set of c with positive Toeplitz matrix. Recall that a subset of $\mathbb{R}^n$ is *semialgebraic* provided it can be defined by a finite number of polynomial equations, inequations, and inequalities. For example, $\mathcal{C}_n$ is a semialgebraic subset of $\mathbb{R}^n$, being defined by polynomial inequalities. A subset of $\mathbb{R}^n$ is algebraic provided it can be defined by a finite number of polynomial equations. Finally, a property of points in $\mathbb{R}^n$ is said to be *generic* if the set of points which enjoy this property is nonempty, with its complement being contained in an algebraic set.

Theorem 7.3 *Let p be any integer such that $\frac{n}{2} \leq p \leq n$. Then the subset of $c \in \mathcal{C}_n$ having positive degree p is a semialgebraic set containing a nonempty open subset of $\mathbb{R}^n$. Consequently, the positive degree has no generic value.*

The proof of this theorem, which has important consequences for so-called subspace identification algorithms [61, 51], can be found in [14]. We shall now illustrate this result by considering the special case $n = 2$. Instead of using the representation $c = (c_1, c_2)'$ for the covariance data, we shall use (γ_0, γ_1), which of course is equivalent. It is easy to check that $p = 0$ if and only if $\gamma_0 = \gamma_1 = 0$, $p = 1$ if and only if

$$|\gamma_1| < \frac{|\gamma_0|}{1 + |\gamma_0|}, \tag{109}$$

and $p = 2$ otherwise [28, 29] (also see [14]). It can be seen that $p = 1$ if and only if the line

$$\alpha_2 = \gamma_1 (1 - \frac{1}{\gamma_0^2})(\gamma_0 \alpha_1 + \gamma_1) \tag{110}$$

intersects the strictly positive real region $\mathcal{P}_2$. These are precisely the points for which $a(z)$ and $b(z)$ have a common factor. All other α correspond to $v(z)$ of degree two. For example, if $\gamma_0 = 0.5$ and $\gamma_1 = 0.2$, condition (109) is satisfied, and the line (110) intersects the positive real region as depicted to the left in Figure 7.1. All $v(z)$ corresponding to points on the interval defined by this intersection have degree one. The positive degree $p = 2$ for those c for which $\gamma \neq 0$ and condition (109) is violated. The situation corresponding to such a point, $\gamma_0 = 0.5$ and $\gamma_1 = 0.4$, is illustrated to the right in Figure 7.1. Here the intersection between $\mathcal{P}_2(\gamma)$ and the line (110) is empty.

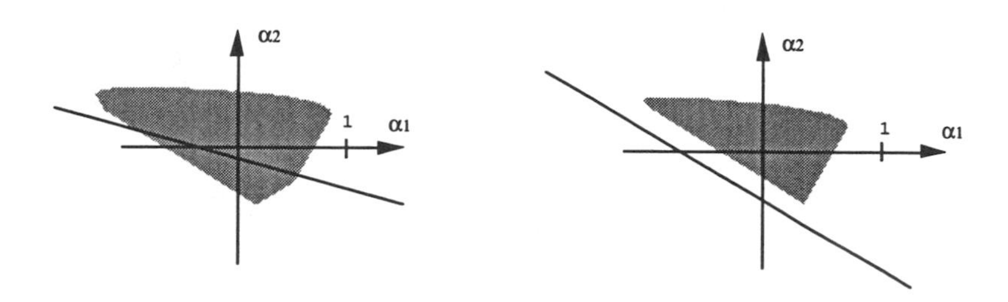

Figure 7.1

For $n > 2$ the situation is more complicated, but we have a sufficient condition for the positive degree p to be strictly less than n, which is similar to (109).

Corollary 7.4 *Suppose $n \geq 2$. Any partial covariance sequence satisfying the condition*

$$|\gamma_{n-1}| < \frac{|\gamma_{n-2}|}{1 + |\gamma_{n-2}|} \tag{111}$$

has a positive degree $p < n$. If $n = 2$, the condition is also necessary.

From this result we see that the occurrence of partial covariance data for which there exists a modeling filter with $a(z)$ and $b(z)$ having common factors is not a "rare event."

8 Simulations

In this section we shall briefly present some computational results about determining modeling filters from partial covariance and zero data. For this we have two different computational procedures: one based on the Covariance Extension Equation presented in Section 6 and another involving a convex optimization problem, details about which we present elsewhere [9].

Let us start by depicting in Figure 8.1 the zeros (marked as $\circ$) and poles (marked as $\times$) of the sixth degree modeling filters whose spectra are illustrated in Figures 3.1 and 3.2 respectively. The left unit circle corresponds to the maximum entropy filter and hence all the zeros are at $z = 0$. To obtain the better fit of the modeling filter of Figure 3.2 we must move the zeros close to the unit circle. Those zeros and the corresponding poles are depicted to the right in Figure 8.1.

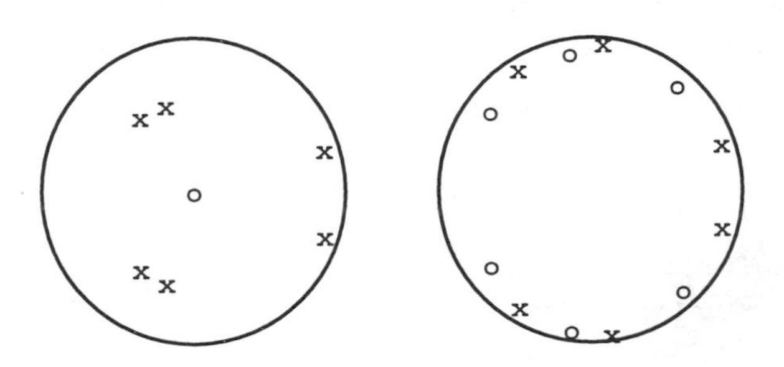

Figure 8.1

With modeling filters of this low degree we can of course not model the fine structure of spectra corresponding to for example 20 ms windows of speech. However, we can handle much larger problems, and next we show an example corresponding to a partial covariance sequence

$$1, c_1, c_2, c_3, \ldots, c_{50}$$

determined from such data in the manner described in Section 3. Figure 8.2 depicts both 50 suitably chosen zeros and the corresponding poles computed by our methods.

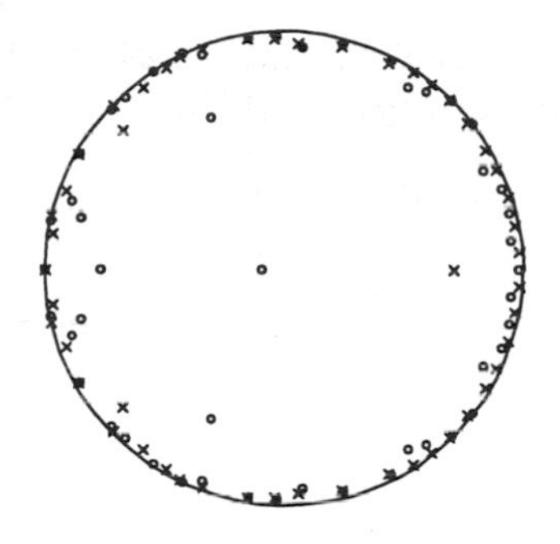

Figure 8.2

The corresponding modeling filter of degree 50 produces an output with the rather rugged spectrum depicted in Figure 8.3.

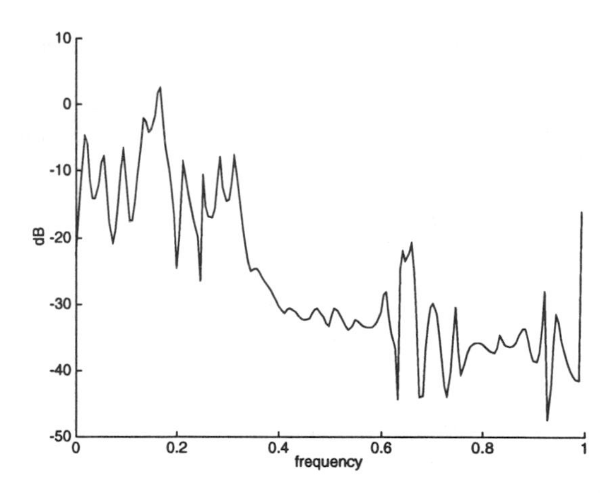

Figure 8.3

These examples show that by choosing sufficiently many zeros close to the unit circle we obtain a more realistic spectrum which does not exhibit the "flatness" of the maximum entropy filter.

Acknowledgment

We would like to thank Per Enqvist for his help with simulations.

References

[1] N.I. Akhiezer. *The Classical Moment Problem.* New York: Hafner, 1965.

[2] B.D.O. Anderson.The inverse problem of stationary convariance generation. *J. Statistical Physics* **1** (1969), 133-147.

[3] M. Aoki. *State Space Modeling of Time Series.* New York: Springer-Verlag, 1987.

[4] K.J. Åström. *Introduction to Stochastic Realization Theory.* New York: Academic Press, 1970.

[5] M.S. Berger. *Nonlinearity and Functional Analysis.* New York: Academic Press, 1977.

[6] G.E. Bredon. Introduction to compact transformation groups. *Pure and Applied Mathematics.* Vol. 46. New York: Academic Press, 1972.

[7] W.A. Boothby. *A Introduction to Differentiable Manifolds and Riemannian Geometry.* New York: Academic Press, 1975.

[8] R.W. Brockett. The geometry of the partial realization problem. *Proc. 1978 IEEE Decision and Control Conference.* 1048-1052.

[9] C.I. Byrnes, S.V. Gusev and A. Lindquist. A convex optimization approach to the rational covariance extension problem. Submitted.

[10] C.I. Byrnes and A. Lindquist. The stability and instability of partial realizations. *Systems and Control Letters* **2** (1982), 2301-2312.

[11] C.I. Byrnes and A. Lindquist. On the geometry of the Kimura-Georgiou parameterization of modelling filter. *Inter. J. of Control* **50** (1989), 99-105.

[12] C.I. Byrnes and A. Lindquist. Toward a solution of the minimal partial stochastic realization problem. *Comptes Rendus Acad. Sci. Paris* **319** (1994), 1231-1236.

[13] C.I. Byrnes and A. Lindquist. Some recent advances on the rational covariance extension problem. *Proc. IEEE European Workshop on Computer-Intensive Methods in Control and Signal Processing 1994.*

[14] C.I. Byrnes and A. Lindquist. On the partial stochastic realization problem. submitted to *IEEE Trans. Automatic Control*

[15] C.I. Byrnes, A. Lindquist, S.V. Gusev and A.V. Matveev. A complete parameterization of all positive rational extensions of a covariance sequence. *IEEE Trans. Automatic Control* **AC-40** (1995), 1841-1857.

[16] C.I. Byrnes, A. Lindquist and T. McGregor. Predictability and unpredictability in Kalman filtering. *IEEE Transactions Auto. Control* **AC-36** (1991), 563-579.

[17] C.I. Byrnes, A. Lindquist and Y. Zhou. Stable, unstable and center manifolds for fast filtering algorithms. *Modeling, Estimation and Control of Systems with Uncertainty.* (G.B. Di Masi, A. Gombani, and A. Kurzhanski, Eds.). Boston: Birkhauser, 1991.

[18] C.I. Byrnes, A. Lindquist and Y. Zhou. On the nonlinear dynamics of fast filtering algorithms. *SIAM J. Control and Optimization* **32** (1994), 744-789.

[19] J.A. Cadzow. Spectral estimation: An overdetermined rational model equation approach. *Proceedings IEEE* **70** (1982), 907-939.

[20] C. Carathéodory. Über den Variabilitätsbereich der Koeffizienten von Potenzreihen, die gegebene Werte nicht annehmen. *Math. Ann.* **64** (1907), 95-115.

[21] C. Carathéodory. Über den Variabilitätsbereich der Fourierschen Konstanten von positiven harmonischen Functionen. *Rend. di Palermo* **32** (1911), 193-217.

[22] J.B. Conway. *Functions of One Complex Variable.* New York: Springer-Verlag, 1993.

[23] Ph. Delsarte, Y. Genin, Y. Kamp and P. van Dooren. Speech modelling and the trigonometric moment problem. *Philips J. Res.* **37** (1982), 277-292.

[24] C.J. Demeure and C.T. Mullis. The Euclid algorithm and the fast computation of cross-covariance and autocovariance sequences. *IEEE Transactions Acoustics, Speech and Signal Processing* **ASSP-37** (1989), 545-552.

[25] L. Euler. De Fractionibus Continuis Dissertatio. In Latin: *Opera Omnia* **1**(14) (1744). In English: *Mathematical Systems Theory* **18** (1985), 295-328.

[26] P. Faurre, M. Clerget and F. Germain. *Opérateurs Rationnels Positifs.* Paris: Dunod, 1979.

[27] F.R. Gantmacher. *The Theory of Matrices.* New York: Chelsea, 1959.

[28] T.T. Georgiou. Partial realization of covariance sequences. *CMST.* Univ. Florida, Gainesville. 1983.

[29] T.T. Georgiou. Realization of power spectra from partial covariance sequences. *IEEE Transactions Acoustics, Speech and Signal Processing* **ASSP-35** (1987), 438-449.

[30] K. Glover. All optimal Hankel norm approximations of linear multivariable systems and their L^∞ error bounds. *Intern. J. Control* **39** (1984), 1115-1193.

[31] W.B. Gragg and A. Lindquist. On the partial realization problem. *Linear Algebra and its Applications* **50** (1983), 277-319.

[32] U. Grenander and G. Szegö. *Nonlinear Methods of Spectral Analysis.* Berkeley: Univ. California Press, 1958.

[33] J. Hadamard. Sur les transformations planes. *C. R. Acad. Sci. Paris* **142** (1906), 74.

[34] J. Hadamard. Sur les transformations ponctuelles. *Bull. Soc. Math. France* **34** (1906), 71-84.

[35] J. Hadamard. Sur les correspondances ponctuelles. Paris: Oeuvres, Editions du Centre Nationale de la Researche Scientifique, 1968. 383-384.

[36] S. Haykin. *Toeplitz Forms and their Applications.* New York: Springer-Verlag, 1979.

[37] N. Jacobson. *Lectures in Abstract Algebra.* Vol. III. New York: Van Nostrand, 1964.

[38] R.E. Kalman. Realization of covariance sequences. *Proc. Toeplitz Memorial Conference 1981.*

[39] R.E. Kalman. On minimal partial realizations of a linear input/output map. *Aspects of Network and System Theory.* (R.E. Kalman and N. de Claris, Eds.). New York: Holt, Rinehart and Winston, 1971. 385-408.

[40] R.E. Kalman. On partial realizations, transfer functions and canonical forms. *Acta Polytech. Scand.* **MA31** (1979), 9-39.

[41] S.M. Kay and S.L. Marple,Jr. Spectrum Analysis-A modern perspective. *Proceedings IEEE* **69** (1981), 1380-1419.

[42] H. Kimura. Positive partial realization of covariance sequences. *Modelling, Identification and Robust Control.* (C.I. Byrnes and A. Lindquist, Eds.). Amsterdam: North-Holland, 1987. 499-513.

[43] L. Kronecker. *Zur Teorie der Elimination einer Variabeln aus zwei algebraischen Gleichnungen.* Berlin: Preuss. Akad. Wiss., 1881.

[44] S. Lang. *Algebra.* New York: Addison & Wesley, 1970.

[45] C. Lanczos. An iteration method for the solution of the eigenvalue problem of linear differential and integral operators. *J. Res. Nat. Bur. Standards* **45** (1950), 255-282.

[46] A. Lindquist. A new algorithm for optimal filtering of discrete-time stationary processes. *SIAM J. Control* **12** (1974), 736-746.

[47] A. Lindquist. Some reduced-order non-Riccati equations for linear least-squares estimation: the stationary, single-output case. *Int. J. Control* **24** (1976), 821-842.

[48] A. Lindquist. On Fredholm integral equations, Toeplitz equations and Kalman-Bucy filtering. *Applied Mathematics and Optimization* **1** (1975), 355-373.

[49] A. Lindquist and G. Picci. On the stochastic realization problem. *SIAM J. Control Optim.* **17** (1979), 365-389.

[50] A. Lindquist and G. Picci. On "subspace method identification" and stochastic model reduction. *Proceedings of the 10th IFAC Symposium on Systems Identification 1994.* 397-403.

[51] A. Lindquist and G. Picci. Canonical correlation analysis, aproximate covariance extension, and identification of stationary time series. *Automatica* **32** (1996), 709-733.

[52] A. Magnus. Certain continued fractions associated with the Padé table. *Math. Z.* **78** (1962), 361-374.

[53] J. Makhoul. Linear prediction: A tutorial review. *Proceedings IEEE* **63** (1975), 561-580.

[54] S.L. Marple, Jr. *Digital Spectral Analysis and Applications.* Englewood Clifs, NJ: Prentice-Hall, 1987.

[55] J.W. Milnor. *Lectures in Modern Mathematics.* Vol. 1. (T.L. Saaty, Ed.). New York: John Wiley and Sons, 1964. 165-183.

[56] J.W. Milnor. *Singular Points of Complex Hypersurfaces.* Princeton, NJ: Princeton University Press, 1969.

[57] B.P. Molinari. The time-invariant linear-quadratic optimal-control problem. *Automatica* **13** (1977), 347-357.

[58] J. Rissanen. Recursive identification of linear systems. *SIAM J. Control* **9** (1971), 420-430.

[59] I. Schur. On power series which are bounded in the interior of the unit circle I and II. *Journal fur die reine und angewandte Mathematik* **148** (1918), 122-145.

[60] O. Toeplitz. Über die Fouriersche Entwicklung positiver Funktionen. *Rendiconti del Circolo Matematico di Palermo* **32** (1911), 191-192.

[61] P. van Overschee and B. De Moor. Subspace algorithms for stochastic identification problem. *IEEE Trans. Automatic Control* **AC-27** (1982), 382-387.

[62] A.S. Willsky. *Digital Signal Processing and Control and Estimation Theory.* Cambridge, MA: MIT Press, 1979.

Department of Systems Science and Mathematics, Washington University, St. Louis, Missouri 63130

Department of Optimization and Systems Theory, Royal Institute of Technology, 100 44 Stockholm, Sweden

Controlling Nonlinear Systems by Flatness

M. Fliess, J. Levine, P. Martin, F. Ollivier, and P. Rouchon[1]

1 Introduction

Many nonlinear control systems encountered in practice are *(differentially) flat*, i.e., linearizable by a special type of dynamic feedbacks called *endogenous*. The main feature of differential flatness is the presence of a fictitious output, called *flat*, or *linearizing*, output, $y = (y_1, \ldots, y_m)$, such that

(i) every system variable may be expressed as a function of the components of y and of a finite number of their time-derivatives;

(ii) every component of y may be expressed as a function of the system variables and of a finite number of their time-derivatives; and

(iii) the number m of components of y is equal to the number of independent input channels.

Those properties, which are enhanced most often by the existence of a flat output with a clear engineering and physical meaning, makes their control rather easy. They permit a straightforward open loop path tracking. The equivalence of the system with a linear controllable one yields a feedback stabilization around the desired trajectory. This new setting has already been illustrated by numerous realistic case-studies such as the motion planning and the stabilization of several nonholonomic mechanical systems [10, 11], the control of a crane [8, 20], of a towed cable [28], of an aircraft [21, 22, 31], of a PVTOL [12, 14, 25], of electric drives [5, 24], of magnetic bearings [19], of chemical reactors [35, 36, 39].

The notion of differential flatness was introduced in 1992 (see [7]) by means of differential algebra. As this setting has already been published in a journal [10], these notes are devoted to a Lie-Bäcklund differential geometric approach which has already been announced by the authors in [6, 8, 9, 12] and by Pomet [34], where jets and prolongations of infinite order play a key role. It permits an intrinsic definition of flatness, which is independent not only of any state space representation, but also of any distinction between system variables. It also yields a proof of the fact that any dynamic feedback linearizable system is (locally) differentially flat. Shadwick [40] noticed the relationship between linearization via dynamic feedback and Cartan's notion of absolute equivalence. Sluis, in his thesis [41], Murray and coworkers [32] have followed this alternative viewpoint where Cartan's approach via differential forms is the main ingredient.

[1]Work partially supported by the G.D.R. "MEDICIS" of the C.N.R.S.

The second part of this paper is devoted to some mechanical control examples. The first one shows that conservative nonholonomic systems are quite often differentially flat: their control may then be viewed as an obvious extension of the elementary computed torque method, which is so familiar in robotics. The second one concerns the one-trailer car where the trailer is not directly hitched to the car (see [38, 23]). Differential flatness is proved via some elliptic integrals. A stabilization feedback around a reference trajectory and at rest is proposed which utilizes time-scaling (see [11]) and the Frénet Formulae. Some simulations are provided.

2 Systems and Diffieties

2.1 Infinite-dimensional Fréchet Manifolds

Let I be a denumerable set of cardinality ℓ, finite or not. Denote by $\mathbb{R}^\ell$ the set of mappings $I \to \mathbb{R}$, where $\mathbb{R}$ is the real line; $\mathbb{R}^\ell$ is equipped with the product topology, which is Fréchet (see, e.g., [15, 46]). For any open subset $\mathcal{V} \subset \mathbb{R}^\ell$, let $C^\infty(\mathcal{V})$ be the set of functions $\mathcal{V} \to \mathbb{R}$, which only depend on a finite number of variables and are C^∞. A C^∞ $\mathbb{R}^\ell$-manifold may be defined like in finite dimension via $\mathbb{R}^\ell$-valued charts. The notions of functions, vector fields, differential forms of class C^∞ on an open subset are clear. If $\{x^i \mid i \in I\}$ are some local coordinates, let us notice that a vector field may be given by an infinite expression $\sum_{i\in I} \zeta^i \frac{\partial}{\partial x^i}$, whereas a differential form $\sum_{finie} \omega_{i_1 \dots i_p} dx^{i_1} \wedge \dots \wedge dx^{i_p}$ is always finite (the ζ^i's and $\omega_{i_1 \dots i_p}$'s are C^∞ functions). The notion of a (local) C^∞ morphism C^∞ between two C^∞ $\mathbb{R}^\ell$ and $\mathbb{R}^{\ell'}$-manifolds, where ℓ and are not necessarily equal, is obvious as well as the notion of (local) isomorphism. On the contrary, the non-validity of the implicit function theorem in these infinite-dimensional Fréchet spaces is forbidding the usual equivalence between various characterizations of (local) submersions and immersions between two finite-dimensional manifolds (see, e.g., [46]). Let us choose the following definition, due to Zharinov [46]: A (local) C^∞ *submersion* (resp. *immersion*) is a C^∞-morphism such that there exist local coordinates where it is a projection (resp. injection).

2.2 Diffieties

A *diffiety*[2] $\mathcal{M}$ is a C^∞ $\mathbb{R}^\ell$-manifold which is equipped with a *Cartan distribution* $CT\mathcal{M}$, i.e., a finite-dimensional and involutive distribution. The dimension n of $CT\mathcal{M}$ is the *Cartan dimension* of $\mathcal{M}$. Any (local) section of $CT\mathcal{M}$ is a (local) *Cartan field* of $\mathcal{M}$. The diffiety is called *ordinary* (resp.

[2]This terminology is due to Vinogradov [44, 45]. Diffieties are a very convenient tool for dealing with jets and prolongations of infinite order, which are so important in the *formal theory* of partial differential equations. More details may be found in [18, 43, 44, 45, 46].

partial) if, and only if, $n = 1$ (resp. $n \not\geq 1$). A *differential equation* is a diffiety. A C^∞ (local) morphism between diffieties is said to be *Lie-Bäcklund*[3] if, and only if, it is compatible with the Cartan distributions. From now on, we will restrict ourselves to ordinary diffieties, i.e., to ordinary differential equations.

Example. Take the nonlinear dynamics

$$\dot{x} = F(x, u) \tag{2.1}$$

where the state $x = (x_1, \ldots, x_n)$ and the control $u = (u_1, \ldots, u_m)$ belong to open subsets of $\mathbb{R}^n$ and $\mathbb{R}^m$; $F = (F_1, \ldots, F_n)$ is a m-tuple of C^∞-functions of their arguments. Associate to (2.1) the infinite-dimensional manifold $\mathcal{D}$ given by the local coordinates $\{t, x_1, \ldots, x_n, u_i^{(\nu_i)} \mid i = 1, \ldots, n; \nu_i \geq 0\}$. The Cartan distribution is spanned by the Cartan field

$$\frac{d}{dt} = \frac{\partial}{\partial t} + \sum_{k=1}^{n} F_k \frac{\partial}{\partial x_k} + \sum_{i=1}^{m} \sum_{\nu_i \geq 0} u_i^{(\nu_i+1)} \frac{\partial}{\partial u_i^{(\nu_i)}}.$$

It is denoted $\frac{d}{dt}$ since when applied to a local real-valued C^∞-function $h(t, x, u, \dot{u}, \ldots, u^{(\nu)})$ on $\mathcal{D}$ it yields its total derivative with respect to t.

Example. A fundamental role is played by the diffiety with global coordinates $\{t, w^\alpha_{i_1 \ldots i_n} \mid \alpha = 1, \ldots, m; i_1, \ldots, i_n \geq 0\}$ and Cartan field

$$\frac{d}{dt} = \frac{\partial}{\partial t} + \sum_{\beta=1}^{m} \sum_{i \geq 0} w^\beta_{i_1, \ldots, i+1, \ldots, i_n} \frac{\partial}{\partial w^\beta_{i_1, \ldots, i, \ldots, i_n}}.$$

It is written $\mathbb{R} \times \mathbb{R}^\infty_m$ and called a *trivial* diffiety since it corresponds to the trivial equation $0 = 0$.

Remark. Spanning the Cartan distribution by another Cartan field, *i.e.*, necessarily of the form $\gamma \frac{d}{dt}$, where $\gamma \neq 0$ is a (local) C^∞ function, yields a *time-scaling*. Thus, our formalism does not preclude an "absolute time" which is fixed once for all.

A (local) *Lie-Bäcklund fiber bundle* (*cf.* [46]) is a triple $\sigma = (\mathcal{X}, \mathcal{B}, \pi)$, where $\pi : \mathcal{X} \to \mathcal{B}$ is a (local) Lie-Bäcklund submersion between two diffieties. For any $b \in \mathcal{B}$, $\pi^{-1}(b)$ is a *fiber*.

2.3 Systems

A *system* [13] is a (local) Lie-Bäcklund fiber bundle $\sigma = (S, \mathbb{R}, \tau)$, where

[3] This terminology is due to Ibragimov [1, 16] (see, also, [46]). It has been criticized by several authors (see, e.g., [33]). Vinogradov writes *C-morphism* (see [18, 44, 45]).

- S is a diffiety where a given Cartan field ∂_S has been chosen once for all[4]

- $\mathbb{R}$ is given a canonical structure of a diffiety, with global coordinate t, and Cartan field $\frac{\partial}{\partial t}$; and

- the Cartan fields ∂_S and $\frac{\partial}{\partial t}$ are τ-related.

A (local) *Lie-Bäcklund morphism* (resp. *immersion, submersion, isomorphism*) $\varphi : (S, \mathbb{R}, \tau) \to (S', \mathbb{R}, \tau')$ between two systems is a Lie-Bäcklund morphism (resp. immersion, submersion, isomorphism) between S and S' such that

- $\tau = \tau'\varphi$,

- ∂_S and $\partial_{S'}$ are φ-related.

A *dynamics* is a (local) Lie-Bäcklund submersion $\delta \; : (S, \mathbb{R}, \tau) \to (U, \mathbb{R}, \mu)$ between two systems, such that the Cartan fields ∂_S and ∂_U are δ-related. In general, U will be an open subset of a trivial diffiety $\mathbb{R} \times \mathbb{R}^m$: it plays the role of input and m is the number of independent input channels. Replace, with a slight abuse of notations, ∂_S and ∂_U, which play the role of total derivation with respect to t, by $\frac{d}{dt}$.

Example. Take a time-dependent dynamics

$$\dot{x} = G(t, x, u, \dot{u}, \ldots, u^{(\alpha)}) \tag{2.2}$$

which may contain moreover derivatives of the control variables and an analogous infinite-dimensional manifold $\mathcal{E}$ given by the local coordinates $\{t, x_1, \ldots, x_n, u_i^{(\nu_i)}\}$ where

$$\frac{d}{dt} = \frac{\partial}{\partial t} + \sum_{k=1}^{n} G_k \frac{\partial}{\partial x_k} + \sum_{i=1}^{m} \sum_{\nu_i \geq 0} u_i^{(\nu_i+1)} \frac{\partial}{\partial u_i^{(\nu_i)}}.$$

The above sumersion τ is given by the projection $\{t, x_1, \ldots, x_n, u_i^{(\nu_i)}\} \mapsto \{t, , u_i^{(\nu_i)}\}$. Notice that the state-space is nothing else that the corresponding fiber.

3 Equivalence and Flatness

Two systems $(S, \mathbb{R}, \tau)$ and $(S', \mathbb{R}, \tau')$ are said to be (locally) *differentially equivalent* if, and only if, they are (locally) Lie-Bäcklund isomorphic. They

[4]Moreover, S should be of *finite type* (see [13]), i.e., finitely generated.

are said to be (locally) ***orbitally equivalent*** if, and only if, S and S' are (locally) Lie-Bäcklund isomorphic. The first definition preserves time, whereas the second one does not: it introduces a time-scaling.

The triple $(\mathbb{R} \times \mathbb{R}^m, \mathbb{R}, \mathrm{pr})$, where $\mathbb{R} \times \mathbb{R}^m = \{t, y_i^{(\nu_i)}\}$ is a trivial diffiety and pr denotes the projection $\{t, y_i^{(\nu_i)}\} \mapsto t$, is called a ***trivial system***. The system $(S, \mathbb{R}, \tau)$ is said to be (locally) ***differentially flat*** if, and only if, it is (locally) differentially equivalent to a trivial system; it is said to be (locally) ***orbitally flat*** if, and only if, it is (locally) orbitally equivalent to a trivial diffiety. The set $y = (y_1, \dots, y_m)$ is called a ***flat***, or ***linearizing, output***.

Arguments based on dimension theory (*cf.* [10, 12, 13, 34]) lead to the

Proposition. *The number of components of a flat output is equal to the number of independent input channels.*

4 Differential Flatness and Dynamic Feedback Linearization

4.1 Bäcklund Correspondences

A (local) *Bäcklund correspondence* between two systems $(S_\iota, \mathbb{R}, \tau_\iota)$, $\iota = 1, 2$, is given by two (local) Lie-Bäcklund submersions $\phi_\iota : (S, \mathbb{R}, \tau) \to (S_\iota, \mathbb{R}, \tau_\iota)$, where $(S, \mathbb{R}, \tau)$ is another system (compare with [46]). This very weak equivalence relation is called a *dynamic feedback*. In order to show that this abstract definition encompasses the usual picture of dynamic feedback (cf. [17, 30]), assume that the fiber corresponding to ϕ_1 is finite-dimensional. As a matter of fact, taking local coordinates $z = (z_1, \dots, z_q)$ on a fiber yields

$$\begin{aligned} \dot{z}_i &= f_i(z, \omega) \quad i = 1, \dots, q \\ \psi &= g(z, \omega') \end{aligned}$$

where $\psi : S_2 \to \mathbb{R}$ is a function, ω and ω' denote finite sets of functions $S_1 \to \mathbb{R}$.

4.2 Dynamic Feedback Linearizability

Adapting the definition of dynamic feedback linearizability in [4] yields: A system $(S, \mathbb{R}, \tau)$ is said to be (locally) ***linearizable by dynamic feedback*** if, and only if, there exists a (local) Lie-Bäcklund submersion $\phi : (\mathbb{R} \times \mathbb{R}^\mu_\infty, \mathbb{R}, \mathrm{pr}) \to (S, \mathbb{R}, \tau)$, where $(\mathbb{R} \times \mathbb{R}^\mu, \mathbb{R}, \mathrm{pr})$ is a trivial system. The next result relates differential flatness to dynamic feedback linearizability [6, 8].

Theorem. *Any system which is locally dynamic feedback linearizable is locally differentially flat.*

In a more traditional language, Lie-Bäcklund isomorphisms and Bäcklund correspondences are respectively related to *endogenous* and *exogenous* dynamic feedbacks (see [12, 14]) for further details). Then, the above theorem reads:

Corollary. *Any system which is locally linearizable by an exogenous dynamic feedback is locally differentially flat, i.e., is locally linearizable by an endogenous dynamic feedback.*

4.3 Sketch of the Proof

Set $\mathbb{R} \times \mathbb{R}^{\mu}_{\infty} = \{t, z_k^{(\nu_k)} \mid k = 1, \dots, \mu; \nu_k \geq 0\}$. Consider the ring $\mathcal{O}(V)$ of C^{∞}-functions on a suitable open subset V of S, and $\mathcal{O}_r(V)$ the subring corresponding to functions of order at most r, i.e., functions of $\{t, z_k^{(\nu_k)} \mid k = 1, \dots, \mu; 0 \leq \nu_k \leq r\}$.

Let f_0 be a minimal set of generators, modulo t, of $\mathcal{O}_0(V_0)$ (up to functional independence), where V_0 is some open subset of V. Complete $f_0, \dot{f}_0$ with f_1 for obtaining a set of generators of $\mathcal{O}_1(V_1)$, where V_1 is some open subset of V_0. We go on completing $f_0 \cup \ldots \cup f_0^{(s)} \cup f_1 \cup \ldots \cup f_1^{(s-1)} \cup \ldots, f_s \cup \dot{f}_s$ with f_{s+1} for getting a set of generators of $\mathcal{O}_{s+1}(V_{s+1})$, where V_{s+1} is some open subset of V_s. The set f_s will be empty for s great enough, all diffieties being of finite type. Dimension-theoretic arguments permit us to prove that the union of the f_s's is a *functionally differentially independent* family; i.e., the elements of the f_i's and their derivatives of arbitrary order are functionally independent.

4.4 Orbital Flatness and Controlling the Clock

Associate to the system $(S, \mathbb{R}, \tau)$ the new one $(S \times \mathbb{R}_{\infty}, \mathbb{R}, \tau')$, where the absolute derivation is given by

$$v\frac{d}{dt} + \frac{\partial}{\partial t'} + \sum_{\ell \geq 0} v^{(\ell+1)} \frac{\partial^{\ell}}{\partial v^{(\ell)}}.$$

The variable v should be understood as a *control of the clock*[5]. The next result is a consequence of the preceeding theorem

Proposition. *The system $(S, \mathbb{R}, \tau)$ is (locally) orbitally flat if, and only if, the system $(S \times \mathbb{R}_{\infty}, \mathbb{R}, \tau')$ is (locally) differentially flat.*

In other words, orbital flatness is "equivalent" to differential flatness via controlling the clock.

[5]The definition given here of controlling the clock differs slightly from a previous one in [8].

5 Nonholonomic Mechanical Systems

Consider $n-m$ *nonholonomic constraints*, with m degrees of freedom (see [13] for further details and related references)

$$A_k = \sum_{i=1}^{n} f_k^i(q)\,\dot{q}_i = 0, \quad k = 1,\dots,n-m \tag{5.1}$$

where $q = (q_1,\dots,q_n)$ are the configuration variables (see, e.g., [2, 29]). Let $\mathcal{I}$ be the *differential ideal* of the trivial diffiety $C^\infty(\mathbb{R}\times\mathbb{R}^\infty_{n-m})$ generated by $A_1,\dots,A_{n-m}$, i.e., the ideal generated by $A_1,\dots,A_{n-m}$, which is closed for the derivation $\frac{d}{dt}$. Assume that the set of zeros of $\mathcal{I}$ is a subdiffiety $\mathcal{N}$ of $\mathbb{R}\times\mathbb{R}^\infty_m$, which is called the *infinite prolongation* associated to (5.1) [6] (this is the case when the rank of the A_k's with respect to $\dot{q}$ is maximum, i.e., when the constraints are independent): $\mathcal{N}$ is called the *configuration diffiety* corresponding to the nonholonomic constraint (5.1). The associated triple $(\mathcal{N},\mathbb{R},\mathrm{pr})$, where pr is the canonical projection, is the *configuration system.*

Take a *conservative* mechanical system, which is subject to (5.1), with external control forces $u = (u_1,\dots,u_m)$, described by (see, [3, equation (12), p. 17]):

$$\begin{aligned} \frac{d}{dt}\left(\frac{\partial L}{\partial \dot{q}_i}\right) - \frac{\partial L}{\partial q_i} &= \textstyle\sum_{k=1}^{n-m} f_k^i\lambda_k + \sum_{j=1}^{m} g_i^j(q)u_j, \quad i = 1,\dots n \\ \textstyle\sum_{i=1}^{n} f_k^i(q)\dot{q}_i &= 0, \quad k = 1,\dots,n-m. \end{aligned} \tag{5.2}$$

$L(q,\dot{q})$ is the Lagrangian; $\lambda = (\lambda_1,\dots,\lambda_{n-m})$ are the Lagrangian multipliers. Assume that (5.2) defines a subsystem $(\mathcal{M},\mathbb{R},\mathrm{pr})$ of $(\mathcal{N},\mathbb{R},\mathrm{pr})$, which is called the *mechanical system.*

Assume that the $n\times n$ matrix $M = (F', G)$ ($'$ denotes transposition) where

$$F = (f_k^i)_{\substack{1\le k\le n-m\\ 1\le i\le n}} \qquad G = (g_i^j)_{\substack{1\le i\le n\\ 1\le j\le m}} \tag{5.3}$$

is invertible for all q. This just means that the external control "forces" u and the constraint "forces" λ are independent.

The nonholonomic constraints (5.1) are said to be *differentially flat* if, and only if, the associated configuration system $(\mathcal{N},\mathbb{R},\mathrm{pr})$ is differentially flat. The reader is invited to check that almost all examples of nonholonomic constraints given in famous treatises such as [2] or [29] are indeed differentially flat.

If $(\mathcal{N},\mathbb{R},\mathrm{pr})$ is differentially flat with a flat output y, the configuration variables q may be expressed as functions of the components of y and their

[6] See [33] and [46] for further information.

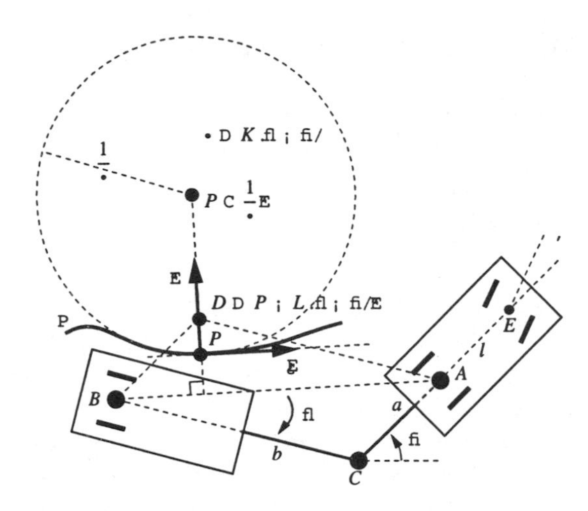

Figure 1: the general 1-trailer system is flat with linearizing point P and with L and K are defined by (6.3) and (6.5).

derivatives. The rank property of M implies that the components of u and λ enjoy the same property. We have proved the

Proposition. *If the configuration system* $(\mathcal{N}, \mathbb{R}, \mathrm{pr})$ *is differentially flat, then the mechanical system* $(\mathcal{M}, \mathbb{R}, \mathrm{pr})$ *is also differentially flat.*

Such nonholonomic mechanical system can be said *fully actuated*: there is the same number of degrees of freedom as independent input channels. The parallel with the *computed torque* method for fully actuated holonomic systems, where the configuration variables q are a flat output, is made clear by replacing q with the new flat output y. The only price to be paid is the utilization of higher order derivatives.

The above analysis may easily be extended to time-varying constraints and additional non-conservative effects such as friction.

6 The General One-trailer System

We will now consider the car with one trailer displayed on figure 1: here the trailer is not directly hitched to the car at the center of the rear axle, but more realistically at a distance a of this point. We have shown in [38, 23] that the nonholonomic constraints are flat and that arc-length parametrization remove the singularity at rest. We assume that the control is the steering angle φ and the car velocity v. In practice, fast and low-level regulators controlling the motor and steering mechanism use the velocity and the steering angle as set-points, respectively. The equations of the high-level dynamics are then as follows:

$$
\begin{aligned}
\dot{x} &= v\cos\alpha \\
\dot{y} &= v\sin\alpha \\
\dot{\alpha} &= \frac{v}{l}\tan\varphi \\
\dot{\beta} &= \frac{-v}{b}\Big(\sin(\beta-\alpha) + \frac{a}{l}\tan\varphi\cos(\beta-\alpha)\Big)
\end{aligned}
\tag{6.1}
$$

where (x, y) are the Cartesian coordinates of point A. Parameters l, a and b are positive lengths. The case $a < 0$ is similar to $a > 0$ and is not treated here.

Proposition 1 *System (6.1) is flat. A possible linearizing output $y = (y_1, y_2)$ is given by the formulae*

$$
\begin{aligned}
y_1 &= x - b\cos\beta - L(\beta-\alpha)\frac{b\sin\beta + a\sin\alpha}{\sqrt{a^2+b^2+2ab\cos(\beta-\alpha)}} \\
y_2 &= y - b\sin\beta + L(\beta-\alpha)\frac{a\cos\alpha + b\cos\beta}{\sqrt{a^2+b^2+2ab\cos(\beta-\alpha)}}
\end{aligned}
\tag{6.2}
$$

with the elliptic function L

$$
L(\beta-\alpha) = ab\int_0^{\beta-\alpha}\frac{\cos\sigma}{\sqrt{a^2+b^2+2ab\cos\sigma}}\,d\sigma \tag{6.3}
$$

Geometrically (y_1, y_2) are the Cartesian coordinates of the point P displayed on figure 1.

Contrary to the car, the explicit derivation of the linearizing output is far from being obvious there. It results from [23]. It is derived via the flag $I^{(2)} \subset I^{(1)} \subset I^{(0)}$ associated to the Pfaffian system $I^{(0)}$ obtained from (6.1) by elimination of v. $I^{(0)}$ is generated thus by the three 1-forms

$$
\begin{aligned}
&\sin\alpha\, dx - \cos\alpha\, dy \\
&l\, d\alpha - \tan\varphi(\cos\alpha\, dx + \sin\alpha\, dy) \\
&b\, d\beta + \Big(\frac{a}{l}\tan\varphi\cos(\beta-\alpha) + \sin(\beta-\alpha)\Big)\ (\cos\alpha\, dx + \sin\alpha\, dy).
\end{aligned}
$$

Then $I^{(2)}$ is generated by a single 1-form that is a combination of dy_1 and dy_2 only (see the proof of the main theorem in [23]). Another way to derive the flat output and close to this one, is to use [26] for transforming the system via static feedback and change of coordinates into a special chained form [27]. Then, the change of coordinates is obtained from the derived flag and can be expressed with the flat output via $(y_1, y_2, \frac{dy_2}{dy_1}, \frac{d^2y_2}{dy_1^2})$ by using the rule $\frac{dy_2}{dy_1} = \dot{y}_2/\dot{y}_1, \ldots$) (see [42] for the precise calculations of these coordinates).

Using (6.2) and (6.1), one can see that

1. The velocity of P is colinear to direction AB; more precisely we have

$$\frac{dP}{dt} = vJ(\alpha - \beta, \varphi)\vec{\tau} \tag{6.4}$$

where

$$\begin{aligned}
&\vec{\tau} = \frac{1}{\|AB\|}\vec{AB} \\
&\rho = \sqrt{a^2 + b^2 + 2ab\cos(\beta - \alpha)} \\
&\rho^2 \; J(\alpha - \beta, \varphi) = \left(b + a\cos(\beta - \alpha) - \frac{a^2}{l}\sin(\beta - \alpha)\tan\varphi\right) \cdot \\
&\qquad \cdot (\cos(\beta - \alpha)\rho + \sin(\beta - \alpha)L(\beta - \alpha)) \,.
\end{aligned}$$

2. The curvature κ of $\mathcal{P}$, obtained by the classical relation

$$\kappa = \frac{\begin{vmatrix} \dot{y}_1 & \ddot{y}_1 \\ \dot{y}_2 & \ddot{y}_2 \end{vmatrix}}{((\dot{y}_1)^2 + (\dot{y}_2)^2)^{3/2}},$$

is given by

$$K(\beta - \alpha) = \frac{-\sin(\beta - \alpha)}{\cos(\beta - \alpha)\; \sqrt{a^2 + b^2 + 2ab\cos(\beta - \alpha)} + L(\beta - \alpha)\; \sin(\beta - \alpha)}. \tag{6.5}$$

Routine calculations show that there exists a unique real $\gamma \in [\pi/2, \pi]$ such that K is a decreasing bijection between $]-\gamma, \gamma[$ and $\mathbb{R}$.

When $a = 0$, previous formulae simplify: $P \equiv B$, $\gamma = \pi/2$ and $K = -\tan(\beta - \alpha)/b$. We recognize here the basic relationships used for the standard 1-trailer system (see [37, 10]).

Geometrically, the above relations can be summarized as follows. Denote by s (resp. s_A) the arc length of $\mathcal{P}$ (resp. of the curve followed by A). Then

$$\begin{aligned}
\frac{dP}{ds} &= \vec{\tau} = \frac{\vec{AB}}{\|AB\|} \\
ds &= J(\alpha - \beta, \varphi)\; ds_A \qquad (6.6)\\
\frac{d^2P}{ds^2} &= \kappa\, \vec{\nu} = K(\beta - \alpha)\, \vec{\nu}
\end{aligned}$$

where $\vec{\nu}$ is the unitary vector with $(\vec{\tau}, \vec{\nu})$ orthogonal and admitting direct orientation. These relations can be inverted: A, α and β can be deduced from P, $\vec{\tau}$ and κ, since K is a bijection form $]-\gamma, \gamma[$ to $\mathbb{R}$. Thus there is a one-to-one correspondence between the set made up with the point A and the two configuration angles α and β such that $|\beta - \alpha| < \gamma$, and the set composed by the linearizing point P, the unitary tangent vector $\vec{\tau}$ and the curvature κ.

An additional derivation with respect to s leads to

$$\frac{dk}{ds} = \left(\frac{\rho}{\cos(\beta-\alpha)\rho + L(\beta-\alpha)\sin(\beta-\alpha)}\right)^3 \cdot \\ \cdot \frac{\sin(\beta-\alpha)/b + (a\cos(\beta-\alpha)/bl + 1/l)\tan\varphi}{b + a\cos(\beta-\alpha) - a^1 \sin(\beta-\alpha)\tan\varphi/l} \tag{6.7}$$

with $\rho = \sqrt{a^2 + b^2 + 2ab\cos(\beta-\alpha)}$.

This relation always gives the direction φ that can be eventually $\pm\pi/2$. In this case the velocity of A becomes zero even if the velocity of P is not zero.

The above consideration leads thus to the following proposition:

Proposition. *Consider (6.1) and two different configurations:* $\tilde{q} = (\tilde{x}, \tilde{y}, \tilde{\alpha}, \tilde{\beta})$ *and* $\overline{q} = (\overline{x}, \overline{y}, \overline{\alpha}, \overline{\beta})$. *Assume that the angles* $\tilde{\beta} - \tilde{\alpha}$ *and* $\overline{\beta} - \overline{\alpha}$ *belong to* $]-\gamma, +\gamma[$. *Then, there exists a smooth open-loop control* $[0,T] \ni t \to (v(t), \varphi(t))$ *steering the system from* $\tilde{p}$ *at time* 0 *to* $\overline{p}$ *at time* $T > 0$, *such that the angle* $\beta - \alpha$ *always remains in* $]-\gamma, \gamma[$ *and such that* $v(t) = 0$ *for* $t = 0, T$.

The initial and final configurations lead for the linearizing curve $\mathcal{P}$ to initial and final position, tangent and curvature. Consider now a smooth mapping $[0,T] \ni t \mapsto \sigma(t) \in [0, S]$ where S is the length of $\mathcal{P}$. The steering control is then given by the following formulae

$$\begin{aligned} \tan\varphi &= l\frac{(b+a\cos(\beta-\alpha))(\cos(\beta-\alpha)+L(\beta-\alpha)\sin(\beta-\alpha)/\rho)^3\frac{dk}{ds} - \sin(\beta-\alpha)/b)}{1+a\cos(\beta-\alpha)/b+a^2\sin(\beta-\alpha)(\cos(\beta-\alpha)+L(\beta-\alpha)\sin(\beta-\alpha)/\rho)^3\frac{dk}{ds}} \\ v &= \frac{\rho^2 \quad \dot{\sigma}(t)}{(b+a\cos(\beta-\alpha)-a^2\sin(\beta-\alpha)\tan\varphi/l)(L(\beta-\alpha)\sin(\beta-\alpha)+\rho\cos(\beta-\alpha))} \end{aligned} \tag{6.8}$$

where $\beta - \alpha = K^{-1}(\kappa(\sigma(t)))$ and $\rho = \sqrt{a^2 + b^2 + 2ab\cos(\beta-\alpha)}$.

The above open-loop control (6.8) is expressed via arc-length parametrization. Such parametrization can be difficult to obtain in practice. A straightforward analysis reveals that any regular parametrization is sufficient for computing explicitly the steering control defined by (6.8).

The fact that $(\beta - \alpha)$ and φ depend only on $(\kappa, d\kappa/ds)$ results directly from the invariance of the problem with respect to the group of planar Euclidian transformations. Such physical and symmetry considerations are essential here in order to keep simple formulae.

This geometric construction can be used for solving the steering problem. As in [37, 10], a simple steering program can be directly deduced from such developments. The MATLAB simulations [7] of figure 2 illustrate the interest of flatness combined with such geometric constructions.

[7]The simulation program can be obtained upon request at the electronic address

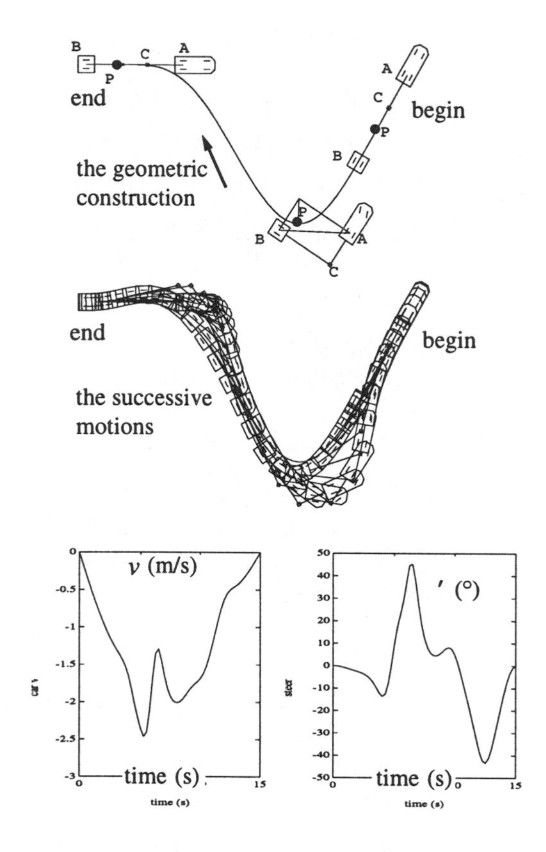

Figure 2: example of backward motions with $l = 1$ m, $a = 1.5$ m, $b = 2.5$ m and $T = 15$ s.

Everyday life shows us that backward motions with a trailer are unstable (contrary to forward motions). This instability comes from the fact that, in the s-scale, the backwards open-loop equations admit (around any straightforward trajectory corresponding to $\dot{\alpha} = 0$ and $\dot{\beta} = 0$) an unstable pole equal to $-1/b$ (remind that $\dot{s} < 0$), the corresponding eigenvector being β the angle of the trailer. This leads to an exponential divergence with respect to s. Thus, trajectory stabilization is strongly needed in practice. Similarly to the open-loop design, the closed-loop design will make ue of the flatness relationships expressed in an intrinsic manner via the Frenet Formulae (6.6,6.7). Such intrinsic computations make the formulae of the dynamic compensator particularly simple.

Denote by subscript c the reference trajectory $\mathcal{P}_c$ for P: $s_c \mapsto P_c(s_c)$ where s_c is the arc length of $\mathcal{P}_c$. Consider also the smooth parametrization

rouchon@cas.ensmp.fr.

$t \mapsto s_c = \sigma(t)$. Then (6.4) implies that, if we set

$$v = [J(\alpha - \beta, \varphi)]^{-1} \xi \, \dot{\sigma}(t) \, . \tag{6.9}$$

Then

$$\frac{dP}{ds_c} = \xi \vec{\tau}$$

with ξ a new control variable close to 1 when P follows P_c. Moreover, we have, denoting by s the real arc length of the curve generated by P,

$$\frac{d}{ds_c} = \xi \frac{d}{ds}.$$

In the sequel we will denote by $'$ the derivation with respect to s_c.

We have

$$P' = \xi \vec{\tau} \quad \text{and} \quad P'' = \xi' \vec{\tau} + \xi \vec{\tau}'.$$

But $\vec{\tau}' = \xi d\vec{\tau}/ds = \xi K(\beta - \alpha)\vec{\nu}$, according to (6.5). Thus we have

$$P'' = \xi' \vec{\tau} + \xi^2 K(\beta - \alpha)\vec{\nu}.$$

Similarly, we have

$$P''' = (\xi'' - \xi^3 K^2(\beta - \alpha)) \, \vec{\tau} + (3\xi\xi' K(\beta - \alpha) + \xi^3 \frac{dK}{ds}) \, \vec{\nu}$$

where dK/ds is given by (6.7). Thus the dynamic compensator

$$\xi' = \zeta, \quad \zeta' = u + \xi^3 K^2(\beta - \alpha)$$

that reads in the t-scale

$$\dot{\xi} = \zeta \, \dot{\sigma}(t), \quad \dot{\zeta} = (u + \xi^3 K^2(\beta - \alpha)) \, \dot{\sigma}(t) \tag{6.10}$$

and the static feedback on φ obtained from (6.8) by replacing $\dfrac{dk}{ds}$ by $\dfrac{w - 3\xi\zeta K(\beta - \alpha)}{\xi^3}$, yields the following dynamics

$$P''' = u \, \vec{\tau} + w \, \vec{\nu}$$

where u and w are the new control.

We can chose u and w for imposing now a linear and stable closed-loop dynamics,

$$\begin{aligned}(P - P_c)''' = &-\left(\tfrac{1}{d_1} + \tfrac{1}{d_2} + \tfrac{1}{d_3}\right) (P - P_c)'' \\ &- \left(\tfrac{1}{d_1 d_2} + \tfrac{1}{d_2 d_3} + \tfrac{1}{d_3 d_1}\right) (P - P_c)' - \left(\tfrac{1}{d_1 d_2 d_3}\right) (P - P_c),\end{aligned} \tag{6.11}$$

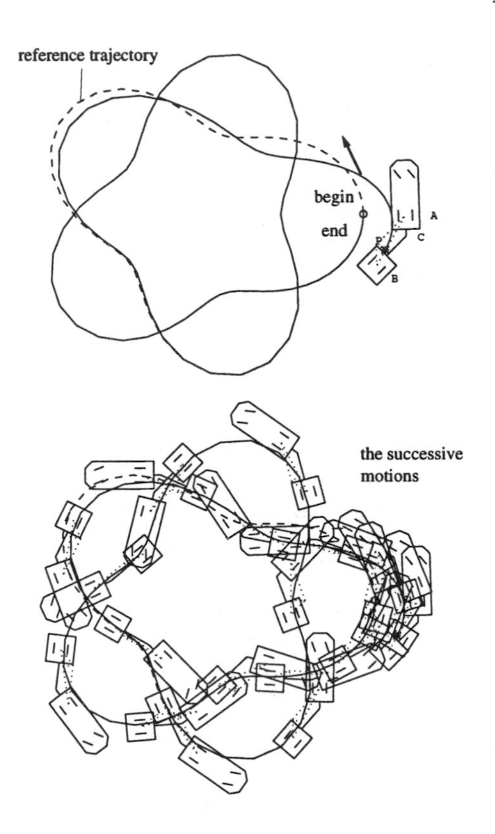

Figure 3: Example of closed-loop motions with $l = 1.0$ m, $a = 0.5$ m, $b = 1.0$ m and $T = 15.$ s (polar equation of the linearizing curve $r = \cos(5\theta/2) + 3$).

where the d_i's correspond to positive lengths.

A simulation using this dynamic feedback is presented in figure 3. It corresponds to a simple MATLAB program that can be obtained upon request at the electronic address `rouchon@cas.ensmp.fr`. The reference trajectory of polar equation $r = \cos(5\theta/2)+3$ is followed with $d_1 = a+b+l$, $d_2 = d_1/1.5$ and $d_3 = d_2/1.5$. The initial error for the state (x, y, α, β) is of $(+l, -l, -\pi/6, -\pi/6)$. We remark that after a distance greater that d_1 (the total length), the error is divided by more than 2.

References

[1] R.L. Anderson and N.H. Ibragimov. *Lie-Bäcklund Transformations in Applications*. Philadelphia: SIAM, 1979.

[2] P. Appell. *Traité de Mécanique Rationnelle. Vol. 2.* Paris: Gauthier-Villars, 1953.

[3] V.I. Arnold. *Dynamical Systems III.* New York: Springer, 1988.

[4] B. Charlet, J. Lévine and R. Marino. Sufficient conditions for dynamic state feedback linearization. *SIAM J. Control Optimiz.* **29** (1991), 38-57.

[5] K. Chelouah, E. Delaleau, P. Martin and P. Rouchon. Differential flatness and control of induction motors. *Proc. CESA 1996.* 80-85.

[6] M. Fliess, J. Lévine, P. Martin, F. Ollivier and P. Rouchon. Flatness and dynamic feedback linearizability: two approaches. *Proc.* 3rd *European Control Conf. 1995.* 649-654.

[7] M. Fliess, J. Lévine, P. Martin and P. Rouchon. Sur les systèmes non linéaires différentiellement plats. *C.R. Acad. Sci. Paris* **I-315** (1992), 619-624.

[8] M. Fliess, J. Lévine, P. Martin and P. Rouchon. Linéarisation par bouclage dynamique et transformations de Lie-Bäcklund. *C.R. Acad. Sci. Paris* **I-317** (1993), 981-986.

[9] M. Fliess, J. Lévine, P. Martin and P. Rouchon. Nonlinear control and Lie-Bäcklund transformations: towards a new differential geometric standpoint. *Proc.* 33rd *IEEE Decision Control Conf. 1994.* 981-986.

[10] M. Fliess, J. Lévine, P. Martin and P. Rouchon. Flatness and defect of nonlinear systems: introductory theory and applications. *Internat. J. Control.* **61** (1995), 1327-1361.

[11] M. Fliess, J. Lévine, P. Martin and P. Rouchon. Design of trajectory stabilizing feedback for driftless flat systems. *Proc.* 3rd *European Control Conf. 1995.* 1882-1887.

[12] M. Fliess, J. Lévine, P. Martin and P. Rouchon. A Lie-Bäcklund approach to dynamic feedback equivalence and flatness. *Robust Control via Variable Structure and Lyapunov Techniques.* (F. Garofalo and L. Glielmo, Eds.). *Lect. Notes Control Inform. Sci.* Vol. 217. London: Springer, 1996. 245-268.

[13] M. Fliess, J. Lévine, P. Martin and P. Rouchon. Deux applications de la géométrie locale des diffiétés. To appear *Ann. Inst. H. Poincaré Phys. Théor.*

[14] M. Fliess, J. Lévine, P. Martin and P. Rouchon. A Lie-Bäcklund approach to equivalence and flatness of nonlinear systems. Preprint, 1996.

[15] M. Golubitsky and V. Guillemin. *Stable Mappings and their Singularities.* New York: Springer, 1973.

[16] N.H. Ibragimov. *Transformation Groups Applied to Mathematical Physics.* Dordrecht, The Netherlands: Reidel, 1985.

[17] A. Isidori. *Nonlinear Control Systems.* New York: Springer, 1989.

[18] I.S. Krasil'shchik, V.V. Lychagin and A.M. Vinogradov. *Geometry of Jet Spaces and Nonlinear Partial Differential Equations.* New York: Gordon and Breach, 1986.

[19] J. Lévine, J. Lottin and J.C. Ponsart. A nonlinear approach to the control of magnetic bearings. *IEEE Trans. on Control Systems Technology* **5** (1996), 524-544.

[20] J. Lévine, P. Rouchon, G. Yuan, C. Grebogi, B.R. Hunt, E. Kostelich, E. Ott and J.A. Yorke. On the control of US Navy cranes. To appear *Proc. European Control Conf. 1997.*

[21] Ph. Martin. *Contribution à l'étude des Systèmes Diffèrentiellement Plats.* Ph.D. Dissertation. École des Mines. Paris. 1992.

[22] Ph. Martin. Aircraft control using flatness. *Proc. CESA 1996*, 194-199.

[23] Ph. Martin and P. Rouchon. Feedback linearization and driftless systems. *Math. Control Signal Syst.* **7** (1994), 235-254.

[24] Ph. Martin and P. Rouchon. Flatness and sampling control of induction motors. To appear *Proc. IFAC World Congress 1996.*

[25] Ph. Martin, S. Devasia and B. Paden. A different look at output tracking: control of a VTOL aircraft. *Automatica* **32** (1995), 101-108.

[26] R.M. Murray. Nilpotent bases for a class on nonintegrable distributions with applications to trajectory generation for nonholonomic systems. *Math. Control Signal Syst.* **7** (1994), 58-75.

[27] R.M. Murray and S.S. Sastry. Nonholonomic motion planning: Steering using sinusoids. *IEEE Trans. Automat. Control* **38** (1993), 700-716.

[28] R.M. Murray. Trajectory generation for a towed cable system using differential flatness. To appear *Proc. IFAC World Congress 1996.*

[29] J.I. Neĭmark and N.A. Fufaev. *Dynamics of Nonholonomic Systems.* Providence, RI: Amer. Math. Soc., 1972.

[30] H. Nijmeijer and A.J. van der Schaft. *Nonlinear Dynamical Control Systems.* New York: Springer, 1990.

[31] M.J. van Nieuwstadt and R.M. Murray. Fast mode switching for a thrust vectored aircraf. *Proc. CESA 1996.* 86-91.

[32] M. van Nieuwstadt, M. Ratinam and R.M. Murray. Differential flatness and absolute equivalence. *Proc.* 33rd *IEEE Conf. Decision Control 1994.* 326-332.

[33] P.J. Olver. *Applications of Lie Groups to Differential Equations.* New York: Springer, 1993.

[34] J.B. Pomet. A differential geometric setting for dynamic equivalence and dynamic linearization. *Geometry in Nonlinear Control and Differential Inclusions.* (B. Jakubczyk, W. Respondek and T. Rzeżuchowski, Eds.). Warsaw: Banach Center, 1995. 319-339.

[35] R. Rothfuss, J. Rudolph and M. Zeitz. Controlling a chemical reactor model using its flatness. To appear *Proc. IFAC World Congress 1996.*

[36] P. Rouchon. Vibrational control and flatness of chemical reactors. *Proc. CESA 1996.* 211-212.

[37] P. Rouchon, M. Fliess, J. Lévine and Ph. Martin. Flatness and motion planning: the car with n-trailers. *Proc. ECC 1993.* 1518-1522.

[38] P. Rouchon, M. Fliess, J. Lévine and Ph. Martin. Flatness, motion planning and trailer systems. *Proc. 32nd IEEE Conf. Decision and Control 1993.* 2700-2705.

[39] J. Rudolph. Flatness of some examples of chemical reactors. *Proc. CESA 1996.* 206-210.

[40] W.F. Shadwick. Absolute equivalence and dynamic feedback linearization. *Systems Control Letters* **15** (1990), 35-39.

[41] W.M. Sluis. *Absolute Equivalence and its Application to Control Theory.* Ph.D. Dissertation. University of Waterloo. Waterloo, Ontario. 1992.

[42] D.M. Tilbury. *Exterior Differential Systems and Nonholonomic Motion Planning.* Ph.D. Dissertation. University of California. Berkeley, California. 1994.

[43] T. Tsujishita. Formal geometry of systems of differential equations. *Sugaku Expos.* **3** (1990), 25-73.

[44] A.M. Vinogradov. Local symmetries and conservation laws, *Acta Appl. Math.* **2** (1984), 21-78.

[45] A.M. Vinogradov, Ed. *Symmetries of Partial Differential Equations.* Dordrecht, The Netherlands: Kluwer, 1989.

[46] V.V. Zharinov. *Geometrical Aspects of Partial Differential Equations.* Singapore: World Scientific, 1992.

Laboratoire des Signaux et Systèmes, CNRS-Supélec, Plateau de Moulon, 91192 Gif-sur-Yvette, France

Centre Automatique et Systèmes, École des Mines de Paris, 35 rue Saint-Honoré, 77305 Fontainebleau, France

Centre Automatique et Systèmes, École des Mines de Paris, 35 rue Saint-Honoré, 77305 Fontainebleau, France

GAGE-CNRS, Centre de Mathématiques, École polytechnique, 91128 Palaiseau, France

Centre Automatique et Systémes, École des Mines de Paris, 60 bld. Saint-Michel, 75272 Paris Cedex 06, France

How Set-Valued Maps Pop Up in Control Theory

H. Frankowska

Abstract

We describe four instances where set-valued maps intervene either as a tool to state the results or as a technical tool of the proof. The paper is composed of four rather independent sections:

1. Set-Valued Optimal Synthesis and Differential Inclusions
2. Viability Kernel
3. Nonsmooth Solutions to Hamilton-Jacobi-Bellman Equations
4. Interior and Boundary of Reachable Sets

1 Optimal Synthesis

We define optimal synthesis in two cases: for the Mayer problem with locally Lipschitz value function and for the time optimal control problem with lower semicontinuous time optimal function.

1.1 Mayer Problem with Lipschitz Value Function

Consider a complete separable metric space U, a continuous $f : R^n \times U \mapsto R^n$, a locally Lipschitz $\varphi : R^n \mapsto R$ and the minimization problem

$$\min \{\varphi(x(1)) \mid x \text{ is solves (1)},\ x(0) = \xi_0\}$$

$$x'(t) \;=\; f(x(t), u(t)),\ \ u(t) \;\in\; U \tag{1}$$

The value function of this problem is defined by

$$V(t_0, x_0) \;=\; \inf \{\varphi(x(1)) \mid x \text{ solves (1)},\ x(t_0) = x_0\}$$

The value function generates the optimal synthesis since it is constant along optimal trajectories. It is well known that in general it is nonsmooth. If

$$\left\{ \begin{array}{ll} i) & \forall\, R > 0,\ \exists\, c_R > 0,\ \forall\, u,\ f(\cdot, u) \text{ is } c_R - \text{Lipschitz on } B_R \\ ii) & \exists\, k > 0,\ \forall\, x,\ \sup_{u \in U} \|f(x,u)\| \;\leq\; k(1 + \|x\|) \end{array} \right. \tag{2}$$

where B_R denotes the closed ball of center zero and radius R, then the value function is locally Lipschitz (see for instance [25, FLEMING & RISHEL]). The optimal feedback set-valued map is given by

$$U(t,x) = \left\{ u \in U \mid \frac{\partial V}{\partial(1,v)}(t,x) = 0,\ v = f(x,u) \right\}$$

where $\frac{\partial V}{\partial(1,v)}$ denotes the directional derivative of V in the direction $(1,v)$. The sets $U(t,x)$ may be empty at points where V is not differentiable. The "optimal" control system can be described then in the following way

$$x'(t) = f(x(t),u(t)), \;\; u(t) \in U(t,x(t)), \;\; x(t_0) = x_0$$

A natural question arises : What are the solutions of the above closed loop system? A possible answer comes from the theory of differential inclusions: Solutions are absolutely continuous functions such that

$$x'(t) \in f(x(t),U(t,x(t))) \;\; \text{a.e.} \;\; \& \;\; x(t_0) = x_0$$

Let us introduce the set-valued map of "optimal dynamics"

$$G(t,x) := f(x,U(t,x)) = \bigcup_{u \in U(t,x)} \{f(x,u)\}$$

Theorem 1.1 ([30, H.F.]) *Assume that V is locally Lipschitz. Then the following two statements are equivalent:*
i) x solves the differential inclusion

$$x'(t) \;\in\; G(t,x(t)), \;\; x(t_0) = x_0 \tag{3}$$

ii) x is optimal: $V(t_0,x_0) = \varphi(x(1))$

Proof — The proof is extremely simple. Fix a trajectory x of (3) and set $\psi(t) = V(t,x(t))$. Then ψ is absolutely continuous and for almost all t

$$\psi'(t) \;=\; \frac{\partial V}{\partial(1,x'(t))}(t,x(t))$$

If $i)$ holds true, then $\psi'(t) = 0$ a.e. and thus ψ is constant equal to $\varphi(x(1))$. If $ii)$ is satisfied, then, $\psi' = 0$ and thus $\frac{\partial V}{\partial(1,x'(t))}(t,x(t)) \;=\; 0$ a.e.. □

It was proved in [12, CANNARSA & FRANKOWSKA] that for smooth problems G is upper semicontinuous but its values are not convex. We recall next the definition of upper semicontinuous maps.

Let X, Y be metric spaces and $F : X \rightsquigarrow Y$ be a set-valued map, i.e., $\forall\, x \in X,\; F(x) \subset Y$. The (Painlevé–Kuratowski) upper limit is defined by

$$\text{Limsup}_{x \to x_0} F(x) := \left\{ \lim_{n\to\infty} y_n \,|\, x_n \to x_0,\, y_n \in F(x_n) \right\}$$

If Y is compact, then F is upper semicontinuous on X if and only if

$$\forall\, x_0 \in X, \;\; \text{Limsup}_{x \to x_0} F(x) \subset F(x_0)$$

When the data f, φ are smooth enough, then the value function V has "regular" directional derivatives and therefore the map G inherits upper semicontinuity, but in the same time the function

$$a \mapsto \frac{\partial V}{\partial a}(t, x(t))$$

is concave. If it is both concave and convex, then V is differentiable at $(t, x(t))$. So the values of G may be nonconvex at points where V is not differentiable. We would like to underline that qualitative theory of differential inclusions is build for upper semicontinuous set-valued maps with convex values. Most of its results are not valid without convexity assumptions. Because of that one should not expect optimal trajectories to have a nice structure when V is nonsmooth.

1.2 Time Optimal Feedback

We describe next the problem for which the value function is in general discontinuous. Consider a complete separable metric space U and a continuous map $f : R^n \times U \mapsto R^n$. Let $y(\cdot; x, u)$ denote the solution to

$$y'(t) = f(y(t), u(t)) \qquad t \geq 0, \;\; y(0) = x \tag{4}$$

where $u(t) \in U$ almost everywhere and let $K \subset R^n$ be a closed set. Consider the *time optimal control problem for system (4) with target* K:

$$T(x) = \inf_u \; \inf\{t \geq 0 \mid y(t; x, u) \in K\}.$$

By the usual convention $T(x) = +\infty$ when no trajectory starting at x reaches K. A vector $p \in R^n$ is called a (proximal) *normal* to $S \subset R^n$ at a point $x \in \bar{S}$ if

$$dist_S(x + p) = \|p\|$$

Proximal normals were introduced in [9, BONY]. The Hamiltonian associated to the above control system is defined by $H(x, p) = \sup_{u \in U} \langle p, f(x, u) \rangle$.

To define time optimal synthesis we need the following extension of directional derivative. For $\varphi : R^n \mapsto R \cup \{+\infty\}$ the upper contingent derivative of φ at x_0 in the direction v is defined by

$$D_\downarrow \varphi(x_0)(v) \;\; = \limsup_{h \to 0+, \; v' \to v} \frac{\varphi(x_0 + hv') - \varphi(x_0)}{h}$$

See [3, AUBIN & FRANKOWSKA] for properties of contingent derivatives. In the two results below we impose assumptions (2) and that for all $x \in R^n$, $f(x, U)$ is closed and convex.

Theorem 1.2 ([13, CANNARSA, H.F. & SINESTRARI]) *Let $\bar{u}(\cdot)$ be a fixed control such that the corresponding trajectory $\bar{y}(\cdot) = y(\cdot; x_0, \bar{u})$ satisfies*

$$\bar{y}(t) \notin K, \ \forall\, t \in [0, T_0[; \ \bar{y}(T_0) \in K; \ H(\bar{y}(T_0), \nu) > 0$$

for some normal ν to $R^n \backslash K$ at $\bar{y}(T)$. Then $\bar{u}$ is time optimal if and only if, for every $t \in [0, T_0[$,

$$D_{\downarrow} T(\bar{y}(t))(v) = -1, \qquad \forall\, v \in Dy(t) := \mathrm{Limsup}_{s \downarrow t} \frac{\bar{y}(s) - \bar{y}(t)}{s - t}$$

The proof is not as straightforward as in the previous section, since $T(\cdot)$ may be merely lower semicontinuous. It is shown first that the co-state $p(\cdot)$ of Pontryagin's maximum principle verifies an adjoint inclusion. Then the Viability Theorem from [2, AUBIN] is applied to show that $t \mapsto V(t, \bar{y}(t))$ is Lipschitz even when V is discontinuous. The above result suggests to define the **time optimal synthesis** in the following way:

$$U(x) = \{u \in U \mid D_{\downarrow} T(x)(f(x, u)) = -1\}$$

The associated set-valued map of "optimal dynamics" is

$$G(x) = f(x, U(x))$$

In view of Theorem 1.2 it is natural to expect optimal trajectories to solve the following closed loop system

$$y'(t) = f(y(t), u(t)), \ \ u(t) \in U(y(t)), \ \ y(0) = x$$

Consider the differential inclusion

$$y'(t) \in G(y(t)), \ \ y(0) = x \tag{5}$$

The time optimal function $T(\cdot)$ being in general discontinuous, the arguments from the proof of Theorem 1.1 are not valid any longer. For this reason we have to change the notion of solution.

Definition 1.3 ([37, MARCHAUD]) *A continuous map $y : [0, T_0] \to R^n$ is a contingent solution of* (5) *if*

$$Dy(t) \cap G(y(t)) \neq \emptyset, \quad \forall\, t \in [0, T_0[\ \ \& \ \ y(0) = x$$

We already know that every time optimal solution is a contingent solution of (5) under all assumptions of Theorem 1.2. Conversely,

Theorem 1.4 ([13, CANNARSA, H.F. & SINESTRARI]) *Suppose that $y(\cdot)$ is a contingent solution of* (5) *in $[0, T_0]$ satisfying*

$$y(t) \notin K, \ \forall\, t \in [0, T_0[; \ \ y(T_0) \in K\,.$$

Then y is time optimal.

2 Viability Kernel

We provide next three examples leading to the notion of viability kernel.

Example 1: Implicit Control System

$$f(x, x', u) = 0, \quad u \in U$$

The way to make it "explicit" is to define the set-valued map

$$F(x) = \{v \mid \exists\, u \in U, \quad f(x, v, u) = 0\}$$

and to study the differential inclusion

$$x'(t) \in F(x(t)) \quad \text{a.e.}$$

But in general F is not defined on the whole space but only on a subset $\mathrm{Dom}(F) := \{x \mid F(x) \neq \emptyset\}$. Furthermore there are $x_0 \in \mathrm{Dom}(F)$ from where no trajectory defined over R_+ of the control system starts. □

Example 2: Control System with State Constraints

$$x'(t) = f(x(t), u(t)), \quad h(x(t), u(t)) \leq 0, \quad u(t) \in U$$

To "get rid" from the constraints let us introduce the set-valued map

$$U(x) = \{u \in U \mid h(x, u) \leq 0\}$$

The new control system is

$$x' = f(x, u), \quad u \in U(x)$$

Again U may be defined only over a subset $K = \{x \mid U(x) \neq \emptyset\}$. Furthermore, there are $x_0 \in K$ from where no trajectory of the control system satisfying state constraints starts. □

Example 3: Bounded Chattering

The problem is to find solutions to the control system

$$x'(t) = f(x(t), u(t)), \quad x(0) = x_0, \quad u(t) \in U(x(t)), \quad \|u'(t)\| \leq M$$

That is u has to be absolutely continuous and, in particular, it can not have jumps. Define a new dynamical system

$$\begin{cases} x'(t) = f(x(t), u(t)), \quad x(0) = x_0, \quad u(t) \in U(x(t)) \\ u'(t) \in B_M \end{cases}$$

Naturally there may exist $x_0 \in \mathrm{Dom}(U)$ from where no trajectory of the above system starts. □

In all three cases we reduced the control system under investigation to the following so called **viability problem**

$$\left\{ \begin{array}{lcl} x'(t) & \in & F(x(t)), \text{ for almost all } t \geq 0, \\ x(t) & \in & K, \ \forall t \geq 0, \\ x(0) & = & x_0 \in K \end{array} \right. \tag{6}$$

The **viability kernel** $Viab(K)$ of K (under F) is the set of **all initial conditions** $x_0 \in K$ from which starts at least one solution (defined over R_+) of the differential inclusion (6). The notion of viability kernel was introduced in [1, AUBIN]. If

$$\left\{ \begin{array}{ll} i) & F \text{ is upper semicontinuous with closed convex images} \\ ii) & \exists\, k > 0, \ \sup_{v \in F(x)} \|v\| \leq k(\|x\| + 1) \end{array} \right.$$

then the viability kernel Viab(K) is closed and enjoys some stability properties. Algorithms were obtained to compute the viability kernel which for low dimensions run on PC's. See [36, FRANKOWSKA & QUINCAMPOIX], [38, SAINT-PIERRE], [15, 16, 17, CARDALIAGUET, QUINCAMPOIX & SAINT-PIERRE]. These (global) algorithms were inspired by "zero-dynamics" of [10, BYRNES & ISIDORI].

Since the notion of viability kernel revealed to be very useful in "computing" Lyapunov functions, time-optimal function, solving the target problem and in some applied problems (see [8, BONNEUIL & MULLERS], [19, CARTELIER & MULLERS], [22, DOYEN & GABAY], [23, DOYEN, GABAY & HOURCADE] and also [2, AUBIN] and its bibliography) the research is carried out in Universit Paris-Dauphine by P. Cardaliaguet, L. Doyen, M. Quincampoix, P. Saint-Pierre and N. Seube to perfection algorithms for computing the viability kernel.

3 Solutions to HJB Equations

We address here the Hamilton-Jacobi-Bellman equation of optimal control. Consider again the Mayer Problem from Section 1.1. As we already observed for locally Lipschitz data the value function is locally Lipschitz. It is easy to understand how the generalized (bilateral) solutions arise in the Lipschitz case.

Definition 3.1 ([21]) *Let $\psi : X \mapsto R \cup \{+\infty\}$ be an extended function and $x_0 \in X$ be such that $\psi(x_0) \neq \infty$. The* subdifferential *of ψ at x_0 is :*

$$\partial_-\psi(x_0) = \left\{ p \mid \liminf_{x \to x_0} \frac{\psi(x) - \psi(x_0) - \langle p, x - x_0 \rangle}{\|x - x_0\|} \geq 0 \right\}$$

The value function V is nondecreasing along solutions of the control system and is constant along optimal solutions. For this reason the following statement follows easily by classical arguments (see for instance [25, FLEMING & RISHEL]): If V is differentiable at (t, x), then

$$-\frac{\partial V}{\partial t}(t,x) \;+\; H\left(x, -\frac{\partial V}{\partial x}(t,x)\right) \;=\; 0$$

$$\forall\,(p_t, p_x) \in \mathrm{Limsup}_{(t',x')\to(t,x)} \frac{\partial V}{\partial x}(t',x'), \quad -p_t + H(x,-p_x) \;=\; 0$$

The Hamiltonian $H(x, \cdot)$ being convex, we have

$$\forall\,(p_t, p_x) \in \partial V(t,x), \;\; -p_t + H(x,-p_x) \;\leq\; 0$$

where $\partial V(t, x)$ denotes Clarke's generalized gradient.

On the other hand V is a viscosity solution to the HJB equation (see [21, CRANDALL, EVANS & LIONS]). In particular,

$$\forall\,(p_t, p_x) \in \partial_- V(t,x), \;\; -p_t + H(x,-p_x) \geq 0$$

But $\partial_- V(t, x) \subset \partial V(t, x)$ and therefore V is a bilateral solution:

$$\forall\,(p_t, p_x) \in \partial_- V(t,x), \;\; -p_t + H(x,-p_x) = 0$$

This notion of solution is valid as well for lower semicontinuous functions [7, BARRON & JENSEN], [32, 33, FRANKOWSKA]. In [7] the authors extended the maximum principle of PDE's to lower semicontinuous functions. The alternative approach proposed in [33] is based on viability theory.

The "geometry" behind the method of proving uniqueness of nonsmooth solutions is the following one (details can be found in [4, AUBIN & FRANKOWSKA]): Consider the reachable set $R(t)$ at time t of

$$\begin{cases} x' \;=\; -f(x, u(t)) & x(1) = x_1 \qquad u(t) \in U \\ z' \;=\; 0 & z(1) = \varphi(x_1) \end{cases}$$

for all possible choices of x_1. Then $R(t)$ is the epigraph of $V(t, \cdot)$:

$$R(t) = \{(x,r) \mid r \geq V(t,x)\}$$

The semigroup properties of reachable sets are used to investigate tangents to the epigraph of V. Namely consider the differential inclusion

$$x'(t) \in F(x(t)), \;\; x(0) = x_0$$

where F is a locally Lipschitz set-valued map with convex compact images, and define its reachable map

$$R(t, x_0) \;=\; \{x(t) \mid x \text{ solves the above inclusion}\}$$

Then by well known results from [24, FILIPPOV]

$$R(t, x_0) \; = \; x_0 + tF(x_0) + o(t)B$$

where B denotes the closed unit ball. The fact that V is a bilateral solution follows from monotonicity properties of the value function. Proofs of the converse are based on Viability Theory [2, AUBIN].

In conclusion, we have to underline that for problems with lower semicontinuous cost function, it is natural to use subdifferentials rather than upper directional derivatives, because subdifferentials are related to tangents to the epigraph of V: $Epi(V) = \{(t, x, r) \mid r \geq V(t, x)\}$ which is closed. On the other hand to construct optimal synthesis via subdifferentials one needs extra assumptions which may be difficult to check. We used here upper directional derivatives, related to tangents to the hypograph of V: $\{(t, x, r) \mid r \leq V(t, x)\}$ not closed in general.

Next we discuss briefly the method of characteristics of HJB equations. Since the Hamiltonian of Mayer's problem is not differentiable at $(x, 0)$, we consider the Bolza problem:

$$(P) \qquad \text{minimize} \int_{t_0}^{T} L(x(t), u(t))dt \; + \; \varphi(x(T))$$

over solution-control pairs (x, u) of control system

$$\begin{cases} x'(t) & = & f(x(t)) + g(x(t))u(t), \quad u \in L^1 \\ x(t_0) & = & x_0 \end{cases} \tag{7}$$

The *Hamiltonian* H in this case is defined by

$$H(x, p) = \sup_u \left(\langle p, f(x) + g(x)u\rangle - L(x, u)\right)$$

and the value function is given by

$$V(t_0, x_0) \; = \; \inf_u \int_{t_0}^{T} L(x(t; t_0, x_0, u), u(t))dt + \varphi(x(T; t_0, x_0, u))$$

where $x(\cdot; t_0, x_0, u)$ denotes the solution to (7) corresponding to the control u. The HJB equation is

$$-V_t \; + \; H(x, -V_x) \; = \; 0, \quad V(T, \cdot) = \varphi(\cdot)$$

If H is smooth, then the characteristic system of this equation is the following Hamiltonian system

$$\begin{cases} x'(t) & = & H'_p(x(t)p(t)) & \quad x(T) = x_T \\ -p'(t) & = & H'_x(x(t), p(t)) & \quad p(T) = -\nabla\varphi(x_T) \end{cases}$$

It is well known that for nonconvex problems it is natural to expect shocks for such system:

$$\exists\, x_1(T) \neq x_2(T),\ \exists\, t_0 < T \ \text{ such that } \ x_1(t_0) = x_2(t_0)$$

This implies that for some initial conditions and some initial time t_0 we have multiple optimal trajectories or equivalently

$$\exists\, x_0 \quad \text{such that } \operatorname{Limsup}_{x \to x_0} \left\{ \frac{\partial V}{\partial x}(t_0, x) \right\}$$

is not a singleton.

In the two results below we impose the following assumptions:

H1) f, g are locally Lipschitz; f, g, $L(\cdot, u)$ are differentiable, $\varphi \in C^1$
H2) $\forall\, (t_0, x_0) \in [0, T] \times \mathbf{R}^n$ an optimal solution of (P) does exist and $V : [0, T] \times \mathbf{R}^n \mapsto \mathbf{R}$ is locally Lipschitz
H3) $L(x, \cdot)$ is continuous, convex and $\exists\, c > 0$ such that $L(x, u) \geq c\, \|u\|^2$
H4) For all $r > 0$, there exists $k_r \geq 0$ such that

$$\forall\, u \in \mathbf{R}^m, \ \ L(\cdot, u) \ \text{ is } \ k_r - \text{Lipschitz on } \ B_r(0)$$

H5) H' is locally Lipschitz and the Hamiltonian system is complete.

Theorem 3.2 *Every solution (x, p) to the Hamiltonian system*

$$\begin{cases} x'(t) & = H'_p(x(t), p(t)) \quad x(t_0) = x_0 \\ -p'(t) & = H'_x(x(t), p(t)) \quad p(t_0) \in -\operatorname{Limsup}_{x \to x_0} V'_x(t_0, x) \end{cases}$$

is so that x is optimal.

Theorem 3.3 (BYRNES & H.F.) *Assume that $H(x, \cdot)$ is strictly convex and let $(\overline{x}, \overline{u})$ be a trajectory-control pair. If $\overline{x}$ is an optimal trajectory of the Bolza problem, then for all $t \in]t_0, T]$, V is differentiable at $(t, \overline{x}(t))$.*

The above extends earlier results of [14, CANNARSA & SONER]) of calculus of variations. Further study of shocks is continued in [18, CAROFF & FRANKOWSKA].

4 Interior and Boundary of Reachable Sets

4.1 Local Controllability

Consider the control system

$$x'(t) = f(x(t), u(t)), \ \ u(t) \in U, \ x(0) = x_0 \tag{8}$$

where f verifies (2) and $f(x,U)$ are closed and convex. Its reachable set at time $t \geq 0$ is given by

$$R(t) = \{\, x(t) \mid x \text{ is solves } (8)\}$$

We address the following question: When $x_0 \in \mathrm{Int}(R(t))$ for all $t > 0$?

Let us first recall the Graves theorem (1947): if $f : X \mapsto Y$ is C^1 and $f'(x_0)$ is surjective, then $\forall\, \varepsilon > 0,\ f(x_0) \in \mathrm{Int}(f(B_\varepsilon(x_0)))$, where $B_\varepsilon(x_0)$ denotes the closed ball of center x_0 and radius ε.

A very similar result holds true also for set-valued maps. Here we apply it to the reachable map $R(\cdot)$. But in order to get such extension of Graves' theorem, one needs to differentiate set-valued maps on metric spaces. Recall first the notion of Painlevé–Kuratowski lower limit of sets. Let $F : X \rightsquigarrow Y$ be a set-valued map. The lower limit is given by

$$\mathrm{Liminf}_{x\to x_0} F(x) := \left\{ \lim_{x\to x_0} y_x \mid y_x \in F(x)\right\}$$

We introduce k-order variations of reachable sets:

$$R^k(0) := \mathrm{Liminf}_{\,t\to 0+} \frac{R(t) - x_0}{t^k}$$

Notice that for all $k \geq 1,\ \ R^k(x_0) \subset R^{k+1}(x_0)$.

Theorem 4.1 **([31, H.F.])** *If* $0 \in f(x_0, U)$ *and for some* $v_1, ...v_p \in R^k(0)$

$$0 \in \mathrm{Int}\ \mathrm{co}\,\{v_1, ..., v_p\}$$

then $x_0 \in \mathrm{Int}(R(t))$ *for all* $t > 0$. *Furthermore there exist* $L > 0,\ \varepsilon > 0$ *such that for all small* $t > 0$, *all* $y_1 \in B_\varepsilon(x_0)$ *and* $y \in R(t)$ *there exists* t_1 *such that*

$$y_1 \in R(t_1)\ \ \&\ \ |t_1 - t| \ \leq\ L\, \sqrt[k]{\|y_1 - y\|}$$

4.2 Lipschitz Behavior of Controls

Consider again the control system (8) and let $(z, \overline{u})$ be its trajectory control pair. We impose assumptions (2) and that $f(\cdot, u) \in C^1$ for all u. The linearized control system is given by

$$\left\{ \begin{array}{rcl} w'(t) & = & f'_x(z(t), \overline{u}(t))w(t) + y(t),\ \ y(t) \in \overline{co}\, f(z(t), U) - z'(t) \\ w(0) & = & 0 \end{array}\right. \tag{9}$$

and the corresponding reachable set by $R^L(T) = \{w(T) \,|\, w \text{ solves } (9)\}$.

Theorem 4.2 **([31, H.F.])** *Assume that* $0 \in \mathrm{Int}\,(R^L(T))$. *Then* $z(T) \in \mathrm{Int}(R(T))$ *and there exist* $\varepsilon > 0,\ L > 0$ *such that for all* $b \in B_\varepsilon(z(T))$ *we can find a control* $u(\cdot)$ *satisfying*

$$x_u(T) = b,\ \ \mu(\{t \in [0,T] \mid u(t) \neq \overline{u}(t)\}) \leq L\, \|b - z(T)\|$$

4.3 Nonsmooth Maximum Principle

Consider the control system (1) and assume (2). Let $g : R^n \to R^k$ be a locally Lipschitz function and $K_0,\ K_1 \subset R^n$ be closed. We impose the following end-point constraints:

$$x(0) \in K_0, \quad x(1) \in K_1 \tag{10}$$

Define the reachable set at time one : $R(1) = \{x(1) \mid x \text{ solves } (1), (10)\}$. Let z be a trajectory of (1), (10). It is well known (see for instance [20, CLARKE], [39, WARGA] etc.) that if $g(z(1))$ is a boundary point of $g(R(1))$, then a maximum principle holds true. The aim of this section is to make evident that behind there is an "alternative" inverse mapping theorem, which is much more than the characterization of boundary of reachable sets. Recall that generalized Jacobian of a locally Lipschitz function $\varphi : R^n \mapsto R^m$ (see [20, CLARKE]) is defined by:

$$\partial\varphi(x_0) \;=\; \overline{co}\ \left(\text{Limsup}_{x\to x_0}\ \varphi'(x)\right)$$

Theorem 4.3 **([31, H.F.])** *Let $(z,\overline{u})$ be a trajectory-control pair of (1), (10). Then at least one of the following two statements holds true:*
i) $\exists\ \lambda \in R^k$ and an absolutely continuous $p : [0,1] \to R^n$ not both equal to zero, satisfying the maximum principle

$$-p'(t) \;\in\; \partial_x f(z(t),\overline{u}(t))^\star p(t) \quad \text{a.e. in } [0,1]$$

$$\max_{u\in U}\ \langle p(t),\ f(z(t),u)\rangle \;=\; \langle p(t),\ f(z(t),\overline{u}(t))\rangle \ \text{ a.e.}$$

$$p(0) \in N_{K_0}(z(0)), \quad -p(1)) \in \partial g(z(1))^\star\lambda + N_{K_1}(z(1))$$

where $N_K(x)$ denotes the Clarke normal cone to K at x and $\partial_x f$ the generalized Jacobian with respect to x.
ii) $\exists\ L > 0,\ \varepsilon > 0$ such that for all $(a,b,c) \in R^k \times R^n \times R^n$ satisfying

$$\|a - g(z(1))\| + \|b\| + \|c\| \le \varepsilon$$

there exists a trajectory-control pair (x_u, u) such that

$$g(x_u(1)) = a, \quad x_u(0) \in b + K_0, \quad x_u(1) \in c + K_1$$

and $\mu(\{t \mid u(t) \neq \bar{u}(t)\}) \le L\,(\|a - g(x_{\bar{u}}(1))\| + \|b\| + \|c\|)$.

In particular, if $g(z(1))$ is a boundary point of $g(R(1))$, then the statement i) holds true.

The above results from the set-valued inverse mapping theorem on metric spaces. Denote by $\mathcal{U}$ the set of all measurable functions $u : [0,1] \to U$. Let

$x(\cdot; u, x_0)$ be the solution of (8) corresponding to the control u and define the set-valued map $G : R^n \times \mathcal{U} \leadsto R \times R^n \times R^n$ by

$$G(x_0, u) = (g(x(1; u, x_0)), x_0, x(1; u, x_0)) - \{0\} \times K_0 \times K_1$$

The "strategy" of the proof is the following one:

1. Approximate G via "smooth" maps by regularizing f and g.
2. Use the inverse mapping theorem on approximations.
3. Go to the limit.

Regularization technics implying nonsmooth maximum principle go back to [39, WARGA]. In [26, FRANKOWSKA] it was shown that Warga's scheme may be refined to get smaller objects than the derivatives containers. The inverse mapping theorem used on approximations is Theorem 4.4 below. Finally Stability Theorem 4.5 is applied to take limits.

Consider $G : X \leadsto Y$, where X is a complete separable metric space and Y is a Banach space with the norm Gâteaux differentiable away from zero. Let $y_0 \in G(x_0)$. The graph of G is defined by

$$\mathrm{Graph}(G) = \{(x, y) \mid y \in G(x)\}$$

The first order "contingent" variation is defined by

$$G^{(1)}(x_0, y_0) = \mathrm{Limsup}_{h \to 0+} \frac{G(B_h(x_0)) - y_0}{h}$$

Theorem 4.4 **([31, H.F.])** *If for some $\varepsilon > 0$, $\rho > 0$, $M > 0$*

$$\rho B \subset \bigcap_{\substack{(x,y) \in \mathrm{Graph}(G) \\ (x,y) \in B_\varepsilon(x_0, y_0)}} \overline{co}\, (G^{(1)}(x, y) \cap MB) \qquad (11)$$

then for all $(x_1, y_1, y_2) \in \mathrm{Graph}(G) \times Y$ near (x_0, y_0, y_0)

$$\mathrm{dist}\,(x_1,\ G^{-1}(y_2)) \leq \frac{1}{\rho} \|y_1 - y_2\|, \quad \text{where } G^{-1}(y) = \{x \mid y \in G(x)\}$$

Theorem 4.5 **([31, H.F.])** *Consider set-valued maps $\{G_i\}_{i \geq 0}$ from a complete metric space X to a Banach space Y having closed graphs. Let $y_0 \in G_0(x_0)$. We assume that for some $\delta > 0$ and for every $\lambda > 0$ there exists an integer I_λ such that for all $i \geq I_\lambda$ and all $x \in B_\delta(x_0)$*

$$G_i(x) \subset G_0(x) + \lambda B$$

If G_i have "a Lipschitz inverse" on a neighborhood of (x_0, y_0) with the same Lipschitz constant, then so does G.

References

[1] AUBIN J.-P. (1987) *Smooth and Heavy Solutions to Control Problems*, in NONLINEAR AND CONVEX ANALYSIS, Eds. B.-L. Lin & Simons S., Proceedings in honor of Ky Fan, Lecture Notes in Pure and Applied Mathematics, June 24-26, 1985

[2] AUBIN J.-P. (1991) VIABILITY THEORY, Birkhäuser, Boston, Basel, Berlin

[3] AUBIN J.-P. & FRANKOWSKA H. (1990) SET-VALUED ANALYSIS, Birkhäuser, Boston, Basel, Berlin

[4] AUBIN J.-P. & FRANKOWSKA H. (to appear) *Set-valued solutions to the Cauchy problem for hyperbolic systems of partial differential inclusions*, NODEA

[5] AUBIN J.-P. & FRANKOWSKA H. (to appear) *The viability kernel algorithm for computing value functions of infinite horizon optimal control problems*, JMAA

[6] AUBIN J.-P. & NAJMAN L. (1994) *L'algorithme des montagnes russes pour l'optimisation globale*, Comptes-Rendus de l'Acadmie des Sciences, Paris, 319, 631-636

[7] BARRON E.N. & JENSEN R. (1990) *Semicontinuous viscosity solutions for Hamilton-Jacobi equations with convex hamiltonians*, Comm. Partial Diff. Eqs., 15, 113-174

[8] BONNEUIL N. & MULLERS K. (to appear) *Viable populations in a predator-prey system*, J. Mathematical Biology

[9] BONY J.M. (1969) *Principe du maximum, inégalité de Harnack et unicité du problème de Cauchy pour les opérateurs elliptiques dégénérés*, Ann. Inst. Fourier, Grenoble, 19, 277–304.

[10] BYRNES C.I. & ISIDORI A. (1984) *A frequency domain philosophy for nonlinear systems with applications to stabilization and adaptive control*, in Proc. 23rd CDC, Las Vegas, NV, 1569-1573

[11] BYRNES Ch. & FRANKOWSKA H. (1992) *Unicité des solutions optimales et absence de chocs pour les équations d'Hamilton-Jacobi-Bellman et de Riccati*, Comptes-Rendus de l'Académie des Sciences, t. 315, Série 1, Paris, 427-431

[12] CANNARSA P. & FRANKOWSKA H. (1991) *Some characterizations of optimal trajectories in control theory*, SIAM J. on Control and Optimization, 29, 1322-1347

[13] CANNARSA P., FRANKOWSKA H. & SINESTRARI C. (to appear) *Properties of minimal time function for target problem*, J. Math. Systems, Estimation and Control

[14] CANNARSA P. & SONER H. (1987) *On the singularities of the viscosity solutions to Hamilton-Jacobi-Bellman equations*, Indiana University Math. J., 36, 501-524

[15] CARDALIAGUET P., QUINCAMPOIX M. & SAINT-PIERRE P. (1994) *Some algorithms for differential games with two players and one target*, MAM, 28, 441-461

[16] CARDALIAGUET P., QUINCAMPOIX M. & SAINT-PIERRE P. (1995) *Contribution l'tude des jeux diffrentiels quantitatifs et qualitatifs avec contrainte sur l'tat*, Comptes-Rendus de l'Acadmie des Sciences, 321, 1543-1548

[17] CARDALIAGUET P., QUINCAMPOIX M. & SAINT-PIERRE P. (to appear) *Numerical methods for optimal control and differential games*, AMO

[18] CAROFF N. & FRANKOWSKA H. (to appear) *Conjugate points and shocks in nonlinear optimal control*, Trans. Amer. Math. Soc.

[19] CARTELIER J. & MULLERS K. (1994) *An elementary Keynesian model: A preliminary approach*, IIASA WP 94-095

[20] CLARKE F.H. (1983) OPTIMIZATION AND NONSMOOTH ANALYSIS, Wiley-Interscience

[21] CRANDALL M.G., EVANS L.C. & LIONS P.L. (1984) *Some properties of viscosity solutions of Hamilton-Jacobi equation*, Trans. Amer. Math. Soc., 282(2), 487-502

[22] DOYEN L. & GABAY D. (1996) *Risque climatique, technologie et viabilit*, Actes des Journes Vie, Environnement et Socits

[23] DOYEN L., GABAY D. & HOURCADE J.-C. (1996) *Economie des ressources renouvelables et viabilit*, Actes des Journes Vie, Environnement et Socits

[24] FILIPPOV A.F. (1967) *Classical solutions of differential equations with multivalued right hand side*, SIAM J. on Control, 5, 609-621

[25] FLEMING W.H. & RISHEL R.W. (1975) DETERMINISTIC AND STOCHASTIC OPTIMAL CONTROL, Springer-Verlag, New York

[26] FRANKOWSKA H. (1984) *The first order necessary conditions for nonsmooth variational and control problems*, SIAM J. on Control and Optimization, 22, 1-12

[27] FRANKOWSKA H. (1986) *Théorème d'application ouverte pour des correspondances*, Comptes-Rendus de l'Académie des Sciences, PARIS, Série 1, 302, 559-562

[28] FRANKOWSKA H. (1987) *Théorèmes d'application ouverte et de fonction inverse*, Comptes Rendus de l'Académie des Sciences, PARIS, Série 1, 305, 773-776

[29] FRANKOWSKA H. (1987) *L'équation d'Hamilton-Jacobi contingente*, Comptes-Rendus de l'Académie des Sciences, PARIS, Série 1, 304, 295-298

[30] FRANKOWSKA H. (1989) *Optimal trajectories associated to a solution of contingent Hamilton-Jacobi equations*, AMO, 19, 291-311

[31] FRANKOWSKA H. (1990) *Some inverse mapping theorems*, Ann. Inst. Henri Poincar, Analyse Non Linaire, 7, 183-234

[32] FRANKOWSKA H. (1991) *Lower semicontinuous solutions to Hamilton-Jacobi-Bellman equations*, Proceedings of 30th CDC, Brighton, December 11-13

[33] FRANKOWSKA H. (1993) *Lower semicontinuous solutions of Hamilton-Jacobi-Bellman equation*, SIAM J. on Control and Optimization, 31, 257-272

[34] FRANKOWSKA H., PLASKACZ S. & RZEZUCHOWSKI T. (1995) *Measurable viability theorems and Hamilton-Jacobi-Bellman equation*, J. Diff. Eqs., 116, 265-305

[35] FRANKOWSKA H. & PLASKACZ S. (1996) *A measurable upper semicontinuous viability theorem for tubes*, J. of Nonlinear Analysis, TMA, 26, 565-582

[36] FRANKOWSKA H. & QUINCAMPOIX M. (1991) *Viability kernels of differential inclusions with constraints: algorithm and applications*, J. Math. Systems, Estimation and Control, 1, 371-388

[37] MARCHAUD H. (1934) *Sur les champs de demi-cônes et les équations différentielles du premier ordre*, Bull. Sc. Math., 62, 1-38

[38] SAINT-PIERRE P. (to appear) *Newton's method for set-valued maps*, Set-Valued Analysis

[39] WARGA J. (1976) *Derivate containers, inverse functions and controllability*, Calculus of Variations and Control Theory, Academic Press, 13-46

CNRS URA 749, CEREMADE, Université Paris-Dauphine, 75775 Paris Cedex 16 France

Circuit Simulation Techniques Based on Lanczos-Type Algorithms

R.W. Freund

1 Introduction

A circuit is a network of electronic devices, such as resistors, capacitors, inductors, diodes, and transistors. Today's integrated circuits are extremely complex, with up to hundreds of thousands or even millions of devices, and prototyping of such circuits is no longer possible. Instead, computational methods are used to simulate and analyze the behavior of the electronic circuit at the design stage. This allows us to correct the design before the circuit is actually fabricated in silicon. It is due to extensive circuit simulation that first-time correct circuits in silicon are almost the norm.

Circuits are characterized by three types of equations: Kirchhoff's current law (KCLs), Kirchhoff's voltage law (KVLs), and the branch constitutive relations (BCRs); see, e.g., [28]. The unknowns in these equations are the currents through the branches of the network, the voltage drops along the branches, and the voltages at the nodes of the network. The KCLs and KVLs are linear algebraic equations that only depend on the topology of the circuit. The KCLs state that, at each node $\mathcal{N}$ of the network, the currents flowing in and out of $\mathcal{N}$ need to sum up to zero. The KVLs state that, for each closed loop $\mathcal{L}$ of the network, the voltage drops along $\mathcal{L}$ sum up to zero. The BCRs are equations that characterize the actual circuit elements. For example, the BCR of a linear resistor is simply Ohm's law. The BCRs are linear equations for simple devices, such as linear resistors, capacitors, and inductors, and they are nonlinear equations for more complex devices, such as diodes and transistors. Furthermore, in general, the BCRs involve time-derivatives of the unknowns, and thus they are ordinary differential equations.

The KCLs, KVLs, and BCRs can be summarized as a system of first-order differential-algebraic equations (DAEs),

$$\frac{d}{dt}q(x,t) + f(x,t) = 0, \tag{1.1}$$

together with suitable initial conditions. In (1.1), q and f are given vector-valued functions, and $x = x(t)$ is the unknown vector of circuit variables. Traditional SPICE-like circuit simulators are based on the numerical solution of the system of DAEs (1.1). However, for very large circuits, such a time-domain integration of (1.1) is often inefficient or even prohibitive.

A much more efficient approach is to generate reduced-order models, based on approximations to the frequency-domain transfer function of the linearized circuit or some large linear subcircuit. Instead of the general system (1.1), we now start with the DAEs for a linear(ized) time-invariant (sub)circuit:

$$\begin{aligned} C\frac{dx}{dt} &= -Gx + Bu(t), \\ y(t) &= E^T x(t). \end{aligned} \tag{1.2}$$

Here, $C, G \in \mathbb{C}^{N\times N}$, $B \in \mathbb{C}^{N\times m}$, and $E \in \mathbb{C}^{N\times p}$ are given matrices, m and p is the number of inputs and outputs, respectively, the components of the given vector-valued function $u\colon [0,\infty) \mapsto \mathbb{C}^m$ are the inputs, and $y\colon [0,\infty) \mapsto \mathbb{C}^p$ is the unknown function of outputs. Denoting by

$$\hat{x}(z) = \int_0^\infty x(t)e^{-zt}\,dt, \quad \hat{u}(z) = \int_0^\infty u(t)e^{-zt}\,dt, \quad \hat{y}(z) = \int_0^\infty y(t)e^{-zt}\,dt$$

the Laplace transform of x, u, and y, respectively, and assuming, for simplicity, zero initial conditions, the system of DAEs (1.2) can be stated in frequency-domain as follows:

$$\begin{aligned} zC\hat{x} &= -G\hat{x} + B\hat{u}(z), \\ \hat{y}(z) &= E^T\hat{x}(z). \end{aligned} \tag{1.3}$$

From (1.3), one obtains the input-output relation $\hat{y}(z) = H(z)\hat{u}(z)$, where $H\colon \mathbb{C} \mapsto (\mathbb{C}\cup\{\infty\})^{p\times m}$ is a rational matrix-valued function defined by

$$H(z) = E^T\,(G + zC)^{-1}\,B. \tag{1.4}$$

The function is called the *transfer function* of the m-input p-output time-invariant linear dynamical system given by (1.2). Note that, in general, the transfer function (1.4) has N poles, namely the eigenvalues of the pencil $G + zC$. In particular, if N is large, then the rational function H is of very high order and working directly with H becomes prohibitive. Instead, one approximates H by a rational function H_n of low order $n \ll N$ that describes a reduced-order model of the linear dynamical system (1.2). One possible approach to constructing suitable reduced-order models is Padé approximation; see, e.g. [2, 3].

For the large linear networks arising in the simulation of electronic circuits, Padé-based reduced-order models have proven to be especially efficient. This was first observed by Huang, Pillage, and Rohrer [19, 24] who, for the single-input single-output case ($m = p = 1$), advocated the use of *asymptotic waveform evaluation* (AWE). The AWE method is based on Padé approximations of the transfer function H, which is now scalar-valued since $m = p = 1$. Furthermore, in AWE, the Padé approximations are generated in the usual way via explicit computation of the underlying moments of the network transfer function. However, this is a numerically unstable procedure. Indeed, in practice, AWE is limited to generating Padé approximants of very low order n only. On the other hand, the numerical problems with AWE can easily be remedied by exploiting the well-known connection [15, 16, 18] between Padé approximants of scalar-valued transfer functions and the classical Lanczos process [23] and generating the Padé approximants via the Lanczos algorithm; see [4, 5, 13]. Feldmann and Freund [6, 8] have extended this approach to the general multi-input multi-output case and shown how matrix Padé approximants to matrix-valued transfer functions can be obtained from a novel Lanczos-type algorithm that was recently developed by Aliaga, Boley, Freund, and Hernández [1].

In this paper, we describe the computation of Padé-based reduced-order models of large linear networks via Lanczos-type algorithms.

2 Padé Approximants of Transfer Functions

Let H be a given matrix-valued transfer function of the form (1.4), where $C, G \in \mathbb{C}^{N\times N}$, $B \in \mathbb{C}^{N\times m}$, and $E \in \mathbb{C}^{N\times p}$ are matrices. Furthermore, we assume that $G+zC$ is a *regular* pencil; i.e., the matrix $G+zC$ is nonsingular for at least one $z \in \mathbb{C}$.

Next, we introduce our notion of an n-th matrix Padé approximant of H. First, we choose a suitable expansion point $z_0 \in \mathbb{C}$ such that the matrix $G+z_0C$ is nonsingular. Let L and U be nonsingular $N \times N$ matrices such that

$$L \cdot U = G + z_0 C, \tag{2.1}$$

and set

$$A = -L^{-1}CU^{-1}, \quad R = L^{-1}B, \quad \text{and} \quad L = U^{-T}E. \tag{2.2}$$

Using (2.1) and (2.2), and setting $z = z_0 + \sigma$ in (1.4), we can rewrite (1.4) as follows:

$$H(z_0+\sigma) = L^T\left(I - \sigma A\right)^{-1} R. \tag{2.3}$$

We remark that in circuit simulation, the matrices G and C are sparse. A factorization (2.1) can then be computed by sparse Gaussian elimination, resulting in factors L and U that are usually also sparse. Although L, C, U are sparse, the matrix A in (2.2) is in general dense. However, Lanczos-type algorithms only use A in the form of matrix-vector products $A \cdot v$ and $A^T \cdot w$, and for computing these, there is no need to explicitly form the matrix A. Indeed, be means of

$$A \cdot v = -L^{-1}\left(C\left(U^{-1}v\right)\right),$$

the product $A \cdot v$ can be obtained by a sparse backsolve with U, a multiplication with the sparse matrix C, and a sparse backsolve with A^T. Computing products $A^T \cdot w$ is done similarly.

We say that an $p \times m$-matrix-valued function $H_n \colon \mathbb{C} \mapsto (\mathbb{C} \cup \{\infty\})^{p \times m}$ is a rational function of *type* $(n-1, n)$ if H_n can be represented as

$$H_n(z_0 + \sigma) = L_n^T \left(I - \sigma A_n\right)^{-1} R_n, \tag{2.4}$$

with matrices $A_n \in \mathbb{C}^{n \times n}$, $R_n \in \mathbb{C}^{n \times m}$, and $L_n \in \mathbb{C}^{n \times p}$. An nth *Padé approximant* (about the expansion point z_0) $H_n(z_0 + \sigma)$ of the $p \times m$-matrix-valued transfer function $H(z_0 + \sigma)$ is a rational $p \times m$-matrix-valued function of type $(n-1, n)$ whose Taylor expansion about $\sigma = 0$ matches as many leading Taylor coefficients of H as possible, i.e.,

$$H_n(z_0 + \sigma) = H(z_0 + \sigma) + \mathcal{O}(\sigma^{m(n)}),$$

where the integer $m(n)$ is maximal. Note that, by (2.3),

$$H(z_0 + \sigma) = \sum_{i=0}^{\infty} M_i \sigma^i, \tag{2.5}$$

where the Taylor coefficients M_i are $p \times m$ matrices given by

$$M_i = L^T A^i R, \quad i = 0, 1, \dots . \tag{2.6}$$

The coefficients (2.6) are called the *moments* of the transfer function H.

We remark that a matrix-valued rational function of type $(n-1, n)$ can always be written in terms of a pair of matrix polynomials. However, since matrix polynomials do not commute and also for other reasons, such a representation involves a number of difficulties (see the discussion in [29]) that can be avoided by using (2.4). Furthermore, the representation (2.4) is also natural, in that it reflects the intimate connection between matrix Padé approximants of transfer functions and the minimal partial realization problem in systems theory; see, e.g., [20, 27].

3 The Single-Input Single-Output Case

In this section, we consider the single-input single-output case, $m = p = 1$. Then, in (2.3), $R, L \in \mathbb{C}^N$, and H is a scalar rational function. Furthermore, the nth Padé approximant H_n can be represented as

$$H_n(z_0 + \sigma) = \frac{\varphi_{n-1}(\sigma)}{\psi_n(\sigma)}, \tag{3.1}$$

where φ_{n-1} and ψ_n are polynomials of degree at most $n-1$ and n, respectively. The standard approach (and the one employed in AWE [24]) to computing H_n is to first generate the $2n$ moments $M_0, M_1, \ldots, M_{2n-1}$, followed by the solution of a linear system with a Hankel matrix whose entries are the moments, to obtain the coefficients of the polynomials φ_{n-1} and ψ_n in (3.1). Unfortunately, this approach is numerically unstable due to the typical extreme ill-conditioning of the Hankel matrix, and also because, in finite-precision arithmetic, the moments M_i with increasing i convey only information about the extreme poles of the transfer function H; see, e.g., [4, 5, 13]. As a result, moment-based approaches, such as AWE, can only be used to compute Padé approximants of very moderate order n, and typically, they are limited to $n \leq 10$.

Next, we describe the *Padé-via-Lanczos* (PVL) algorithm [4, 5] to compute H_n. Instead of (3.1), the PVL approach uses the following well-known Lanczos-Padé connection (see, e.g., [15, 16, 18]):

$$H_n(z_0 + \sigma) = \left(\eta D_n^T e_1\right)^T (I - \sigma T_n)^{-1} (\rho e_1), \tag{3.2}$$

where $e_1 = [1 \quad 0 \quad \cdots \quad 0]^T \in \mathbb{R}^n$. All the quantities involved in (3.2) are generated in the course of the first n steps of the Lanczos algorithm [23]. Starting with $v_1 = R/\rho$ and $w_1 = L/\eta$, where $\rho, \eta > 0$ are arbitrary normalization factors, the first n Lanczos steps produce two sequences of vectors,

$$v_1, v_2, \ldots, v_n \quad \text{and} \quad w_1, w_2, \ldots, w_n, \tag{3.3}$$

such that

$$\begin{aligned} \operatorname{span}\{ v_1, v_2, \ldots, v_n \} &= \operatorname{span}\{ R, AR, \ldots, A^{n-1} R \}, \\ \operatorname{span}\{ w_1, w_2, \ldots, w_n \} &= \operatorname{span}\{ L, A^T L, \ldots, (A^T)^{n-1} L \}, \end{aligned} \tag{3.4}$$

and the vectors (3.3) are biorthogonal:

$$w_i^T v_l = 0 \quad \text{for all} \quad i \neq l, \ i, l = 1, 2, \ldots, n. \tag{3.5}$$

Setting

$$V_n = [\, v_1 \quad v_2 \quad \cdots \quad v_n \,] \quad \text{and} \quad W_n = [\, w_1 \quad w_2 \quad \cdots \quad w_n \,], \tag{3.6}$$

the matrices D_n and T_n in (3.2) are now defined as

$$D_n = W_n^T V_n \quad \text{and} \quad T_n = D_n^{-1} W_n^T A V_n.$$

Note that, by (3.5), the matrix D_n is diagonal. Moreover, the vectors (3.3) can be generated by three-term recurrences, and the entries of T_n are just the coefficients of these recursions. Consequently, the matrix T_n is tridiagonal. Finally, we remark that, in the classical Lanczos process [23], so-called breakdowns or near-breakdowns due to division by zero or a near-zero number can occur. However, these problems can be remedied by incorporating

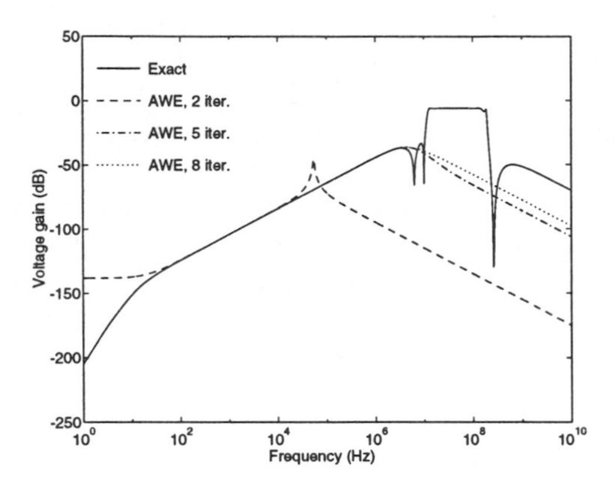

Figure 1: Results for simulation of voltage gain with AWE.

look-ahead techniques into the Lanczos process; see, e.g., [11]. The basic idea is to relax the biorthogonality (3.5) slightly whenever a breakdown or near-breakdown would occur in the classical Lanczos algorithm. To keep the exposition as simple as possible, in this paper, we always assume that breakdowns will not occur in the Lanczos process. However, we stress that both the Lanczos-Padé connection (3.2) and the result in Theorem 5.1 below also hold true if look-ahead Lanczos steps are performed to avoid breakdowns or near-breakdowns. Only the structure of the matrices D_n and T_n are affected by look-ahead steps. More precisely, in (3.2), D_n is block diagonal and T_n is block tridiagonal when look-ahead steps occur.

Next, we present two numerical examples (taken from [5]) from circuit simulation to illustrate the behavior of the PVL algorithm for computing nth Padé approximants (3.1). In both cases, we plot the absolute values

$$|H_n(2\pi \mathrm{i}\omega)| \quad \text{and} \quad |H(2\pi \mathrm{i}\omega)|$$

of H_n and the original transfer function H, respectively, over some relevant frequency range $0 \leq \omega_{\min} \leq \omega \leq \omega_{\max}$; here $\mathrm{i} = \sqrt{-1}$.

Example 3.1 The circuit simulated here is a voltage filter, where the frequency range of interest is $1 \leq \omega \leq 10^{10}$. This example was first run with AWE, and in Figure 1, we show the computed $|H_n|$ for $n = 2, 5, 8$, together with the exact $|H|$. Note that H_8 clearly has not yet converged to H. On the other hand, it turned out that the H_n's practically did not change anymore for $n \geq 8$, and in particular, AWE never converged in this example. Next, we ran this example with PVL, and in Figure 2, we show the computed results for $n = 2, 8, 28$, together with the exact $|H|$. Note that the results for $n = 8$ (the dotted curves) in Figures 1 and 2 are vastly different, although they both correspond to the same function H_8. The

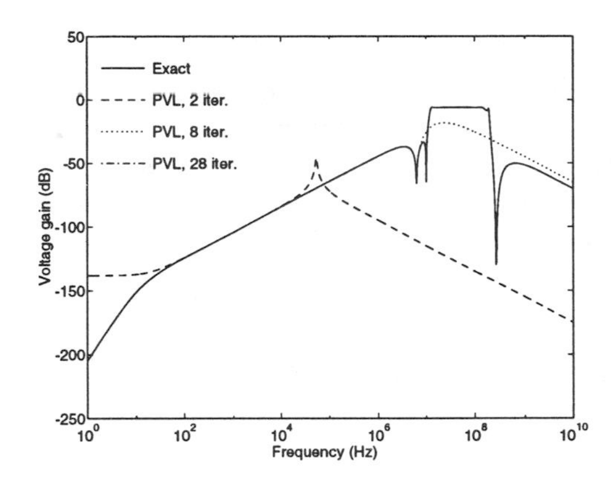

Figure 2: Results for simulation of voltage gain with PVL.

reason for this is that AWE is numerically unstable, while PVL generates H_8 stably. Furthermore, note that PVL converges, with the computed 28th Padé approximant being practically identical to H.

Example 3.2 The second example is a larger circuit (the so-called PEEC circuit) with 2100 capacitors, 172 inductors, and 6990 inductive couplings. Here the frequency range of interest is $0 \leq \omega \leq 5 \times 10^9$. In Figure 3, we show the exact transfer function, together with the 60-th Padé approximant as computed by PVL.

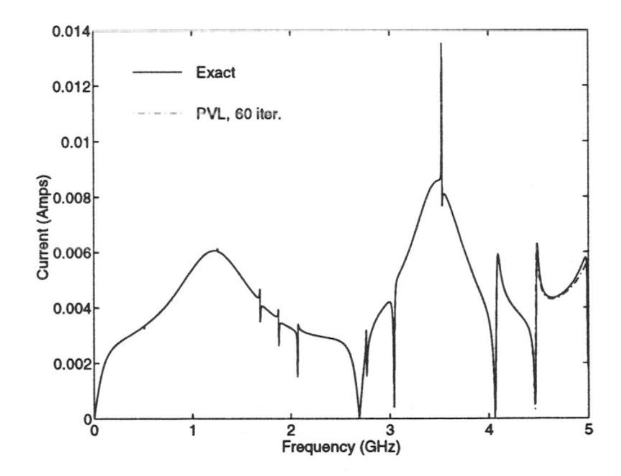

Figure 3: Results for the PEEC circuit, 60 PVL iterations.

Remark 3.3 It is well known that, in general, Padé-based reduced-order models need not be stable, even though the original transfer function H was stable. Recall that the function H is stable if all its poles have non-positive real parts and possible poles on the imaginary axis are simple. However, stability of Padé-based reduced-order models can be guaranteed for important special classes of linear networks, namely RC, RL, and LC circuits consisting of only resistors and capacitors, resistors and inductors, and inductors and capacitors, respectively; see [10].

Remark 3.4 In circuit simulation, it is often crucial to compute sensitivities of the transfer function or its poles or zeros, with respect to certain circuit parameters. An extension of PVL that allows such sensitivity computations is described in [9].

4 A Lanczos-Type Algorithm for Multiple Starting Vectors

We now return to the general m-input p-output case, with *arbitrary* $m, p \geq 1$. A Lanczos-type process connected to matrix Padé approximants of transfer functions H given by (2.3) needs to be able to handle multiple starting vectors, namely the m, respectively p, columns of the matrices R, respectively L, in (2.3), where, in general, m and p can be different. Such an algorithm was recently developed by Aliaga, Boley, Freund, and Hernández [1], while all earlier algorithms (see, e.g., [21, 22, 25, 26]) are restricted to the special case $m = p$. In this section, we briefly describe the method proposed in [1].

Like the classical Lanczos process, the algorithm for multiple starting vectors generates two sequences of vectors (3.3) that are again constructed to fulfill the biorthogonality relations (3.5). However, instead of (3.4), the vectors (3.3) now span the subspaces generated by the first n linearly independent columns of the *Krylov matrices*

$$\begin{aligned} K^{(r)} &= [\,R \quad AR \quad A^2R \quad \cdots \quad A^{N-1}R\,] \\ \text{and} \quad K^{(l)} &= [\,L \quad A^TL \quad (A^T)^2L \quad \cdots \quad (A^T)^{N-1}L\,], \end{aligned} \tag{4.1}$$

respectively. In particular, the process of constructing the vectors (3.3) now also needs to be able to detect linearly dependent columns in (4.1) and to delete these vectors. We refer to this process as *deflation*. In our Lanczos process, a linearly dependent vector occurs if, and only if, $v_n = 0$ or $w_n = 0$, and deflation simply consists of deleting v_n or w_n, respectively. As a result, the Lanczos vectors (3.3) now span the spaces generated by the first n columns of the *deflated* Krylov matrices

$$\begin{aligned} K^{(r)}_{\text{defl}} &= [\,R_1 \quad AR_2 \quad A^2R_3 \quad \cdots \quad A^{j_{\max}-1}R_{j_{\max}}\,] \\ \text{and} \quad K^{(l)}_{\text{defl}} &= [\,L_1 \quad A^TL_2 \quad (A^T)^2L_3 \quad \cdots \quad (A^T)^{k_{\max}-1}L_{k_{\max}}\,], \end{aligned} \tag{4.2}$$

respectively. The matrices (4.2) are obtained from (4.1) by deleting all linearly dependent vectors that one encounters when scanning the columns of (4.1) from left to right. In (4.2), $R_j \in \mathbb{C}^{N \times m_j}$ and $L_k \in \mathbb{C}^{N \times p_k}$ are submatrices of R and L, respectively. Moreover, R_j is a submatrix of R_{j-1}, and L_j is a submatrix of L_{j-1}, so that $m_1 \geq m_2 \geq \cdots \geq m_{j_{\max}} \geq 1$ and $p_1 \geq p_2 \geq \cdots \geq p_{k_{\max}} \geq 1$.

We now set $N_0 = \min\{\operatorname{rank} K^{(r)}, \operatorname{rank} K^{(l)}\}$. Then, for each integer n with $\max\{m_1, p_1\} \leq n \leq N_0$, we define $j = j(n)$ and $k = k(n)$ as the maximal integers such that

$$m_1 + m_2 + \cdots + m_j \leq n \quad \text{and} \quad p_1 + p_2 + \cdots + p_k \leq n, \tag{4.3}$$

respectively. The integers $j(n)$ and $k(n)$ will be used in Section 4 to state our main result. We remark that if no deflation occurs, then $j(n) = \lfloor n/m \rfloor$ and $k(n) = \lfloor n/p \rfloor$.

After these preliminaries, we can now present a sketch of the actual algorithm. First, the starting vectors are biorthogonalized, resulting in relations of the following form:

$$R = [\, v_1 \quad v_2 \quad \cdots \quad v_{m_1} \,]\, \rho \quad \text{and} \quad L = [\, w_1 \quad w_2 \quad \cdots \quad w_{p_1} \,]\, \eta,$$

where $\rho \in \mathbb{C}^{m_1 \times m}$ and $\eta \in \mathbb{C}^{p_1 \times p}$. Moreover, we initialize the indices $m_c = m_1$ and $p_c = p_1$, which are used to keep track of the current block sizes; m_c respectively p_c are reduced by 1 every time deflation occurs.

Then, for $n = 1, 2, \ldots, N_0 - \max\{m_c, p_c\}$, we repeat the following steps:

$$\text{Set} \quad v_{n+m_c} = Av_n - \sum_{i=n-p_c}^{n+m_c-1} v_i \, \frac{w_i^T A v_n}{w_i^T v_i};$$

If $v_{n+m_c} = 0$, then delete v_{n+m_c} and set $m_c = m_c - 1$;

$$\text{Set} \quad w_{n+p_c} = A^T w_n - \sum_{i=n-m_c}^{n+p_c-1} w_i \, \frac{v_i^T A^T w_n}{w_i^T v_i};$$

If $w_{n+p_c} = 0$, then delete w_{n+p_c} and set $p_c = p_c - 1$.

Using the matrices V_n and W_n introduced in (3.6), the above recurrences to generate the Lanczos vectors (3.3) can be summarized in compact matrix form as follows:

$$\begin{aligned} AV_n &= V_n T_n + [\, \underbrace{0 \quad 0 \quad \cdots \quad 0}_{n-m_c} \quad \underbrace{\star \quad \star \quad \cdots \quad \star}_{m_c} \,], \\ A^T W_n &= W_n \tilde{T}_n^T + [\, \underbrace{0 \quad 0 \quad \cdots \quad 0}_{n-p_c} \quad \underbrace{\star \quad \star \quad \cdots \quad \star}_{p_c} \,]. \end{aligned} \tag{4.4}$$

Here, T_n and $\tilde{T}_n$ are $n \times n$ banded matrices with lower bandwidth $m+1$ and upper bandwidth $p+1$. Moreover, each deflation step reduces the lower or upper bandwidth by 1. Finally, the matrices T_n and $\tilde{T}_n$ are connected by the relation

$$\tilde{T}_n = D_n T_n D_n^{-1}, \quad \text{where} \quad D_n = W_n^T V_n. \tag{4.5}$$

Remark 4.1 In the above description of the algorithm, we assumed, for simplicity, that only zero vectors are deflated. Of course, in finite-precision arithmetic, also nonzero vectors that are in some sense close to zero need to deflated. Such *inexact* deflation is described in detail in [1].

5 A Connection to Matrix Padé Approximation

We continue to use the notation introduced in Section 4. In particular, ρ, η, D_n, and V_n are the quantities generated by the Lanczos-type process. We denote by $0_{i \times l}$ the $i \times l$ zero matrix. We now establish the following connection between the Lanczos-type process and matrix Padé approximants.

Theorem 5.1 *Let* $\max\{m_1, p_1\} \leq n \leq N_0$, *and let* $j = j(n)$ *and* $k = k(n)$ *be the integers defined in* (4.3). *Then,*

$$H_n(z_0 + \sigma) = \begin{bmatrix} D_{p_1}^T \eta \\ 0_{(n-p_1) \times p} \end{bmatrix}^T (I - \sigma T_n)^{-1} \begin{bmatrix} \rho \\ 0_{(n-m_1) \times m} \end{bmatrix} \tag{5.1}$$

is an n*th Padé approximant of the function* H *defined in* (2.3), *and*

$$H_n(z_0 + \sigma) = H(z_0 + \sigma) + \mathcal{O}(\sigma^{j+k}). \tag{5.2}$$

Moreover, if $n = m_1 + m_2 + \cdots + m_j$ *or* $n = p_1 + p_2 + \cdots + p_k$, *then* H_n *is a Padé approximant of minimal order* n *that matches the first* $j + k$ *moments* (2.6), $M_0, M_1, \ldots, M_{j+k-1}$, *of* H.

Proof. Using (4.4), the band structure of T_n and $\tilde{T}_n$, the definition of j and k in (4.3), and (4.5), one can verify the following relations:

$$\begin{aligned} A^{i_1} R &= V_n T_n^{i_1} \begin{bmatrix} \rho \\ 0_{(n-m_1) \times m} \end{bmatrix}, \quad i_1 = 0, 1, \ldots, j-1, \\ L^T A^{i_2} &= [\, \eta^T \quad 0_{p \times (n-p_1)} \,] \tilde{T}_n^{i_2} W_n^T, \\ &= \begin{bmatrix} D_{p_1}^T \eta \\ 0_{(n-p_1) \times p} \end{bmatrix}^T T_n^{i_2} D_n^{-1} W_n^T, \quad i_2 = 0, 1, \ldots, k-1. \end{aligned}$$

These relations, together with (2.6) and $D_n = W_n^T V_n$, imply that

$$M_{i_1+i_2} = L^T A^{i_1+i_2} R = \begin{bmatrix} D_{p_1}^T \eta \\ 0_{(n-p_1) \times p} \end{bmatrix}^T T_n^{i_1+i_2} \begin{bmatrix} \rho \\ 0_{(n-m_1) \times m} \end{bmatrix} \tag{5.3}$$

for all $i_1 + i_2 = 0, 1, \ldots, j + k - 2$. Using (4.4) and the fact that the Lanczos vectors are biorthogonal, it is possible to show that (5.3) even holds true for $i_1 + i_2 = j + k - 1$. Thus, by (5.3), the first $j + k$ coefficients of the Taylor expansions about $\sigma = 0$ of the functions (2.5) and (5.1) match, and this concludes the proof of (5.2). In remains to verify that the number $j+k$ of matched coefficients is maximal so that H_n is indeed an nth Padé approximant. This can be done by exploiting the connection between matrix Padé approximants and the partial minimal realization problem and by showing that the criterion given in [14, Theorem 0.1] is satisfied.

6 Some Related Work

For the single-input single-output case, Grimme, Sorensen, and Van Dooren [17] proposed to use an implicitly restarted Lanczos algorithm to remedy the possible instability of Padé-based reduced-order models. The basic idea here is to adaptively choose a polynomial ψ that dampens out unstable poles of the reduced-order model, and then implicitly run n steps of the Lanczos algorithm applied to A with modified starting vectors $\psi(A)R$ and $\psi(A^T)L$. This results in some Padé-type approximation

$$\begin{aligned} H(z_0+\sigma) &= L^T\,(I-\sigma A)^{-1}\,R \\ &\approx (\psi(A^T)L)^T\,(I-\sigma A)^{-1}\,(\psi(A)R) \\ &\approx L^T R\, e_1^T\,(I-\sigma T_n)^{-1}\,e_1 =: \tilde{H}_n(z_0+\sigma), \end{aligned}$$

where H and $\tilde{H}_n$ match in the $2n$ *modified* moments

$$L^T A^j\,(\psi(A))^2\,R, \quad j = 0, 1, \ldots, 2n-1.$$

With the same amount of computational effort, PVL would generate a true Padé approximation that matches the $2(n + \deg\psi)$ moments LA^jR, $j = 0, 1, \ldots, 2(l + \deg\psi) - 1$.

An alternative to Padé approximation is interpolation of $H(z_0 + \sigma_j)$ at, say s, carefully selected nonzero points $\sigma_j \in \mathbb{C}$, $j = 1, 2, \ldots, n$. This approach now requires the computation of

$$H(z_0+\sigma_j) = -\frac{1}{\sigma_j} L^T\,(A-(1/\sigma_j)I)^{-1}\,R = -\frac{1}{\sigma_j} L^T X^{(j)}, \quad j = 1, 2, \ldots, s,$$

where $X^{(j)}$ is the solution the block system

$$(A-(1/\sigma_j)I)\,X^{(j)} = R. \tag{6.1}$$

Freund and Malhotra [12] developed a block-QMR (BL-QMR) algorithm for the iterative solution of block systems such as (6.1). In fact, BL-QMR is also based on the Lanczos-type algorithm sketched in Section 4. Note that s systems of the form (6.1) need to be solved. These systems all have the same right-hand side, R, and their coefficient matrices differ from the fixed matrix A only by different shifts $(1/\sigma_j)I$. The solution of such shifted systems requires only one run of the underlying Lanczos-type algorithm (applied to A), resulting in considerable computational savings when BL-QMR is used to solve the s shifted block systems (6.1); see, e.g., [7].

References

[1] J.I. Aliaga, D.L. Boley, R.W. Freund, and V. Hernández. A Lanczos-type algorithm for multiple starting vectors. Numerical Analysis Manuscript. No. 96-18. Bell Laboratories. Murray Hill, NJ. 1996.

[2] G.A. Baker, Jr. and P. Graves-Morris. *Padé Approximants, Second Edition.* New York: Cambridge University Press, 1996.

[3] A. Bultheel and M. Van Barel. Padé techniques for model reduction in linear system theory: a survey. *J. Comput. Appl. Math.* **14** (1986), 401-438.

[4] P. Feldmann and R. W. Freun. Efficient linear circuit analysis by Padé approximation via the Lanczos process. *Proc. EURO-DAC 1994 with EURO-VHDL 1994.* 170-175.

[5] P. Feldmann and R. W. Freund. Efficient linear circuit analysis by Padé approximation via the Lanczos process. *IEEE Trans. Computer-Aided Design* **14** (1995), 639-649.

[6] P. Feldmann and R.W. Freund. Reduced-order modeling of large linear subcircuits via a block Lanczos algorithm. *Proc. 32nd Design Automation Conference 1995.* 474-479.

[7] R.W. Freund. Solution of shifted linear systems by quasi-minimal residual iterations. *Numerical Linear Algebra.* (L. Reichel, A. Ruttan, and R. S. Varga, W. de Gruyter, Eds.) Berlin: Springer-Verlag, 1993. 101-121.

[8] R.W. Freund. Computation of matrix Padé approximations of transfer functions via a Lanczos-type proces. *Approximation Theory VIII, Vol. 1: Approximation and Interpolation.* (C. K. Chui and L. L. Schumaker, Eds.). Singapore: World Scientific Publishing Co., 1995. 215-222.

[9] R.W. Freund and P. Feldmann, Small-signal circuit analysis and sensitivity computations with the PVL algorithm. *IEEE Trans. Circuits and Systems–II: Analog and Digital Signal Processing* **43** (1996), 577-585.

[10] R.W. Freund and P. Feldmann. Reduced-order modeling of large passive linear circuits by means of the SyPVL algorithm. To appear *Tech. Dig. 1996 IEEE/ACM International Conference on Computer-Aided Design 1996.*

[11] R.W. Freund, M.H. Gutknecht, and N.M. Nachtigal. An implementation of the look-ahead Lanczos algorithm for non-Hermitian matrices. *SIAM J. Sci. Comput.* **14** (1993), 137-158.

[12] R.W. Freund and M. Malhotra. A block-QMR algorithm for non-Hermitian linear systems with multiple right-hand sides. Numerical Analysis Manuscript No. 95-09. AT&T Bell Laboratories. Murray Hill, NJ, 1995. To appear *Linear Algebra Appl.*

[13] K. Gallivan, E. Grimme, and P. Van Dooren. Asymptotic waveform evaluation via a Lanczos method. *Appl. Math. Lett.* **7** (1994), 75-80.

[14] I. Gohberg, M. A. Kaashoek, and L. Lerer. On minimality in the partial realization problem. *Systems Control Lett.* **9** (1987), 97-104.

[15] W.B. Gragg. Matrix interpretations and applications of the continued fraction algorithm. *Rocky Mountain J. Math.* **4** (1974), 213-225.

[16] W.B. Gragg and A.Lindquist. On the partial realization problem. *Linear Algebra Appl.* **50** (1983), 277-319.

[17] E.J. Grimme, D.C. Sorensen, and P. Van Dooren. Model reduction of state space systems via an implicitly restarted Lanczos method. *Numer. Algorithms* **12** (1996), 1-31.

[18] M.H. Gutknecht. A completed theory of the unsymmetric Lanczos process and related algorithms, part I. *SIAM J. Matrix Anal. Appl.* **13** (1992), 594-639.

[19] X. Huang. *Padé Approximation of Linear(ized) Circuit Responses.* Ph.D. Dissertation. Carnegie Mellon University. Pittsburgh, Pennsylvania. 1990.

[20] T. Kailath. *Linear Systems*, Englewood Cliffs: Prentice-Hall, 1980.

[21] H.M. Kim and R.R. Craig, Jr. Structural dynamics analysis using an unsymmetric block Lanczos algorithm. *Internat. J. Numer. Methods Engrg.* **26** (1988), 2305-2318.

[22] H.M. Kim and R.R. Craig, Jr. Computational enhancement of an unsymmetric block Lanczos algorithm. *Internat. J. Numer. Methods Engrg.* **30** (1990), 1083-1089.

[23] C. Lanczos. An iteration method for the solution of the eigenvalue problem of linear differential and integral operators. *J. Res. Nat. Bur. Standards* **45** (1950), 255-282.

[24] L.T. Pillage and R.A. Rohrer. Asymptotic waveform evaluation for timing analysis. *IEEE Trans. Computer-Aided Design* **9** (1990), 352-366.

[25] T.-J. Su. A decentralized linear quadratic control design method for flexible structures. Ph.D. Dissertation. The University of Texas at Austin. Austin, Texas. 1989.

[26] T.-J. Su and R.R. Craig, Jr. Model reduction and control of flexible structures using Krylov vectors. *J. Guidance Control Dynamics* **14** (1991), 260-267.

[27] A.J. Tether. Construction of minimal linear state-variable models from finite input-output data. *IEEE Trans. Automat. Control* **AC-15** (1970), 427-436.

[28] J. Vlach and K. Singhal. *Computer Methods for Circuit Analysis and Design, Second Edition.* New York: Van Nostrand Reinhold, 1993.

[29] G.-L. Xu and A. Bultheel. Matrix Padé approximation: definitions and properties. *Linear Algebra Appl.* **137/138** (1990), 67-136.

Bell Laboratories, Lucent Technologies, Room 2C–420, 700 Mountain Avenue, Murray Hill, New Jersey 07974-0636

Dynamical Systems Approach to Target Motion Perception and Ocular Motion Control

B.K. Ghosh[1], E.P. Loucks, C.F. Martin[2], and L. Schovanec [3]

1 Introduction

In this paper, we introduce a dynamical systems approach to two problems. The first problem is to dynamically estimate the motion parameters of a moving target. The second problem is to dynamically control the orientation of the visual system. In the first problem we consider a planar textured surface undergoing a rigid or an affine motion. The visual system is a CCD camera which is assumed to be held fixed in space. The observation model for the camera is assumed to be a perspective projection model. We show the underlying dynamical system is a "perspective system." In the second problem, the visual system is assumed to be the oculomotor system. A dynamic model of the associated control system is developed that is particularly appropriate for modeling saccadic and smooth pursuit eye movements. The eye is controlled by extraocular muscles and the control signals are generated by motoneuronal activity. The problem of orientation control of a visual system to track a given moving target is an example of perspective control which has been introduced in this paper. Perspective control of ocular motion would be a subject of future investigation.

2 Dynamic Model Based Motion Estimation

2.1 Introduction to Perspective Systems Theory

The first part of this paper presents a summary of recent results in the area of "perspective systems theory" that have been reported in [7], [8] and [9]. The theory is motivated from problems in visually guided motion and shape estimation and from perspective control. The motion and shape estimation is well known in computer vision (see for example [20], [23]) whereas the perspective control problem has connection with various visually guided control problems in robotics [15] and with the control of the motion of the human eye [16] (see [17]), which has been detailed in the next section.

---

[1]The research of the first two authors was supported in part by the DOE grant No. DE-FG02-90ER14140

[2]Supported by NSF grant DMS-9628558 and NASA grant NAG 2-902

[3]Supported by the Texas Advanced Research Program Grant No. 003644-123 and NSF grant DMS-9628558

To introduce a sample problem, consider a linear system with two state variables described as follows

$$\begin{pmatrix} \dot{x}_1(t) \\ \dot{x}_2(t) \end{pmatrix} = \begin{pmatrix} a_1 & a_2 \\ a_3 & a_4 \end{pmatrix} \begin{pmatrix} x_1(t) \\ x_2(t) \end{pmatrix} \tag{2.1.1}$$

where we assume that the parameters a_1, a_2, a_3, a_4 are unknown constants. We now consider the output function

$$Y : \mathbb{R}^2 - \{(0,0)\} \rightarrow \mathbb{RP}^1 \tag{2.1.2}$$

defined as

$$(x_1, x_2) \longmapsto [x_1, x_2]$$

where $[x_1, x_2]$ describes a point in $\mathbb{RP}^1$ in homogeneous coordinates where $\mathbb{RP}^1$ refers to the real projective space of all homogeneous lines in $\mathbb{R}^2$ (see [1] and [12] for details). In other words, we assume that the non-zero state (x_1, x_2) of the dynamical system (2.1.1) is observed up to a homogeneous line, i.e., a line through (x_1, x_2) passing through the origin. If we now consider the coordinate chart

$$\{[x_1, x_2] : x_2 \neq 0\}$$

of $\mathbb{RP}^1$ defined by the coordinate

$$y_t = \frac{x_1(t)}{x_2(t)}$$

we obtain the following recursion on y_t given by

$$\dot{y_t} = a_2 + (a_1 - a_4)y_t - a_3 y_t^2. \tag{2.1.3}$$

Note in particular that the right hand side of (2.1.3) is a quadratic function in y_t. Hence (2.1.3) is a Riccati equation in one variable. We now introduce the following problem.

Problem 2.1.1 (Sample Identification Problem) Consider the Riccati dynamical system (2.1.3), where we assume that we observe y_t, $t \geq 0$. The problem is to identify parameters a_1, a_2, a_3, a_4 to the extent possible from this data.

It is clear that at best one can hope to identify a_2, $a_1 - a_4$ and a_3. It would be important to understand, however, under what condition on the parameters a_1, a_2, a_3, a_4 and initial condition $x_1(0), x_2(0)$ would the identification indeed be possible. It may be remarked that the problem 2.1 is motivated classically from the following Riccati's question.

Question 2.1.2 (Riccati's Question) If points on a plane satisfying a linear dynamical system in continuous time are observed up to the slope of the line which the point makes with respect to the origin, what can be said about the dynamics of the slope?

In order to generalize the above example, we now consider a linear system

$$\dot{x}_t = Ax_t, \quad z_t = Cx_t \tag{2.1.4}$$

where $x_t \in \mathbb{R}^n$, $z_t \in \mathbb{R}^p$ for $t \geq 0$. Furthermore, we assume that $p > 1$ and that z_t is observed only up to a homogeneous line; i.e., we have as in (2.1.2), the observation function

$$Y : \mathbb{R}^p - \{0\} \to \mathbb{RP}^{p-1} \tag{2.1.5}$$

$$z_t \longmapsto [z_t] = [Cx_t]$$

where $[z_t]$ is a point in the projective space $\mathbb{RP}^{p-1}$ of all homogeneous lines in $\mathbb{R}^p$. The pair (2.1.4), (2.1.5) would be referred to as a "Perspective Dynamical System." We now consider the following observability problem.

Problem 2.1.3 (Perspective Observability Problem) The dynamical system (2.1.4)), (2.1.5) is said to be perspectively observable if for every pair of initial conditions x_0, x_0' where $[x_0] \neq [x_0']$, there exists some T_1 and T_2 such that $[Ce^{At}x_0]$ and $[Ce^{At}x_0']$ are defined and $[Ce^{At}x_0] \neq [Ce^{At}x_0']$ for all $t \in [T_1, T_2]$. We seek condition on A, C such that the dynamical system (2.1.4), (2.1.5) is perspectively observable.

The following result is essentially due to Dayawansa, Ghosh, Martin and Wang [4] (see also Ghosh and Rosenthal [10]).

Theorem 2.1.1 *The perspective system* (2.1.4), (2.1.5) *is perspectively observable if for any set of eigenvalues* λ_0, λ_1 *of* A *(possibly repeated) one has*

$$\text{rank} \begin{pmatrix} (A - \lambda_0 I)(A - \lambda_1 I) \\ C \end{pmatrix} = n. \tag{2.1.6}$$

Moreover if the eigenvalues of A *are in* $\mathbb{R}$, *the condition* (2.1.6) *is also necessary.*

Remark 2.1.5 Theorem 2.1.1 is true even over the base field $\mathbb{C}$ of complex numbers. In fact since every eigenvalue of A is in $\mathbb{C}$, the condition (2.1.6) would always be necessary and sufficient for perspective observability over $\mathbb{C}$. Note that the rank condition (2.1.6) is an immediate generalization of the Popov-Belevitch-Hautus condition [13], for checking the observability of a linear dynamical system. If the parameters A and C are unknown, one introduces the following identification problem.

Problem 2.1.6 (Perspective Identification Problem) Consider the dynamical system (2.1.4), (2.1.5). Assume that $x_0 \neq 0$ and the parameters A, C are unknown. For some T_1, T_2 where $0 \leq T_1 < T_2$, assume that

$[Ce^{At}x_0]$ has been observed for all $t \in [T_1, T_2]$. The problem is to compute A, C and $[x_0]$ from this data to the extent possible.

It may be noted that the parameters A, C, $[x_0]$ cannot be completely recovered from $[z_t] = [Ce^{At}x_0]$. The main result of this paper is to show that the nonuniqueness in C, A, $[x_0]$ which produces the same $[z_t]$ can be described as an orbit of a perspective group $\mathcal{G}$. The orbit of the perspective group $\mathcal{G}$ is described as follows:

1. $P \in GL(n)$ acting on (C, A, x_0):

$$(C, A, x_0) \longmapsto (CP, P^{-1}AP, P^{-1}x_0); \tag{2.1.7}$$

2. λ_1, $\lambda_3 \in \mathbb{R} - \{0\}$ acting on (C, A, x_0):

$$(C, A, x_0) \longmapsto (\lambda_1 C \;\; A \;\; \lambda_3 x_0); \tag{2.1.8}$$

3. $\lambda_2 \in \mathbb{R}$ acting on (C, A, x_0) :

$$(C, A, x_0) \longmapsto (C \;\; \lambda_2 I + A \;\; x_0). \tag{2.1.9}$$

The action (2.1.7) is already well known in linear system theory and is obtained as a result of change of basis in the state space. The scaling action (2.1.8), (2.1.9) is new and is a result of the perspective observation function (2.1.5).

The collective actions (2.1.7), (2.1.8), (2.1.9) will be referred to as the action due to a perspective group $\mathcal{G}$. The following important result for the case when the number of outputs p is greater than or equal to the number of state variables n has been shown in [8] (see also [11]).

Theorem 2.1.2 *Let us consider triplets (C, A, x_0) such that the vectors Cx_0, CAx_0, ..., $CA^{n-1}x_0$ are linearly independent. Under this generic condition on the triplet (C, A, x_0), it is possible to locally identify the parameters of the system* (2.1.4), (2.1.5) *up to orbits of the perspective group $\mathcal{G}$.*

We now propose to describe the perspective control problem. To introduce the problem, let us consider a linear dynamical system in continuous time defined by

$$\dot{x} = Ax + Bu \tag{2.1.10}$$

where we assume that $x \in \mathbb{R}^n$ and $u \in \mathbb{R}^m$. We assume furthermore that the state function $x(t)$ is not observed entirely but is observed up to an unknown nonzero scale factor. In particular, the output function $X(t)$ is defined as

$$\begin{aligned} X : \mathbb{R}^n - (0, 0, \ldots, 0) &\to \mathbb{RP}^{n-1} \\ x &\longmapsto [x] \end{aligned} \tag{2.1.11}$$

where $[x]$ is the homogeneous line defined by the non-zero vector x. The pair (2.1.10), (2.1.11) is an example of a perspective control system. Let us now consider the following controllability problem.

Problem 2.1.8 Let x_0 and x_T be two non-zero vectors in $\mathbb{R}^n$. Does there exist a non-negative scalar $T \geq 0$ together with a piece wise continuous function $u(t), t \in [0, T]$ such that the state $X(t)$ can be controlled starting from $X(0) = [x_0]$ to $X(T) = [x_T]$ with the chosen control $u(t)$?

Definition 2.1.9 The perspective dynamical system (2.1.10), (2.1.11) is said to be perspectively controllable if Problem 2.1 has an affirmative answer for every pair of nonzero vectors in $\mathbb{R}^n$.

In order to write down the main result on perspective controllability, we define the following:

$$\begin{array}{l} H_0 = span\left\{x_0, B, AB, \ldots, A^{n-1}B\right\} \\ H_T = span\left\{x_T, B, AB, \ldots, A^{n-1}B\right\} \end{array} . \tag{2.1.12}$$

The main theorem on perspective controllability is now stated.

Theorem 2.1.3 *The dynamical system* (2.1.10), (2.1.11) *is perspectively controllable in between the two non-zero vectors x_0 and x_T iff the following two conditions are satisfied.*

1. *H_0 and H_T have the same dimension.*
2. *$\exists T > 0 : e^{AT} H_0 = H_T$.*

For a proof of the above theorem and for many other controllability and observability results, we would like to refer to [17].

In the next section, we introduce perspective dynamical system that arises in a specific motion estimation problem for a textured plane undergoing an affine motion. We shall also consider the special case when the motion is rigid.

2.2 Shape Dynamics of a Surface Patch

In this section, we shall consider a plane described as

$$Z = pX + qY + r \tag{2.2.1}$$

where p, q, r are shape parameters that are changing in time as a result of an affine motion field given by

$$\dot{\mathcal{X}} = A\mathcal{X} + b \tag{2.2.2}$$

where

$$A = [a_{ij}], b = col[b_1, b_2, b_3] \tag{2.2.3}$$

are respectively a 3×3 matrix and a 3×1 vector and where $\mathcal{X} = col\,[X, Y, Z]$. We would propose to derive a differential equation that describes the motion of the shape parameters p, q, r. This is done as follows.

Let us homogenize the vector (X, Y, Z) as $X = \bar{X}/\bar{W}$, $Y = \bar{Y}/\bar{W}$, $Z = \bar{Z}/\bar{W}$ and the vector (p, q, r) as

$$p = \bar{p}/\bar{s}, q = \bar{q}/\bar{s}, r = \bar{r}/\bar{s}. \tag{2.2.4}$$

and rewrite (2.2.1) as $\begin{pmatrix} \bar{p}, & \bar{q}, & -\bar{s}, & \bar{r} \end{pmatrix} \bar{\mathcal{X}} = 0$ and (2.2.2) as $\dot{\bar{\mathcal{X}}} = -\mathcal{A}^T \bar{\mathcal{X}}$ where
$\bar{\mathcal{X}} = \begin{pmatrix} \bar{X}, & \bar{Y}, & \bar{Z}, & \bar{W} \end{pmatrix}^T$ and

$$-\mathcal{A}^T = \begin{pmatrix} A & b \\ 0 & 0 \end{pmatrix}. \tag{2.2.5}$$

It follows that

$$\frac{d}{dt} \begin{pmatrix} \bar{p}, & \bar{q}, & -\bar{s}, & \bar{r} \end{pmatrix}^T = \mathcal{A} \begin{pmatrix} \bar{p}, & \bar{q}, & -\bar{s}, & \bar{r} \end{pmatrix}^T \tag{2.2.6}$$

where $\mathcal{A}$ is the 4×4 matrix in (2.2.5) and is defined up to addition by a scalar multiple of the identity matrix. If we assume initial condition to be $\bar{s}(0) = 1$, $\bar{p}(0) = p(0)$, $\bar{q}(0) = q(0)$, $\bar{r}(0) = r(0)$, it may be concluded that the dynamical system (2.2.6) describes the motion of the shape parameters p, q, r. In fact from (2.2.4) and (2.2.6) the dynamics of p, q, r can be written as the following Riccati Equation

$$\begin{aligned} \dot{p} &= (a_{33} - a_{11})p - a_{21}q + a_{31} - a_{13}p^2 - a_{23}pq \\ \dot{q} &= (a_{33} - a_{22})q - a_{12}p + a_{32} - a_{13}pq - a_{23}q^2 \\ \dot{r} &= -(a_{33} + a_{23}q + a_{13}p)r + (b_3 - b_2q - b_1p). \end{aligned} \tag{2.2.7}$$

In general, the Riccati Equation (2.2.7) propagates in time, the relationship between co-ordinates X, Y and Z expressed via the plane (2.2.1). Note that the equation (2.2.7) is parameterized by 12 motion parameters and 3 initial conditions on shape parameters. Thus there is a total of 15 parameters describing the shape dynamics (2.2.6) for the affine motion.

An important special case of the affine motion (2.2.2) is the case when A is a skew symmetric matrix given by

$$\begin{pmatrix} 0 & \omega_1 & \omega_2 \\ -\omega_1 & 0 & \omega_3 \\ -\omega_2 & -\omega_3 & 0 \end{pmatrix} \triangleq \Omega. \tag{2.2.8}$$

Under this assumption, the motion field (2.2.2) describes a rigid motion. The shape dynamics (2.2.6) can be written as

$$\frac{d}{dt} \begin{pmatrix} \bar{p} \\ \bar{q} \\ -\bar{s} \\ \bar{r} \end{pmatrix} = \begin{pmatrix} \Omega & 0 \\ -b^T & 0 \end{pmatrix} \begin{pmatrix} \bar{p} \\ \bar{q} \\ -\bar{s} \\ \bar{r} \end{pmatrix}. \tag{2.2.9}$$

Note that the shape dynamics (2.2.7) reduces to $\dot{p} = -\omega_2(1+p^2)+\omega_1 q - \omega_3 pq$, $\dot{q} = -\omega_3(1+q^2)-\omega_1 p-\omega_2 pq$ and $\dot{r} = b_3-b_1p-b_2q-r(\omega_3 q+\omega_2 p)$ which is parameterized by a total of 6 motion parameters and 3 initial conditions on shape parameters. Thus there is a total of 9 parameters describing the shape dynamics (2.2.9) for the rigid motion.

Assume that the planar surface is textured and the intensity $E(X, Y, Z, t)$ of a point (X, Y, Z) on the surface at time t does not change along the integral curves of (2.2.2). Let (x, y) be the coordinates of the image plane obtained under the projection of a point (X, Y, Z) on the surface of the object. We define

$$x = \frac{fX}{Z+\delta}, \quad y = \frac{fY}{Z+\delta} \tag{2.2.10}$$

where $\delta \in [0, f]$ and f is the focal length of the camera. Note that if $\delta = 0$ we obtain a viewer centered projection. If $\delta = f$ we obtain an image centered projection. These two projections have been described in [14]. Finally note that if $\delta = f$ and $f \rightarrow \infty$, we obtain

$$x = X, \quad y = Y \tag{2.2.11}$$

which is known in the literature [14] as the "orthographic projection."

Let us now restrict attention to a planar surface (2.2.1) with affine motion (2.2.2) and assume a generalized projection (2.2.10). The "optical flow" equation for this special case can be written as follows

$$\begin{aligned} \dot{x} &= d_1 + d_3 x + d_4 y + \tfrac{1}{f}\left(d_7 x^2 + d_8 xy\right) \\ \dot{y} &= d_2 + d_6 y + d_5 x + \tfrac{1}{f}\left(d_8 y^2 + d_7 xy\right) \end{aligned} \tag{2.2.12}$$

where

$$\begin{aligned} d_1 &= f(a_{13}+c_1), d_2 = f(a_{23}+c_2), d_3 = (a_{11}-a_{33}) - (c_3 + pc_1) \\ d_4 &= a_{12} - qc_1, d_5 = a_{21} - pc_2, \\ d_6 &= (a_{22}-a_{33}) - (c_3 + qc_2), d_7 = pc_3 - a_{31}, d_8 = qc_3 - a_{32} \end{aligned} \tag{2.2.13}$$

and where

$$c_i = (b_i - a_{i3}\delta)/(r+\delta), i = 1, 2, 3. \tag{2.2.14}$$

Assume that the intensity function is smooth so that all its partial derivatives exist and can be computed. If the motion field is affine given by (2.2.2), the intensity dynamics is given by

$$\frac{\partial e}{\partial t} + F(x,y)\frac{\partial e}{\partial x} + G(x,y)\frac{\partial e}{\partial y} = 0 \tag{2.2.15}$$

where $e(x, y, t)$ is the observed intensity function on the image plane and

$$\begin{aligned} F(x,y) &= d_1 + d_3 x + d_4 y + \tfrac{1}{f}\left(d_7 x^2 + d_8 xy\right) \\ G(x,y) &= d_2 + d_6 y + d_5 x + \tfrac{1}{f}\left(d_8 y^2 + d_7 xy\right). \end{aligned} \tag{2.2.16}$$

The parameters $d_1, \dots, d_8$ can be defined from (2.2.13). Combining (2.2.15) and (2.2.16) we now write

$$v^T d = -\frac{\partial e}{\partial t} \tag{2.2.17}$$

where

$$v^T = \left(e_x, e_y, xe_x, ye_x, xe_y, ye_y, \frac{1}{f}\left(x^2 e_x + xye_y\right), \frac{1}{f}\left(xye_x + y^2 e_y\right) \right) \tag{2.2.18}$$

and

$$d = (d_1, \dots, d_8)^T. \tag{2.2.19}$$

In order to compute an estimate of the coefficient vector d, we proceed as follows. Let us choose $n \geq 8$ points on the image plane denoted by $(x_i, y_i), i = 1, \cdots, n$. From the observed data $e(x, y, t)$ we now form the matrices

$$V = \begin{pmatrix} v(x_1, y_1) & v(x_2, y_2) & \cdot & \cdot & \cdot & v(x_n, y_n) \end{pmatrix} \tag{2.2.20}$$

and

$$u = \begin{pmatrix} -e_t(x_1, y_1) & -e_t(x_2, y_2) & \cdot & \cdot & \cdot & -e_t(x_n, y_n) \end{pmatrix}^T. \tag{2.2.21}$$

From (2.2.17) it follows that $V^T d = u$. If the points (x_i, y_i) are chosen in such a way that rank $V = 8$, we compute

$$\hat{d} = (VV^T)^{-1} V u \tag{2.2.22}$$

as an estimate of d. We therefore have the following theorem.

Theorem 2.2.1 *Assume that the function $e(x, y, t)$ is such that all its partial derivatives are available and can be measured. Assume furthermore that the points $(x_i, y_i), i = 1, \dots, n$ are such that rank $V = 8$, where V is given by (2.2.20). It is possible to obtain an unique estimate of d.*

We shall now consider identifying parameters of (2.2.6) by considering the output equation given by (2.2.13). However since (2.2.13) is nonlinear in the parameters, we would like to homogenize the vector $(d_1, \dots, d_8)^T$ as follows. Let us define

$$d_j = \frac{y_j}{y_9}, \quad j = 1, \dots, 8 \tag{2.2.23}$$

so that the vector

$$(y_1, \dots, y_9) \tag{2.2.24}$$

is a homogenization of the essential parameters. Equation (2.2.13) can be written as

$$\begin{pmatrix} y_1 \\ y_2 \\ y_3 \\ y_4 \\ y_5 \\ y_6 \\ y_7 \\ y_8 \\ y_9 \end{pmatrix} = \begin{pmatrix} 0 & 0 & -fb_1 & fa_{13} \\ 0 & 0 & -fb_2 & fa_{23} \\ -b_1' & 0 & b_3 - \delta a_{11} & a_{11} - a_{33} \\ 0 & -b_1' & -\delta a_{12} & a_{12} \\ -b_2' & 0 & -\delta a_{21} & a_{21} \\ 0 & -b_2' & b_3 - \delta a_{22} & a_{22} - a_{33} \\ -b_3' & 0 & -\delta a_{31} & a_{31} \\ 0 & -b_3' & -\delta a_{32} & a_{32} \\ 0 & 0 & -\delta & 1 \end{pmatrix} \begin{pmatrix} \bar{p} \\ \bar{q} \\ -\bar{s} \\ \bar{r} \end{pmatrix} \tag{2.2.25}$$

where $\bar{p}$, $\bar{q}$, $\bar{s}$ and $\bar{r}$ have been defined as given by (2.2.4) and

$$b' = (b_1 - a_{13}\delta \; b_2 - a_{23}\delta \; b_3 - a_{33}\delta) \triangleq (b_1' \; b_2' \; b_3'). \tag{2.2.26}$$

The dynamical system (2.2.6) together with the output function (2.2.25) constitutes an example of a perspective system as has been described in (2.1.4), (2.1.5). The motion estimation problem is to identify the parameters A, b given by (2.2.3) and the initial conditions $\bar{p}(0), \bar{q}(0), \bar{s}(0), \bar{r}(0)$ to the extent possible from the essential parameter vector $(d_1, ...d_8)$.

The main result that we now describe is a complete answer to the motion and shape estimation problem. Note in particular that the perspective system (2.2.6), (2.2.25) is parameterized by a set of 12 motion parameters A, b and a set of 3 shape parameters p, q, r. We shall show that not all the 15 parameters are identifiable, i.e., there is a non-unique choice of parameters for which the observation described by (2.2.25) is the same. The nonunique choice is parameterized by the orbit of a group action and is described as follows.

Theorem 2.2.2 *Under a suitable generic condition on the set of 15 parameters of the perspective system (2.2.6), (2.2.25), the following parameters or functions of parameters are locally identifiable. They are*

$$(A, p, q, c_1, c_2, c_3) \tag{2.2.27}$$

where c_1, c_2, c_3 is defined in (2.2.14).

To sum up, in this section we have shown the extent to which the motion parameters of a textured planar patch can be recovered by a single CCD camera. The problem, we show, reduces to a suitable parameter identification problem of a perspective system. These have been detailed in [8] and [9].

3 Dynamics and Control of Eye Movement

3.1 Introduction to the Oculomotor System

In the study of the oculomotor system and the mathematical description of this plant as a pursuit or tracking system, one must address the mechanics of the extraoculor muscles as well as the role of motoneuronal activity in producing a desired eye movement. The objective of this section of the paper is to formulate a realistic mathematical representation of the eye movement control system. In particular, we develop a model of the human eye-positioning mechanism which accurately depicts the dynamics of the eye movement mechanism once it has been given an appropriate control signal that corresponds to a specific type of eye movement. In this paper, attention will be focused upon saccadic and smooth pursuit eye movements. Saccadic eye movements are among the fastest voluntary muscle movements the human body is capable of producing. They are characterized by a rapid shift of gaze from one point to another and are used to bring a new part of the visual field into the foveal region. Smooth pursuit eye moments are associated with the attempt to move the eye at the same velocity as the target. There is evidence [22] that in attempting to follow an unpredictable target, the eye performs either a saccadic or pursuit movement or some combination of the two. It should be noted that there are several additional eye movements, for example, vestibular or vergence movements, that will not be addressed here.

A vast amount of literature deals with the study of eye movements. One of the first models was developed by Descarte [5] in 1630 based on the principal of reciprocal innervation, a notion of paired muscular activity in which a contraction of one muscle is associated with the relaxation of the other. In 1954 Westheimer [21] developed a linear second order approximation of eye dynamics during a saccade in which the input to the model was assumed to be a step of muscle force. The model worked well for 10° saccades but not for larger such movements. In addition, the model predicted the unphysical results that the time of saccade duration would be independent of saccade magnitude and that the peak velocity would be directly proportional to saccade magnitude. A more realistic representation of eye movement was advanced by Robinson [19]. His linear fourth-order model could simulate saccades between 5° and 40° but the velocity profiles predicted by this model were not physically realistic. It was recognized by Westheimer and Robinson that the eye movement mechanism was inherently nonlinear, issues not addressed by their work. Roughly speaking, the nonlinear features of the system can be attributed to the geometry of the system as well as the nonlinear physiological behavior of certain components that describe the extraocular muscle. Martin and Lu [18] advanced a model of the eye system that assumed a linear model of muscle behavior but accounted for the nonlinear effects associated with forces that develop when the recti muscles act in a nontangential fashion on the eyeball. They were able to construct a control law that enabled the eye to track a target through a range of both large and small displacements. The model that was utilized omitted some physiological features of muscle and did not distinguish the effects of passive and active muscle behavior, a notion that will be elaborated upon later. Another group of investigators have concentrated upon ocular models that emphasize the effects of muscular physiology upon system performance. Along these lines, a sixth-order nonlinear model proposed by Cook and Stark [3] and subsequently modified in [2] produced realistic position, velocity and acceleration profiles. This Cook-Clark-Stark model addressed the nonlinear relationship between force and velocity but ignored the force-length characteristics of muscle. This assumption was tantamount to assuming that the medial and lateral rectus muscles operate near the primary position that corresponds to looking straight ahead. The model that is investigated here incorporates both length-velocity and length-tension characteristics of muscle. It does not account for the geometric nonlinearities addressed in [18]. A justification for this assumption is that for saccadic movements, when motion is typically less than 30°, the nontangential forces associated with the recti muscles do not occur. Before proceeding to the development of the model, a digression on the physiology of muscle models is presented.

3.2 The Plant Model

Figure 1 shows a top view of the left eyeball. Of the six extraocular muscles, the lateral and medical recti are used primarily for rotations about the z-axis, the superior and inferior recti for rotations about the y axis and the superior and inferior oblique for rotations about the x axis. Only the medial and lateral recti are shown since the investigation here is restricted to horizontal movements that correspond to rotations about the z axis (see Figure 1). For displacements from straight ahead that are less than 30° to 40°, the moment arms of the lateral and medial rectus are approximately equal to the radius of the eyeball. The mechanics is considerably simplified under these assumptions and the derivations presented here are restricted to these range of motions since the primary intent of this work is to model a saccadic movement.

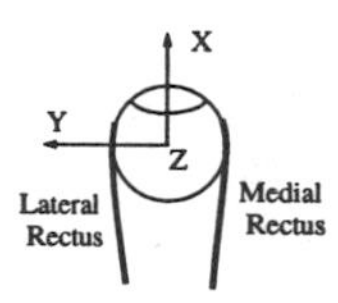

Figure 1. Top view of globe with medial and lateral recti muscles responsible for horizontal rotations about z axis.

Figure 2 shows the basic arrangement of the elements of the plant, that is, the eyeball or globe and extraocular muscle. For the sake of convenience, the muscles are shown unwrapped from the globe as though they were translating it. In Figure 2, J is the moment of inertia of the eyeball which is acted upon by three forces: the tendon force of the two muscles which constitute the agonist-antagonist pair, T_1, T_2, and the force of the nonmuscular passive tissues.

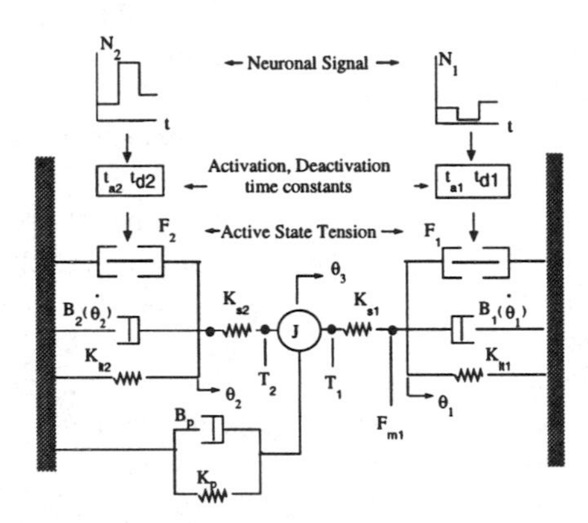

Figure 2. The organization of mechanical elements of the eye model. The agonist and antagonist elements are denoted by subscripts 1,2 respectively.

A so-called Hill model is utilized to describe the agonist and antagonist muscles. This phenomenological model is composed of viscoelastic elements which represent the active forces that are developed in response to neural stimulation in addition to forces that correspond to passive elastic and viscoelastic structure. Here the parallel passive effects have been incorporated into the forces of the nonmuscular passive tissues since data is available for the viscous and elastic elements B_b and K_p under these assumptions. The series passive effects are primarily due to the tendon and its elastic behavior and correspond to the series springs with constants K_{s1}, K_{s2}. The active contractile element is conceptually divided into two parts, one described by a force-velocity relationship and the other by a force-length dependence. In Figure 2, the parallel viscous elements B_1, B_2 are used to describe force-velocity relationship that acts like a viscosity in that it produces a force that is related to the speed of contraction. The dependence of the active contractile element on length is modeled by the parallel elastic element with spring constant K_{lt1}, K_{lt2}. The active state forces are determined by innervation, denoted by N_1, N_2. There are activation dynamics that are interposed between, say N_1 and the active state tension F_1. The details associated with these model elements will be expounded upon in the subsequent discussion

The manner in which the force of the contractile element depends on the rate of change in length is described by the well-known Hill equation,

$$v[T(v) + a] = b[T_0 - T(v)].$$

Here v is the velocity of contraction, $T(v)$ is the instantaneous tension in the series elastic component, T_0 is the maximal isometric tension and a, b are constants. Hill's equation may be written in the form $T(v) = T_0 - B(v)v$ where $B(v)$ corresponds to the damping coefficients in the parallel viscous elements. There are several experimentally determined forms of $B(v)$ and the expressions that will be used here are taken from [2] and are given by

$$B_1 = \frac{1.25}{\dot{\theta}_1 + 900} F_1, \quad B_2 = \frac{F_2 - 40}{\dot{\theta}_2 - 900}. \tag{3.2.1}$$

The length-tension muscle effects are incorporated into the model by identifying the muscle force developed in the tendon with a linear approximation of the experimentally determined force-length curve. This is achieved by considering the force applied on the globe by the agonist T_1 in a state of static equilibrium. In this case it is simple matter to show that

$$\begin{aligned} T_1 &= \frac{K_{lt1} K_{s1} \theta_3}{K_{lt1} + K_{s1}} + \frac{K_{s1} F_1}{K_{lt1} + K_{s1}} \\ &= K' \theta_3 + \frac{K_{s1} F_1}{K_{lt1} + K_{s1}}. \end{aligned}$$

The above equation provides a linear approximation to the experimentally determined force-length curve from which K' may be determined. Since K_{s1} is readily available from a quick-release experiment, one may then determine K_{lt1} and similarly K_{lt2}. The values for K_{lt1}, K_{lt2} that are used in this model are taken from [6].

3.3 The Dynamics of Eye Motion

The differential equations that describe eye movement are derived by considering a summation of active and passive forces that act on the eyeball and the activation dynamics that describe the innervation due to the neuronal signal. In this development one must distinguish between the force that is generated in the active muscle denoted by F_{m1}, F_{m2} and that which is transmitted through the tendon and is exerted on the globe, T_1, T_2 (see Figure 2). Henceforth we assume that the spring constants for the series and parallel elastic elements are identical and denote them by K_s and K_{lt} respectively. For the agonist we have that the force at the tendon satisfies

$$T_1 = K_{se}(\theta_1 - \theta_3)$$

while the actively generated muscle force is

$$F_{m1} = F_1 - B_1 \dot{\theta}_1 - K_{lt} \theta_1 = K_s(\theta_1 - \theta_3).$$

If we define $K_{st} = (K_s + K_{lt})/K_s$ some simple manipulations give

$$T_1 = \frac{K_s}{K_{st}} F_1 - \frac{K_s K_{lt}}{K_{st}} \theta_3 - \frac{K_s B_1}{K_{st}} \dot{\theta}_1. \tag{3.3.2}$$

In the same way one has for the antagonist that

$$T_2 = \frac{K_s}{K_{st}} F_2 - \frac{K_s K_{lt}}{K_{st}} \theta_3 + \frac{K_s B_2}{K_{st}} \dot{\theta}_2. \tag{3.3.3}$$

By Newton's second law it follows by a balance of forces on the globe that

$$K_s(\theta_1 - \theta_2) - K_s(\theta_3 - \theta_2) = K_p \theta_3 + B_p \dot{\theta}_3 + J \ddot{\theta}_3. \tag{3.3.4}$$

The inputs into the model are the neural control signals N_1, N_2. These signals are transformed into active state tensions by what are known as activation dynamics (see Figure 3). Here we adopt the approach of [2] and assume that these processes can be depicted as a low pass filter with activation-deactivation time constants $t_{1a}, t_{1d}, t_{2a}, t_{2d}$. (For a more complete discussion and more sophisticated models of activation dynamics see [24].) In particular we take

$$\tau_1 \dot{F}_1 - F_1 = N_1 \tag{3.3.5}$$
$$\tau_2 \dot{F}_2 - F_2 = N_2 \tag{3.3.6}$$

where

$$\tau_1 = t_{1a}[u(t) - u(t - T_1)] + t_{1d} u(t - T_1)$$
$$\tau_2 = t_{2a}[u(t) - u(t - T_2)] + t_{2d} u(t - T_2)$$

and T_1, T_2 correspond to the duration of the agonist and antagonist pulse.

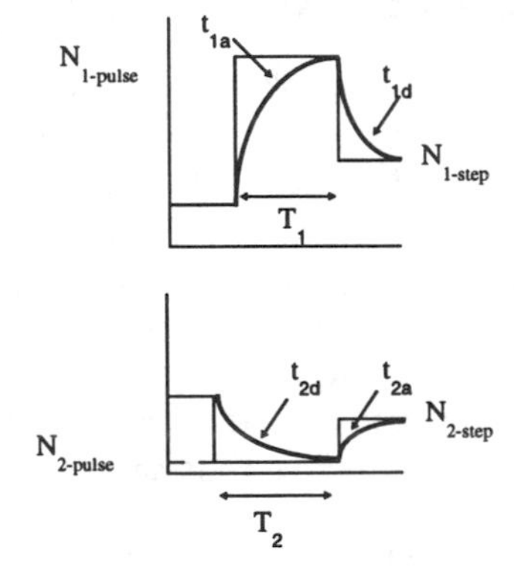

Figure 3. For saccadic movements a pulse step produced by the motoneurons is converted into active state tension by first order activation and deactivation dynamics.

The state variables that will be utilized are the three positions $\theta_1, \theta_2, \theta_3$, the eye velocity $\dot{\theta}_3$ and the two active state tensions F_1, F_2. Define

$$
\begin{aligned}
x_1 &= \theta_1 = \text{ position of agonist node} \\
x_2 &= \theta_2 = \text{ position of antagonist node} \\
x_3 &= \theta_3 = \text{ position of eye} \\
x_4 &= \dot{\theta}_3 = \text{ eye velocity} \\
x_5 &= F_1 = \text{ agonist active state} \\
x_6 &= F_2 = \text{ antagonist active state.}
\end{aligned}
$$

By solving equations (3.3.2-3.3.6) for the state variables and substituting the velocity dependent viscosities in (3.2.1) into these equations one obtains the state equations

$$\dot{x_1} = x_4$$

$$\dot{x_2} = \frac{900((K_{st} - K_{lt})x_1 - K_{st}x_2 + x_5)}{-(K_{st} - K_l)x_1 + K_{st}x_2 + .25x_2}$$

$$\dot{x_3} = \frac{-900((K_{st} - K_{lt})x_1 - K_{st}x_3 - x_6)}{-(K_{st} - K_{lt})x_1 + K_{st}x_3 + 2x_6 - 40}$$

$$\dot{x_4} = \frac{-2(K_{se} + K_p)x_1 + K_{se}(x_2 + x_3) - B_p x_4}{J_p}$$

$$\dot{x_5} = \frac{N_1 - x_5}{\tau_1}$$

$$\dot{x_6} = \frac{N_2 - x_6}{\tau_2}.$$

The initial conditions reflect the fact that the recti muscles are not at rest in the primary position that corresponds to looking straight ahead and are given by

$$
\begin{aligned}
x_1(0) &= x_4(0) = 0 \\
x_2(0) &= -x_3(0) = 5.6^\circ \\
x_5(0) &= x_6(0) = 20.6\text{gm.}
\end{aligned}
$$

Again, it should be noted that the nonlinearities in this system are due to the force-velocity dependence of the active muscle. As a matter of comparison we note that in [18] the muscle model that is adopted corresponds to a constant viscosity parameter and inextensible tendon (and so the parallel elastic element with spring constant K_s is omitted). The nonlinear nature of their system is due to the geometrical implications that arise when the moment arms of the lateral and medial rectus are not equal to the radius of the globe.

3.4 Simulations

A variety of sources provide experimental values for the parameters that appear in the model that has been presented in this paper. The data used here is taken from [3, 6]. The parameter values are $K_s = 2.5$ gm/°, $K_{lt} = 1.2$ gm/°, $K_p = .5$ gm/°, $J = 4.3(10^{-5})$ gm/s^2, $B_p = .06$ gm/°. There is not universal agreement on the relationship between motoneuronal firing and saccade duration and amplitude. The following expressions used to represent the nervous activity are obtained from electromyographic studies and have been converted into units of force [6]: $N_{1-pulse} = 55 + 11\theta$ gm, $N_{1-step} = 20.6 + 2.35\theta$ gm, $N_{2-pulse} = .5$ gm, $N_{2-step} = 20.6 - .74\theta$ gm, $t_{1a} = 11.7 - .2\theta$ ms, $t_{1d} = .2$ ms, $t_{2a} = 2.4$ ms, $t_{2d} = 1.9$ ms, $T_1 = 10 + \theta$ ms, $T_2 = T_1 + 6$ ms. In the above expression, θ denotes the magnitude of the saccade.

The system of 6 coupled nonlinear differential equations for the state variables $x_1, \cdots, x_6$ is solved numerically when the control input corresponds to agonist activity that begins at time $t = 1$. Simulations were computed for saccades of 10° to 50°. Figure 4 shows that the model provides very good agreement of the eye rotation for saccades up to 30°. A comparison with actual physiological data for saccades of greater magnitude suggests that the control signal is most likely not a single pulse-step [19]. The discrepancy for larger saccades could also be attributed to the nonlinear effects that are associated with the nontangential forces that have been described earlier and which were addressed by [18].

The phase portraits for these movements are illustrated in Figure 5. Again, when compared with the experimental results of [19], the velocities predicted by the model are in good agreement for saccades of 30° or less. The speeds are somewhat underestimated for the larger movements.

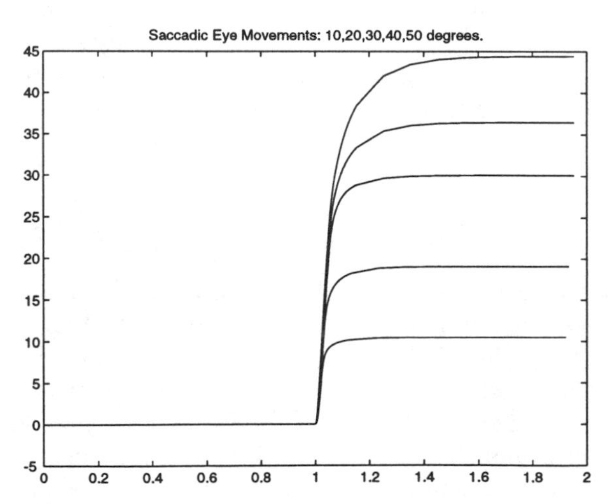

Figure 4. Simulations of saccades of 10, 20, 30, 40 and 50 degrees.

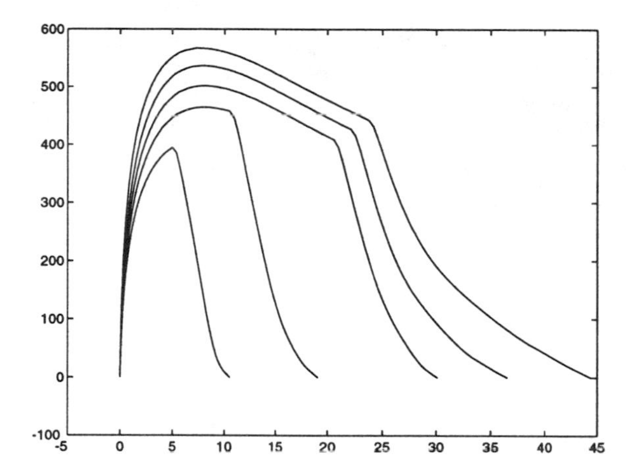

Figure 5. Phase portraits for saccades of 10, 20, 30, 40 and 50 degrees.

The agonist force that is applied to the globe through the tendon may be expressed as

$$T_1 = \frac{K_{se}}{K_{st}} F_1 - \frac{K_{se} K_{lt}}{K_{st}} \theta_3 - \frac{K_{se} 1.25}{K_{st}(\dot{\theta}_1 + 900)} \dot{\theta}_1.$$

By utilizing the approximate solution of the state equations this force is readily computed and the results are illustrated in Figure 6. The predicted forces are all within the known range of tensions that have been measured experimentally.

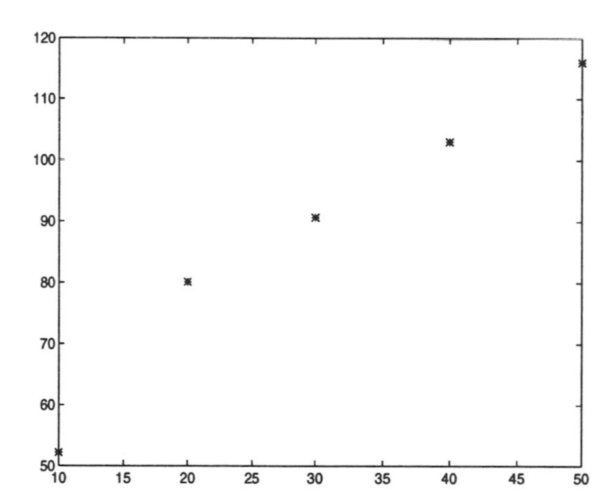

Figure 6. The developed tension on the globe versus saccade magnitude.

Several issues need to be addressed to improve upon the model that has been presented here. Data suggests that the series elastic element should be replaced by a component that represents a nonlinear dependence of force on muscle length. More importantly, it is believed that this element is parametrically modulated by the active state tension and as such, the spring stiffness K_s should be a function of nervous activity and active force. The use of a linear approximation of the force-length curve to determine the parameter K_{lt} undoubtedly introduces some error. Data suggests that the muscle force-length characteristics are better approximated by a hyperbolic form [19]. Finally, the role of feedback and the stretch reflex needs to be explored within in the context of ocular control and movement.

4 Conclusion

In this paper, we have introduced a dynamical systems approach in order to identify the motion parameters of a moving target. We also introduced a new model for eye movement by considering the mechanics of the extraocular muscles. The problem of directing "gaze" towards a moving target may be viewed as a problem of tracking, up to direction, the motion of a moving target (exosystem) and is a subject of future research.

References

[1] W.M. Boothby. *An Iintroduction to Differential Manifolds and Riemannian Geometry.* New York: Academic Press, 1975.

[2] M.R. Clark and L. Stark. Control of human eye movements; I. Modelling of extraocular muscles; II. A model for the extraocular plant mechanism; III. Dynamic characteristics of the eye tracking mechanism. *Math. Biosc.* **20** (1974), 191-265.

[3] G. Cook and L. Stark. The human eye movement mechanism: Experiments, modeling and model testing. *Arch. Opthalmol.* **79** (1968), 428-436.

[4] W.P. Dayawansa, B.K. Ghosh, C. Martin and X. Wang. A necessary and sufficient condition for the perspective observability problem. Systems *&* Control Let. **25**(1) (1995), 159-166.

[5] R. Descartes. *Treatise of Man.* Translation by T.S. Hall. Cambridge, MA: Harvard Univ. Press, 1972.

[6] J.D. Enderle. The fast eye movement control system. *Biomedical Engineering Handbook.* (J.D. Bronsino, Ed.). Boca Raton, Florida : CRC Press, 1995. 2473-2493.

[7] B.K. Ghosh, M. Jankovic and Y.T. Wu. Perspective problems in system theory and its application to machine vision. *Journal of Mathematical Systems, Estimation and Control* **4**(1) (1994), 3-38.

[8] B.K. Ghosh and E.P. Loucks. A perspective theory for motion and shape estimation in machine vision. *SIAM Journal on Control and Optimization* **33**(5) (1995), 1530-1559.

[9] B.K. Ghosh and E.P. Loucks. A realization theory for perspective systems with applications to parameter estimation problems in machine vision. *IEEE Trans. on Aut. Contr.* **41**(12) (1996), 154-162.

[10] B.K. Ghosh and J. Rosenthal. A generalized Popov-Belevitch-Hautus test of observability. *IEEE Trans. on Automatic Control* **40**(1) (1995), 176-180.

[11] B.K. Ghosh and E.P. Loucks. Identification of parameters of linear dynamical systems with perspective observation. To appear *Syst. & Contr. Let.*

[12] P. Griffiths and J. Harris. *Principles of Algebraic Geometry.* New York: Wiley-Interscience Publication, 1978.

[13] M.L.J. Hautus. Controllability and observability condition of linear autonomous systems. *Ned. Akad. Wetenschappen, Proc. Ser. A* **72** (1969), 443-448.

[14] K. Kanatani. *Group-Theoretical Methods in Image Understanding.* New York: Springer Verlag, 1990.

[15] M. Lei. *Vision Based Robot Tracking and Manipulation.* Ph.D. Dissertation. Washington University. St. Louis, Missouri. 1994.

[16] P. Lockwood and C. Martin. Binocular observablity. *Computation and Control IV.* (K. Bowers and J. Lund, Eds.). *Progress in Systems and Control.* Boston: Birkhäuser, 1995. 219-228.

[17] E.P. Loucks, C. Martin and B.K. Ghosh. Controllability, observability and realizability of perspective dynamical systems. To appear *IEEE Trans. on Aut. Contr.*.

[18] S.Y. Lu and C.F. Martin. Dynamics of ocular motion. *Computation and Control IV.* (K. Bowers and J. Lund, Eds.). *Progress in Systems and Control.* Boston: Birkhäuser, 1995. 229-245.

[19] D.A. Robinson. The mechanics of human saccadic eye movement. *J. of Physiol.* **174** (1964), 245-264.

[20] A. M. Waxman and U. Ullman. Surface structure and 3-d motion from image flow: kinematic analysis. *Intl. J. Robotics Research* **4** (1985), 72-94.

[21] G. Westheimer. Mechanisms of saccadic eye movements. *Amer. Med. Assoc. Arch. Opthamol.* **52** (1954), 710-724.

[22] L.R. Young and L. Stark. Variable feedback experiments testing a sampled data model for eye tracking movements. *IEEE Trans. on Human Factors in Electronics* **4** (1963), 38-51.

[23] R.Y. Tsai and T.S. Huang. Estimating three dimensional motion parameters of a rigid planar patch. *IEEE Trans. on ASSP* **29** (1981), 1147-1152.

[24] F. Zajac. Muscle and tendon: properties, models, scaling and application to biomechanics and motor control. *Critical Reviews in Biomedical Engineering* **17** (1989), 359-441

Department of Systems Science and Mathematics, Washington University, St. Louis, Missouri 63130

Chiron Corporation, 4777 LeBourget Drive, St. Louis, Missouri 63134

Department of Mathematics, Texas Tech University, Lubbock, Texas 79409

Department of Mathematics, Texas Tech University, Lubbock, Texas 79409

The Jacobi Method: A Tool for Computation and Control

U. Helmke and K. Hüper[1]

1 Introduction

The interaction between numerical linear algebra and control theory has crucially influenced the development of numerical algorithms for linear systems in the past. Since the performance of a control system can often be measured in terms of eigenvalues or singular values, matrix eigenvalue methods have become an important tool for the implementation of control algorithms. Standard numerical methods for eigenvalue or singular value computations are based on the QR-algorithm. However, a number of computational problems in control and signal processing are not amenable to standard numerical theory or cannot be easily solved using current numerical software packages. Various examples can be found in the digital filter design area. For instance, the task of finding sensitivity optimal realizations for finite word length implementations requires the solution of highly nonlinear optimization problems for which no standard numerical solution of algorithms exist.

There is thus the need for a new approach to the design of numerical algorithms that is flexible enough to be applicable to a wide range of computational problems as well as has the potential of leading to efficient and reliable solution methods. In fact, various tasks in linear algebra and system theory can be treated in a unified way as optimization problems of smooth functions on Lie groups and homogeneous spaces. In this way the powerful tools of differential geometry and Lie group theory become available to study such problems. With Brockett's paper (1988) [3] as the starting point there has been ongoing success in tackling difficult computational problems by geometric optimization methods. We refer to Helmke and Moore (1994) [10] and the recent PhD theses by Smith (1993) [20], Mahony (1994) [16], Dehaene (1995) [4], and Hüper (1996) [11] for more systematic and comprehensive state of the art descriptions. Some of the further application areas where our methods are potentially useful include diverse topics such as frequency estimation, principal component analysis, perspective motion problems in computer vision, pose estimation, system approximation, model reduction, computation of canonical forms and feed-

[1]This work was partially supported by grant I-0184-078.06/91 from the German-Israeli-Foundation for Scientific Research and Development.

back controllers, balanced realizations, Riccati equations, and structured eigenvalue problems.

In this paper a generalization of the classical Jacobi method for symmetric matrix diagonalization, see Jacobi (1846) [13], is considered that is applicable to a wide range of computational problems. In recent years, Jacobi-type methods have gained increasing interest, due to superior accuracy properties and inherent parallelism as compared to QR-based methods. The classical Jacobi method successively decreases the sum of squares of the off-diagonal elements of a given symmetric matrix to compute the eigenvalues. Similar extensions exist to compute eigenvalues or singular values of arbitrary matrices. Instead of using a special cost function such as the off-diagonal norm in Jacobi's method, other classes of cost functions are feasible as well. We consider a class of perfect Morse-Bott functions on homogeneous spaces that are defined by unitarily invariant norm functions or by linear trace functions. In addition to gaining further generality, our choice of functions leads to an elegant theory as well as yielding improved convergence properties for the resulting algorithms.

Rather than trying to develop the new Jacobi method in full generality on arbitrary homogeneous spaces, we demonstrate its applicability by means of examples from linear algebra and system theory. New classes of Jacobi-type methods for symmetric matrix diagonalization, balanced realization, and sensitivity optimization are obtained. In comparison with standard numerical methods for matrix diagonalization, the new Jacobi-method has the advantage of achieving automatic sorting of the eigenvalues. This sorting property is particularly important towards applications in signal processing; i.e., frequency estimation or estimation of dominant subspaces.

This is a survey paper. Complete proofs of the results reported here have been or will be published elsewhere.

2 Optimization on Homogeneous Spaces

In this section, following Helmke (1993) [8], we describe a rather general Lie group approach to optimization problems of interest in linear algebra as well as in control theory.

Let G be a real reductive Lie group and $K \subset G$ a maximal compact subgroup. Let

$$\alpha : G \times V \to V, \quad (g, x) \mapsto g \cdot x \tag{2.1}$$

be a linear algebraic action of G on a finite dimensional vector space V. Given an element $x \in V$, the subset of V

$$G \cdot x = \{g \cdot x | g \in G\}$$

is called an *orbit* of G. Each orbit $G\cdot x$ of such a real algebraic group action then is a smooth submanifold of V that is diffeomorphic to the homogeneous space G/H, with $H := \{g \in G | g\cdot x = x\}$ the stabilizer subgroup. We are interested in understanding the structure of critical points of a smooth proper function $f: G\cdot x \to \mathbb{R}_+$ defined on orbits $G\cdot x$. Some of the interesting cases actually arise when f is defined by a norm function on V. Thus given a positive definite inner product $\langle\, ,\, \rangle$ on V let $\|x\|^2 = \langle x, x\rangle$ denote the associated Hermitian norm. An Hermitian norm on V is called $K-$invariant if

$$\langle k\cdot x, k\cdot y\rangle = \langle x, y\rangle \tag{2.2}$$

holds for all $x, y \in V$ and all $k \in K$, for K a maximal compact subgroup of G. Fix any such $K-$invariant Hermitian norm on V. For any $x \in V$ we consider the smooth distance function on $G\cdot x$ defined as

$$\phi: G\cdot x \to \mathbb{R}_+, \quad \phi(g\cdot x) = \|g\cdot x\|^2. \tag{2.3}$$

We then have the following result due to Kempf and Ness (1979) [14]. For an important generalization to plurisubharmonic functions on complex homogeneous spaces, see Azad and Loeb (1990) [2].

Theorem 2.1 *(a) The norm function $\phi: G\cdot x \to \mathbb{R}_+$, $\phi(g\cdot x) = \|g\cdot x\|^2$, has a critical point if and only if the orbit $G\cdot x$ is a closed subset of V.*

(b) Let $G\cdot x$ be closed. Every critical point of $\phi : G\cdot x \to \mathbb{R}_+$ is a global minimum and the set of global minima is a single uniquely determined $K-$orbit.

(c) If $G\cdot x$ is closed, then $\phi : G\cdot x \to \mathbb{R}_+$ is a perfect Morse-Bott function. The set of global minima is connected.

Theorem 2.1 completely characterizes the critical points of $K-$invariant Hermitian norm functions on $G-$orbits $G\cdot x$ of a reductive Lie group G. Similar results are available for compact groups. We describe such a result in a special situation which suffices for the subsequent examples. Thus let G now be a compact semisimple Lie group with Lie algebra $\mathfrak{g}$. Let

$$\alpha: G \times \mathfrak{g} \to \mathfrak{g}, \quad (g, x) \mapsto g\cdot x = \mathrm{Ad}\,(g)x \tag{2.4}$$

denote the adjoint action of G on its Lie algebra. Let $G\cdot x$ denote an orbit of the adjoint action and let

$$(x, y) := -\mathrm{tr}\,(\mathrm{ad}_x \circ \mathrm{ad}_y) \tag{2.5}$$

denote the Killing form on $\mathfrak{g}$. Then for any element $a \in \mathfrak{g}$ the trace function

$$f_a: G\cdot x \to \mathbb{R}_+,\; f_a(g\cdot x) = -\mathrm{tr}\,(\mathrm{ad}_a \circ \mathrm{ad}_{g\cdot x}) \tag{2.6}$$

defines a smooth function on $G \cdot x$. For a proof of the following result, formulated for orbits of the co-adjoint action, we refer to Atiyah (1982) [1], Guillemin and Sternberg (1982) [6].

Theorem 2.2 *Let G be a compact, connected, and semisimple Lie group over $\mathbb{C}$ and let $f_a : G \cdot x \to \mathbb{R}_+$ be the restriction of a linear function on a co-adjoint orbit, defined via evaluation with an element a of the Lie algebra. Then*

(a) $f_a : G \cdot x \to \mathbb{R}$ is a perfect Morse-Bott function.

(b) If $f_a : G \cdot x \to \mathbb{R}$ has only finitely many critical points, then there exists a unique local=global minimum. All other critical points are saddle points or maxima.

Suppose now in an optimization exercise we want to compute the set of critical points of a smooth function $\phi : G \cdot x \to \mathbb{R}_+$, defined on an orbit of a Lie group action. Thus let G denote a Lie group acting smoothly on a finite dimensional vector space V. For $x \in V$ let $G \cdot x$ denote an orbit. Let $\Omega_1, \ldots, \Omega_N$ denote a basis of the Lie algebra $\mathfrak{g}$ of G, with $N = \dim G$. Denote by $G_i(t) = \exp(t\Omega_i), t \in \mathbb{R}$, the associated one parameter subgroups of G. We refer to $G_1(t), \ldots, G_N(t)$ as the *basic* transformations of G. The proposed algorithm for minimizing a smooth proper function $\phi : G \cdot x \to \mathbb{R}_+$ then consists of a recursive application of so-called sweep operations. The algorithm is termed a *Jacobi-type* algorithm.

Jacobi Sweep. Define

$$\begin{aligned} x_k^{(1)} &:= G_1(t_*^{(1)}) \cdot x_k \\ x_k^{(2)} &:= G_2(t_*^{(2)}) \cdot x_k^{(1)} \\ &\vdots \\ x_k^{(N)} &:= G_N(t_*^{(N)}) \cdot x_k^{(N-1)} \end{aligned}$$

where for $i = 1, \ldots, N$

$$t_*^{(i)} := \arg\min_{t \in \mathbb{R}} \{\phi(G_i(t) \cdot x_k^{(i-1)})\}$$

if

$$\phi(G_i(t) \cdot x_k^{(i-1)}) \not\equiv \phi(x_k^{(i-1)}),$$

and

$$t_*^{(i)} := 0$$

otherwise.

Thus $x_k^{(i)}$ is recursively defined as the minimum of the smooth function $\phi : G\cdot x \to \mathbb{R}$, when restricted to the i–th curve $\{G_i(t)\cdot x_k^{(i-1)} | t \in \mathbb{R}\}$ containing $x_k^{(i-1)}$. The algorithm then consists of the iteration of sweeps.

Jacobi-type Algorithm on the Homogeneous Space $G\cdot x$.

- Let $x_0, \ldots, x_k \in G\cdot x$ be given for $k \in \mathbb{N}$.
- Define the recursive sequence $x_k^{(1)}, x_k^{(2)}, \ldots, x_k^{(N)}$ as above (sweep).
- Set $x_{k+1} := x_k^{(N)}$. Proceed with the next sweep.

Proposition 2.1 *Let $\phi : G\cdot x \to \mathbb{R}_+$ be a smooth proper function and let $x_k \in G\cdot x$, $k \in \mathbb{N}$, denote the sequence generated by the Jacobi algorithm. Then x_k converges to the intersection of the set of critical points of $\phi : G\cdot x \to \mathbb{R}_+$ with a level set.*

The proof is by a familiar Lyapunov-type argument. Better results can be obtained in special cases, such as for the Hermitian norm or trace functions, but we will not go into this here. Although this algorithm, or actually the variants discussed below, may look naive at first glance, we will show that fast competitive numerical algorithms can be obtained in this way.

3 Applications to Linear Algebra and Control

The purpose of this section is to show that important questions from linear algebra and system theory can be solved by specialization of the previous general theory. Thus the tasks of symmetric matrix diagonalization, the singular value decomposition, balancing and sensitivity minimization are reformulated equivalently as the solutions to matrix optimization problems.

3.1 Symmetric Matrix Diagonalization

A core problem in numerical linear algebra is the diagonalization of a real symmetric matrix by orthogonal transformations. Here we show that this problem is equivalent to solving a matrix least squares problem. Let

$$\Lambda := \operatorname{diag}(\lambda_1 I_{n_1}, \ldots, \lambda_r I_{n_r})$$

be a real diagonal $n\times n$ matrix with eigenvalues $\lambda_1 > \cdots > \lambda_r$ and multiplicities satisfying $n_1 + \cdots + n_r = n$. Let $\mathcal{O}(\Lambda)$ denote the isospectral set

$$\mathcal{O}(\Lambda) = \{X \in \mathbb{R}^{n\times n} \mid X = \Theta\Lambda\Theta', \Theta \in O(n)\}$$

of real symmetric $n \times n$ matrices with eigenvalues $\lambda_1,\ldots,\lambda_r$. Then $\mathcal{O}(\Lambda)$ is a smooth, compact, and connected manifold. Moreover, $\mathcal{O}(\Lambda)$ is the Λ–orbit of the $O(n)$–action on real symmetric matrices $(\Theta, X) \mapsto \Theta X \Theta'$. Given any symmetric matrix $N \in \mathbb{R}^{n\times n}$ consider the smooth least squares distance function

$$f_N : \mathcal{O}(\Lambda) \to \mathbb{R}, \; f_N(X) = \|N - X\|^2 \tag{3.1}$$

with $\|X\|^2 = \operatorname{tr}(XX')$. Choose $N = \operatorname{diag}(\mu_1,\ldots,\mu_n)$, with $\mu_1 > \cdots > \mu_n$ for matrix diagonalization. Theorem 3.1 is a slight extension of a result of [3].

Theorem 3.1 *([10]). Let N, Λ as above. Then*

(a) $f_N : \mathcal{O}(\Lambda) \to \mathbb{R}$ has exactly $\dfrac{n!}{n_1! \cdots n_r!}$ critical points. These are diagonal matrices $X = \operatorname{diag}(\lambda_{\pi(1)}, \ldots, \lambda_{\pi(n)})$, for π a permutation.

(b) There is a unique local = global minimum $X = \operatorname{diag}(\lambda_1, \ldots, \lambda_n)$, and a unique local = global maximum $X = \operatorname{diag}(\lambda_n, \ldots, \lambda_1)$ with $\lambda_1 \geq \cdots \geq \lambda_n$. All other critical points are saddle points.

3.2 Singular Value Decomposition

The approach of the last section can be easily adapted to the singular value decomposition of rectangular matrices. Let $N, \Sigma \in \mathbb{R}^{n\times m}$, $n \leq m$, be matrices of the form

$$N = (\operatorname{diag}(\mu_1, \ldots, \mu_n), 0_{n\times(m-n)}), \quad \Sigma = (\operatorname{diag}(\sigma_1 I_{n_1}, \ldots, \sigma_r I_{n_r}), 0_{n\times(m-n)})$$

with $\sigma_1 > \cdots > \sigma_r \geq 0$, $\mu_1 > \cdots > \mu_n > 0$, and $n_1 + \cdots + n_r = n$. The Lie group $O(n) \times O(m)$ acts on $\mathbb{R}^{n\times m}$ by

$$\alpha : O(n) \times O(m) \times \mathbb{R}^{n\times m} \to \mathbb{R}^{n\times m}, \quad ((U, V), X) \to UXV'. \tag{3.2}$$

The set $\mathcal{O}(\Sigma)$ of $n \times m$ real matrices X with singular values $\sigma_1, \ldots, \sigma_n$

$$\mathcal{O}(\Sigma) = \{X = U\Sigma V' | U \in O(n), \, V \in O(m)\} \tag{3.3}$$

is an orbit of α and therefore $\mathcal{O}(\Sigma)$ is a smooth, compact manifold. Consider the task of minimizing the smooth least squares distance function

$$F_N : \mathcal{O}(\Sigma) \to \mathbb{R}, \quad F_N(X) = \|N - X\|^2. \tag{3.4}$$

We assume $n < m$, in which case $\mathcal{O}(\Sigma)$ is connected. For a proof of Theorem 3.2 see [9], for a complete discussion of the general case see [11].

Theorem 3.2 *Let $n < m$.*

(a) A matrix $X \in \mathcal{O}(\Sigma) = \{U\Sigma V' | U \in O(n),\ V \in O(m)\}$ *is a critical point of* $F_N : \mathcal{O}(\Sigma) \to \mathbb{R}$ *if and only if* $NX' = XN'$ *and* $N'X = X'N$ *hold.* X *is a critical point if and only if*

$$X = (\operatorname{diag}(\varepsilon_1 \sigma_{\pi(1)}, \ldots, \varepsilon_n \sigma_{\pi(n)}), 0_{n\times(m-n)})$$

with $\varepsilon_i \in \{\pm 1\}$ *and* π *a permutation.*

(b) There is a unique local=global minimum

$$X = (\operatorname{diag}(\sigma_1 I_{n_1}, \ldots, \sigma_r I_{n_r}), 0_{n\times(m-n)}).$$

All other critical points are saddle points (or maxima).

3.3 Balancing and Sensitivity Optimization

Turning to applications in control theory we consider the tasks of computing balanced and sensitivity optimal realizations of a linear system. A general cost function approach for balancing and sensitivity is presented which is based on [7], [8], and [10].

We begin with a general discussion of finding realizations that minimize Hermitian norm functions.

3.3.1 Norm Balancing

Consider the real vector space of matrix triples (A, B, C)

$$L(n, m, p) = \{(A, B, C) \in \mathbb{R}^{n\times n} \times \mathbb{R}^{n\times m} \times \mathbb{R}^{p\times n}\}.$$

The Lie group $GL(n)$ of invertible $n \times n$-matrices acts on $L(n, m, p)$ via the similarity action

$$\sigma : GL(n) \times L(n, m, p) \to L(n, m, p),\ (T, (A, B, C)) \mapsto (TAT^{-1}, TB, CT^{-1}).$$

The orbits of σ

$$\mathcal{O}(A, B, C) = \{(TAT^{-1}, TB, CT^{-1}) | T \in GL(n)\} \tag{3.5}$$

are smooth submanifolds of the Euclidean space $L(n, m, p)$. Given a positive definite inner product $\langle\ ,\ \rangle$ on $L(n, m, p)$ let

$$\|(A, B, C)\| = \langle (A, B, C), (A, B, C) \rangle^{\frac{1}{2}} \tag{3.6}$$

denote the associated Hermitian norm on $L(n, m, p)$. A Hermitian norm on $L(n, m, p)$ is called *unitarily invariant* if

$$\|(TAT^{-1}, TB, CT^{-1})\| = \|(A, B, C)\| \tag{3.7}$$

holds for all $(A,B,C) \in L(n,m,p)$ and all orthogonal $T \in O(n)$. Every Hermitian norm then defines a smooth function $f : \mathcal{O}(A,B,C) \to \mathbb{R}$,

$$f(TAT^{-1}, TB, CT^{-1}) = \|(TAT^{-1}, TB, CT^{-1})\|^2. \tag{3.8}$$

We are interested in characterizing the critical points of such norm functions on $\mathcal{O}(A,B,C)$. We call a realization $(F,G,H) \in \mathcal{O}(A,B,C)$ *norm balanced* and *norm minimal*, respectively, if the norm function $f : \mathcal{O}(A,B,C) \to \mathbb{R}$, $f(F,G,H) = \|(F,G,H)\|^2$, has a critical point or a global minimum at (F,G,H), respectively. For a proof of the following results we refer to [10].

Theorem 3.3 *Denote by $\|\cdot\|$ a unitarily invariant Hermitian norm on $L(n,m,p)$. Let $f : \mathcal{O}(A,B,C) \to \mathbb{R}$ be defined as $f(F,G,H) = \|(F,G,H)\|^2$. Then*

(a) there exists a global minimum $f(F,G,H) \in \mathcal{O}(A,B,C)$ if and only if (A,B,C) is similar to

$$\left(\begin{bmatrix} A_{11} & 0 \\ 0 & A_{22} \end{bmatrix}, \begin{bmatrix} B_1 \\ 0 \end{bmatrix}, \begin{bmatrix} C_1 & 0 \end{bmatrix} \right)$$

where (A_{11}, B_1, C_1) is controllable and observable and A_{22} is diagonalizable over $\mathbb{C}$.

(b) Suppose a global minimum exists. Then the critical points of $f : \mathcal{O}(A,B,C) \to \mathbb{R}$ coincide with the global minima. Any two norm minimal realizations $(A_1,B_1,C_1), (A_2,B_2,C_2) \in \mathcal{O}(A,B,C)$ are orthogonally similar via $T \in O(n)$ as $(A_2, B_2, C_2) = (TA_1T^{-1}, TB_1, C_1T^{-1})$.

Theorem 3.3 can be extended in various directions, see [10]. First, the norm function can be replaced by an arbitrary smooth unitarily invariant strictly plurisubharmonic function $f : \mathcal{O}(A,B,C) \to \mathbb{R}$. For any such function part (b) remains in force. This covers, for example, the trace function

$$f_N : \mathcal{O}(A,B,C) \to \mathbb{R},\ \mho_N(\mathbb{A},\mathbb{B},\mathbb{C}) := \mathrm{tr}\,(\mathbb{N}(\mathbb{W} + \mathbb{W}_{\rtimes})) \tag{3.9}$$

where W_c and W_o denote the controllability and observability Gramians of (A,B,C) and $N = N' \in \mathbb{R}^{\ltimes \times \ltimes}$. A second generalization is towards norm minimal realizations with prescribed symmetry properties, such as, e.g., for Hamiltonian or signature symmetric realizations

$$(AJ)' = AJ, \qquad C' = JB, \qquad J = \begin{bmatrix} 0 & I \\ -I & 0 \end{bmatrix}$$

$$(AI_{pq})' = AI_{pq}, \quad C' = I_{pq}B, \quad I_{pq} = \mathrm{diag}\,(I_p, -I_q).$$

In these cases the Lie group $GL(n)$ has to be replaced by the symplectic group $Sp(n,\mathbb{R})$ or the pseudo-orthogonal group $O(p,q)$, respectively. Again, the above result can be extended to such situations, see [10].

Theorem 3.4 *(a) For the standard Euclidean norm on $L(n,m,p)$*

$$\|(A,B,C)\|^2 := \operatorname{tr}(AA' + BB' + C'C). \qquad (3.10)$$

a realization $(F,G,H) \in \mathcal{O}(A,B,C)$ is norm minimal if and only if $FF' + GG' = F'F + H'H$.

(b) Two norm minimal realizations $(F_1,G_1,H_1),(F_2,G_2,H_2) \in \mathcal{O}(A,B,C)$ are equivalent via an orthogonal transformation $T \in O(n)$

$$(F_1, G_1, H_1) = (TF_2T^{-1}, TG_2, H_2T^{-1}).$$

In particular, a norm minimal realization (F,G,H) exists if and only if (A,B,C) satisfies part (a) of Theorem 3.3. Moreover, for the special case where $B = C = 0$ this implies a classical result by Schur that the similarity orbit $\mathcal{O}(A) = \{TAT^{-1} | T \in GL(n)\}$ of a matrix A has an element $F \in \mathcal{O}(A)$ of minimal Frobenius norm if and only if A is similar to a normal matrix.

3.3.2 Balanced Realizations

We consider the task of computing balanced realizations for asymptotically stable discrete-time systems

$$x_{k+1} = Ax_k + Bu_k, \quad y_k = Cx_k. \qquad (3.11)$$

Thus A is assumed to be Schur-stable; i.e., $\rho(A) < 1$ holds for the spectral radius. The continuous-time case can be treated similarly. While numerical methods for computing balanced realizations are well-known, see Laub *et al.* (1987) [15], Safonov *et al.* (1989) [19], our purpose here is to show that efficient algorithms for balancing can also be obtained using Jacobi-type methods.

Recall a discrete-time stable realization (A,B,C) with associated controllability and observability Gramians

$$W_c = \sum_{k=0}^{\infty} A^k BB'(A')^k, \quad W_o = \sum_{k=0}^{\infty} (A')^k C'CA^k, \qquad (3.12)$$

respectively, is called balanced if

$$W_c = W_o = \operatorname{diag}(\sigma_1 I_{n_1}, \ldots, \sigma_r I_{n_r}) \qquad (3.13)$$

with $\sigma_1 > \cdots > \sigma_r \geq 0$ being the singular values and multiplicities satisfying $n_1 + \cdots + n_r = n$. For (A,B,C) a stable realization with Gramians W_c, W_o and $(TAT^{-1}; TB, CT^{-1})$ an equivalent balanced realization the Gramians satisfy

$$TW_cT' = T'^{-1}W_oT^{-1} = \text{diagonal}.$$

The matrix $T \in GL(n)$ is called a balancing transformation. Sufficient but not necessary conditions for the existence of balancing transformations are that (A, B, C) is controllable and observable.

Instead of defining balanced realizations as having equal and diagonal controllability and observability Gramians, one can equivalently characterize them as global minima of a cost function. This is the variational approach to balancing as developed by Mullis and Roberts (1976) [17], [8], and [10]. For this consider the smooth weighted trace function

$$f_N : \mathcal{O}(A, B, C) \to \mathbb{R}_+, \ \mho_{\mathbb{N}}(\mathbb{A}, \mathbb{B}, \mathbb{C}) = \operatorname{tr}(\mathbb{N}(\mathbb{W} + \mathbb{W}_{\rtimes})) \tag{3.14}$$

with $N = N'$. We choose $N = \operatorname{diag}(\mu_1, \ldots, \mu_n)$ with $0 < \mu_1 < \cdots < \mu_n$.

Theorem 3.5 *Let (A, B, C) be controllable and observable. Consider the weighted trace function $f_N : \mathcal{O}(A, B, C) \to \mathbb{R}$ defined by $f_N(A, B, C) = \operatorname{tr}(N(W_c + W_o))$ where W_c and W_o are the controllability and observability Gramians, respectively, and N as above. Then*

(a) $f_N : \mathcal{O}(A, B, C) \to \mathbb{R}$ has compact sublevel sets and a minimum exists.

(b) A realization $(F, G, H) \in \mathcal{O}(A, B, C)$ is a critical point of f_N if and only if (F, G, H) is balanced. The local minima coincide with the global minima of f_N. The global minima are characterized by $W_c = W_o = \operatorname{diag}(\sigma_1, \ldots, \sigma_n)$ with $\sigma_1 \geq \cdots \geq \sigma_n$.

Similarly, if $N = I_n$ is chosen as the identity matrix, then the critical points of $f : \mathcal{O}(A, B, C) \to \mathbb{R}$, $f(A, B, C) = \operatorname{tr}(W_c + W_o)$ are exactly the pre-balanced realizations for which the Gramians W_c and W_o are equal (but not necessarily diagonal).

3.3.3 L^2–Sensitivity Minimization

An important issue in robust digital filter design is that of minimizing sensitivity measures of the transfer function subject to coefficient errors of the realizations. A natural sensitivity measure of a realization is the L^2–index or related indices as in Thiele (1986) [21]; see also Gevers and Li (1993) [5]. For simplicity we focus on stable SISO systems ($\rho(A) < 1$)

$$x_{k+1} = Ax_k + bu_k, \quad y_k = cx_k \tag{3.15}$$

with $(A, b, c) \in L(n, 1, 1)$ and associated transfer function $G(z) = c(zI - A)^{-1}b$. The L^2–sensitivity measure of (A, b, c) then is defined as

$$\mathcal{S}(A, b, c) = \left\| \frac{\partial G}{\partial A}(z) \right\|^2 + \left\| \frac{\partial G}{\partial b}(z) \right\|^2 + \left\| \frac{\partial G}{\partial c}(z) \right\|^2$$

where

$$\|X(z)\|^2 := \frac{1}{2\pi i} \oint \operatorname{tr}\,(X(z)X^*(z))\, \frac{\mathrm{d}\,z}{z}$$

denotes the L^2−norm of a stable rational function on the unit circle. Using standard rules of matrix calculus, we have the well-known formulae for the partial derivatives of transfer functions with respect to A, b, c as

$$\begin{aligned}\left(\frac{\partial G}{\partial A}\right)'(z) &= (zI-A)^{-1}bc(zI-A)^{-1} =: \mathcal{A}(z)\\ \left(\frac{\partial G}{\partial b}\right)'(z) &= c(zI-A)^{-1} =: \mathcal{C}(z)\\ \left(\frac{\partial G}{\partial c}\right)'(z) &= (zI-A)^{-1}b =: \mathcal{B}(z).\end{aligned}$$

Therefore, after a simple manipulation of terms

$$\begin{aligned}\mathcal{S}(A,b,c) &= \|\mathcal{A}(z)\|^2 + \|\mathcal{C}(z)\|^2 + \|\mathcal{B}(z)\|^2\\ &= \frac{1}{2\pi i}\oint \operatorname{tr}\left(\mathcal{A}(z)\mathcal{A}^*(z)\right)\frac{\mathrm{d}\,z}{z} + \operatorname{tr} W_c + \operatorname{tr} W_o\end{aligned} \tag{3.16}$$

where W_c and W_o are the controllability and observability Gramians, respectively. In particular, the cost function $\operatorname{tr}(N(W_c+W_o))$, $N=N'>0$, for balancing is seen as the weighted sensitivity measure subject to errors in the input and output vectors b and c. The occurrence of the first term $\|\mathcal{A}(z)\|^2$ in the above formula makes the minimization problem much more complicated. The above L^2−index defines a smooth function $\mathcal{S} : \mathcal{O}(A,b,c) \to \mathbb{R}_+$ defined by

$$\mathcal{S}(TAT^{-1}, Tb, cT^{-1}) = \|T\mathcal{A}(z)T^{-1}\|^2 + \operatorname{tr}(TW_cT') + \operatorname{tr}(T'^{-1}W_oT^{-1}),$$

on the space of realizations of $G(z)$. We refer to $\mathcal{S}$ as the L^2−*sensitivity* function. A realization $(F,g,h) \in \mathcal{O}(A,b,c)$ is called *sensitivity minimal* if it minimizes the L^2−sensitivity function on the orbit $\mathcal{O}(A,b,c)$. The basic existence and uniqueness properties of sensitivity optimal realizations are summarized in the following result, see [5], [7], and [10].

Proposition 3.1 *Let (A,b,c) be a discrete-time stable controllable and observable realization of a transfer function $G(z)$. Then the L^2−sensitivity function $\mathcal{S}:\mathcal{O}(A,b,c) \to \mathbb{R}_+$ achieves a minimum. The global minima are the only critical points of $\mathcal{S}$. Moreover, two realizations (A_1,b_1,c_1),(A_2,b_2,c_2) $\in \mathcal{O}(A,b,c)$ of $G(z)$ that minimize the L^2−sensitivity function are similar by a unique orthogonal similarity transformation $T \in O(n)$*

$$(A_2,b_2,c_2) = (TA_1T^{-1}; Tb_1, c_1T^{-1}).$$

Although the L^2−sensitivity index differs substantially from the cost index used for balancing, there is a similar interpretation of sensitivity minimal realizations in terms of *sensitivity* Gramians. The *controllability* and *observability sensitivity* Gramians $\widetilde{W}_c$ and $\widetilde{W}_o$ are defined as

$$\widetilde{W}_c := \frac{1}{2\pi i}\oint (\mathcal{A}(z)\mathcal{A}^*(z) + \mathcal{B}(z)\mathcal{B}^*(z))\frac{\mathrm{d}z}{z},$$

$$\widetilde{W}_o := \frac{1}{2\pi i}\oint (\mathcal{A}^*(z)\mathcal{A}(z) + \mathcal{C}^*(z)\mathcal{C}(z))\frac{\mathrm{d}z}{z}.$$

These sensitivity Gramians generalize the standard controllability and observability Gramians. In contrast to the simple covariance properties

$$(W_c, W_o) \mapsto (TW_cT', T'^{-1}W_oT^{-1})$$

of the Gramians under state space transformations the sensitivity Gramians $\widetilde{W}_c, \widetilde{W}_o$ do not transform similarly. In analogy to the familiar notion of balancing, we introduce the notion of a sensitivity balanced realization.

Definition 2.1 A realization (A, b, c) of a transfer function $G(z)$ is called *sensitivity* balanced, if the sensitivity Gramians are equal and diagonal:

$$\widetilde{W}_c = \widetilde{W}_o = \widetilde{\Sigma} = \mathrm{diag}\,(\tilde{\sigma}_1, \ldots, \tilde{\sigma}_n) > 0.$$

Thus $\tilde{\sigma}_1, \ldots, \tilde{\sigma}_n$ are called the sensitivity singular values of $G(z)$.

Given a sensitivity balanced realization one can now do sensitivity balanced model reduction based on sensitivity balanced truncation. These sensitivity balanced truncations seem to have some advantages compared with standard balanced truncation. However, additional problems arise due to the fact that the truncated subsystems of a sensitivity balanced realization are not in general sensitivity balanced, see Yan and Moore (1992) [22].

4 Algorithms

Using the cost function approach developed in Section 3, we can now put the Jacobi-type algorithm in concrete terms. For space reasons we focus on the tasks of diagonalization, balancing, and sensitivity optimization.

4.1 Sort-Jacobi Algorithm for Symmetric Matrix Diagonalization

In this example the theory is particularly simple and elegant. We apply the Jacobi method introduced in Section 2 to minimize the cost function $f_N : \mathcal{O}(\Lambda) \to \mathbb{R}$, $f_N(X) = \|N - X\|^2$ on the homogeneous space $\mathcal{O}(\Lambda)$. Let

$\Omega_{pq} = e_q e_p' - e_p e_q'$, for $1 \leq p < q \leq n$, denote the standard basis vectors of the Lie algebra $\mathfrak{so}(n)$. Consider the associated orthogonal rotations in $\mathbb{R}^{n}$

$$G_{pq}(t) := \exp(t\Omega_{pq})$$

with (p,q)–th submatrix $\begin{bmatrix} \cos t & -\sin t \\ \sin t & \cos t \end{bmatrix}$. Using an arbitrary ordering of $\{(p,q)\in\mathbb{N}^2 \mid 1\leq p<q\leq n\}$ we denote by $G_1(t),\ldots,G_N(t)$ the $N=n(n-1)/2$ different orthogonal matrices. The iterates of the Jacobi algorithm then are recursively defined as the minima of the cost function $f_N : \mathcal{O}(\Lambda) \to \mathbb{R}$, restricted to the closed geodesics $\{G_{pq}(t)XG_{pq}(-t) | t\in\mathbb{R}\}$. The convergence properties of this algorithm are established by Theorem 4.1, see [11], [12].

Theorem 4.1 *Let $X_0 = X_0' \in \mathbb{R}^{n\times n}$ be given and $N = \mathrm{diag}\,(\mu_1,\ldots,\mu_n)$ with $\mu_1 > \cdots > \mu_n$. Let $f_N : \mathcal{O}(X_0) \to \mathbb{R}$, $f_N(X) = \|N - X\|^2$ be the smooth function to be minimized on $\mathcal{O}(X_0)$. Let X_k, $k = 0,1,2,\ldots$, denote the sequence of isospectral symmetric matrices generated by the Jacobi algorithm. Then*

(a) global convergence holds, that is $\lim_{k\to\infty} X_k = \mathrm{diag}\,(\lambda_1,\ldots,\lambda_n)$ with $\lambda_1 \geq \cdots \geq \lambda_n$.

(b) The algorithm is quadratically convergent.

Similar results hold for the singular value decomposition, see [11] and [12] for details. An advantage of our method, in comparison to the conventional approach based on off-norm minimization, is that the algorithm converges quadratically fast even in the case of multiple eigenvalues or singular values. Furthermore, for the singular value decomposition of non-square matrices, no preparatory QR-decomposition is needed in order to ensure global convergence. The proposed algorithm incorporates an additional sorting property which is particularly useful for applications. Thus eigenvalues or singular values are automatically sorted by the algorithm with respect to *any* preassigned ordering. The new Jacobi-type algorithm, having an elegant convergence theory, appears to be the first efficient scheme developed by geometric optimization methods. Recently, classes of conjugate gradient and Newton-type algorithms for symmetric matrix diagonalization have been proposed by [16] and [20], using geometric optimization methods. The bottleneck of their algorithms, however, is the necessity of online computations of matrix exponentials for dense skew-symmetric matrices. Since the computational complexity of evaluating such matrix exponentials is more or less equivalent to the diagonalization of skew symmetric matrices there is no advantage in using their methods.

4.2 A Jacobi-type Algorithm for System Balancing

We now apply the Jacobi algorithm presented in Section 2 to minimize the cost function $f_N : \mathcal{O}(A,B,C) \to \mathbb{R}_+$, $f_N(A,B,C) = \mathrm{tr}\,(N(W_c + W_o))$. A basis of $\mathfrak{gl}(n)$ is chosen, defined via the standard basis vectors $e_1, \ldots, e_n$ of $\mathbb{R}^n$. Thus the basic transformations for successive minimization of f_N are $G_{pq}(t)$, for $1 \leq p, q \leq n$ and $t \in \mathbb{R}$, where

$$G_{pq}(t) := \exp(t(e_q e_p' - e_p e_q')), \qquad \text{for } p < q \tag{4.1}$$

with (p,q)–th submatrix $\begin{bmatrix} \cos t & -\sin t \\ \sin t & \cos t \end{bmatrix}$ as above,

$$G_{pq}(t) := \exp(t(e_q e_p' + e_p e_q')), \qquad \text{for } p > q \tag{4.2}$$

is a hyperbolic rotation with (p,q)–th submatrix $\begin{bmatrix} \cosh t & -\sinh t \\ \sinh t & \cosh t \end{bmatrix}$, and

$$G_{pp}(t) := \exp(t e_p e_p'), \qquad \text{for } p = q \tag{4.3}$$

differs from I_n by the (pp)–th entry which is equal to e^t. Using an arbitrary ordering of $\{(p,q) \in \mathbb{N}^2 | 1 \leq p, q \leq n\}$ we denote by $G_1(t), \ldots, G_N(t)$ the $N = n^2$ different basic transformations. The Jacobi iterates are thus recursively defined as minima of the cost function $f_N : \mathcal{O}(A,B,C) \to \mathbb{R}$, when restricted to a finite sequence of closed geodesics diffeomorphic to S^1 and noncompact curves diffeomorphic to hyperbolae. It is easy to present computer codes for the algorithms. The first one, Algorithm 4.2, defines a function that diagonalizes symmetric 2×2–matrices such that the eigenvalues appear in decreasing order. *Any* ordering of the eigenvalues $\lambda_1, \ldots, \lambda_n$ can be achieved by specifying for each index pair (p,q) an appropriate order in the last if-else structure of Algorithm 4.2. The second and third functions determine hyperbolic and diagonal congruence transformations that minimize the trace of the sum of two symmetric positive definite 2×2–matrices. Algorithms 4.2 and 4.2 present two alternatives of a complete Jacobi scheme for computing balanced realizations with sorted Hankel singular values. In Algorithm 4.2 the usual cyclic by row scheme is used, the generalization to arbitrary cyclic schemes is straightforward. In Algorithm 4.2 a sweep is built up by three subsweeps, each consisting only of orthogonal, hyperbolic, or diagonal transformations, respectively. MATLAB-like algorithmic language and notation is used. The matrices G_{pq} appearing below differ from the identity matrix only by the (pq)–th submatrix $\begin{bmatrix} g_{pp} & g_{pq} \\ g_{qp} & g_{qq} \end{bmatrix}$ with $1 \leq p,q \leq n$. To simplify the notation we write X and Y instead of W_c and W_o, respectively. Accumulation of the congruence transformations is possible, but is not explicitly formulated in the algorithms.

Algorithm 4.1 Given $X = X' \in \mathbb{R}^{n \times n}$ and $p, q \in \mathbb{N}$ with $1 \leq p < n$ and $p+1 \leq q \leq n$, a (cos, sin)–pair is computed such that if $\widetilde{X} = G_{pq} X G_{pq}'$ then $\tilde{x}_{pq} = 0$ and $\tilde{x}_{pp} \geq \tilde{x}_{qq}$.

function: $(g_{pp}, g_{pq}, g_{qp}, g_{qq})$ = **sym.schur2.sorted** (X, p, q)

if $x_{pq} \neq 0$
  if $x_{qq} \neq x_{pp}$

$$\tau = \frac{x_{qq} - x_{pp}}{2x_{pq}};\ tan = \frac{\operatorname{sign}(\tau)}{|\tau| + \sqrt{1+\tau^2}}$$

$$cos = \frac{1}{\sqrt{1+tan^2}};\ sin = tan \cdot cos$$

  else

$$cos = \frac{1}{2}\sqrt{2};\ sin = -\operatorname{sign}(x_{pq}) \cdot cos$$

  end
else
  $cos = 1;\ sin = 0$
end
if $x_{pp} \geq x_{qq}$
  $g_{pp} = cos;\ g_{pq} = -sin;\ g_{qp} = sin;\ g_{qq} = cos$
else
  $g_{pp} = sin;\ g_{pq} = cos;\ g_{qp} = -cos;\ g_{qq} = sin$
end
end sym.schur2.sorted

Algorithm 4.2 Given two symmetric positive definite matrices $X, Y \in \mathbb{R}^{\kappa \times \kappa}$ and $p, q \in \mathbb{N}$ with $1 \leq p < n$ and $p+1 \leq q \leq n$, a $(cosh, sinh)-$pair is computed such that if $\widetilde{X} = G_{pq} X G'_{pq}$ and $\widetilde{Y} = {G'}_{pq}^{-1} X G_{pq}^{-1}$ then $\tilde{x}_{pp} + \tilde{x}_{qq} + \tilde{y}_{pp} + \tilde{y}_{qq}$ is minimized.

function: $(g_{pp}, g_{pq}, g_{qp}, g_{qq}) =$ **sym2** (X, Y, p, q)

$$a = 2(y_{pq} - x_{pq});\ b = x_{pp} + x_{qq} + y_{pp} + y_{qq}$$

$$cosh = \frac{\sqrt{b+a} + \sqrt{b-a}}{2\sqrt[4]{b^2 - a^2}};\ sinh = \frac{\sqrt{b+a} - \sqrt{b-a}}{2\sqrt[4]{b^2 - a^2}}$$

$g_{pp} = cosh;\ g_{pq} = sinh;\ g_{qp} = sinh;\ g_{qq} = cosh$
end sym2

Algorithm 4.3 Given two symmetric positive definite matrices $X, Y \in \mathbb{R}^{\kappa \times \kappa}$ and $p \in \mathbb{N}$ with $1 \leq p \leq n$, a $g_{pp} > 0$ is computed such that if $\widetilde{X} = G_{pp} X G'_{pp}$ and $\widetilde{Y} = {G'}_{pp}^{-1} Y G_{pp}^{-1}$ then $\tilde{x}_{pp} + \tilde{y}_{pp}$ is minimized.

function: $g_{pp} =$ **diag1** (X, Y, p)

$$g_{pp} = \sqrt[4]{\frac{y_{pp}}{x_{pp}}}$$

end diag1

Algorithm 4.4: (Cyclic by Row Jacobi for Balancing) Given two symmetric positive definite matrices $X, Y \in \mathbb{R}^{\kappa \times \kappa}$, an $N = \operatorname{diag}(\mu_1, \ldots, \mu_n) \in \mathbb{R}^{\kappa \times \kappa}$ with $0 < \mu_1 < \cdots < \mu_n$, and a tolerance $\epsilon \geq 0$, this algorithm overwrites X, Y with diagonal $\Sigma = GXG' = G'^{-1}YG^{-1} = \operatorname{diag}(\sigma_1, \cdots, \sigma_n)$, where $\sigma_1 \geq \cdots \geq \sigma_n$, $G \in GL^+(n)$, and $\Delta f = \operatorname{tr} N(X + Y - 2\Sigma) \leq \epsilon$.

$fNew = trace((N(X+Y))(1:n, 1:n))$
$fOld = fNew - 2\epsilon$ (to ensure starting the while loop)
while $(fNew - fOld) > \epsilon$
  for $p = 1 : n-1$
    $g_{pp} =$ **diag1**(X, Y, p); $X = G_{pp} X G'_{pp}$; $Y = G'^{-1}_{pp} Y G^{-1}_{pp}$
      for $q = p+1 : n$
        $(g_{pp}, g_{pq}, g_{qp}, g_{qq}) =$ **sym.schur2.sorted**$(X + Y, p, q)$
        $X = G_{pq} X G'_{pq}$; $Y = G'^{-1}_{pq} Y G^{-1}_{pq}$
        $(g_{pp}, g_{pq}, g_{qp}, g_{qq}) =$ **sym2**(X, Y, p, q)
        $X = G_{pq} X G'_{pq}$; $Y = G'^{-1}_{pq} Y G^{-1}_{pq}$
      end
  end
  $g_{nn} =$ **diag1**(X, Y, n); $X = G_{nn} X G'_{nn}$; $Y = G'^{-1}_{nn} Y G^{-1}_{nn}$
  $fOld = fNew$; $fNew = trace((N(X+Y))(1:n, 1:n))$
end

Algorithm 4.5: (Jacobi for Balancing via Subsweeps) Given two symmetric positive definite matrices $X, Y \in \mathbb{R}^{\kappa \times \kappa}$, an $N = \operatorname{diag}(\mu_1, \ldots, \mu_n) \in \mathbb{R}^{\kappa \times \kappa}$ with $0 < \mu_1 < \cdots < \mu_n$, and a tolerance $\epsilon \geq 0$, this algorithm overwrites X, Y with diagonal $\Sigma = GXG' = G'^{-1}YG^{-1} = \operatorname{diag}(\sigma_1, \cdots, \sigma_n)$, where $\sigma_1 \geq \cdots \geq \sigma_n$, $G \in GL^+(n)$, and $\Delta f = \operatorname{tr} N(X + Y - 2\Sigma) \leq \epsilon$.

$fNew = trace((N(X+Y))(1:n, 1:n))$
$fOld = fNew - 2\epsilon$ (to ensure starting the while loop)
while $(fNew - fOld) > \epsilon$
  for $p = 1 : n-1$
      for $q = p+1 : n$
        $(g_{pp}, g_{pq}, g_{qp}, g_{qq}) =$ **sym.schur2.sorted**$(X + Y, p, q)$
        $X = G_{pq} X G'_{pq}$; $Y = G'^{-1}_{pq} Y G^{-1}_{pq}$
      end
  end
  for $p = 1 : n$
    $g_{pp} =$ **diag1**(X, Y, p); $X = G_{pp} X G'_{pp}$; $Y = G'^{-1}_{pp} Y G^{-1}_{pp}$

```
    end
    for p = 1 : n − 1
        for q = p + 1 : n
            (g_pp, g_pq, g_qp, g_qq) = sym2(X, Y, p, q)
            X = G_pq X G'_pq; Y = G'^{-1}_pq Y G^{-1}_pq
        end
    end
    fOld = fNew; fNew = trace((N(X + Y))(1 : n, 1 : n))
end
```

The convergence properties of Algorithms 4.2 and 4.2 are established by Theorem 4.2. By the explicit description of Algorithms 4.2-4.2, the overall schemes 4.2 and 4.2 do not depend on the specific values of the diagonal entries of N, but only on the ordering they define. This is in sharp contrast to differential equation approaches for balancing; see [10], where the entries of N in fact crucially influence the transient behavior of the flows.

Theorem 4.2 *Let $(A_0,B_0,C_0) \in L(n,m,p)$ be controllable and observable with Gramians $(W_c)_0, (W_o)_0$. Let $N = \operatorname{diag}(\mu_1, \dots, \mu_n)$ with $0 < \mu_1 < \cdots < \mu_n$. Let $((W_c)_k, (W_o)_k)$, $k = 0, 1, 2, \dots$, denote the sequence of positive definite symmetric matrices generated by the Algorithms 4.2 or 4.2. Then*

(a) global convergence holds, i.e.,

$$\lim_{k\to\infty} ((W_c)_k, (W_o)_k) = (\operatorname{diag}(\sigma_1, \dots, \sigma_n), \operatorname{diag}(\sigma_1, \dots, \sigma_n))$$

with $\sigma_1 \geq \cdots \geq \sigma_n > 0$. The σ_i are the Hankel singular values of (A_0,B_0,C_0).

(b) The convergence of the Gramians to $\operatorname{diag}(\sigma_1, \dots, \sigma_n))$ *is quadratically fast.*

Proof. We only give a sketch of the proof. Consider the homogeneous space $\mathcal{O}(\Sigma,\Sigma) = \{(T\Sigma T', T'^{-1}\Sigma T^{-1}) | T \in GL(n)\}$, with $\Sigma = \operatorname{diag}(\sigma_1, \dots, \sigma_n)$. Consider the smooth cost function $g_N : \mathcal{O}(\Sigma,\Sigma) \to \mathbb{R}$ defined by $g_N(W_c,W_o) = \operatorname{tr}(N(W_c + W_o))$. By Theorem 3.5 the only local=global minimum of g_N is given by (Σ, Σ) with $\Sigma = \operatorname{diag}(\sigma_1, \dots, \sigma_n))$ and $\sigma_1 \geq \cdots \geq \sigma_n > 0$. It is then easily seen by inspection that the unique fixed point of the algorithm is the global minimum. This proves global convergence.

To prove quadratic convergence it can be shown using the implicit function theorem that a sweep $s : \mathcal{O}(\Sigma,\Sigma) \to \mathcal{O}(\Sigma,\Sigma)$ defines a smooth mapping on an open neighborhood $U \subset \mathcal{O}(\Sigma,\Sigma)$ of $(\Sigma,\Sigma) \in \mathcal{O}(\Sigma,\Sigma)$. Furthermore, the first derivative of s at (Σ,Σ) is equal to zero. Thus the result follows by a standard Taylor expansion argument.

The above result shows convergence of the Gramians but not necessarily for the realizations. The additional freedom obtained by passing from the Gramians to the realizations may be used in order to enforce canonical structures on (A, B, C). In this way it might be possible to develop algorithms that compute balanced canonical forms such as those described by Ober (1987) [18].

4.3 Sensitivity Minimization

Our goal now is to develop a numerical algorithm to compute sensitivity balanced realizations. This has been first done by [22] where it was found that model reduction based on sensitivity balanced truncation can in fact have desirable properties in comparison to standard balanced truncation. However, in addition to rather slow linear convergence of the proposed algorithm being observed by [22], quadratically convergent algorithms have not been obtained so far. The algorithm which we propose is defined by the iteration of generalized Jacobi sweeps. Each generalized sweep consists of two subsweeps, by first applying a hyperbolic subsweep that minimizes the L^2- sensitivity function

$$\mathcal{S}(A,b,c) = \frac{1}{2\pi i}\oint \mathrm{tr}\,(\mathcal{A}(z)\mathcal{A}^*(z))\,\frac{\mathrm{d}\,z}{z} + \mathrm{tr}\,(W_c) + \mathrm{tr}\,(W_o)$$

and then performing an orthogonal subsweep that minimizes the trace function

$$\widetilde{\mathcal{S}}_N(A,b,c) := \frac{1}{2\pi i}\oint \mathrm{tr}\,(N\,(\mathcal{A}(z)\mathcal{A}^*(z) + \mathcal{B}(z)\mathcal{B}^*(z)))\,\frac{\mathrm{d}\,z}{z}$$

for $N = \mathrm{diag}\,(\mu_1, \ldots, \mu_n)$, and $0 < \mu_1 < \cdots < \mu_n$. Each hyperbolic subsweep consists of successive application of the $n(n+1)/2$ hyperbolic transformations (4.2) and (4.3), whereas the orthogonal subsweep consists of the $n(n-1)/2$ Jacobi rotations (4.1). Thus each generalized sweep is built up by a total number of n^2 basic coordinate transformations. The next theorem shows global convergence of the resulting algorithm to the class of sensitivity balanced realizations. Note that the sensitivity singular values become automatically sorted in decreasing order. We conjecture that the sensitivity Gramians actually converge quadratically fast.

Theorem 4.3 *Let (A_k, b_k, c_k), $k \in \mathbb{N}$, denote the sequence of realizations of a stable transfer function $G(z)$ generated by the above Jacobi-type algorithm. Then global convergence holds; i.e., the sequence (A_k, b_k, c_k) converges to the set of sensitivity balanced realization $(A_\infty, b_\infty, c_\infty)$ with sensitivity singular values sorted as $\tilde{\sigma}_1 \geq \cdots \geq \tilde{\sigma}_n$. Moreover, each $(A_\infty, b_\infty, c_\infty)$ minimizes the L^2-sensitivity function.*

Proof. (sketch) Using hyperbolic rotations forces the sequence to converge to the compact orbit of sensitivity minimal realizations for which the sensitivity Gramians are equal (but not necessarily diagonal). On this limit set the orthogonal rotations then force the sequence to converge to the set of $(A_\infty, b_\infty, c_\infty)$, as stated. The result then follows using global smoothness of the hyperbolic sweeps. Full details will appear elsewhere.

5 Numerical Experiments

In this section we present a few numerical experiments with the algorithms for balancing. In Figure 1 Algorithms 4.2 and 4.2 are applied to a 15–th order system with multiple and clustered Hankel singular values. The Gramians W_c, W_o are formed by a randomly generated congruence transformation T, i.e., $((W_c)_0, (W_o)_0) = (T\Sigma_1 T', T'^{-1}\Sigma_1 T^{-1})$ with $\Sigma_1 = \text{diag}\,(1{,}1+\varepsilon{,}1+\varepsilon{,}2{,}2{,}2{,}2{,}5{,}5{,}5{,}10{,}10{,}10{,}10{,}10)$ and $\varepsilon = 10^{-7}$. Figure 2 shows Algorithms 4.2 and 4.2 applied to a system with Hankel singular values $\Sigma_2 = \text{diag}\,(1, \ldots, 15)$. In both figures the values of the normalized cost function $f = \text{tr}\,(N((W_c)_k + (W_o)_k - 2\Sigma))/\text{tr}\,(2N\Sigma)$ are plotted against the number k of sweeps. Local quadratic convergence of the cost function is obvious from the plots. Many more simulations not presented here let us conjecture that in general Algorithm 4.2 is faster than Algorithm 4.2.

We use the same Gramians as in Figures 1 and 2 as the starting points for our simulations in Figures 3 and 4, respectively. There we compare the balancing algorithm proposed by [15] with a modification where the conventional Kogbetliantz algorithm for the SVD used in [15] is replaced by the Sort-Jacobi algorithm for SVD. Note that the Kogbetliantz algorithm is generally not quadratically convergent for matrices with multiple singular values. In contrast, the Sort-Jacobi algorithm is always quadratically convergent. This is in agreement with the simulation experiment in Figure 3 which shows improved performance by the Sort-Jacobi algorithm. Thus replacing the Kogbetliantz algorithm by the Sort-Jacobi algorithm can lead to an improvement of the algorithm proposed by [15]. The values of the cost function $g = \text{off}\,((L_o' L_c)_k)$ are plotted against the number k of sweeps. The algorithm of [15] computes the singular value decomposition of the product $L_o' L_c$ where L_o, L_c are the Cholesky factors of the Gramians W_o and W_c, respectively. From the SVD $L_o' L_c = U\Sigma V'$ a balancing transformation is then immediately obtained as $T = L_c V \Sigma^{-\frac{1}{2}}$, see [15].

Finally, in Figure 5 we consider a situation where some system modes are nearly uncontrollable or unobservable. Our Algorithms 4.2 and 4.2 are applied to a system taken from [19]. It is a 10–th order system with 2 inputs and 2 outputs. Due to the large gap between the dominant singular value and the subdominant ones the dominant singular value is separated already after 1 sweep by Algorithm 4.2 (2 sweeps by Algorithm 4.2) up to an accuracy of 5 significant digits. The algorithms then slow down and local quadratic convergence is not observed due to finite computer accuracy of about 10^{-16}.

In all figures plotted points are joined by linear interpolation.

References

[1] M.F. Atiyah. Convexity and commuting Hamiltonians. *Bull. London Math. Soc.* **14** (1992), 1-15.

[2] H. Azad and J.J. Loeb. On a theorem of Kempf and Ness. *Ind. Univ. Math. J.* **39**(1) (1990), 61-65.

[3] R.W. Brockett. Dynamical systems that sort lists, diagonalize matrices, and solve linear programming problems. *Proc. IEEE of the 27th Conference on Decision and Control 1988.* 799-803. See also *Lin. Algebra & Applic.* **146** (1991), 79-91.

[4] J. Dehaene. *Continuous-time Matrix Algorithms Systolic Algorithms and Adaptive Neural Networks.* Ph.D. Dissertation. Katholieke Universiteit. Leuven, Germany. 1995.

[5] M. Gevers and G. Li. *Parametrizations in Control, Estimation and Filtering Problems.* London: Springer, 1993.

[6] V. Guillemin and S. Sternberg. Convexity properties of the moment mapping. *Inventiones Math.* **67** (1982), 491-513.

[7] U. Helmke. A several complex variables approach to sensitivity analysis and structured singular values. *J. of Mathematical Systems, Estimation, and Control* **2**(3) (1992), 339-351.

[8] U. Helmke. Balanced realizations for linear systems: a variational approach. *SIAM J. Control and Optimization* **31**(1) (1993), 1-15.

[9] U. Helmke and J.B. Moore. Singular-value decomposition via gradient and self-equivalent flows. *Lin. Algebra & Applic.* **169** (1992), 223-248.

[10] U. Helmke and J.B. Moore. *Optimization and Dynamical Systems.* London: Springer, 1994.

[11] K. Hüper. *Structure and convergence of Jacobi-type methods for matrix computations.* Ph.D. Dissertation. Technical University of Münich. Munich, Germany. 1996.

[12] K. Hüper and U. Helmke. Structure and convergence of Jacobi-type methods. To appear *Num. Math.*

[13] C.G.J. Jacobi. Über ein leichtes Verfahren, die in der Theorie der Säcularstörungen vorkommenden Gleichungen numerisch aufzulösen. *Crelle's J. für die reine und angewandte Mathematik* **30** (1846), 51-94.

[14] G. Kempf and L. Ness. The length of vectors in representation spaces. *Lect. Notes in Math.* **732** (1979), 233-243.

[15] A.J. Laub, M.T. Heath, C.C. Paige, and R.C. Ward. Computation of system balancing transformations and other applications of simultaneous diagonalization algorithms. *IEEE Transactions on Automatic Control* **32**(2) (1987), 115-122.

[16] R. Mahony. *Optimization algorithms on homogeneous spaces.* Ph.D. Dissertation. Australian National University. Canberra, Australia. 1994.

[17] C.T. Mullis and R.A. Roberts. Synthesis of minimum roundoff noise fixed point digital filters. *IEEE Transactions on Circuits and Systems* **23** (1976), 551-562.

[18] R.J. Ober. Balanced realizations: canonical form, parametrization, model reduction. *Int. J. Control* **46**(2) (1987), 643-670.

[19] M.G. Safonov and R.Y. Chiang. A Schur method for balanced-truncation model reduction. *IEEE Transactions on Automatic Control* **34**(7) (1989), 729-733.

[20] S.T. Smith. *Geometric optimization methods for adaptive filtering.* Ph.D. Dissertation. Harvard University. Cambridge, Massachusetts. 1993.

[21] L. Thiele. On the sensitivity of state-space systems. *IEEE Transactions on Circuits and Systems* **33** (1986), 502-510.

[22] W. Yan and J.B. Moore. On L^2–sensitivity minimization of linear state-space systems. *IEEE Transactions on Circuits and Systems* **39** (1992), 641-648.

Department of Mathematics, University of Würzburg, D-97074, Würzburg, Germany

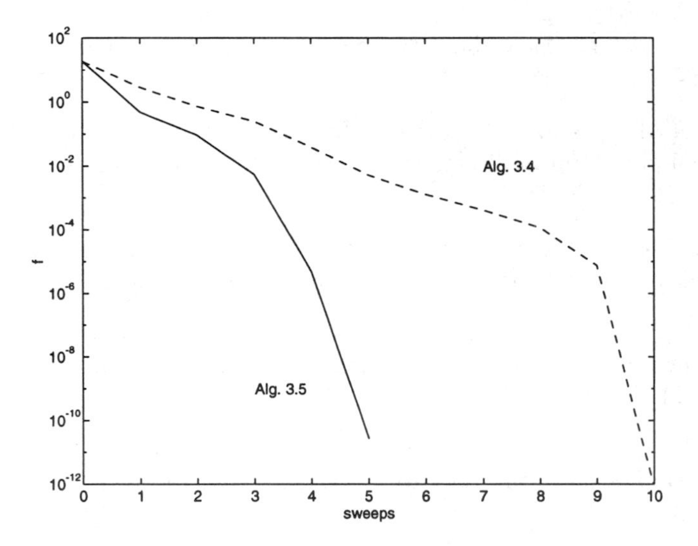

Figure 1: Balancing of a 15−th order system with clustered Hankel singular values. $\Sigma_1 = \mathrm{diag}\,(1,1+\varepsilon,1+\varepsilon,2,2,2,2,5,5,5,10,10,10,10,10)$, $T \in GL(n)$, $((W_c)_0,(W_o)_0) = (T\Sigma_1 T', T'^{-1}\Sigma_1 T^{-1})$, $\varepsilon = 10^{-7}$, $N = \mathrm{diag}\,(1,\ldots,15)$, and normalized $f = \mathrm{tr}\,(N((W_c)_k + (W_o)_k - 2\Sigma_1))/\mathrm{tr}\,(2N\Sigma_1)$.

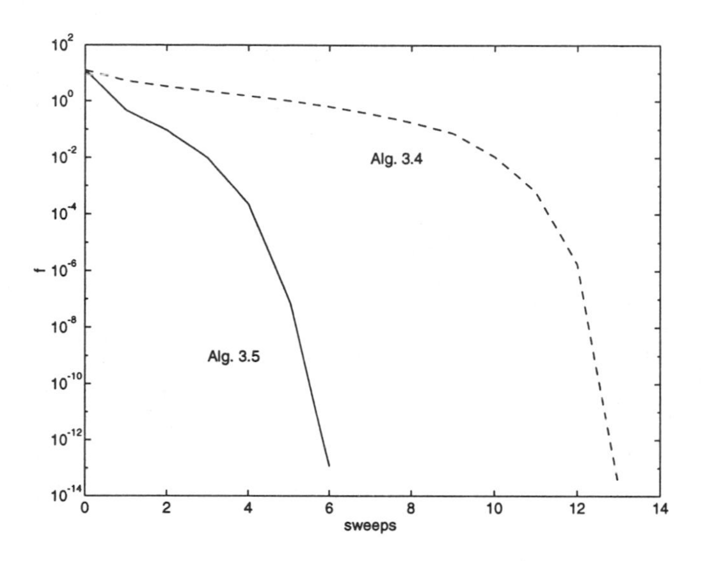

Figure 2: Balancing of a 15−th order system with distinct Hankel singular values. $((W_c)_0, (W_o)_0) = (T\Sigma_2 T', T'^{-1}\Sigma_2 T^{-1})$, $\Sigma_2 = N = \text{diag}(1, \ldots, 15)$, $T \in GL(n)$, and normalized $f = \text{tr}(N((W_c)_k + (W_o)_k - 2\Sigma_2))/\text{tr}(2N\Sigma_2)$.

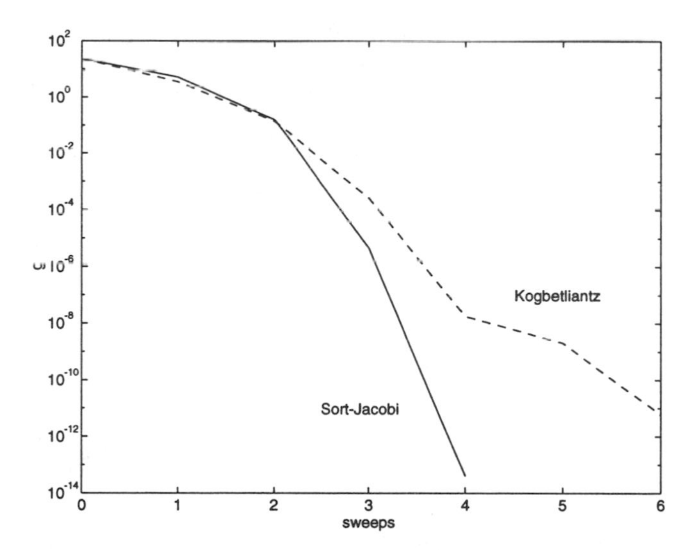

Figure 3: SVD of $L_o' L_c$. L_o, L_c are the Cholesky factors of the Gramians W_o, W_c. $\Sigma_1 = \text{diag}(1{,}1+\varepsilon{,}1+\varepsilon{,}2{,}2{,}2{,}2{,}5{,}5{,}5{,}10{,}10{,}10{,}10{,}10)$, $\varepsilon = 10^{-7}$, $((W_c)_0, (W_o)_0) = (T\Sigma_1 T', T'^{-1}\Sigma_1 T^{-1})$, $T \in GL(n)$, and $g = \text{off}((L_o' L_c)_k)$.

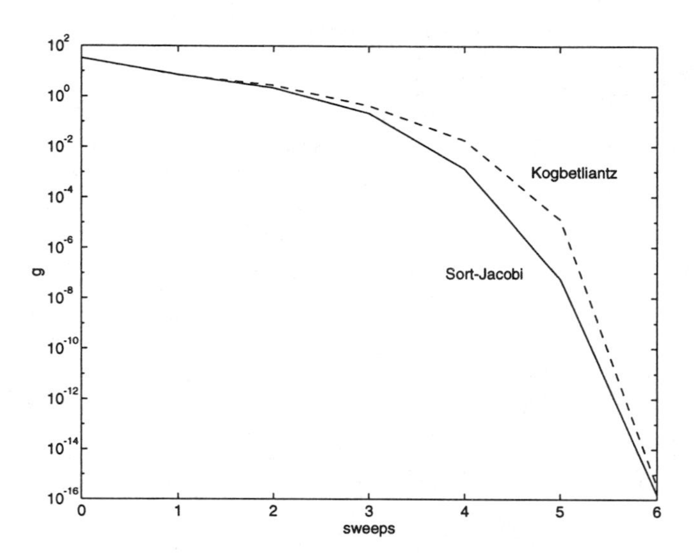

Figure 4: SVD of $L_o'L_c$. L_o, L_c are the Cholesky factors of the Gramians W_o, W_c. $((W_c)_0, (W_o)_0) = (T\Sigma_2 T', T'^{-1}\Sigma_2 T^{-1})$, $\Sigma_2 = \operatorname{diag}(1, \ldots, 15)$, $T \in GL(n)$, and $g = \operatorname{off}((L_o'L_c)_k)$.

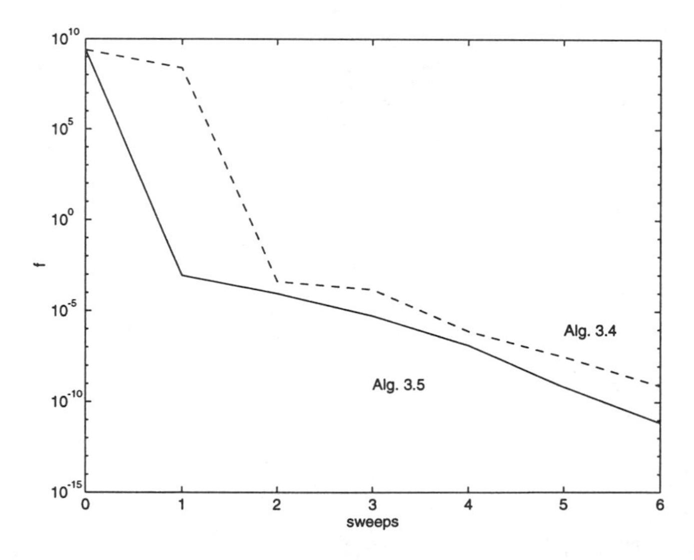

Figure 5: Balancing of the 10–th order system taken from [19].

Ellipsoidal Calculus for Estimation and Feedback Control

A.B. Kurzhanski

1 Introduction

The emphasis of the present paper is to overview the constructive techniques for modeling and analyzing an array of problems in uncertain dynamics and control, as collected in monograph [17] and in some further investigations. It deals with problems of *guaranteed control synthesis and set-valued estimation* for systems that operate under "set-membership uncertainty" - unknown but bounded inputs and disturbances and presents a unified approach to these topics based on descriptions involving the notions of set-valued calculus.

In this paper we deal with *linear time-variant systems*

$$\dot{x} = A(t)x + u + f(t), \quad x(t_0) = x_0, \tag{1.1}$$

with *magnitude* ("hard") bounds on the controls u and the uncertain items $f(t), x_0$:

$$u(t) \in \mathcal{P}(t), \quad f(t) \in \mathcal{Q}(t), \quad x^0 \in X_0, \tag{1.2}$$

where $\mathcal{P}(t), \mathcal{Q}(t)$ are multifunctions with values in convex compact sets, continuous in time. We shall further refer to system (1.1), (1.2) as a *linear-convex* system.

The constructive schemes treated here are based on presenting the set-valued solution elements for the respective problems in terms of *ellipsoidal-valued representations* as introduced in papers [15], and monograph [17]. This "ellipsoidal" move leads to rather effective algorithms with possibility of further computer animation. For the problem of control synthesis it particularly allows us to present the solutions in terms of analytical designs rather than algorithms as required in the "exact" theory. [1]

In Sections 3 and 4 we shall presume the constraints on u, f, x^0 to be ellipsoidal-valued. Namely by introducing the notation $\mathcal{E}(a, K) = \{x : (x-a, K^{-1}(x-a) \leq 1\}, x \in \mathbb{R}^n, \; K > 0$, in terms of inclusions, we shall then have:

$$u \in \mathcal{E}(p(t), P(t)) \; , f \in \mathcal{E}(q(t), Q(t)) \; , x^0 \in \mathcal{E}(x^*, X_0), \tag{1.3}$$

[1] The ellipsoidal calculus mentioned here is aimed at *exact representation* of the respective convex sets by parameterized families of ellipsoids rather than at approximating them by only one or several ellipsoids, as done in other publications.

where the continuous functions $p(t), q(t)$ and the vector x^* are given together with continuous matrix functions $P(t) > 0, Q(t) > 0$ and matrix $X_0 > 0$. We also use notation $\rho(l|\mathcal{P}) = \max\{(l,x)|x \in \mathcal{P})\}$ for the support function of a compact set $\mathcal{P}$.

2 Guaranteed Control Synthesis for Uncertain Systems

In this section we present some basic relations for the problem of control synthesis, presenting them in the form appropriate for further ellipsoidal approximations. Consider system (1.1), (1.2) and a "terminal set" $\mathcal{M} \in$ *comp*$\mathbb{R}^n$ - the variety of convex compact sets in $\mathbb{R}^n$.

Definition 2.1 The problem of *guaranteed control synthesis* consists in specifying *a solvability set* $\mathcal{W}(\tau, t_1, \mathcal{M})$ and a set-valued **feedback control strategy** $u = \mathcal{U}(t,x)$, such that all the solutions to the differential inclusion

$$\dot{x} \in A(t)x + \mathcal{U}(t,x) + \mathcal{Q}(t) \tag{2.1}$$

that start from any given position $\{\tau, x_\tau\}$, $x_\tau \in \mathcal{W}(\tau, t_1, \mathcal{M})$, $\tau \in [t_0, t_1)$, would reach the terminal set $\mathcal{M}$ at time $t_1 : x(t_1) \in \mathcal{M}$.

This obviously means that strategy $\mathcal{U}(t,x)$ is supposed to steer system (1.1), $u = \mathcal{U}(t,x)$, to set $\mathcal{M}$, *whatever is the disturbance* $f(t) \in \mathcal{E}(p(t), P(t))$. Definition 1.1 is nonredundant if $\mathcal{W}(\tau, t_1, \mathcal{M}) \neq \emptyset$. Taking the set-valued function $\mathcal{W}[t] = \mathcal{W}(t, t_1, \mathcal{M})$, $\tau \leq t \leq t_1$, we come to *the solvability tube* $\mathcal{W}[\cdot]$. (Note that $\mathcal{W}(\tau, t_1, \mathcal{M})$ is the largest set of states, relative to inclusion, from which the solution does exist at all).

The formulated problem does not have any optimization criteria and requires us to find just a feasible solution. Nevertheless, we shall seek for such a solution by applying some mini-maximization schemes. We shall first deal with a Dynamic Programming interpretation of the problem. Considering equation (1.1), (1.2) and target set $\mathcal{M}$; introduce the value function

$$\mathcal{V}(t,x) = \min_{\mathcal{U}} \max_{x(\cdot)} \{\mathcal{I}(t,x) | \mathcal{U}(\cdot,\cdot) \in U_{\mathcal{P}}^{c}, x(\cdot) \in \mathcal{X}_{\mathcal{U}}(\cdot)\},$$

where

$$\mathcal{I}(t,x) = d^2(x[t_1], \mathcal{M})$$

and

$$d^2(x, \mathcal{M}) = \min\{(x-z, x-z) | z \in \mathcal{M}\} = h_+^2(x, \mathcal{M}),$$

is the square of the Euclid distance from x to set $\mathcal{M}$;

$$h_+(\mathcal{Q}, \mathcal{M}) = \max_{x} \min_{z} \{(x-z, x-z)^{1/2} | x \in \mathcal{Q}, z \in \mathcal{M}\}$$

is the Hausdorff semidistance between sets $\mathcal{Q}, \mathcal{M}$; $\mathcal{X}_{\mathcal{U}}(\cdot)$ is the variety of all trajectories $x(\cdot)$ of inclusion (1.1) generated by given strategy $\mathcal{U}$ and $U^c_{\mathcal{P}}$ stands for the class of strategies that ensure the existence and unicity of solutions to the differential inclusion (2.1) (this, for example, is the class of set-valued functions with values in *comp*$\mathbb{R}^n$, upper semicontinuous in x and continuous in t).

The formal H-J-B-I (Hamilton-Jacobi-Bellman-Isaacs) equation for the value $\mathcal{V}(t,x)$ looks as follows

$$\frac{\partial \mathcal{V}}{\partial t} + \min_u \max_f \left(\frac{\partial \mathcal{V}}{\partial x}, A(t)x + u + f \right) = 0 \tag{2.2}$$

with boundary condition

$$\mathcal{V}(t_1, x) = h_+^2(x, \mathcal{M}). \tag{2.3}$$

The last equation may not have a classical solution though, and its treatment may therefore require us to use the generalized notion of viscosity solution [5] or its equivalent - the minmax solution [22]. However, in the specific problems of this paper, the value function $\mathcal{V}(t,x)$ shall turn to be directionally differentiable for any direction $(1,h), h \in \mathbb{R}^n$. Then, in terms of subdifferential calculus, equation (2.2) could be presented as

$$\min_u \max_f \{D\mathcal{V}(t,x)|(1, A(t)x + u + f)\} = 0, \tag{2.4}$$

with boundary condition (2.3), where $D\mathcal{V}(t,x)|(1,h)$ is the *directional derivative* of function $\mathcal{V}$ along the direction $(1,h)$ at point (t,x), and $u \in \mathcal{P}(t)$, $f \in \mathcal{Q}(t)$. Function $\mathcal{V}$ is a solution to (2.4) if it satisfies this equation for all $\{t,x\}$ in the domain $[t_0,t_1] \times \mathbb{R}^n$. We shall have this remark in mind while continuing to write the H-J-B -I equation in the formal "classical" transcription (2.2). [2]

We further deal with the following nondegenerate case.

Assumption 2.1 . The solvability tube $\mathcal{W}[\cdot]$ is such that there exists an $\epsilon > 0$ and a continuous function $w_*(t)$, that yield the inclusion $w_*(t) + \epsilon\mathcal{S} \in \mathcal{W}[t]$, $t_0 \le t \le t_1$, where $\mathcal{S} = \{x : (x,x) \le 1\}$.

Lemma 2.1 *Under Assumption 1.1 equation (2.2) (interpreted as (2.4)), with boundary condition (2.3), has a unique solution* $\mathcal{V}(t,x)$ *in the domain* $[t_0,t_1] \times \mathbb{R}^n$.

Denote $\mathcal{W}_*[t] = \{x : \mathcal{V}(t,x) \le 0\}$. An important fact is given by

[2]In this paper all the value functions shall turn to be either convex or quasi-convex in x and all the solutions to the respective H-J-B or H-J-B-I equations shall not lead beyond the notions of viscosity or minmax solutions.

Lemma 2.2 *Under Assumption 1.1 the solvability set* $\mathcal{W}[t]$ *of Definition 1.1 may be presented as*

$$\mathcal{W}[t] \ = \ \mathcal{W}_*[t], \ \ t_0 \leq t \leq t_1. \tag{2.5}$$

If equation (2.3) (2.4) would be solved, one could obtain the solution strategy as

$$\mathcal{U}_*(t,x) = \arg\min\{(\partial\mathcal{V}(t,x)/\partial x, u)|u \in \mathcal{P}(t)\} \tag{2.6}$$

(if the gradient $\partial V_*(t,x)/\partial x$ does exist at $\{t,x\}$), or, more generally, as

$$\mathcal{U}_*(t,x) = \{u : \max_f\{dh_+^2(x,\mathcal{W}_*[t])/dt | f \in \mathcal{Q}(t)\} \leq 0\}. \tag{2.7}$$

To specify the control strategy $\mathcal{U}_*$ it therefore suffices to know only the cut $\mathcal{W}_*[t]$ of the value function $\mathcal{V}(t,x)$. [3] Since the direct integration of the H-J-B-I equation may require rather subtle numerical techniques, we shall further investigate and approximate the cuts of the value function rather than the value function itself. By dealing only with the cuts, we shall thus *avoid the integration* of the H-J-B-I equation for the class of problems considered in this paper. Particularly, we shall introduce an array of parameterized internal and external *ellipsoidal approximations* for these cuts. It is therefore important to be able to find the cuts $\mathcal{W}_*[t]$ without calculating $\mathcal{V}$. Thus, it is not uninteresting to observe that there exists a set-valued integral whose value happens to be precisely the set $\mathcal{W}[t]$ and therefore also the cut $\mathcal{W}_*[t]$ of the value function $\mathcal{V}(t,x)$. This is the "alternated integral" introduced by L.S.Pontryagin [19], (see [17], Section 1.7).

Lemma 2.3 *Under Assumption 1.1 set* $\mathcal{W}_*[t]$ *may be presented as*

$$\mathcal{W}_*[t] \ = I(t,t_1,\mathcal{M}), \ t_0 \leq t \leq t_1, \tag{2.8}$$

where $I(t,t_1,\mathcal{M})$ *is the set-valued alternated integral of Pontryagin.*

Finally, the set-valued function $\mathcal{W}[t]$ is a solution to the following evolution equation of the "funnel type":

Lemma 2.4 *Under Assumption 1.1 the set-valued function* $\mathcal{W}[t]$ *satisfies for all* $t \in [t_0,t_1]$ *the evolution equation*

$$\lim_{\sigma\to 0}\sigma^{-1}h_+(\mathcal{W}[t-\sigma]+\sigma\mathcal{Q}(t),\ \mathcal{W}[t]-\sigma\mathcal{P}(t)) = 0 \ \ ,\mathcal{W}[t_1] = \mathcal{M} \ \ . \tag{2.9}$$

The map $\mathcal{W}[t] = \mathcal{W}(t,t_1,\mathcal{M})$ *satisfies the semigroup property*

$$\mathcal{W}(t,t_1,\mathcal{M}) \ = \ \mathcal{W}(t,\tau,\mathcal{W}(\tau,t_1,\mathcal{M})), \ t \leq \tau \leq t_1. \tag{2.10}$$

[3]Here the properties of the value function $\mathcal{V}(t,x)$ are such that the respective strategy $\mathcal{U}(t,x) \in U_c$ is always feasible.

The set-valued function $\mathcal{W}[\cdot]$ also turns to be a "stable bridge," as defined in [9], [10]. The basic property of the solution is as follows:

Theorem 2.1 *Suppose Assumption 1.1 holds. Then, with $x_\tau = x(\tau) \in \mathcal{W}_*[t]$, all the solutions $x[t] = x(t, \tau, x_\tau)$ to the differential inclusion*

$$\dot{x} \in \mathcal{U}_*(t, x) + f(t), \quad x[\tau] = x_\tau, \tag{2.11}$$

satisfy the relation $x[t] \in \mathcal{W}_*[t]$, $\tau \leq t \leq t_1$, and therefore reach the terminal set $\mathcal{M}$: $x[t_1] \in \mathcal{M}$, whatever is the unknown disturbance $f(t)$.

Among the feasible set-valued strategies $\mathcal{U}(t, x)$ that satisfy the inclusion $\mathcal{U}(t, x) \subseteq \mathcal{U}_*(t, x)$ and therefore ensure the result of theorem 1.1, if substituted in (1.11) instead of $\mathcal{U}_*(t, x)$, is $\mathcal{U}_e(t, x)$ - the "extremal aiming strategy" ([9], [10]), given by relation

$$\mathcal{U}_e(t, x) = \partial_l k(t, -l^0) \;=\; \arg\min\{(l^0, u\}|u \in \mathcal{P}(t)\}. \tag{2.12}$$

Here $\partial_l k$ stands for the subdifferential of function $k(t, l)$ in the variable l, $k(t, l) = \rho(l|\mathcal{P}(t))$, and $l^0 = l^0(t, x) \neq 0$ is the maximizer for the problem

$$d_*[\tau, x] = \max\{(l, x) - \rho(l|\mathcal{W}_*[\tau])|\ \|l\| \leq 1\} > 0 \tag{2.13}$$

with $l^0(t, x) = 0$ when $d_*[\tau, x] = 0$. [4]

Lemma 2.5 *With $x_\tau \in \mathcal{W}^*[\tau]$ strategy $\mathcal{U}_e(t, x)$ ensures the inclusion $X_{\mathcal{W}^*}(t, \tau, x_\tau) \subseteq \mathcal{W}^*[t]$, $\quad \tau \leq t \leq t_1$.*

Here $X_*(t, \tau, x_\tau)$ is the tube of all solutions to system (2.1), $x[\tau] = x_\tau$, with $\mathcal{U}(t, x) = \mathcal{U}_e(t, x)$. The results of the above may be summarized into

Theorem 2.2 *The synthesizing strategy $\mathcal{U}_e(t, x)$ resolves the problem of guaranteed control synthesis under uncertainty.*

Under Assumption 1.1 the last theorems do not require any matching conditions for the constraints on the inputs u, f, (of type $\mathcal{P}(t) = \alpha\mathcal{Q}(t)$, $0 < \alpha < 1$).

Finally, we also mention that the cut $\mathcal{W}_*[t]$ may be obtained from the following value function

$$\mathcal{V}_*(\tau, x) = \min_{\mathcal{U}} \max_{f} \max_{x(\cdot)} \{\mathcal{I}_*(\tau, x)|\mathcal{U}(\cdot, \cdot), f(\cdot)\},$$

as the cut $\mathcal{W}_*[\tau] = \{x : \mathcal{V}_*(\tau, x) \leq 0\}$, where

$$\mathcal{I}_*(\tau, x) = \int_\tau^{t_1} (h_+^2(u, \mathcal{P}(t) - h_+^2(f, \mathcal{Q}(t))dt + h_+^2(x(t_1), \mathcal{M}),$$

[4] A justification of the last formula, which reflects N.N. Krasovski's "extremal aiming" rule is given in [9], [10], see also [17].

and $x(\cdot)$ varies over all the trajectories of system (2.11) generated by strategy $\mathcal{U}$ and function f. Value function $\mathcal{V}_*(t,x)$ reflects the so-called H_∞ approach to the problem of control synthesis (see [3], [8], [11]).

The application of the described techniques requires to calculate the tube $\mathcal{W}_*[\cdot]$ in one or another way, producing finally the desired strategy $\mathcal{U}_*$ or $\mathcal{U}_e$ *in the form of an algorithm.* The aim of this paper is further to emphasize an ellipsoidal technique that would allow us to produce a computational algorithm for the solvability tubes and further on, to generate the "guaranteed " control synthesis in the form of a design - an "analytical" controller.

Before passing to the ellipsoidal technique, we briefly treat the case when $f(t) = q(t)$ is given which means there is *no input uncertainty.* Let us also suppose that $\mathcal{P}(t)$ is an ellipsoid given by (1.3). Then (2.3) reduces to the formal H-J-B equation (see Remark 2.1)

$$\frac{\partial V}{\partial t} + \left(\frac{\partial V}{\partial x}, A(t) + p(t) + q(t)\right) + \frac{1}{4}\left(\frac{\partial V}{\partial x}, P(t)\frac{\partial V}{\partial x}\right)^{1/2} = 0, \qquad (2.14)$$

under boundary condition (2.3), where $\mathcal{V} = V$. (Here $p(t), q(t)$ are taken to be continuous). With

$$\mathcal{M} = \{x : (x-m, M^{-1}(x-m)) \leq 1\} = \mathcal{E}(m, M), \;\; M > 0, \qquad (2.15)$$

also being an ellipsoid, the boundary condition (2.3) transforms into

$$V(t_1, x) = (x-m, M(x-m))(1-(x-m, M(x-m))^{-1/2})^2, \; x \notin \mathcal{E}(m, M),$$

$$V(t_1, x) = 0, \;\; x \in \mathcal{E}(m, M).$$

The respective solvability set shall be denoted as $W[t]$ with $W[t] = \{x : V(t,x) \leq 0\}$. [5]

The value function $V(t,x)$ may now be calculated through duality methods of convex analysis. This yields (see [17], Section 5).

Lemma 2.6 *With* $\mathcal{P}(t) = \mathcal{E}(p(t), P(t)), \mathcal{M} = \mathcal{E}(m, M)$, *the value function*

$$V(t,x) = \max\{\Phi(\tau, x, l) | l \in \mathbb{R}^n\}, \qquad (2.16)$$

where

$$\Phi(\tau, x, l) = (l, x - x^*(\tau)) - H(t, l) - \frac{1}{4}(l, S(t_1, \tau) M S'(t_1, \tau) l).$$

[5] From here on we distinguish the case with no uncertainty ($f(t)$ given) by using notations V, W, U for value function, solvability tube and solution strategy instead of $\mathcal{V}, \mathcal{W}, \mathcal{U}$ as in the general case.

Here

$$H(\tau,l) = \int_{\tau}^{t_1} (l, S(t,\tau)P(t)S'(t,\tau)l)^{1/2}dt + (l, S(t_1,\tau)MS'(t_1,\tau)l)^{1/2};$$

function $S(t,\tau)$ is the solution to the adjoint matrix equation

$$\partial S(t,\tau)\partial t = -S(t,\tau)A(t), \; S(\tau,\tau) = I;$$

symbol I stands for the unit matrix, and $x^*(t)$ is the solution to equation

$$\dot{x}^* \;=\; A(t)x^* \;+\; p(t) \;+\; q(t), \tag{2.17}$$

with $x^*(t_1) = m$.

The maximizing vector $l = l^0(t,x)$ in (2.16) is unique. The function $V(t,x)$ satisfies the H-J-B equation (2.15) (see Remark 2.1) and the synthesizing strategy

$$U_e(t,x) \;=\; \arg\min\left\{ \left(\frac{\partial V(t,x)}{\partial x}, u\right) | u \in \mathcal{P}(t) \right\}.$$

Here $\partial \mathcal{V}(t,x)/\partial x \;=\; l^0(t,x)$. Particularly, with $V(t,x) \;=\; 0$, this yields $U_e(t,x) = \mathcal{E}(p(t),P(t))$.

The support function of the cut $W[t] = \{x : V(t,x) \leq 0\}$ may now be calculated directly. This gives $\rho(l|W[t]) = (l,x^*(t)) + H(t,l)$. Now the strategy U_e may be also calculated through the last relation with $l^0(t,x)$ substituted for vector $l^*(t,x)$ - the maximizer for the problem

$$\max\{(l,x) - \rho(l|W[t]) \mid (l,l) \leq 1\} = d(x, W[t]). \tag{2.18}$$

In the absence of uncertainty ($f(t)$ given), this value function solves the minimization problem

$$V(t,x) = \min_{u} \mathcal{I}(t,x),$$

and *is the same, whether the problem is solved over open-loop controls* $u(t) \in \mathcal{P}(t)$ *or closed-loop strategies* $\mathcal{U}(t,x) \in U_c$.

The respective synthesizing control strategy is now given by relation (2.12), (2.13) with $\mathcal{W}_*[t]$ substituted by $W[t]$ or by the "extremal aiming strategy" (in the absence of uncertainty), which is $\mathcal{U}_e(t,x)$ as in (2.12), (2.18), but with $l^*(t,x) = l^0(t,x)$ taken from either (2.16) or (2.18).

We shall now pass to the main part of the paper which describes the ellipsoidal approximations for the solvability tubes $\mathcal{W}[\cdot], W[\cdot]$ and their crossections $\mathcal{W}[t], W[t]$ - the cuts of the value functions $\mathcal{V}(t,x), V(t,x)$.

3 Ellipsoidal Solvability Tubes

In this section and in the next one we further assume all the sets $\mathcal{P}, \mathcal{Q}, X^0, \mathcal{M}$ to be ellipsoidal-valued as given by (1.3), (2.15). We shall start with the simplest case of systems without uncertainty ($f(t)$ given), considering the problem of control synthesis under Assumption 2.1. In this case one may observe that $W[t]$ is given by a set-valued "Aumann" integral which is

$$W[t] = \mathcal{E}(m, M) \;-\; \int_t^{t_1} S(t_1, s)\mathcal{E}(p(s) + f(s), P(s))ds. \tag{3.1}$$

Introducing the notation

$$F(H, P, X) = (HXH')^{\frac{1}{2}}(HPH')^{\frac{1}{2}} \;+\; (HPH')^{\frac{1}{2}}(HXH')^{\frac{1}{2}},$$

consider the differential equations (2.18) and

$$\dot{X}_+ = A(t)X_+ + X_+A(t) - \pi(t)X_+ - \pi^{-1}(t)P(t), \tag{3.2}$$

$$\dot{X}_- \;=\; A(t)X_- \;+\; X_-A(t) \;-H^{-1}(t)F(H(t), X_-, H'(t))H'^{-1}(t), \tag{3.3}$$

with boundary conditions

$$x^*(t_1) = m \;,\; X_+(t_1) = M,\; X_-(t_1) = M. \tag{3.4}$$

Denote the solutions to (3.2), (3.3) with boundary conditions (3.4) as $x(t) = x(t, t_1, m)$, $X_+(t) = X_+(t, t_1, M)$, $X_-(t) = X(t, t_1, M)$ respectively. Then, following the approximation schemes of [17], we come to

Theorem 3.1 *The following inclusions are true*

$$E_-[t] = \mathcal{E}(x^*(t), X_-(t)) \subseteq W[t] \subseteq \mathcal{E}(x^*(t), X_+(t)) = E_+[t], \tag{3.5}$$

whatever the solutions are to differential equations (3.2)-(3.4) with measurable functions $\pi(t) > 0, H(t) = H'(t)$, ($H(t)$ is symmetrical).

As in [17], the last assertion develops into exact representations

Theorem 3.2 *The following "external" representation is true*

$$W[t] \;=\; \cap\{\mathcal{E}(x^*(t), X_+(t))|\pi(\cdot)\} \tag{3.6}$$

where the intersection is taken over all measurable functions $\pi(t)$ that satisfy the inequalities

$$\min\{(l, P(t)l)|(l, l) = 1\} \leq \pi(t) \leq \max\{(l, P(t)l)|(l, l) = 1\}, \tag{3.7}$$

The following "internal " representation is true

$$W[t] \;=\; \overline{\cup\{\mathcal{E}(x^*(t), X_-(t))|H(\cdot)\}} \tag{3.8}$$

where the union is taken over all measurable functions $H(t) = H'(t)$.

Here the dash stands for the closure of the respective set. For given pairs of functions $\pi(\cdot), H(\cdot)$ and boundary values m^*, M^* denote

$$E_+(t,\tau,\mathcal{E}(m^*,M^*)) = \mathcal{E}(x^*(t,\tau,m^*), X_+(t,\tau,M^*)), \tag{3.9}$$

$$E_-(t,\tau,\mathcal{E}(m^*,M^*)) = \mathcal{E}(x^*(t,\tau,m^*), X_-(t,\tau,M^*)). \tag{3.10}$$

Then, obviously, $E_-(t,\tau,\mathcal{E}(m^*,M^*)) \subseteq W[t] \subseteq E_+(t,\tau,\mathcal{E}(m^*,M^*))$.

Lemma 3.1 *The following identities are true with* $t \leq \tau \leq t_1$ *:*

$$E_-(t,\tau,E_-(\tau,t_1,\mathcal{E}(m,M))) \equiv E_-(t,t_1,\mathcal{E}(m,M)) \tag{3.11}$$

$$E_+(t,t_1,\mathcal{E}(m,M)) \equiv E_+(t,\tau,E_+(\tau,t_1,\mathcal{E}(m,M))). \tag{3.12}$$

Relations (3.11), (3.12) thus define, in backward time, the "lower" and "upper" semigroup properties of the corresponding mappings.

Let us now return to the uncertain system (2.1.), (2.2) with $f(t)$ unknown but bounded. As indicated above, the cut $\mathcal{W}_*[t]$ of the value function $\mathcal{V}[t]$ is now described by the alternated integral of Pontryagin and its evolution in time is due to the funnel equation (2.8). Consider the differential equations (3.2) and also

$$\dot{X}_+ = A(t)X_+ + X_+A(t) - \pi(t)X_+ - \pi^{-1}(t)P(t) + \tag{3.13}$$

$$+H^{-1}(t)F(H(t),X_+,Q(t))H'^{-1}(t),$$

$$\dot{X}_- = A(t)X_- + X_-A(t) + \pi(t)X_- + \pi^{-1}(t)Q(t) - \tag{3.14}$$

$$-H^{-1}(t)F(H(t),X_-,P(t))H'^{-1}(t),$$

which have to be taken with boundary conditions (3.4). Denote the solutions to (3.13), (3.14) under boundary condition (3.4) as $X_+(t|\pi(\cdot),H(\cdot))$, $X_-(t|\pi(\cdot),H(\cdot))$. Following the techniques of [15], [17] it is possible to prove the following assertion:

Theorem 3.3 *For every vector* $l \in \mathbb{R}^n$*, the following inclusions are true for any measurable functions* $\pi(t) > 0$*, and* $H(t) = H'(t)$ *:*

$$\mathcal{E}_-[t] = \mathcal{E}(x^*(t), X_+(t|\pi(\cdot),H(\cdot)) \supseteq \mathcal{W}[t] \supseteq \mathcal{E}(x^*(t), X_+(t|\pi(\cdot),H(\cdot)) = \mathcal{E}_+[t]. \tag{3.15}$$

Moreover, the following relations hold:

$$\rho(l|W[t]) = \inf\{\rho(l|\mathcal{E}(x^*(t), X_+(t|\pi(\cdot),H(\cdot))|\pi(\cdot),H(\cdot)\}, \tag{3.16}$$

and

$$\mathcal{W}[t] = \cap\{\mathcal{E}(x^*(t), X_+(t|\pi(\cdot),H(\cdot))\}, \tag{3.17}$$

as well as

$$\rho(l|W[t]) = \sup\{\rho(l|\mathcal{E}(x(t), X_-(t|\pi(\cdot), H(\cdot)))|\pi(\cdot), H(\cdot)\}, \quad (3.18)$$

and

$$W[t] = \overline{\cup\{\mathcal{E}(x^*(t), X_-(t|\pi(\cdot), H(\cdot))|\pi(\cdot), H(\cdot)\}}, \quad (3.19)$$

where $\pi(t) > 0$, $H(t) = H'(t)$, *are measurable functions.*

We emphasize, once more, that $\mathcal{E}_-[t]$ is an internal approximation of Pontryagin's alternated integral $I(t, t_1, \mathcal{M})$ of (2.8), while $E_-[t]$ is a similar approximation of the Aumann integral (3.1). One may also observe from the previous theorem that the upper and lower ellipsoidal estimates $\mathcal{E}_+[t] = \mathcal{E}(x(t), X_+(t|\pi(\cdot), H(\cdot))$ and $\mathcal{E}_-[t] = \mathcal{E}(x(t), X_-(t, |\pi(\cdot), H(\cdot))$ are respectively *inclusion-minimal and inclusion-maximal* as compared to any other external or internal ellipsoidal approximations of set $\mathcal{W}[t]$.

For a given pair of functions $\pi(\cdot), H(\cdot)$ and a given pair of boundary values m^*, M^* consider the mappings (3.9), (3.10), where the respective functions $x(t, \tau, m^*)$, $X_+(t, \tau, M^*)$, $X_-(t, \tau, M^*)$ now satisfy (3.13), (3.14), with boundary condition (3.4). Then the mappings (3.9), (3.10) again satisfy the upper and lower semigroup properties (3.11), (3.12), now for the case of uncertain dynamics. The property of being *inclusion-maximal* as well as *the semigroup property* of the internal approximation $E_-(t, \tau, \mathcal{E}(m^*, M^*))$ are crucial for using ellipsoidal techniques in the analytical design of the solution strategies for the problem of control synthesis. [6]

4 Control Synthesis Through Ellipsoidal Techniques

Let us return to the original problem of control synthesis. There the strategy $\mathcal{U}_e(t, x)$ was synthesized such that it ensured all solutions $x[t] = x(t, \tau, x_\tau)$ of the differential inclusion (2.11), $x_\tau = x[\tau] \in \mathcal{W}[\tau]$, to satisfy the relation $x[t] \in \mathcal{W}[t]$, $\tau \leq t \leq t_1$, and therefore to ensure $x[t_1] \in \mathcal{M} = W[t_1]$, *whatever the disturbance* $f(t)$ is. Here $\mathcal{W}[t]$ is the *solvability set* described in Section 1. The exact solution requires, as indicated above, to calculate the tube $\mathcal{W}[\cdot]$ and then, for each instant t, to solve an extremal problem of type (2.13) whose solution finally yields the desired strategy $\mathcal{U}_e(t, x)$, which is therefore actually defined *as an algorithm*.

[6] For the problem of control synthesis we need only the internal approximations of the solvability set. Nevertheless, we present also the external approximations of this set, since, with a reversal of time, the results of this section give the external and internal approximations of attainability domains (when there is no uncertainty), as well as of the more complicated domains of "attainability under uncertainty" in the general case, see [17].

In order to obtain a simpler scheme, we will now substitute $\mathcal{W}[t]$ by one of its *internal ellipsoidal approximations* $\mathcal{E}_-[t] = \mathcal{E}(x^*, X(t))$. The conjecture is that once $\mathcal{W}[t]$ is substituted by $\mathcal{E}_-[t]$, we should just copy the scheme of Section 1, constructing a strategy $\mathcal{U}_-(t,x)$ such that for every solution $x[t] = x(t,\tau,x_\tau)$ to equation

$$\dot{x}[t] \in \mathcal{U}_-(t, x[t]) + f(t), \quad \tau \le t \le t_1, \quad x[\tau] = x_\tau, \quad x_\tau \in \mathcal{E}_-[\tau], \tag{4.1}$$

the inclusion $x[t] \in \mathcal{E}_-[t], \quad \tau \le t \le t_1$, would be true and therefore $x[t_1] \in \mathcal{E}(m, M) = \mathcal{M} = \mathcal{E}_-[t]$, whatever the disturbance $f(t)$ is.

In constructing the "ellipsoidal synthesis," we shall imitate the strategy $\mathcal{U}_e$ given by (2.12), applying the respective scheme to the internal approximation $\mathcal{E}_-[t]$ of $\mathcal{W}[t]$. It was indicated in [17] that once approximation $\mathcal{E}_-[t]$ is selected "appropriately," (that is due to (3.13), (3.14)), the respective "ellipsoidal-based" extremal strategy $\mathcal{U}_-(t,x)$ does solve the problem. More precisely, we have the same relations (2.12), (2.13), except that $\mathcal{W}[t]$ will now be substituted by $\mathcal{E}_-[t]$. Namely, this gives

$$\mathcal{U}_-(t,x) = \begin{cases} \mathcal{E}(p(t), P(t)) & \text{if } x \in \mathcal{E}_-[t] \\ p(t) - P(t)l^0(l^0, P(t)l^0)^{-1/2} & \text{if } x \notin \mathcal{E}_-[t], \end{cases} \tag{4.2}$$

where $l^0 = l^0(t,x)$ is the unit vector that solves the problem

$$d[t,x] = (l^0, x) - \rho(l^0|\mathcal{E}_-[t]) = \max\{(l,x) - \rho(l|\mathcal{E}_-[t])| \|l\| \le 1\} \tag{4.3}$$

and $d[t,x] = h_+(x, \mathcal{E}_-[t])$.

One may readily observe that relation (4.3) coincides with (2.12), if set $W[t]$ is substituted for $\mathcal{E}_-[t]$ and $\mathcal{P}(t)$ for $\mathcal{E}(p(t), P(t))$; however here the maximization problem (4.3) may be solved in more detail than its more general analogue (2.13), (since $\mathcal{E}_-[t]$ is an ellipsoid).

If s^0 is the solution to the minimization problem

$$s^0 = \arg\min\{\|(x - s)\| | s \in \mathcal{E}_-[t], \ x = x(t)\}, \tag{4.4}$$

then in (4.3) one may take $l^0 = k(x(t) - s^0), \quad k > 0$, in (4.3), so that l^0 will be *the gradient* of the distance $d[t,x] = h_+(x, \mathcal{E}_-[t])$ with t fixed. (This can be verified by differentiating either (4.3) or (4.4) in x).

Lemma 4.1 *Consider a nondegenerate ellipsoid* $\mathcal{E} = \mathcal{E}(x^*, X)$ *and a vector* $x \notin \mathcal{E}(a, Q)$. *Then the gradient*

$$l^0 = \partial h_+(x, \mathcal{E}(x^*, X))/\partial x$$

may be expressed as

$$l^0 = (x - s^0)/\|x - s^0\|, \ s^0 = (I + \lambda X^{-1})^{-1}(x - a) + a, \tag{4.5}$$

where $\lambda > 0$ *is the unique root of the equation* $f(\lambda) = 0$, *where*

$$f(\lambda) = \left((I + \lambda X^{-1})^{-1}(x - a), X^{-1}(I + \lambda X^{-1})^{-1}(x - a)\right)^{-1}.$$

Corollary 4.1 With parameters x^*, X given and x varying, the multiplier λ may be uniquely expressed as a function $\lambda = \lambda(x)$.

Looking at relation (2.12) in view of the fact that $\mathcal{P}(t) = \mathcal{E}(p(t), P(t))$, we observe

$$\arg\min\{(l^0, u) | u \in \mathcal{E}(p(t), P(t))\} = \mathcal{U}_-(t, x). \quad (4.6)$$

The result (4.2) is then a consequence of the following fact:

Lemma 4.2 *Given ellipsoid $\mathcal{E}(p, P)$, the minimizer u^* for the problem*

$$\min\{(l, u) | u \in \mathcal{E}(p, P)\} = (l, u^*) \ , l \neq 0,$$

is the vector $u^* = p \ - \ Pl(l, Pl)^{-\frac{1}{2}}$.

This lemma follows from the formula for the support function of an ellipsoid. The result may now be summarized in the assertion:

Theorem 4.1 *For the set $\mathcal{W}[t]$ define an internal approximation $\mathcal{E}_-[t] = \mathcal{E}_-(x^*(t), X_-(t))$ with fixed parameterizing functions $\pi(t), H(t)$ in (3.14). If $x[\tau] \in \mathcal{E}_-[\tau]$ and the synthesizing strategy is taken as $\mathcal{U}_-(t, x)$ of (4.2), then the following inclusion is true: $x[t] \in \mathcal{E}_-[t], \quad \tau \leq t \leq t_1$, and therefore $x[t_1] \in \mathcal{E}(m, M)$, whatever is the unknown disturbance $f(t)$.*

The "ellipsoidal" synthesis described in this section thus gives a solution strategy $\mathcal{U}_-(t, x)$ for any internal approximation $\mathcal{E}_-[t] = \mathcal{E}_-(x(t), X_-(t))$ of $\mathcal{W}[t]$. With $x \notin \mathcal{E}_-[t]$, the function $\mathcal{U}_-(t, x)$ is single-valued, whilst with $x \in \mathcal{E}_-[t]$ we have $\mathcal{U}_-(t, x) = \mathcal{E}_-[t]$. The overall solution $\mathcal{U}(t, x)$ turns to be upper-semicontinuous in x and measurable in t, ensuring therefore the existence of a solution to the differential inclusion (4.1). Due to Theorem 3.3 (see (3.20)), each element $x \in \text{int}W[t]$ belongs to a certain ellipsoid $\mathcal{E}_-[t]$ and may therefore be steered to the terminal set $\mathcal{M}$ by means of a certain "ellipsoidal-based" strategy $\mathcal{U}_-(t, x)$, under Assumption 2.1.

Relations (4.2),(4.5) indicate that strategy $\mathcal{U}_-(t, x)$ is given explicitly, with the only unknown being the multiplier λ of Lemma 3.1, which can be calculated as the only root of equation $f(\lambda) = 0$. But in view of Corollary 4.1 the function $\lambda = \lambda(t, x)$ may be calculated *in advance*, depending on the parameters $x^*(t), X_-(t)$ of the internal approximation $\mathcal{E}_-[t]$ (which may also be calculated in advance and is such that it ensures the feasibility of $\mathcal{U}_-$). With this specificity, the suggested strategy $U_-(t, x)$ may be considered as an *analytical design.*

In the absence of uncertainty (with $f(t)$ being known), the given scheme works again, but now in a simplified form, since $\mathcal{W}[t]$ should be substituted for $W[t]$ and the internal approximation $\mathcal{E}_-[t] = \mathcal{E}(x^*(t), X_-(t))$ for $E_-[t] = \mathcal{E}(x*(t), X_-(t))$, with $X_-(t)$ calculated due to equation (3.3), $X_-(t_1) = M$.

5 Viability and State Estimation

We shall begin with a discussion of *Dynamic Programming tcchniqucs for the viability problem*, [1], [17]. Consider system (1.1), (1.2) under an additional *viability constraint*:

$$G(t)x(t) \in \mathcal{E}(y(t), K(t)), \quad t_0 \leq t \leq t_1, \; K(t) > 0, \tag{5.1}$$
$$x(t_1) \in \mathcal{E}(m, M), \; M > 0,$$

with function $f(t) \equiv 0$. This constraint may particularly arrive due to a *measurement equation*

$$y(t) = G(t)x + v(t), \quad v(t) \in \mathcal{E}(0, K(t)), \tag{5.2}$$

where $y(t)$ is the observed measurement output.

We shall look for the *viability kernel* $W_v[\tau]$ at given instant τ, namely, the set of all points $x = x(\tau)$ for each of which there exists a control $u = u(t)$ that ensures the viability constraint (5.1). We shall determine $W[\tau]$ as *the level set* (the cut) $W[\tau] = \{x : V_v(\tau, x) \leq 0\}$ of the *the viability function* $V_v(\tau, x)$, which we define as the solution to the following problem:

$$V_v(\tau, x) = \min_u \left\{ \int_\tau^{t_1} (h_+^2(x(t), \mathcal{E}(q(t), Q(t))dt + h_+^2(x(t_1), \mathcal{E}(m, M)) \right\} \tag{5.3}$$

under restrictions $u(t) \in \mathcal{E}(p(t), P(t), \; \tau \leq t \leq t_1; \; x[\tau] = x.$

The external ellipsoidal approximation $\mathcal{E}(x^0(\tau), X(\tau)) \supseteq W[\tau]$ may be achieved through the scheme of paper [2] or through the following relations that follow from [17], Section 5.5 :

$$\dot{x}^0 = A(t)x^0 + p(t) + M(t)(y(t) - G(t)x(t)), \tag{5.4}$$

where

$$\dot{X} = (A(t) - M(t)G(t))X + X(A(t) - G'(t)M'(t)) - \tag{5.5}$$
$$-(\pi(t) + \gamma(t))X - (\gamma(t))^{-1}P(t) - (\pi(t))^{-1}M(t)K(t)M'(t),$$

where $x(\tau) = x, \;\; X(\tau) = X_0$ and $\pi(t) > 0, \gamma(t) > 0, M(t)$ are continuous functions.

What follows is the assertion

Theorem 5.1 *The viability domain $W_v(\tau)$ for system (1.1). (1.2) under restriction (1.3) on $u(t)$ and state constraints (5.1) (with $y(t), K(t)$ continuous) satisfies the inclusion $W_v[\tau] \in \mathcal{E}(x(\tau), X(\tau))$, where $x(t), X(t)$ satisfy the differential equations (5.4), (5.5) within the interval $\tau \leq t \leq t_1$. Moreover, the following relation is true*

$$W_v[\tau] = \cap\{\mathcal{E}(x(t), X(t))|\pi(\cdot), \gamma(\cdot), M(\cdot)\} \tag{5.6}$$

where $\pi(t) > 0, \gamma(t) > 0, M(t)$ are continuous functions.

System (5.4),(5.5) was derived under the assumption that function $y(t)$ is continuous. However, we may as well assume that $y(t)$ is allowed to be piece-wise "right"-continuous or even Lebesgue-measurable. Then the respective relations may follow the lines of paper [6].

A similar problem solved in direct time, through a "forward" H-J-B equation, leads to the description of *"guaranteed state estimation"* under unknown but bounded errors, see [15], Part IV.

References

[1] J.P. Aubin. *Viability Theory.* Boston: Birkhäuser, 1991.

[2] J.S. Baras and A.B. Kurzhanski. Nonlinear filtering: the set-membership (bounding) and the H_∞ approaches. *Proc. of the IFAC NOLCOS Conference, 1995.*

[3] T. Basar and P. Bernhard. H^∞ *Optimal Control and Related Minimax Design Problems.* Boston: Birkhäuser, 1995.

[4] F.L. Chernousko. *State Estimation for Dynamic Systems.* New York: CRC Press, 1994.

[5] M.G. Crandall and P.L. Lions. Viscosity Solutions of Hamilton-Jacobi Equations. *Trans. Amer. Math. Soc.* **277** (1983), 1-42.

[6] T.F. Filippova, A.B. Kurzhanski, K. Sugimoto and I. Vályi. Ellipsoidal calculus, singular perturbations and the state estimation problem for uncertain systems. *Journ. of Math. Systems. Estimation and Control* **6**(3) (1996), 323-338.

[7] W.H. Fleming and H.M. Soner. *Controlled Markov Processes and Viscosity Solutions.* Berlin: Springer-Verlag, 1993.

[8] H. Knobloch, A. Isidori and D. Flockerzi. *Topics in Control Theory.* Boston: Birkhäuser, 1993.

[9] N.N. Krasovskii. In Russian: *Game-Theoretic Problems on the Encounter of Motions.* Moscow: Nauka, 1970. In English: *Rendezvous Game Problems.* Springfield, VA: Nat. Tech. Inf. Serv., 1971.

[10] N.N. Krasovski and A.N. Subbotin. *Positional Differential Games.* Berlin: Springer-Verlag, 1988.

[11] A. Krener. Necessary and sufficient conditions for worst-case H_∞ control and estimation. *Math. Syst. Estim. and Control* **4** (1994).

[12] A.B. Kurzhanski. *Control and Observation Under Uncertainty.* Moscow: Nauka, 1977.

[13] A.B. Kurzhanski and O.I. Nikonov. On the problem of synthesizing control strategies: evolution equations and set-valued integration. *Doklady Akad. Nauk SSSR* **311** (1990), 788-793.

[14] A. Kurzhanski and O.I. Nikonov. Evolution equations for tubes of trajectories of synthesized control systems. *Russ. Acad. of Sci. Math. Doklady* **48**(3) (1994), 606-611.

[15] A.B. Kurzhanski and I. Vályi. Ellipsoidal techniques for dynamic systems: the problems of control synthesis. *Dynamics and Control* **1** (1991), 357-378.

[16] A.B. Kurzhanski and I. Vályi. Ellipsoidal techniques for dynamic systems: control synthesis for uncertain systems. *Dynamics and Control* **2** (1992), 87-111.

[17] A.B. Kurzhanski and I. Vályi. *Ellipsoidal Calculus for Estimation and Control.* Boston: Birkhäuser, 1996.

[18] P.L. Lions and P.E. Souganidis. Differential games, optimal control and directional derivatives of viscosity solutions of Bellman's and Isaac's equations. *SIAM J. Cont. Opt.* **23** (1995), 566-583.

[19] L.S. Pontryagin. Linear differential games of pursuit. *Mat. Sbornik* **112**(154) (1980).

[20] R.T. Rockafellar. *Convex Analysis.* Princeton, N.J.: Princeton University Press, 1970.

[21] F.C. Schweppe. Recursive state estimation: unknown but bounded errors and system inputs. *IEEE, Trans. Aut. Cont.*, **13** (1968).

[22] A.I. Subbotin. *Generalized Solutions of First-Order PDE's. The Dynamic Optimization Perspective.* Boston: Birkhäuser, 1995.

Moscow State University, Russia.

Control and Stabilization of Interactive Structures

I. Lasiecka

1 Introduction

Questions related to stabilizability and control of interactive structures, which often arise in the context of the so called "smart materials technology," have attracted considerable attention in recent years. These models are governed by systems of coupled PDE equations with prescribed transmission conditions. The equations involved are, typically, combinations of parabolic and hyperbolic dynamics with boundary/point controls. Specific examples include: thermoelastic plates reinforced with shape memory fibers, structural acoustic models with piezoceramic actuators, electromagnetic structures with piezoelectric sensors/actuators, etc. In this paper, the emphasis is placed on the mathematical theory which describes-predicts the dynamic response of the structures subject to control actions. The mathematical interest/challenge in studying this kind of model is precisely the effect of coupling, the "mixing" of two different types of dynamics which, in turn, may often produce unexpected effects affecting the entire structure, without being valid for each element separately.

It is known (and it will be briefly reviewed below) that the properties of the feedback control operator corresponding to standard optimization problems will depend critically on the type of dynamics considered. For instance, in the case of boundary or point control (more generally, the so called unbounded control action) it is known that while in the parabolic case the optimal feedback operator has "smoothing" properties and the resulting gain operator is always bounded; this is not the case for hyperbolic-like dynamics. Indeed, in this latter case we have to deal with gain operators intrinsically unbounded and only densely defined on a basic state space. This, of course, has negative implications on the issue of robustness or numerical implementations of such feedback operators. However, if we change the scenario and consider a coupled system of parabolic-hyperbolic dynamics, then the natural question which arises is whether the parabolic or the hyperbolic features of the dynamics will prevail for the overall structure.

Do we get any benefit from the analyticity associated with parabolic problems? Is this effect transfered onto the entire structure? Or perhaps, only partially on some components (of interest to us) of the structure.

It is not always the case that the "parabolic" part of the system is the one "better behaved." Indeed, everything depends on the type of control problem considered. If one considers exact controllability questions or other related inverse type/ recovery type problems, then the dominating effect of dynamics should be hyperbolic-like. Thus, similar questions as the ones raised above should be asked with respect to the possibility of "propagation" of hyperbolicity onto the entire (or part of) dynamics and the parabolic part should be suppressed.

These are all very natural questions to ask, and the answer predictably depends on the type of interaction between various components of dynamics or, in other words, on the transmission conditions or coupling conditions between the two problems, including the geometry of the coupling. What is more interesting, from a control point of view, is that this coupling can be "enhanced" by a "smart" application of control devices (shape memory alloys, rheological fluids, piezoelectric elements, optic fibers, etc) which may produce desirable and sought after effects.

In order to focus our attention, and to convey main ideas, we shall concentrate on a discussion of a rather specific problem of a nonlinear thermoelastic plate, which provides a good example of a model with a strong interplay between "hyperbolicity" and parabolicity of coupled dynamics. Other examples of problems in the same category are models related to structural acoustic problems (see [8], [10],[9] and references therein) which constitute structural combinations of hyperbolic and parabolic-like systems. The effects of parabolicity and hyperbolicity on control theoretic properties of structural acoustic models have been analyzed in [3], [2], [1].

Our paper is organized as follows: in section 1 we provide a brief overview of various results dealing with LQR and Riccati theory for abstract systems with unbounded controls. We shall distinguish the two cases of parabolic and hyperbolic like dynamics. In the same section we shall also briefly discuss general results pertaining to controllability and stabilizability of such dynamics. Particular emphasis will be paid to the issue as to what is and what is not possible to achieve with a given type of actuators.

In section 2 we present stability theory pertinent to nonlinear thermoelastic plates.

Section 3 deals with the boundary control problems and related Riccati Equations addressed for several models of linear thermoelastic plates.

2 Abstract Results

In this section we shall consider an abstract control system with unbounded control actions. Let H (resp. U) be two Hilbert spaces corresponding respectively to the state and control space. Inner product in H (resp. U) will be denoted by $(.,.)$, (resp. $\langle .,. \rangle$). Let A denotes a generator of a strongly continuous semigroup $S(t)$ on H. The domain of the generator A will be denoted by $D(A) \subset H$. The control operator is denoted by $B : U \rightarrow [D(A)]'$ where X' stands for a dual (pivotal) space to X with respect to the topology in H. Note that if the control operator is bounded, we have $B : U \rightarrow H$.

The abstract control system to be considered is

$$y_t(t) = Ay(t) + Bu(t) \text{ on } [D(A)]'; \quad y(0) = y_0 \in H. \tag{2.1}$$

With control system (2.1) we associate the functional cost given by

$$J(y,u) \equiv \int_0^\infty [|Ry(t)|_Z^2 + |u(t)|_U^2]dt \tag{2.2}$$

where the observation operator $R : H \to Z$ is assumed bounded. (For many results this restriction is not necessary. However, in order to simplify the exposition and to focus on the maximal unboundedness of the control operator, we shall assume boundedness of the observation). We assume the usual finite cost condition.

Assumption 2.1 For any $y_0 \in H$, there exist $u \in L_2(0,\infty;U)$ such that $J(u,y(u)) \leq \infty$ for $y(u)$ satisfying (2.1).

By classical optimization methods one shows that

- for every initial data y_0 there exist an optimal control u^0 which minimizes $J(u,y)$ given by (2.2) with respect to all pairs (u,y) subject to the dynamics in (2.1), and
- there exist positive, selfadjoint and bounded operator $P : H \to H$ such that $(Px, x) = \int_0^\infty [|Ry^0(t)|_Z^2 + |u^0(t)|_U^2]dt$.

The main point is feedback characterization of the optimal control u^0 in terms of the optimal state y^0. This is related to appropriate solvability of a weak form of Riccati Equation:

$$(A^*P(t)x, y) + (P(t)Ax, y) + (R^*Rx, y) = \langle B^*P(t)x, B^*P(t)y \rangle, \tag{2.3}$$

where $x, y \in D$, $D \subset H$ is an appropriate dense set in H.

Our goal is to establish the optimal synthesis i.e. the following representation: $u^0(t) = -B^*Py^0(t)$; $t > 0$.

Since the operator B is unbounded, it is not clear a priori what the meaning of a composition operator B^*P is. This is precisely the main point which distinguishes between two types of dynamics: hyperbolic and parabolic. Now we shall consider two cases separately. Accordingly, we shall formulate the following assumptions:

Assumption 2.2 Let A be a generator of an *analytic* semigroup $S(t)$ on H and let the control operator B satisfies the following condition:

$$[A - \lambda I]^{-\gamma} B \in \mathcal{L}(U, H); \text{ for some } \gamma < 1 \tag{2.4}$$

and for some λ in the resolvent set $\rho(A)$.

Assumption 2.3 Let A be a generator of a C_0 semigroup $S(t)$ on H and the control operator B satisfies the following conditions:

$$[A - \lambda I]^{-1} B \in \mathcal{L}(U, H); \tag{2.5}$$

$$\int_0^T |B^* S^*(t) x|_U^2 dt \leq C_T |x|_H^2; \text{ for all } x \in D(A^*). \tag{2.6}$$

The following results, quoted from [20] (see also [22], [21], [30] and references therein), are available for each case

Theorem 2.1 *Assume Assumption 2 and let $\epsilon > 0$ be an arbitrary small constant.. The operator P has the following regularity property:*

$$P \in \mathcal{L}(H, D([A^* - \lambda I]^{1-\epsilon})), \tag{2.7}$$

and, in particular:

$$B^* P \in \mathcal{L}(H, U). \tag{2.8}$$

Moreover, the operator P satisfies Riccati Equation (2.3) with $D \equiv D([A - \lambda I]^\epsilon)$. The optimal feedback operator $A_P \equiv A - BB^* P$ is a generator of an analytic semigroup.

The optimal synthesis can be written in a pointwise manner as:

$$u^0(t) = -B^* P y^0(t); \text{ for all } t \geq 0, y_0 \in H. \tag{2.9}$$

Under the additional, usual "dectability" condition we also have the uniform stabilizability of the feedback semigroup.

$$|e^{A_P t} x|_H \leq C e^{-\omega t} |x|_H ; t > 0, \omega > 0. \tag{2.10}$$

Theorem 2.2 *Let us assume Assumption 2. Instead of regularity properties in (2.8) we only have $B^* P : H \to U$is densely defined.*

$A_P \equiv A - BB^* P$ is a generator of a C_0 semigroup

$$B^* P \in \mathcal{L}(D(A_P), U). \tag{2.11}$$

The gain operator is also defined on $D(A)$, but this requires, in general, a suitable extension of a typically uncloseable operator B^*. Indeed, we have

$$B_e^* P \in \mathcal{L}(D(A), U), \tag{2.12}$$

where B_e^* is a suitable extension of B^*. The optimal synthesis can be written in a pointwise manner as:

$$u^0(t) = -B^* P y^0(t); \text{ a.e. } t > 0, y_0 \in H; \text{ and for all } t > 0, \text{when } y_0 \in D(A_P). \tag{2.13}$$

Moreover, the operator P satisfies Algebraic Riccati Equation with $D \equiv D(A_P)$, and in the case $D \equiv D(A)$ we need to replace B^*P by its extension B_e^*P.

Under the additional "dectability" condition we also have the uniform stabilizability of the feedback semigroup

$$|e^{A_P t}x|_H \leq Ce^{-\omega t}|x|_H; t > 0; \omega > 0. \tag{2.14}$$

Remark 2.1 Theorems 2.1 and 2.2 deal with the infinite horizon problem. For results pertinent to the finite horizon problem with unbounded controls we refer to [20], [21], [13], [28] and references therein.

Remark 2.2 Note that the very basic difference between the two cases, hyperbolic and parabolic, is the fact that the gain operator is bounded in the former case while it is intrinsically *unbounded* in the latter case. Moreover, it was shown (see [20]) that in the case of time reversible dynamics, bounded gain operators B^*P can not provide uniform stabilization. This means that for time reversible and dectable systems, the gain operator B^*P must be *unbounded.*

We conclude this section with few general remarks about stabilizability/ controllability of parabolic and hyperbolic systems.

As it is well known, stabilization of parabolic systems is, essentially, a finite dimensional problem. Indeed, due to the analyticity of the underlying semigroup, the unstable part of the system is confined to finitely many modes. Analyticity also implies that the uniform decays are guaranteed once all the elements of the spectrum are located in the left complex plane. In view of this, the entire issue of stabilization is reduced to a finite dimensional pole placement problem (modulo some technicalities to assure that the spillover effect will not destabilize the already stable infinite-dimensional part of the system).

In contrast, the concept of uniform stability for hyperbolic systems is rather demanding. While *strong* stability is easily achievable (via application of the usual La Salle invariance principle), the *uniform stability* is an all together different matter. To illustrate this point, we recall a negative result, which is classical by now.

Theorem 2.3 *[30] Assume that the semigroup $S(t)$ is time reversible. Let $F : H \to H$ be a relatively bounded and finite rank operator. Then, $A + F$ is uniformly stable, iff A is uniformly stable.*

Remark 2.3 The original version of this result, proved under the assumption that the perturbation F is compact (rather than relatively bounded), was first established by D. Russell [29].

In view of Theorem 2.3, it is clear that finite dimensional and relatively bounded feedback can not uniformly stabilize time reversible systems. Thus there are only two avenues to pursue. Either to abandon altogether a possibility of using finite dimensional controllers or to consider "highly unbounded" feedbacks. In fact, a successful example illustrating applicability of this second alternative is the case of boundary control. Indeed, if one considers one-dimensional problem with boundary stabilizing feedback (as considered by many authors), then such feedback is of finite rank. However, it can be shown that this feedback is *not relatively bounded*, thus the result of Theorem 2.3 does not apply. Many other examples of boundary feedbacks stabilizing various hyperbolic dynamics are given in [24], [18], [20] and references therein. Nevertheless, one needs to be aware of limitations imposed by "hyperbolicity," as far as uniform stabilization is concerned. The inability of using finite dimensional feedbacks in the conventional, distributed manner is of great concern to practitioners.

Finally, few words about exact controllability. In contrast to stabilization, parabolic problems are, typically, not exactly controllable. In fact, this is a simple consequence of smoothing effects of analytic dynamics. Instead, an approximate controllability of parabolic systems is rather common. For the hyperbolic problems, exact controllability is achievable and several results in this direction (including the cases of boundary controllability) are given in [24], [19] [18], [20] and references therein.

The remaining part of the paper will describe in more details an example of a thermoelastic plate which is a classical example of coupling between hyperbolicity and parabolicity. The interplay and trade-off between parabolicity and hyperbolicity is put to evidence and taken advantage of for the purposes of more effective control.

3 Thermoelastic Plates

In what follows, we shall consider a model of thermoelastic plate where it is assumed that in addition to mechanical strains/stresses, the dynamics of the plate is affected by thermal stresses resulting from variations of the temperature. The following PDE model is taken from [18].

$$w_{tt} - \gamma \Delta w_{tt} + \Delta^2 w + \alpha \Delta \theta = \beta[\mathcal{F}(w), w] \tag{3.1}$$

$$\theta_t - \eta \Delta \theta + \sigma \theta - \alpha \Delta w_t = 0 \tag{3.2}$$

where Ω is a two-dimensional, bounded open region with a smooth boundary Γ and $\alpha, \beta, \sigma, \eta$ are positive parameters (for physical interpretation of these parameters consult [18]). The constant γ is proportional to the square of the thickness and it is, therefore, small. For this reason some of the models assume the value of this constant equal to zero. $\mathcal{F}(w)$ is the Airy's stress function satisfying

$$\Delta^2\mathcal{F}(w) = -[w,w] \text{ in } Q,\ \mathcal{F} = \frac{\partial \mathcal{F}}{\partial \nu} = 0 \text{ on } \Gamma. \tag{3.3}$$

The von Karman bracket is given by $[u,v] \equiv u_{x,x}v_{y,y} + u_{y,y}v_{x,x} - 2u_{x,y}v_{x,y}$.

The initial conditions associated with the equation are given by: $w(0) = w_0, w_t(0) = w_1, \theta(0) = \theta_0$ in Ω. With the above equations we associate Robin type of boundary conditions imposed for the temperature

$$\frac{\partial\theta}{\partial\nu} = -\lambda\theta \quad \text{on} \quad \Gamma \times (0,\infty) \equiv \Sigma,\ \lambda \geq 0, \tag{3.4}$$

and we consider several types of boundary conditions satisfied by the displacement w.

- **Clamped boundary conditions**

$$w = \frac{\partial w}{\partial \nu} = 0 \quad \text{on} \quad \Gamma \times (0,\infty) \equiv \Sigma. \tag{3.5}$$

- **Hinged boundary conditions**

$$w = 0; \Delta w + B_1 w + \alpha\theta = 0 \quad \text{on} \quad \Sigma. \tag{3.6}$$

where $B_1 w \equiv (1-\mu)[2n_1n_2w_{xy} - n_1^2 w_{yy} - n_2^2 w_{xx}]$; the constant $\mu \in [0,1]$ is Poisson's modulus and $\nu = [n_1, n_2]$.

- **Simply supported boundary conditions**

$$\Delta w + B_1 w + \alpha\theta = 0; \ \frac{\partial \Delta w}{\partial \nu} + B_2 w - \gamma\frac{\partial w_{tt}}{\partial \nu} + \alpha\frac{\partial\theta}{\partial\nu} = 0 \quad \text{on} \quad \Sigma. \tag{3.7}$$

where $B_2 w \equiv (1-\mu)\frac{\partial}{\partial\tau}[(n_1^2 - n_2^2)w_{xy} + n_1n_2(w_{yy} - w_{xx})]$.

One could also consider various combinations of mixed boundary conditions when one type of boundary condition is prescribed on one portion of the boundary and another type on the second portion of the boundary. For simplicity of exposition, we are not going to do this.

As we see from the above equations, the thermoelastic model consists of coupled von Karman plate with a heat equation. Thus, we have a classical case of coupling between hyperbolic and parabolic dynamics. What is important here is the fact that the coupling is "strong" (it involves terms of higher order that that of the energy level). Therefore, one may anticipate that this may allow for a change of character of dynamics in some components of the structure. To analyze this phenomenon we introduce the energy of the system. This is given by:

$$E(t) = \int_\Omega w_t^2(t) + \gamma|\nabla w_t(t)|^2 + a(w(t),w(t)) + |\theta(t)|^2 + 1/2\beta|\Delta\mathcal{F}(w)|^2 d\Omega \tag{3.8}$$

where $a(w,w)$ is a quadratic form topologically equivalent to $H^2(\Omega)$ norm and its exact structure depends on the type of boundary conditions considered. By a simple integration by parts procedure one easily finds out that the energy of the system is nonincreasing, i.e.:

$$E_t(t) = -\int_\Omega [\eta|\nabla\theta|^2 + \sigma|\theta|^2]d\Omega - \eta\lambda \int_\Gamma \theta^2 d\Gamma. \tag{3.9}$$

In the linear case, standard semigroup arguments provide a proof of existence and uniqueness of finite energy solutions (see [18]). In the nonlinear case ($\beta > 0$), the wellposedness of finite energy solutions follows from the techniques presented in [12]. It should be noted that the uniqueness of solutions for the nonlinear case with $\gamma = 0$ is a rather intricate result (the problem has been open in the literature until recently see [23], [18], etc) and it involves newly discovered in [16], [12], [11] sharp regularity of the Airy stress function. In what follows we shall discuss stability and optimal control problem associated with the thermoelastic plate introduced above. As we shall see, the control theoretic properties will heavily depend on the value of the parameter γ, which, in turn, changes the character of the underlying dynamics.

3.1 Stability

Since the energy of the system is decreasing (see (3.9), a legitimate question to ask is that of asymptotic stability. As it is well known, the heat equation is uniformly stable while the von Karman (or Kirchhoff-in the linear case) component of the system is not. Thus the relevant question is whether the coupling between two components is strong enough to transfer stability from the heat component of the system onto the entire structure. If we do not ask for *uniform* stability, the answer to this question is particularly simple. Indeed, it can be shown, by standard Liapunov method, that the linear system is strongly stable. This is to say that for all initial data of finite energy $E(0) < \infty$ we have:

$$E(t) \to 0; \text{ when } t \to \infty. \tag{3.10}$$

However, if one is interested in a stronger stability requirements, the problem is more complex. In fact, a first result in this direction is due to Kim [17]; here he shows that a clamped, *linear* plate with $\gamma = 0$ is uniformly stable. Later, Lagnese [18] shows that by adding mechanical dissipation on the boundary of a simply supported plate, the energy of the *linear* model with $\gamma > 0$ decays exponentially to zero. This raises an interesting question whether mechanical dissipation is necessary for uniform stability. In discussing this issue we shall consider two separate cases: $\gamma = 0$ and $\gamma > 0$. As we shall see, the value of this parameter has predominant effect on the type of dynamics of the overall structure. This is easily seen by noting that the coupling term Δw_t has anisotropic order equal to four, when $\gamma = 0$, and isotropic order equal to three, when $\gamma > 0$. Thus, in the case of $\gamma = 0$ the coupling is of the order of the principal of the linear operator. Indeed, in this case, it was recently discovered [26] that the linear model (3.1), (3.2) with $\beta = 0$ and boundary conditions either clamped (see (3.5) or hinged (3.6) generates an *analytic* semigroup. (A question whether the same result holds for simply supported boundary conditions, remains still at the level of a conjecture). This simply means that the analyticity generated by the heat component of the structure is transferred onto the entire system. This is possible, of course, due to strong coupling represented by differential operator of the same order as that of principal part.

In this situation the problem of uniform stability of linear part of the model is essentially solved. Indeed, strong stability result stated in (3.10), together with the analyticity of the underlying semigroup imply automatically exponential decay rates for the energy function. In the nonlinear case ($\beta > 0$), the exponential decay rates have been established by pseudodifferential multipliers method. The precise result is formulated below:

Theorem 3.1 *Let w, θ be a solution of (3.1, (3.2) with clamped (3.5) or hinged (3.6) boundary conditions and with the value of the parameter $\gamma = 0$. Then, there exist constants $C(E(0) > 0, \omega(E(0)) > 0$ such that:*

$$E(t) \leq Ce^{-\omega t}E(0); \ \ t > 0, \omega > 0. \tag{3.11}$$

In the linear case, these constants do not depend on the initial energy $E(0)$.

Remark 3.1 The linear version of this result, in the case $\gamma = 0$, has been first shown, in [17], [27]. Still linear version of this result, but valid for the full range of values of the parameter $\gamma \geq 0$ (covering also the non-analytic case) has been proved in [6] by using pseudodifferential multipliers. In fact, this technique extends also to the nonlinear case, yielding the final result of Theorem 3.1 proved in [5].

Remark 3.2 The result of Theorem 3.1 covers the cases of clamped and hinged boundary conditions only. As expected, the case of simply supported boundary conditions is more complicated. (This was recognized already in [27]). There is no problem with strong stability, which property is robust with respect to boundary conditions and values of the parameter γ (see [18]); the issue is of course that of uniform stability. While it is conjectured that the analyticity of the overall structure still holds (with $\gamma = 0$) for simply supported plates, we do not know of any rigorous proof pertaining to this fact. Thus, the only available route to prove the uniform stability, is to carry the estimates. It was noticed by Avalos that the arguments presented in [4] can be modified to yield the sought after stability result stated in Theorem 3.1 valid also for simply sported plates with $\gamma = 0$.

Now, let us turn toward the case $\gamma > 0$. In this case we do not have any longer the analyticity of the underlying linear part of the system and the issue of stability is predictably more complicated. Indeed, in this case strong stability of the system gives no indication about uniform decay rates. In fact, in order to deal with the issue, hard PDE estimates are necessary. By using pseudodifferential multipliers including some elements of microlocal analysis the following result has been established in [5], [4].

Theorem 3.2 *[5], [4]*

- *Let w, θ be a solution of (3.1, (3.2) with clamped (3.5) or hinged (3.6) and with the value of the parameters $\gamma \geq 0, \beta \geq 0$. Then, there exist constants $C(E(0) > 0, \omega(E(0)) > 0$ such that:*

$$E(t) \leq Ce^{-\omega t}E(0); \; t > 0, \omega > 0. \tag{3.12}$$

  where the constants $C(E(0))$ and $\omega(E(0))$ are uniform with respect to $0 \leq \gamma \leq 1$.

- *In the case of simply supported boundary conditions (3.7), the estimate in (3.12) still holds but with the values C, ω depending on $\gamma > 0$.*

- *In the linear case, the constants C, ω do not depend on the initial energy $E(0)$.*

Concluding the discussion of stability issues, we see that in all the cases considered the coupling in the structure is instrumental in transferring stability properties from one component of the system onto the entire structure and, therefore, there is no need for mechanical dissipation.

Finally, a few words about controllability. As easily anticipated, exact controllability is not a property to be expected from thermoelastic systems (due to the existence of analytic component). Indeed, a negative result provided in [15] indicates that this property fails even for a one-dimensional thermoelastic rod. However, several positive results on *partial* exact controllability of thermoelastic plates have been shown in [19] and on *nullcontrollability* of one-dimensional thermoelastic rodes in [15].

3.2 Optimal Control

In this subsection we shall discuss several optimal control problems associated with the linear thermoelastic plate. We will focus on boundary control problems where the control is exercised either via thermal control(controlling temperature or, flux of the temperature through the boundary) or via mechanical control (moments, shears, forces applied to the edge of the plate). We minimize the functional

$$J(w,\theta) = \int_0^\infty [\int_\Omega [|\Delta w|^2 + |w_t|^2 + \gamma|\nabla w_t|^2 + |\theta|^2]d\Omega + \int_\Gamma |u(t)|^2 d\Gamma]dt \tag{3.13}$$

subject to u, w, θ satisfying equations (3.1), (3.2) with $\beta = 0$ and the following boundary conditions:

Problem 3.1

$$\frac{\partial \theta}{\partial \nu} = -\lambda\theta + u; \lambda \geq 0; \; w = \frac{\partial w}{\partial \nu} = 0 \quad \text{on} \quad \Gamma \times (0, \infty) \equiv \Sigma.$$

Problem 3.2 $\theta = u; w = \frac{\partial w}{\partial \nu} = 0$ on Σ.

Problem 3.3 $\frac{\partial \theta}{\partial \nu} = -\lambda\theta,\ \lambda \geq 0;\ w = 0; \Delta w + B_1 w + \alpha\theta = u$ on Σ.

Problem 3.4 $\frac{\partial \theta}{\partial \nu} = -\lambda\theta,\ \lambda \geq 0; \Delta w + B_1 w + \alpha\theta + c_1 \frac{\partial\, w_t}{\partial \nu} = k_1 u;\ \frac{\partial \Delta w}{\partial \nu} + B_2 w - \gamma \frac{\partial w_{tt}}{\partial \nu} + \alpha \frac{\partial \theta}{\partial \nu} - c_2 w_t = k_2 u$ on Σ, where the constants $c_1, c_2 \geq 0$ and $k_i \in R$.

The first two problems correspond to thermal control while the remaining two deal with mechanical control.

Remark 3.3 The analysis provided below remains the same if some of the boundary conditions are prescribed only on a portion of the boundary, and the others on the remaining part.

As before, we shall distinguish two cases, the "analytic" case $\gamma = 0$ and the non analytic case $\gamma > 0$. For the case $\gamma = 0$ we have the following result:

Theorem 3.3 *Let* $\gamma = 0$.

1. *Then the optimal control Problems 3.2, 3.2, and operator* $B^*P \in \mathcal{L}(L_2(\Gamma); H^2(\Omega) \times L_2(\Omega))$ *where the operator* P *and the corresponding analytic, uniformly stable feedback semigroup* $A - BB^*P$ *satisfy all the properties listed in Theorem 2.1.*

2. *In the case of Problem 3.2, we assume in addition that* $k_i \neq 0 \Rightarrow c_i > 0$. *Then the optimal feedback synthesis holds with the gain operator* $B^*P : L_2(\Gamma) \to H^2(\Omega) \times L_2(\Omega))$ *which is densely defined and it complies with the conditions of Theorem*

Proof. It is clear that the Finite Cost Condition (2) and dectability assumption are satisfied by the virtue of the uniform stabilizability result of Theorem as already noted in the previous section, we deal with the analytic case and the first part of the Assumption 2 is satisfied. Therefore, we are in a position to take advantage of the results presented in Theorem 2.1, provided we verify the validity of condition (2.4). To accomplish this we put all three problems into an abstract framework.

Problem 3.5 We define the following state and control spaces: $H \equiv H_0^2(\Omega) \times L_2(\Omega) \times L_2(\Omega)$; $U \equiv L_2(\Gamma)$. Then the functional cost is bounded on the energy space (i.e.: R is bounded). To define the appropriate operators A, B we need to introduce the following auxiliary operators: $\mathcal{A} : L_2(\Omega) \to L_2(\Omega)$ defined by $\mathcal{A} \equiv \Delta^2 w$; $w \in D(\mathcal{A}) \equiv \{w \in H_0^2(\Omega) \cap H^4(\Omega)\}$, $A_N : L_2(\Omega) \to L_2(\Omega)$ defined by $A_N \equiv -\Delta\theta$, $\theta \in D(A_N) \equiv \{w \in H^2(\Omega) : \frac{\partial\theta}{\partial\nu} = -\lambda\theta \text{ on } \Gamma\}$, $N : L_2(\Gamma) \to L_2(\Omega)$ defined by $Nu = v$, iff $\Delta v = 0$, *and* $\frac{\partial v}{\partial \nu} = -\lambda v + u$ *on* Γ. With the above notation we write the generator $A : H \to H$

$$A[w_1, w_2, \theta] \equiv [w_2, -\mathcal{A}w_1 + \alpha A_N\theta, +\alpha\Delta w_2 - (\sigma I + \eta A_N)\theta] \tag{3.14}$$

with $D(A) \equiv D(\mathcal{A}) \times H_0^2(\Omega) \times D(A_N)$, and the operator $B : U \to [D(A)]'$ is defined by

$$Bu = [0, -\alpha A_N N u, \eta A_n N u]. \tag{3.15}$$

To make our calculations somewhat simpler we take $\sigma = 0, \eta = 1$ (this is not essential). Elementary computations show: $A^{-1}Bu = [\alpha\mathcal{A}^{-1}A_N Nu, 0, Nu]$. Application of Green's formula gives: $\mathcal{A}^{-1}A_N N = 0$, hence $A^{-1}Bu = [0, 0, Nu]$. Noting that $D(\mathcal{A}^{1/2}) = H_0^2(\Omega)$ [14], interpolation between $D(A)$ and H yields:

$$D(A^r) = D(\mathcal{A}^{1/2+r/2}) \times D(\mathcal{A}^{r/2}) \times D(A_N{}^r); \ 0 \le r \le 1. \tag{3.16}$$

Since $N : L_2(\Gamma) \to D(A_N{}^{3/4-\epsilon})$ is bounded (see [20]), we conclude from (3.15) and (3.16) the boundedness of the operator

$$A^{-1}B : U \to D(A^{3/4-\epsilon}). \tag{3.17}$$

This implies that the assumption (2.4) in the formulation of Theorem. Thus the conclusion of Theorem 2.1 applies to the Problem 3.1, proving the theorem in this case.

Problem 3.6 Here, the analysis is similar. The spaces H and U and the operator $\mathcal{A}$ are the same as before. In addition we define the following auxiliary operators: $A_D : L_2(\Omega) \to L_2(\Omega)$ defined by $A_D \equiv -\Delta\theta$, $\theta \in D(A_D) \equiv \{w \in H^2(\Omega) : \theta = 0 \text{ on } \Gamma\}$, $D : L_2(\Gamma) \to L_2(\Omega)$ defined by $Du = v$, iff $\Delta v = 0$, *and* $v = u$ *on* Γ. Proceeding as before and replacing operators A_N, N with A_D, D we arrive at the conclusion

$$A^{-1}B : U \to D(A^{1/4-\epsilon}) \text{ is bounded.} \tag{3.18}$$

This implies that the assumption (2.4) in the formulation of Theorem Thus the conclusion of Theorem 2.1 applies to the Problem 3.2.

Problem 3.7 We define $H \equiv H^2(\Omega) \cap H_0^1(\Omega) \times L_2(\Omega) \times L_2(\Omega)$; $U \equiv L_2(\Gamma)$. As before, we easily see that the functional cost is bounded on the energy space (i.e.: R is bounded). To define the appropriate operators A, B we to introduce the following auxiliary operators: $\mathcal{A}_1 : L_2(\Omega) \to L_2(\Omega)$ defined by $\mathcal{A}_1 \equiv \Delta^2 w$; $w \in D(\mathcal{A}) \equiv \{w \in H_0^1(\Omega) \cap H^4(\Omega) : \Delta w + B_1 w = 0 \ on \Gamma\}$, $G_1 : L_2(\Gamma) \to L_2(\Omega)$ defined by $G_1 u = v$, $iff \ \Delta^2 v = 0$, $and \ v = 0, \Delta v + B_1 v = u$; $on \ \Gamma$.

The operators $A : H \to H$ and $B : U \to [D(A)]'$ are given by

$$A[w_1, w_2, \theta] \equiv [w_2, -\mathcal{A}_1 w_1 + \alpha(A_N + \mathcal{A}G_1)\theta, -\alpha A_D w_2 - (\sigma I + \eta A_N)\theta] \tag{3.19}$$

with $D(A) \equiv \{(w_1, w_2) \in [D(\mathcal{A}_1{}^{1/2})]^2, \theta \in D(A_N) : w_1 - \alpha G_1 \theta \in D(\mathcal{A}_1)\}$; (where we have used the fact that $D(\mathcal{A}_1{}^{1/2}) = H^2(\Omega) \cap H_0^1(\Omega)$, [14]);

$$Bu \equiv [0, \mathcal{A}_1 G_1 u, 0]. \tag{3.20}$$

Simple calculations show: $A^{-1}Bu = [-G_1 u, 0, 0]$. Noting that

$$D(A^r) = D(\mathcal{A}_1{}^{1/2+r/2}) \times D(\mathcal{A}_1{}^{r/2}) \times D(A_N{}^r) \ for \ 0 \leq r < 1/8,$$

and that $G_1 : L_2(\Gamma) \to D(\mathcal{A}_1{}^{5/8-\epsilon})$ is bounded (see [20]), we obtain that

$$A^{-1}B \in \mathcal{L}(U, D(A^{1/8-\epsilon})).$$

This implies that the assumption (2.4) in the formulation of Theorem Thus the conclusion of Theorem 2.1 applies to the Problem 3.3, completing the proof of the first part of the Theorem.

Problem 3.8 By similar arguments as above, one can show that the control operator B satisfies: $A^{-\gamma}B \in \mathcal{L}(U, H)$ with the value of $\gamma = 5/8 + \epsilon$. However, the analyticity of the underlying semigroup, in the case of simply supported boundary conditions, is still an open problem. For this reason, we are not in a position to apply the analytic result, and the result stated in the second part of the Theorem is considerably weaker (the gain operator is not guaranteed to be bounded). We shall return to this problem later, and complete the proof of the second part of Theorem, when dealing with the case $\gamma > 0$ which case is discussed below.

Since in the case $\gamma > 0$ we do not have analyticity, we shall work within the "hyperbolic" framework by applying the results of Theorem 2.2. Our main result is

Theorem 3.4 *Let $\gamma > 0$. Then the optimal control Problems 3.2, and 3.2 admit the optimal feedback synthesis with the* **densely defined** *gain operator $B^*P : L_2(\Gamma) \to H^2(\Omega) \times H^1(\Omega))$ where the operator P and the corresponding uniformly stable feedback semigroup $A - BB^*P$ satisfy all the properties listed in Theorem 2.2.*

In the case of Problem 3.2, the same conclusion holds, provided that $c_1 > 0$ if $k_1 \neq 0$. If $k_1 = 0$, then the gain operator is bounded.

Remark 3.4 In the nonanalytic case $\gamma > 0$ the Problem 3.2 does not comply with the assumptions of Theorem 2.2. (The operator B is not admissible). However, this particular problem falls into the category of non-admissible control problems studied by methods of integrated sermigroups in [22].

Proof. The finite cost condition (2) and detectability condition are satisfied due to uniform stability result of Theorem is topologically equivalent to $H^2(\Omega) \times H^1(\Omega) \times L_2(\Omega)$, the observation operator R is bounded. Therefore, we will be in a position to take advantage of the results presented in Theorem 2.2, provided we verify validity of conditions (2.5) and (2.6). To accomplish this we put all three problems into an abstract framework.

Problem 3.9 We define $H \equiv H_0^2(\Omega) \times H_0^1(\Omega) \times L_2(\Omega); U \equiv L_2(\Gamma)$. The operators $\mathcal{A}, A_D, A_N$ are the same as before. With the above notation the operators $A : H \to H$ and $B : U \to [D(A)]'$ are given by:

$$A[w_1, w_2, \theta] \equiv [w_2, A_D{}^{-1}(-\mathcal{A}w_1 + \alpha A_N \theta), +\alpha \Delta w_2 - (\sigma I + \eta A_N)\theta] \quad (3.21)$$

with $D(A) \equiv D(\mathcal{A}^{3/4}) \times H_0^2(\Omega) \times D(A_N)$, where we have used that $D(\mathcal{A}^{1/4}) \sim H_0^1(\Omega)$ and $\mathcal{A}^{1/4} A_D{}^{-1} \mathcal{A}^{1/4}$ is bounded [20].

$$Bu \equiv [0, 0, \eta A_N N u] \quad (3.22)$$

Taking (wlog) $\sigma = 0, \eta = 1$, we calculate : $A^{-1}Bu = [\alpha \mathcal{A}^{-1} A_N N u, 0, Nu]$. Application of Green's formula gives: $\mathcal{A}^{-1} A_N N = 0$, hence $A^{-1}Bu = [0, 0, \eta N u]$ and since $N \in \mathcal{L}(U, H^{3/2}(\Omega))$, we have that $A^{-1}Bu \in \mathcal{L}(U, H)$, as desired for (2.5). Our next task is to show that the condition (2.6) in Theorem $N^* A_N w = w|_\Gamma$, $B^*[w_1, w_2, \theta] = \theta|_\Gamma$. Thus, Condition (2.6) is satisfied as soon as we establish the following inequality:

$$\int_0^T \int_\Gamma \theta^2 d\Gamma dt \le C_T[|w_0|^2_{H^2(\Omega)} + |w_1|^2_{H^1(\Omega)} + |\theta_0|^2_{L_2(\Omega)}] \quad (3.23)$$

for all solutions w, θ corresponding to the homogenous equation (i.e. with $u = 0$).

The inequality in (3.23) follows from energy method combined with the Trace Theorem ([25]). Thus we have verified all the assumptions required by Theorem 2.2, and its conclusion applies to the present case.

Problem 3.10 With $H \equiv H^2(\Omega) \cap H_0^1(\Omega) \times H_0^1(\Omega) \times L_2(\Omega)$; $U \equiv L_2(\Gamma)$. The generator $A : H \to H$ and the boundary operator $B : U \to [D(A)]'$ are given by:

$$A[w_1, w_2, \theta] \equiv [w_2, A_D{}^{-1}(-\mathcal{A}_1 w_1 + \alpha(A_N + \mathcal{A}_1 G_1)\theta), \\ -\alpha A_D w_2 - (\sigma I + \eta A_N)\theta] \quad (3.24)$$

with $D(A) \equiv \{w_1, w_2 \in H^2(\Omega) \cap H_0^1(\Omega); \theta \in D(A_N); w_1 - \alpha G_1 \theta \in D(\mathcal{A}_1{}^{3/4})\}$.

$$Bu \equiv [0, A_D{}^{-1}\mathcal{A}_1 G_1 u, 0] \tag{3.25}$$

Since $G_1 : L_2(\Gamma) \to H^{5/2}(\Omega) \cap H_0^1(\Omega)$ we have $A^{-1}Bu = [-G_1 u, 0, 0] \in \mathcal{L}(U, H)$, and assumption (2.5) in the formulation of Theorem we compute $B^*[w_1, w_2, \theta] = \frac{\partial w_2}{\partial \nu}|_\Gamma$. Thua, (2.6) in Theorem 2.2 is satisfied as soon as we establish the following inequality

Lemma 3.5 *Let w, θ be a solution to (3.1), (3.2) with the homogenous boundary conditions (3.6). Then, the following inequality holds true:*

$$\int_0^T \int_\Gamma |\frac{\partial w_t}{\partial \nu}|^2 d\Gamma dt \leq C_T[|w(0)|^2_{H^2(\Omega)} + |w_t(0)|^2_{H^1(\Omega)} + |\theta(0)|^2_{L_2(\Omega)}]. \tag{3.26}$$

Notice that the result claimed in (3.26) does not follow from interior regularity of the solutions. Indeed, we only have $w_t \in H^1(\Omega)$, so the normal derivative on the boundary may be not even defined. However, it was shown in [4] that this sharp boundary regularity holds true. Thus the conclusion of Theorem 2.2 applies to the Problem 3.3.

Problem 3.11 Before we provide semigroup formulation for the problem, it is helpful to write down explicitly the abstract form of the second order equation. To accomplish this we introduce the following operators: $G_2 : L_2(\Gamma) \to L_2(\Omega)$ defined by: $G_2 u = v \; iff \; \Delta^2 v = 0; \Delta w + B_1 w = 0; \frac{\partial \Delta v}{\partial \nu} + B_2 v = u \; on \Gamma$, the operator $\mathcal{A}_2 : L_2(\Omega) \to L_2(\Omega)$; $\mathcal{A}_2 w = \Delta^2 w; D(\mathcal{A}_2) = \{w \in H^4(\Omega); \Delta w + B_1 w = 0, \partial \nu + B_2 v = 0 \; on \; \Gamma\}$, and the operator $A_0 : L_2(\Omega) \to L_2(\Omega)$ defined by: $A_0 \equiv I - \gamma\Delta$ with the associated Neumann homognous boundary conditions.

With the above notation, the abstract model for the original equation corresponding to Problem 3.2 becomes:

$$\begin{aligned} A_0 w_{tt} + \mathcal{A}_2 w + \mathcal{A}_2 G_1 (c_1 \frac{\partial w_t}{\partial \nu} + \alpha\theta) + \mathcal{A}_2 G_2 (-c_2 \frac{\partial w_t}{\partial \nu} + \alpha \frac{\partial \theta}{\partial \nu}) \\ -\alpha A_N \theta = k_1 \mathcal{A}_2 G_1 u + k_2 \mathcal{A}_2 G_2 u \\ \theta_t + (\eta A_N + \sigma I)\theta - \alpha \Delta w_t = 0. \end{aligned} \tag{3.27}$$

With $H \equiv H^2(\Omega) \times H^1(\Omega) \times L_2(\Omega)$; $U \equiv L_2(\Gamma)$, we obtain the following representation for the generator A and the operator $B : U \to [D(A)]'$:

$$\begin{aligned} A[w_1, w_2, \theta] \equiv [w_2, -A_0{}^{-1}(\mathcal{A}_2(w_1 + G_1(c_1 \frac{\partial w_2}{\partial \nu} + \alpha\theta) + \\ G_2(-c_2 \frac{\partial w_t}{\partial \nu} + \alpha \frac{\partial \theta}{\partial \nu})) + \alpha A_N \theta), \alpha \Delta w_2 - (\sigma I + \eta A_N)\theta] \end{aligned} \tag{3.28}$$

$$Bu \equiv [0, A_0{}^{-1}\mathcal{A}_2(k_1 G_1 u + k_2 G_2 u), 0]. \tag{3.29}$$

If $k_1 = 0$, then it is straightforward to show that the operator B is bounded on the state space. Thus, in this case a classical B-bounded theory [7] applies. We shall concentrate on the unbounded case $k_1 \neq 0$, and, for simplicity, we take $k_2 = 0$. Since $G_1 : L_2(\Gamma) \to H^{5/2}(\Omega)$ we have $A^{-1}Bu = [-k_1 G_1 u, 0, 0] \in \mathcal{L}(U, H)$. This implies that the first assumption (2.5) in the formulation of Theorem Now, we turn to the admissibility of the control operator. We compute B^*. Topologising $H^1(\Omega)$ with the graph norm of $D({A_0}^{1/2})$ and repeating similar computations to those in Problem 3.2 with A_D replaced by A_0 gives: $B^*[w_1, w_2, \theta] = \frac{\partial w_2}{\partial \nu}|_\Gamma$. Hence,the condition (2.6) in Theorem 2.2 is satisfied as soon as we establish the following inequality

Lemma 3.6 *Let w, θ be a solution to (3.1), (3.2) with the homogenous boundary conditions in Problem 3.2 (i.e.: $k_i = 0$). Moreover, we assume that the constant $c_1 > 0$.Then, the following inequality holds true:*

$$\int_0^T \int_\Gamma |\frac{\partial w_t}{\partial \nu}|^2 d\Gamma dt \leq C_T[|w(0)|^2_{H^2(\Omega)} + |w_t(0)|^2_{H^1(\Omega)} + |\theta(0)|^2_{L_2(\Omega)}]. \quad (3.30)$$

Notice that like in Lemma 3.5, the result claimed in (3.30) does not follow from the interior regularity of the solutions. However, the validity of (3.30) can be established by applying energy methods to the abstract form of equations (3.27). In the process of calculation one uses the identifications $G_1^* \mathcal{A}_2 w = \frac{\partial w}{\partial \nu}, G_2^* \mathcal{A}_2 w = -w|_\Gamma$ and topological equivalence of $D({\mathcal{A}_2}^{1/2}) \sim H^2(\Omega), D({A_0}^{1/2}) \sim H^1(\Omega)$. The details are omitted for lack of space.

Thus the conclusion of Theorem 2.2 applies to the Problem

We also note that identical argument to the one presented above gives the statement in the second part of Theorem 3.3 completing the proof of Theorem 3.3.

References

[1] G. Avalos. The exponential stability of a coupled hyperbolic/parabolic system arising in structural acoustics. *Abstract and Applied Analysis* **1** (1996), 203-219.

[2] G. Avalos and I. Lasiecka. The strong stability of a semigroup arising from a coupled hyperbolic/parabolic system. To appear *Semigroup Forum.*

[3] G. Avalos and I. Lasiecka. Differential Riccati equation for the active control of a problem in structural acoustics. To appear *Journal of Optimization Theory and Applications.*

[4] G. Avalos and I. Lasiecka. Exponential stability of a thermoelastic system without mechanical dissipation :the case of simply supported boundary conditions. Submitted to *Siam J. of Mathematical Analysis.*

[5] G. Avalos and I. Lasiecka. Uniform decays in nonlinear thermoelastic systems without mechanical dissipation. Preprint, 1996.

[6] G. Avalos and I. Lasiecka. *Exponential Stability of a Thermoelastic System without Mechanical Dissipation.* To be published by Rendiconti dell Instituto di Matematica dell Universita di Trieste.

[7] A.V. Balakrishnan. *Applied Functional Analysis.* New York: Springer Verlag, 1981.

[8] H.T. Banks and R.C. Smith. Feedback control of noise in a 2-d nonlinear structural acoustic model. *Discrete and Continuous Dynamical Systems* **1** (1995), 119-149.

[9] H.T. Banks and R.C. Smith. Implementation issues regarding pde-based controllers-control of transient and periodic vibrations. Preprint, 1995.

[10] H.T. Banks, R.C. Smith, and Y. Wang. The modeling of piezoceramic patch interactions with shells, plates and beams model. To appear *Quarterly of Applied Mathematics.*

[11] A. Favini, I. Lasiecka, M. A. Horn, and D. Tataru. Addendum to the paper:global existence, uniqueness and regularity of solutions to a von kármán system with nonlinear boundary dissipation. To appear *Differential and Integral Equations.*

[12] A. Favini, I. Lasiecka, M. A. Horn, and D. Tataru. Global existence, uniqueness and regularity of solutions to a von kármán system with nonlinear boundary dissipation. *Differential and Integral Equations* **9**(2) (1996), 267-294.

[13] F. Flandoli, I. Lasiecka, and R. Triggiani. Algebraic Riccati equations with non-smoothing observation arising in hyperbolic and Euler-Bernoulli boundary control problems. *Annali di Mat. Pura et Applicata* **153** (1988), 307-382.

[14] P. Grisvard. Caracterization de quelques espaces d'interpolation. *Arch. Rat. Mech. Anal* **25** (1967), 40-63.

[15] S. Hansen. Boundary control of a one-dimensional linear thermoelastic rode. *SIAM J. on Control* **32** (1994), 1052-1074.

[16] M. A. Horn, I. Lasiecka, and D. Tataru. Wellposedness and uniform decay rates of weak solutions to a von karman system with nonlinear dissipative boundary conditions. *Lecture Notes in Pure and Applied Mathematics, Marcel Dekker* **160** (1994), 133-159.

[17] U. Kim. On the energy decay of a linear thermoelastic bar and plate. *SIAM J. Math. Anal.* **23** (1992), 889-899.

[18] J. Lagnese. *Boundary Stabilization of Thin Plates.* Philadelphia: SIAM, 1989.

[19] J. Lagnese and J.L Lions. *Modeling Analysis and Control of Thin Plates.* Paris: Masson, 1988.

[20] I. Lasiecka and R. Triggiani. *Differential and Algebraic Riccati Riccati Equations with Applications to Boundary/Point Control.* New York: Springer Verlag, 1991.

[21] I. Lasiecka and R. Triggiani. Riccati differential equations with unbounded coefficients and nonsmooth terminal condition-the case of analytic semigroup. *SIAM J. Math. Anal* **23** (1992), 449-481.

[22] I. Lasiecka and R. Triggiani. Algebraic Riccati equations arising from systems with unbounded input-solution operator. *Nonl. Anal. Theory and Applications* **20** (1993), 659-695.

[23] J. Lions. *Quelques Methods do Resolution des Problemes aux Limits Nonlinearies.* Paris: Dunod, 1969.

[24] J. Lions. *Controllabilite Exacte,Perturbations et Stabilization de Systems Distribues.* Paris: Masson, 1989.

[25] J. L. Lions and E. Magenes. *Non-homogenous Boundary Value Problems and Applications.* New York: Springer Verlag, 1972.

[26] Z. Liu and M. Renardy. A note on the equations of thermoelastic plate. *Appl. Math. Letters* **8** (1995), 1-6.

[27] Z. Liu and S. Zheng. Exponential stability of the Kirchhoff plate with termal or viscoelastic damping. To appear *Quarterly of Applied Mathematics.*

[28] G. Da Prato, I. Lasiecka, and R. Triggiani. A direct study of the Riccati equations arising in hyperbolic boundary control problems. *J. Diff. Equations* **64** (1986), 26-47.

[29] D. Russell. Decay rates for weakly damped systems in Hilbert spaces obtained via control-theoretic methods. *J. Diff. Equations* **19** (1975), 344-370.

[30] R. Triggiani. Finite rank, relatively bounded perturbations of semigroups generators. Part III: a sharp result on the lack of uniform stabilization. *Diff. and Int. Equations* **3** (1990), 503-522.

Department of Applied Mathematics, University of Virginia, Charlottesville, Virginia 22903

Risk Sensitive Markov Decision Processes

S.I. Marcus[1], E. Fernández-Gaucherand[2], D. Hernández-Hernández,
S. Coraluppi, and P. Fard

1 Introduction

Risk-sensitive control is an area of significant current interest in stochastic control theory. It is a generalization of the classical, risk-neutral approach, whereby we seek to minimize an exponential of the sum of costs that depends not only on the expected cost, but on higher order moments as well.

Research effort has been directed towards establishing results that parallel those already available in the risk-neutral setting, as well as towards exploring the connections of risk-sensitive control to robust control and differential games. This paper summarizes some contributions to the first of these objectives.

For linear systems and exponential of the sum of quadratic costs, the problem has been studied by [26] in the fully observed setting. Extensions to the partially observed setting are due to [3] and [34]. A somewhat surprising result is that the conditional distribution of the state given past observations does not constitute an information state. The equivalence, in the large risk limit, to a differential game arising in H^∞ control is due to [12] and [20]. Nonlinear systems have been studied in [18, 31, 34]. Partially observed nonlinear systems are treated in [27], where the appropriate information state and dynamic programming equations are presented.

In parallel to the work in the control community, there has been a body of research on risk-sensitive Markov Decision Processes (MDPs) – i.e., discrete-time problems with a finite or countable state space. An early formulation of the risk-sensitive MDP problem is due to [25]. Discounted cost problems are studied in [8]; a surprising result is that, for the infinite horizon problem with discounted costs, the optimal policy is not stationary in general. An information state and dynamic programming equations for the partially observed problem are introduced in [2]. Structural results for the value function in the partially observed setting are provided in [15]. The average cost problem has been studied in [16], [17], [22], [23], [25].

[1]Research supported in part by the National Science Foundation under grant EEC 9402384.

[2]Research supported in part by a grant from the University of Arizona Foundation and the Office of the Vice President for Research; and in part by the National Science Foundation under grant NSF-INT 9201430.

The nonstationarity of the discounted problem has motivated an alternative generalization of the risk-neutral formulation, which preserves the stationarity of the optimal control law. The formulation is developed in [28], and has been studied recently in the linear systems context in [21]. Further results and algorithms for the discounted problem, including the problem with partial observations, are presented in [9].

2 The Risk Sensitive MDP Model

In this section we give the general idea of the problem formulation; specific assumptions will be stated in later sections. We restrict attention to discrete-time stochastic dynamical systems with finite or countable state and observation spaces, and finite or compact action (or control) set. For this class of systems, we can employ an MDP (or controlled Markov chain) description. This is given by $M = (X, Y, U, \{P(u), u \in U\}, \{Q(u), u \in U\})$, where X is the state space, Y is the output space, and U is the set of controls. X_t, Y_t, and U_t denote the state, output, and control at time t. $P(u)$ and $Q(u)$ are the state transition matrix and the output matrix, respectively, for $u \in U$. More precisely, $p_{i,j}(u) := pr(X_{t+1} = j | X_t = i, U_t = u)$ (also denoted $P(j|i,u)$) and $q_{i,y}(u) := pr(Y_t = y | X_t = i, U_{t-1} = u)$; in addition, we define the matrix $\overline{Q}(y,u) := diag(q_{i,y})(u)$.

A policy or control law is a sequence of mappings $\pi = (\pi_0, \pi_1, \ldots)$ such that $u_k = \pi_k(Y^k), Y^k = (Y_1, \ldots, Y_k)$. Let us denote by $\mathbf{\Pi}$ the set of admissible policies. Traditionally, an additive cost structure has been employed, of the form

$$J^0(\pi) = E^\pi[C_M],$$

where $c(i,u)$ is the cost per stage and $C_M = \sum_{t=0}^{M-1} c(X_t, U_t)$.

The *finite horizon risk-sensitive control* problem is to find a policy π to minimize

$$J^\gamma(\pi) := \gamma \log \mathbb{E}^\pi \left[exp(\gamma^{-1} \cdot \mathcal{C}_M) \right]$$

where $\gamma^{-1} \neq 0$ is the *risk factor;* in this paper we will consider the *risk-averse* case in which $\gamma > 0$. Notice that, to first order in γ^{-1},

$$J^\gamma(\pi) \simeq \mathbb{E}^\pi \left[\mathcal{C}_M \right] + \frac{1}{2\gamma} Var^\pi \left[\mathcal{C}_M \right].$$

The minimization of $J^\gamma(\pi)$ is equivalent to the minimization of

$$\overline{J}^\gamma(\pi) := \mathbb{E}^\pi \left[exp(\gamma^{-1} \cdot \mathcal{C}_M) \right].$$

We will also discuss the ***average cost risk-sensitive control*** problem, in which one seeks to minimize

$$J_a^\gamma(\pi) = \limsup_{T\to\infty} \frac{\gamma}{T} \log \mathbb{E}^\pi \exp\{\frac{1}{\gamma}\sum_{t=0}^{T-1} c(X_t, U_t)\}.$$

3 Complete State Observations

In this section, we assume that $Y_t = X_t$ – i.e., that we have complete observations of the state.

3.1 The Finite Horizon Case

This problem was first considered in [25]. Define the value function by

$$S_{k,M}^\gamma(x) := \min_\pi \mathbb{E}^\pi \left[exp(\gamma^{-1} \cdot \mathcal{C}_{k,M}) | X_k = x\right],$$

where $\mathcal{C}_{k,M} = \sum_{t=k}^{M-1} c(X_t, U_t)$. The value function satisfies the dynamic programming recursion:

$$\begin{aligned} S_{N,N}^\gamma &= 1 \\ S_{k,N}^\gamma &= \min_{u\in U}\{\mathcal{D}(u) S_{k+1,N}^\gamma\} \end{aligned}$$

where the minimum is taken separately for each component of the vector equation and $[\mathcal{D}(u)]_{i,j} := p_{i,j}(u) \cdot exp(\gamma^{-1}c(i,u))$ is the "disutility contribution matrix." There exists a Markov policy (i.e., a policy such that π_k depends only on the state X_k) that is optimal.

3.2 Infinite Horizon, Average Cost

For the finite state case, this problem has been studied in [25] and, more recently, in [16]; the latter paper also discusses the relation to robust control problems. We will present results for the countable state case, following [22, 23]. As in [22], we assume first that U is Borel space; for $x \in X$, $U(x)$ is the set of admissible actions. In addition, we will assume:

Assumption A.1.

(i) For each $x \in X, U(x)$ is a compact subset of U.

(ii) The cost function c is nonnegative, continuous and bounded.

(iii) For all $x, z \in X$, the function $u \mapsto P(z|x,u)$ is continuous on $U(x)$.

The average cost risk sensitive optimal control problem is to find a policy $\pi^* \in \mathbf{\Pi}$ that minimizes $J_a^\gamma(\pi)$. Define

$$\Lambda := \inf_{\pi \in \mathbf{\Pi}} J_a^\gamma(\pi).$$

The main objective is to find sufficient conditions to ensure the existence of a stationary optimal policy.

Verification Theorem.

Theorem 3.1 [22]. *Suppose that there exist a number λ and a bounded function $W : X \to \mathbb{R}$ such that*

$$e^{\lambda + W(x)} = \min_{u \in U(x)} \{ e^{\gamma^{-1} c(x,u)} \sum_{z \in X} e^{W(z)} P(z|x,u) \}. \tag{3.1}$$

Then

$$\lambda\gamma \leq J_a^\gamma(\pi) \text{ for all } \pi \in \mathbf{\Pi}.$$

Further, if π^* is a stationary policy, with $\pi^*(x)$ achieving the minimum on the r.h.s. of (3.1) for each $x \in X$, then π^* is optimal, and

$$\lambda = \lim_{T \to \infty} \frac{1}{T} \log \mathbb{E}^{\pi^*} \exp \{ \frac{1}{\gamma} \sum_{t=0}^{T-1} c(X_t, U_t) \}.$$

Remark. Notice that the right hand side of (3.1) looks like a moment generating function; we will use this fact later.

Existence of Solutions.

We turn to the question of existence of a solution to the dynamic programming equation (3.1). The main result is the following:

Theorem 3.2 [22]. *For each $e \in X$ and $\pi \in \mathbf{\Pi}$ define*

$$\tau_e := \min \{ t > 0 : x_t = e \}.$$

If there exist $e \in X$ and $C > 0$ such that

$$\mathbb{E}^\pi (\tau_e | X_0 = x) < C \tag{3.2}$$

for all $\pi \in \mathbf{\Pi}$ and $x \in X$, then there exists a solution (λ, W) to the dynamic programming equation (3.1), with W bounded.

Remarks. (i) Similar results are proved under weaker conditions in [23], and are briefly discussed below.

(ii) One might expect that this theorem can be proved by using the "vanishing discount" approach that has been so successful in the risk neutral case (see, e.g., [1] and the references therein); in this approach, one solves the corresponding discounted problem and obtains the solution of the average cost problem as a limit of discounted problems as the discount factor approaches 1. However, in the risk sensitive case, the optimal policies for the "corresponding" discounted problem with cost

$$\mathbb{E}^{\pi}\left[exp(\gamma^{-1}\sum_{t=0}^{\infty}\beta^{t}c(X_t,U_t))\right] \tag{3.3}$$

are *not stationary* [8]! An intuitive explanation is that the decision maker appears less risk averse, by a factor β, from step to step, approaching risk-neutrality as $t \to \infty$. An alternative approach, sketched here, employs instead a sequence of discounted dynamic games [16, 22].

This approach depends in a fundamental way on a duality result. Let $P(X)$ be the set of probability vectors on X, i.e.

$$P(X) = \{\mu = (\mu^0, \mu^1, \ldots) : \mu^i \geq 0, \sum_{i \in X} \mu^i = 1\}.$$

Fix $\nu \in P(X)$, and define the relative entropy function $I(\cdot||\nu) : P(X) \to \mathbb{R} \cup \{+\infty\}$ by

$$I(\mu||\nu) = \begin{cases} \sum_{x\in X} log(r(x))\mu(x) & \text{if } \mu << \nu \\ +\infty & \text{otherwise} \end{cases}$$

where

$$r(x) = \begin{cases} \frac{\mu(x)}{\nu(x)} & \text{if } \nu(x) \neq 0 \\ 1 & \text{otherwise.} \end{cases}$$

The next lemma establishes, using a Legendre-type transformation, the duality relationship between the relative entropy function and the logarithmic moment generating function.

Lemma 3.3 [10, Proposition II.4.2]. *Let ψ be a bounded function defined on X, and let $\nu \in P(X)$. Then,*

$$\log \sum_{z\in X} e^{\psi(z)}\nu(z) = \sup_{\mu \in P(X)} \{\sum_{z\in X} \psi(z)\mu(z) - I(\mu||\nu)\};$$

the supremum is attained at the unique probability measure μ^* defined by

$$\mu^*(x) = \frac{e^{\psi(x)}}{\int e^{\psi} d\nu} \nu(x), \quad x \in S.$$

Using Lemma 3.3, we rewrite equation (3.1) as

$$\lambda + W(x) = \min_{u \in U(x)} \sup_{\mu \in P(X)} \{\sum W\mu + \frac{1}{\gamma} c(x,u) - I(\mu || P(\cdot|x,u))\} \quad (3.4)$$

This equation corresponds to the Isaacs equation associated with a stochastic dynamic game with average cost per unit time criterion (see [11, 22]).

Theorem 3.1 can then be proved via the vanishing discount approach, by first considering the corresponding infinite horizon discounted cost stochastic dynamic game. Let W_β be the upper value function of this game. Then, once we find a uniform bound for a "differential" discounted value function, i.e. $h_\beta(x) := W_\beta(x) - W_\beta(e)$, with e as in (3.2), the theorem follows by letting $\beta \to 1$.

First we introduce the infinite horizon discounted cost dynamic game.

Stochastic dynamic game. Let X be the state space, U be the control set for Player 1 (minimizer), and $P(X)$ be the control set for Player 2 (maximizer). The reward function is $(x, u, \mu) \mapsto \frac{1}{\gamma} c(x,u) - I(\mu || P(\cdot|x,u))$.

The evolution of the system is as follows (c.f. [16] [22]). At each time $t \in \{0, 1, \ldots\}$ the state of the system is observed, say $X_t = x \in X$. Then, a control $U_t \in U(x)$ is chosen for Player 1, and $\mu_t \in P(X)$ is chosen for Player 2. Then, a reward $\frac{1}{\gamma} c(X_t, U_t) - I(\mu_t || P(\cdot|X_t, U_t))$ is earned, and the state of the system moves to the state X_{t+1} according to the probability distribution μ_t.

Strategies. For each $t \geq 0$, let N_t and K_t be the set of feasible histories up to time t for Player 1 and Player 2, respectively. That is, $N_0 = S$ and $N_t = (S \times P(S))^t \times S$, while $K_0 = K$ and $K_t = K^t \times K$, where $K = \{(x,u) : u \in U(x), x \in X\}$. Generic elements of N_t and K_t are vectors of the form $n_t = (X_0, \mu_0, \ldots, X_{t-1}, \mu_{t-1}, X_t)$ and $K_t = (X_0, U_0, \ldots, X_{t-1}, U_{t-1}, X_t, U_t)$, respectively. A non-randomized strategy for Player 1 is a sequence $\pi = \{\pi_t\}$ of functions π_t from N_t to U, such that $\pi_t(n_t) \in U(X_t)$ for all $n_t \in N_t$. We say that π is stationary if, for all $t \geq 0, \pi_t$ depends only on the current state X_t, and π_t is independent of t. A non-randomized strategy for Player 2 is a sequence $\xi = \{\xi_t\}$ of functions ξ_t from K_t to $P(X)$. Stationarity of ξ is defined similarly.

Given the initial state $x \in X$, let $P_x^{\pi,\xi})$ be the probability induced by the strategies π, ξ , and $\mathbb{E}_x^{\pi,\xi}$ the corresponding expectation operator. Equation (3.3) corresponds to the dynamic progamming (Isaacs) equation of the stochastic dynamic game described above with average cost optimality criterion, defined for each $x \in X, \pi, \xi$ as

$$\Lambda(x,\pi,\xi) := \limsup_{T\to\infty} \mathbb{E}_x^{\pi,\xi} \frac{1}{T} \sum_{t=0}^{T-1} \left[\frac{1}{\gamma} c(X_t, U_t) - I(\xi_t || P(\cdot|X_t, U_t)) \right].$$

The corresponding discounted games are defined via the cost functionals

$$J_\beta(x,\pi,\xi) = \mathbb{E}_x^{\pi,\xi} \sum_{t=0}^{\infty} \beta^t \left[\frac{1}{\gamma} c(x_t, a_t) - I(\xi_t || P(\cdot|X_t, U_t)) \right],$$

where $\beta \in (0,1)$ is the discount factor.

Definition 3.4. When there exist a pair of strategies (π^*, ξ^*) such that

$$J_\beta(x,\pi^*,\xi) \le J_\beta(x,\pi^*,\xi^*) \le J_\beta(x,\pi,\xi^*)$$

for all π, ξ, the value $W_\beta(x) = J_\beta(x,\pi^*,\xi^*)$ is called the value of the game, and (π^*, ξ^*) are referred to as *optimal strategies*.

Lemma 3.5 [22]. *There is a unique bounded solution to the Isaacs equation*

$$W_\beta(x) = \min_{u\in U(x)} \sup_{\mu\in P(X)} \{ \sum_z \beta W_\beta(z)\mu(z) + \frac{1}{\gamma} c(x,u) - I(\mu || P(\cdot|x,u)) \}, \tag{3.5}$$

and it is the value function of the discounted cost stochastic dynamic game. Moreover, stationary strategies are optimal.

Remark. Note that, by Lemma 3.3, equation (3.5) can be rewritten as

$$e^{W_\beta(x)} = \min_{u\in U(x)} \{ e^{\frac{1}{\gamma} c(x,u)} \sum_z e^{\beta W_\beta(z)} P(z|x,u) \}. \tag{3.6}$$

This optimality equation has been studied by Eagle [11] in the context of a particular type of risk-sensitive discounted Markov decision process. Chung and Sobel [8] (see also the references therein), study risk-sensitive discounted Markov decision processes with a very different optimality equation, which results in *nonstationary* optimal policies (see Section 5 below).

Now, to employ the vanishing discount approach, we define $h_\beta(x) := W_\beta(x) - W_\beta(e)$, with e as in (3.2), and write (3.6) as

$$e^{(1-\beta)W_\beta(e)} \cdot e^{h_\beta(x)} = \min_{u\in U(x)} e^{\frac{1}{\gamma} c(x,u)} \sum e^{\beta h_\beta(z)} P(z|x,u) \}. \tag{3.7}$$

Sketch of proof of Theorem 3.1 [22]. It is first proved (using the assumption (3.2) and the boundedness of c) that $(1-\beta)W_\beta(e)$ and h_β

are uniformly bounded. Let $\beta_n \uparrow 1$ be given. Then, boundedness of $(1 - \beta)W_\beta(e)$ and h_β imply that, by a suitable diagonalization, we may pick a subsequence $\{\beta_n\}$ (denoting it again by $\{\beta_n\}$) along which $h_{\beta_n}(x), x \in X$, and $(1-\beta)W_\beta(e)$ converge to some limits $W(x)$ and λ, respectively. Thus, the theorem follows from (3.7) and an application of the Dominated Convergence Theorem.

In [23], this problem is studied under considerably weaker hypotheses, similar to those used in previous literature for the risk-neutral average cost criterion [5]-[7].

Assumption A.2.

(i) For each $x, z \in X$, the mapping $u \to P(z|x,u)$, with $u \in U(x)$ is lower semi-continuous.

(ii) For each $x \in X, U(x)$ is a compact subset of U.

Define

$$J_a^\gamma(x,\pi) = \limsup_{T\to\infty} \frac{\gamma}{T} \log \mathbb{E}^\pi[\exp\{\frac{1}{\gamma}\sum_{t=0}^{T-1} c(X_t,U_t)\}|X^0 = x].$$

Assumption A.3 (a) There exists a stationary policy $\bar{\pi} \in \mathbf{\Pi}$ such that

$$\rho := J_a^\gamma(x,\bar{\pi})$$

is finite and independent of x.

(b)

$$\liminf_{x\to\infty} \min_{u\in U(x)} c(x,u) > \rho.$$

The following theorem, which presents a dynamic programming *inequality*, is proved via the dynamic stochastic game and vanishing discount approach discussed above.

Theorem 3.6 [23]. *Under Assumptions A.2 and A.3, there exist a number ρ^* and a (possibly extended) function W on X such that for all $x \in X$*

$$e^{\rho^*+W(x)} \geq \inf_{u\in U(x)}\{e^{c(x,u)}\sum e^{W(z)}P(z|x,u)\}$$

and the set $H := \{x \in X : W(x)$ is finite$\}$ is not empty. Moreover, there exists an optimal policy $\pi^* \in \mathbf{\Pi}$ whenever the initial state belongs to H, and

$$\rho^* = J_a^\gamma(x,\pi^*)$$

for all $x \in H$.

4 Partial State Observations

In this section, we discuss risk sensitive Markov decision processes with partial state observations, also know as ***hidden Markov models.*** We assume throughout that X, U, and Y are finite with cardinalities N_X, N_U, and N_Y, respectively.

4.1 The Finite Horizon Case

As for the risk-neutral case [1], [4], [29], an equivalent stochastic optimal control problem can be formulated in terms of ***information states*** and ***separated policies.*** Here we follow the work of Baras, Elliott, and James [2], [27]. Let $\mathcal{Y}_t$ be the filtration generated by the available observations up to time t, and let $\mathcal{G}_t$ be the filtration generated by the sequence of states and observations up to that time. Then the probability measure induced by a policy π is equivalent to a canonical distribution $\mathcal{P}^\dagger$, under which $\{Y_t\}$ is independently and identically distributed (i.i.d), uniformly distributed, independent of $\{X_t\}$, and $\{X_t\}$ is a controlled Markov chain with transition matrix $P(u)$. Also,

$$\frac{d\mathcal{P}^\pi}{d\mathcal{P}^\dagger}|_{\mathcal{G}_t} = \lambda_t^\pi := N_Y^t \cdot \Pi_{k=1}^t q_{X_k,Y_k}(U_{k-1}).$$

The cost incurred by using the policy π is given by

$$\overline{J}^\gamma(\pi) = \mathbb{E}^\dagger\left[\lambda_M^\pi exp(\gamma^{-1} \cdot \mathcal{C}_M)\right].$$

Following [2], [27], the information state is given by

$$\sigma_t^\gamma(i) := \mathbb{E}^\dagger\left[I[X_t = i]exp(\gamma^{-1} \cdot \mathcal{C}_t) \cdot \lambda_t^\pi \mid \mathcal{Y}_t\right],$$

where $I[A]$ is the indicator function of the event A, and $\sigma_0^\gamma(i) = p_0$, where p_0 is the initial distribution of the state and is assumed to be known. With this definition of information state, similar results as in the risk-neutral case can be obtained. In particular, one obtains a recursive updating formula for $\{\sigma_t^\gamma\}$, which is driven by the output (observation) path and evolves forward in time. Moreover, the value functions can be expressed in terms of the information state only, and dynamic programming equations give necessary and sufficient optimality conditions for ***separated policies***, i.e., maps $\sigma_t^\gamma \mapsto \tilde{\pi}_t(\sigma_t^\gamma) \in U$; see [2], [27]. In particular we have that:

$$\overline{J}^\gamma(\pi) = \mathbb{E}^\dagger\left[\sum_{i=1}^{N_X} \sigma_M^\gamma(i)\right],$$

where $\{\sigma_M^\gamma\}$ is obtained under the action of policy π.

General Results.

The following lemma gives the recursions that govern the evolution of the information state.

Lemma 4.1 [2], [27]. *The information state process* $\{\sigma_t^\gamma\}$ *is recursively computable as:*

$$\sigma_{t+1}^\gamma = N_Y \cdot M(Y_{t+1}, U_t)\sigma_t^\gamma, \tag{4.1}$$

where

$$M(Y_{t+1}, U_t) := \overline{Q}(Y_{t+1}, U_t)\mathcal{D}^T(U_t),$$

$\overline{Q}(y,u) := diag(q_{i,y})(u)$, *$\mathcal{D}$ is defined in Section 3.1, T denotes transpose.*

Remark. Observe that as $\gamma^{-1} \to 0$, $\mathcal{D}(u) \to P(u)$ (elementwise). Therefore, (4.1) is the "natural" extrapolation of the (unnormalized) conditional probability distribution of the (unobservable) state, given the available observations, which is the standard risk-neutral information state [1], [4], [29].

Define value functions $J^\gamma(\cdot, M-k) : \mathbb{R}_+^{N_X} \to \mathbb{R}$, $k = 1, \ldots, M$, as follows:

$$J^\gamma(\sigma, M-k) := \min_{\pi_{M-k} \cdots \pi_{M-1}} \Big\{ \mathbb{E}^\dagger \{ \sum_{i=1}^{N_X} \sigma_M^\gamma(i) \mid \sigma_{M-k}^\gamma = \sigma \} \Big\}.$$

Lemma 4.2 [2]. *The dynamic programming equations for the value functions are:*

$$\begin{aligned} J^\gamma(\sigma, M) &= \sum_{i=1}^{N_X} \sigma(i); \\ J^\gamma(\sigma, M-k) &= \min_{u \in U} \{ \mathbb{E}^\dagger \left[J^\gamma(N_Y M(u, Y_{M-k+1}) \cdot \sigma, M-k+1) \right] \}, \\ k = 1, 2, \ldots, M. & \end{aligned} \tag{4.2}$$

Furthermore, a separated policy $\pi^* = \{\pi_0^*, \ldots, \pi_{M-1}^*\}$ that attains the minimum in (4.2) is risk-sensitive optimal.

The following generalize similar structural results for the standard risk-neutral case [1], [4], [13], [29], [32].

Lemma 4.3 [15]. *The value functions given by (4.2) are concave and piecewise linear functions of $\sigma \in \mathbb{R}_+^{N_X}$.*

Lemma 4.4 [15]. *Optimal separated policies $\{\pi_t^*\}$ are constant along rays through the origin; i.e., if $\sigma \in \mathbb{R}_+^{N_X}$ then $\pi_t^*(\sigma') = \pi_t^*(\sigma)$, for all $\sigma' = \alpha\sigma$, $\alpha \geq 0$.*

An action $\overline{u} \in U$ is said to be a resetting *action if there exists $j^* \in X$ such that $p_{i,j^*}(\overline{u}) = 1$, for all $i \in X$.*

Using these results and a result of Lovejoy [30, Lemma 1], the next Theorem can be proved.

Theorem 4.5 [15]. *Let $\overline{u} \in U$ be a resetting action. Then $CR^k_{\overline{u}}$, the region in which the control value $\overline{u}$ is optimal at time k, is a convex subset of $\mathbb{R}_+^{N_X}$.*

A Case Study.

In [15], risk sensitive control of a popular benchmark problem is considered; much is known about this problem in the risk-neutral case. This is a two-state replacement problem which models failure-prone units in production/manufacturing systems, communication systems, etc. The underlying state of the unit can either be *working* ($X_t = 0$) or *failed* ($X_t = 1$), and the available actions are to *keep* ($U_t = 0$) the current unit or *replace* ($U_t = 1$) the unit by a new one. The cost function $(x, u) \mapsto c(x, u)$ is as follows: let $R > C > 0$, then $c(0,0) = 0$, $c(1,0) = C$, $c(x,1) = R$. The observations have probability $1/2 < q < 1$ of coinciding with the true state of the unit. The state transition matrices are given as:

$$P(0) = \begin{bmatrix} 1-\theta & \theta \\ 0 & 1 \end{bmatrix}; \quad P(1) = \begin{bmatrix} 1 & 0 \\ 1 & 0 \end{bmatrix},$$

with $0 < \theta < 1$; see [13], [14], [33] for more details. With the above definitions, the matrices used to update the information state vector are given by:

$$M(y,0) = \begin{bmatrix} q_y(1-\theta) & 0 \\ (1-q_y)\theta & (1-q_y)e^{\gamma C} \end{bmatrix}; \quad M(y,1) = \begin{bmatrix} q_y e^{\gamma R} & q_y e^{\gamma R} \\ 0 & 0 \end{bmatrix}, \tag{4.3}$$

where $q_y := q(1-y) + (1-q)y$, $y = 0, 1$. For this case $\sigma = (\sigma(1), \sigma(2))^T \in \mathbb{R}^2_+$. Define the *replace* control region $CR^k_{replace}$ and the *keep* control region CR^k_{keep} in the obvious manner. The next result follows from (4.3), Lemma 4.4, and Theorem 4.5.

Lemma 4.6 [15]. *For all decision epochs the* replace *control region is a (possibly empty) conic segment in $\mathbb{R}^2_+$.*

The next result establishes an important *threshold* structural property of the optimal control policy. This is similar to well known results for the risk neutral case [13], [14], [30], [33].

Theorem 4.7 [15]. If $CR^k_{replace}$ is nonempty, then it includes the $\sigma(2)$-axis, i.e., $\mathbb{R}^2_+$ is partitioned by a line through the origin such that for values of $\sigma \in \mathbb{R}^2_+$ above the line it is optimal to *replace* the unit, and it is optimal to *keep* the unit otherwise.

Further structural results for this example, as well as results on limiting behavior for large and small risk factors, are presented in [15].

4.2 Infinite Horizon, Average Cost

Risk sensitive control of average cost, finite state, partially observed models has been studied in [17], using the approach discussed in Section 3.2 for the completely observed case. However, an information state is used in place of the state. In this paper, we will only consider the risk sensitive average cost version of the two-state replacement problem discussed above in Section 4.1; i.e., with cost functional $J_a^\gamma(\pi)$; for the risk neutral case, this problem has been studied in detail in [13]. As suggested in [17], it is convenient for the infinite horizon problem to use a normalized information state

$$\rho_t^\gamma := \frac{\sigma_t^\gamma}{|\sigma_t^\gamma|} = \begin{bmatrix} 1-\alpha_t^\gamma \\ \alpha_t^\gamma \end{bmatrix}$$

where

$$|\sigma| := \sum_{j=1}^{N_Y} \sigma(j).$$

Thus α_t^γ can be used as the (one-dimensional) information state. Then

$$\rho_{t+1}^\gamma = \frac{M(Y_{t+1},U_t)\rho_t^\gamma}{|M(Y_{t+1},U_t)\rho_t^\gamma|}$$

or

$$\alpha_{t+1}^\gamma = f(\alpha_t^\gamma, Y_{t+1}, U_t),$$

where f is defined implicitly.

The dynamic programming equation corresponding to (3.1) is

$$e^{\lambda + W(\alpha)} = \min_{u\in U}\{\mathbf{E}^\dagger\{|M(y,u)\begin{bmatrix} 1-\alpha \\ \alpha \end{bmatrix}|e^{W(f(\alpha,y,u))}\}, \tag{4.4}$$

and we have the following Verification Theorem.

Theorem 4.8 [24]. *Let (λ, W) be a solution of equation (4.4), with W bounded. Then the separated stationary policy π^*, with $\pi^*(\alpha)$ achieving the minimum in the r.h.s. of (4.4) is optimal and $\lambda\gamma$ is the optimal average cost.*

In order to study the existence of solutions to (4.4), we can take the same approach as in Section 3.2. Again optimal policies for discounted risk sensitive control problems are nonstationary, so we take logarithms in (4.4) and use the duality [10], via the Legendre transformation, between the log moment generating function and the relative entropy, to convert (4.4) into the following optimality equation for an average cost game:

$$\lambda + W(\alpha) = \min_{u \in U} \sup_{\xi \in P(Y)} \sum_{j=0}^{1} \xi^j \{-log(2\xi^j) + log|M(j,u) \begin{bmatrix} 1-\alpha \\ \alpha \end{bmatrix}|$$
$$+W(f(\alpha, j, u))\}.$$

Approximation with the corresponding discounted equations, as in Section 3.2, yields the existence of solutions to (4.4).

Proposition 4.9 [24]. There exist a number λ and a bounded function $W : [0,1] \to \mathbb{R}$ such that (4.4) is satisfied.

Structural Results.
Using arguments similar to those in [13], we can obtain results on the structure of the optimal policy.

Lemma 4.10 [24]. *If every average cost optimal policy replaces at* $\alpha \in [0,1]$*, then every average cost optimal policy replaces in the interval* $[\alpha, 1]$.

Lemma 4.11 [24]. *It is average cost optimal to produce in the interval* $[0, f(0,0,0)]$.

Condition C.1. For each $\beta \in (0,1)$,

$$\frac{1}{\gamma}C \leq (1-\beta)[\frac{1}{\gamma}R + \beta W_\beta(0)].$$

Theorem 4.12 [24]. *a) If (C.1) holds, then it is optimal to produce for all* $\alpha \in [0,1]$.
b) If (C.1) does not hold, then there exists $\alpha_\gamma \in [0,1)$ *such that it is optimal to produce in* $[0, \alpha_\gamma)$ *and repair in* $[\alpha_\gamma, 1]$.

Thus a simple threshold or "bang-bang" is optimal. Simulations are presented in [24] to compare the optimal policies for the risk sensitive and risk neutral cost criteria, and to study how the policies vary as a function of the risk factor γ^{-1}.

5 Alternative Risk Sensitive Approach

The risk sensitive approaches discussed above are generalizations of the risk neutral approach, in the sense that they seek to minimize a cost functional that is a generalization of the risk neutral cost functional. This works well for finite horizon and average cost risk sensitive problems, and results have been presented that correspond to those in the risk neutral case. However, as noted in Section 3.2, the minimization of the risk sensitive discounted cost functional (3.3) results in nonstationary optimal policies [8].

Instead of generalizing the expression for the cost to be minimized, one can alternatively generalize the risk-neutral dynamic programming equations characterizing the value function. This leads to a formulation for which, on the infinite horizon and with discounting and stationary costs, there exists a stationary optimal policy (see [8], [9], [21], [28]). This alternative formulation does *not* involve the optimization of a single cost (or utility) function for the entire path of the process; indeed, there is considerable debate in the decision theory and economics literature about whether the optimization of a single expected utility function is "rational" (see [28] and the references therein). In this alternative formulation, the control at each time is chosen to minimize an immediate cost, plus the discounted "certain equivalent" return from future stages, resulting in the following dynamic programming equation for the value function h^γ in the completely observed, finite state case:

$$h^\gamma(x) = \min_{u \in U} c(x,u) + \beta\gamma \log \sum_{z \in X} P_{xz}(u) \exp(\frac{1}{\gamma} h^\gamma(z)). \tag{5.1}$$

Existence of stationary optimal policies, as well as policy iteration algorithms, are shown in [8]. Extensions of the theory and the partially observed case are discussed in [9].

The discounted dynamic programming equation (5.1) is similar to (3.6) for the discounted game problem of Section 3.2. Indeed, Eagle [11] studied (3.6) in a context quite similar to that discussed in this section. An intriguing question is that of developing a more fundamental understanding of the relationship between the discounted problems discussed in this section and the discounted games discussed in Section 3.2.

References

[1] A. Arapostathis, V.S. Borkar, E. Fernández-Gaucherand, M.K. Ghosh and S.I. Marcus. Discrete-time controlled Markov processes with average cost criterion: a survey. *SIAM J. Control and Optim.* **31** (1993), 282-344.

[2] J.S. Baras and M.R. James. Robust and risk-sensitive output feedback control for finite state machines and hidden Markov models. To appear *J. Math. Systems, Estimation and Control.*

[3] A. Bensoussan and J.H. Van Schuppen. Optimal control of partially observable stochastic systems with exponential of integral performance index. *SIAM J. Control and Optim.* **23** (1985), 599-613.

[4] D.P. Bertsekas. *Dynamic Programming: Deterministic and Stochastic Model.* Englewood Cliffs, NJ: Prentice-Hall, 1987.

[5] V.S. Borkar. On minimum cost per unit of time control of Markov chains. *SIAM J. Cont. and Optim.* **22** (1984), 965-978.

[6] R. Cavazos-Cadena. Weak conditions for the existence of optimal stationary policies in average Markov decision chains with unbounded costs. *Kybernetika* **25** (1989), 145-156.

[7] R. Cavazos-Cadena and L.I. Sennott. Comparing recent assumptions for the existence of average optimal stationary policies. *Oper. Res. Lett.* **11** (1992), 33-37.

[8] K.-J. Chung and M.J. Sobel. Discounted MDP's: Distribution functions and exponential utility maximization. *SIAM J. Control and Optim.* **25** (1987), 49-62.

[9] S. Coraluppi and S.I. Marcus. Risk-sensitive control of Markov decision processes. *Proc. 1996 Conf. on Information Science and Systems.* 934-939.

[10] P. Dupuis and R.S. Ellis. *A Weak Convergence Approach to the Theory of Large Deviations.* To be published by John Wiley & Sons.

[11] J.N. Eagle, II. *A Utility Criterion for the Markov Decision Process.* Ph.D. Dissertation. Stanford University. Stanford, Caliornia. 1975.

[12] C.-H. Fan, J.L. Speyer and C.R. Jaensch. Centralized and decentralized solutions to the linear-exponential-Gaussian problem. *IEEE Transactions on Automatic Control* **39** (1994), 1986-2003.

[13] E. Fernández-Gaucherand, A. Arapostathis, and S.I. Marcus. On the average cost optimality equation and the structure of optimal Policies for partially observable Markov decision processes. *Annals of Operations Research* **29** (1991), 439-470.

[14] E. Fernández-Gaucherand, A. Arapostathis and S.I. Marcus. Analysis of an adaptive control scheme for a partially observed controlled Markov chain. *IEEE Transactions on Automatic Control* **38** (1993), 987-993.

[15] E. Fernández-Gaucherand and S.I. Marcus. Risk-sensitive optimal control of hidden Markov models: structural results. To appear *IEEE Transactions on Automatic Control.*

[16] W.H. Fleming and D. Hernández-Hernández. Risk sensitive control of finite state machines on an infinite horizon I. To appear *SIAM J. Control and Optim.*

[17] W.H. Fleming and D. Hernández-Hernández. Risk sensitive control of finite state machines on an infinite horizon II. Technical Report. Division of Applied Mathematics. Brown University.

[18] W.H. Fleming and W.M. McEneaney. Risk-sensitive control and differential games. *Springer Lecture Notes in Control and Info. Sci.* **184** (1992), 185-197.

[19] W.H. Fleming and W.M. McEneaney. Risk-sensitive control on an infinite horizon. *SIAM J. Control and Optim.* **33** (1995),1881-1915.

[20] K. Glover and J.C. Doyle, State-space formulae for all stabilizing controllers that satisfy an H_∞-norm bound and relations to risk sensitivit. *Systems and Control Lett.* **11** (1988), 167-172.

[21] L.P. Hansen and T.J. Sargent. Discounted linear exponential quadratic Gaussian control. *IEEE Transactions on Automatic Control* **40** (1995), 968-971.

[22] D. Hernández-Hernández and S.I. Marcus. Risk-sensitive control of Markov processes in countable state space. To appear *Systems and Control Lett.*.

[23] D. Hernández-Hernández and S.I. Marcus. Existence of risk sensitive optimal stationary policies for controlled Markov processes. Technical Report. Institute for Systems Research. University of Maryland.

[24] D. Hernández-Hernández, S.I. Marcus, and P. Fard. Analysis of a risk sensitive control problem for hidden Markov chains. Technical Report. Institute for Systems Research. University of Maryland.

[25] R.A. Howard and J.E. Matheson. Risk-sensitive Markov decision processes. *Management Sci.* **18** (1972), 356-369.

[26] D.H. Jacobson. Optimal stochastic linear systems with exponential performance criteria and their relation to deterministic differential games. *IEEE Transactions on Automatic Control* **18** (1973), 124-131.

[27] M.R. James, J.S. Baras and R.J. Elliott. Risk-sensitive control and dynamic games for partially observed discrete-time nonlinear systems. *IEEE Transactions on Automatic Control* **39** (1994), 780-792.

[28] D.M. Kreps and E.L. Porteus. Temporal resolution of uncertainty and dynamic choice theory. *Econometrica* **46** (1978), 185-200.

[29] P.R. Kumar and P. Varaiya. *Stochastic Systems: Estimation, Identification and Adaptive Control.* Englewood Cliffs, NJ: Prentice-Hall, 1986.

[30] W.S. Lovejoy. On the convexity of policy regions in partially observed systems. *Operations Research* **35** (1987), 619-621.

[31] T. Runolfsson. The equivalence between infinite horizon control of stochastic systems with exponential-of-integral performance index and stochastic differential games. *IEEE Transactions on Automatic Control* **39** (1994), 1551-1563.

[32] R.D Smallwood and E.J. Sondik. The optimal control of partially observable Markov processes over a finite horizon. *Operations Research* **21** (1973), 1071-1088.

[33] C.C. White. A Markov quality control process subject to partial observation. *Management Science* **23** (1977), 843-852.

[34] P. Whittle. *Risk-Sensitive Optimal Control.* New York: John Wiley & Sons, 1990.

Electrical Engineering Department and Institute for Systems Research, University of Maryland, College Park, Maryland 20742

Systems and Industrial Engineering Department, University of Arizona, Tucson, Arizona 85721

Department of Mathematics, CINVESTAV IPN, Apartado postal 14-740 Mexico D.F. 07000, Mexico

Electrical Engineering Department and Institute for Systems Research, University of Maryland, College Park, Maryland 20742

Electrical Engineering Department and Institute for Systems Research, University of Maryland, College Park, Maryland 20742

On Inverse Spectral Problems and Pole-Zero Assignment

Y.M. Ram

1 Introduction

The inverse eigenvalue problem of constructing a band symmetric matrix from spectral data may be regarded as a problem of assigning the poles and zeros of a certain frequency response function by passive control. This problem has application in controlling the dynamic response of undamped vibratory systems. Pole assignment is needed to avoid harmonic excitation near resonance frequencies. Particular components of the vibration response may be made to vanish by assigning the zeros appropriately. Alternatively, pole and zero assignment may be achieved by active control.

Consider for example the n degree-of-freedom second order system

$$\mathbf{M}\ddot{\mathbf{x}} + \mathbf{C}\dot{\mathbf{x}} + \mathbf{K}\mathbf{x} = \mathbf{e_n}e^{\lambda t}, \tag{1.1}$$

$\mathbf{e_n}$ the $n-$th unit vector, $\mathbf{M}$ positive definite symmetric *mass* matrix, $\mathbf{C}$ and $\mathbf{K}$ positive semi-definite *damping* and *stiffness* matrices, respectively. Equation (1.1) has the particular solution

$$\mathbf{x}(t) = \mathbf{v}e^{\lambda t}, \tag{1.2}$$

$\mathbf{v} = (v_1, v_2, \ldots, v_n)^T$ a constant vector. Substituting (1.2) in (1.1) gives

$$\mathbf{Q}(\lambda) = \mathbf{e_n}, \tag{1.3}$$

where

$$\mathbf{Q}(\lambda) = \lambda^2\mathbf{M} + \lambda\mathbf{C} + \mathbf{K}. \tag{1.4}$$

The *quadratic pencil* $\mathbf{Q}(\lambda)$ has $2n$ eigenvalues $\lambda_1, \lambda_2, ..., \lambda_{2n}$ which are the roots of

$$det(\mathbf{Q}(\lambda)) = 0. \tag{1.5}$$

Let $\hat{\mathbf{Q}}(\lambda)$ be the leading $n-1$ dimensional principal subpencil of $\mathbf{Q}(\lambda)$, and denote its $2n-2$ eigenvalues by $\mu_1, \mu_2, ..., \mu_{2n-2}$. The Cramer's rule applied to (1.3) gives

$$v_n = c\frac{\prod_{i=1}^{2n-2}(\lambda - \mu_i)}{\prod_{i=1}^{2n}(\lambda - \lambda_i)}, \tag{1.6}$$

c constant. By assigning the *zero* μ_p, $1 \leq p \leq 2n-2$ some integer to λ, the response v_n is made to vanish. By assigning the poles λ_i, $i = 1, 2, ..., 2n$, far away from the exciting frequency λ, excitation near resonance frequencies is avoided.

Section 2 deals with pole and zero assignment for simply connected systems by passive control. An inverse eigenvalue problem for the quadratic pencil (1.4) is solved where $\mathbf{C}$ and $\mathbf{K}$ are constructed to have the prescribed poles and zeros of (1.6). The classical result of reconstructing a tridiagonal *linear* pencil from spectral data (see e.g.. [1] and [3]) is thus generalised to the construction of tridiagonal *quadratic* pencil.

In Section 3 we present an explicit solution of the partial pole assignment problem for the system (1.1). Only a part of the spectrum is assigned while the rest of the spectrum remains unchanged. Hence, *spill-over*, the phenomenon in which poles not intended to be changed are modified by the control, will not occur and the stability of the controlled system can thus be assured. Saad [6] solved the associated problem for first order system by methods of projection and deflation. The explicit solution presented here holds for second order system. Reduction to first order realisation may thus be avoided. A comprehensive study on eigenvalue assignment for quadratic pencil is presented in Lancaster and Maroulas [4].

The problem of assigning both poles and zeros for an undamped system is addressed in the last section. It is shown that certain poles and zeros may be assigned simultaneously by appropriately selecting the position vector and the control function.

2 Inverse Eigenvalue Problem for a Quadratic Pencil

For a simply connected mass-spring-damper system $\mathbf{M}$ is diagonal, $\mathbf{C}$ and $\mathbf{K}$ are Jacobi matrices. Consider the simply connected system with $\mathbf{M}=\mathbf{I}$ and denote

$$\mathbf{C}=\begin{bmatrix} \alpha_1 & \beta_1 & 0 & \dots & 0 \\ \beta_1 & \alpha_2 & \beta_2 & \dots & 0 \\ 0 & \beta_2 & \alpha_3 & \ddots & 0 \\ \vdots & \vdots & \ddots & \ddots & \vdots \\ 0 & 0 & 0 & \dots & \alpha_n \end{bmatrix}, \quad \mathbf{K}=\begin{bmatrix} \gamma_1 & \delta_1 & 0 & \dots & 0 \\ \delta_1 & \gamma_2 & \delta_2 & \dots & 0 \\ 0 & \delta_2 & \gamma_3 & \ddots & 0 \\ \vdots & \vdots & \ddots & \ddots & \vdots \\ 0 & 0 & 0 & \dots & \gamma_n \end{bmatrix}. \tag{2.1}$$

Expanding the determinant of $\mathbf{Q}(\lambda)$ by its last row gives

$$\prod_{i=1}^{2n}(\lambda-\lambda_i)=(\gamma_n+\alpha_n\lambda+\lambda^2)\prod_{i=1}^{2n-2}(\lambda-\mu_i)-(\beta_{n-1}\lambda+\delta_{n-1})^2 q_{2n-4}(\lambda) \tag{2.2}$$

where $q_{2n-4}(\lambda)$ is the characteristic polynomial of the leading $(n-2)\times(n-2)$ subpencil of $\mathbf{Q}(\lambda)$.

¿From the coefficients of λ^{2n-4} in (2.2) we find that

$$\alpha_n = \sum_{i=1}^{2n-2} \mu_i - \sum_{i=1}^{2n} \lambda_i. \tag{2.3}$$

Substituting $\lambda = \mu_j, j = 1, 2, \ldots, 2n-2$ in (2.2) gives

$$\prod_{i=1}^{2n} (\mu_j - \lambda_i) = -p(\mu_j), \quad j = 1, 2, \ldots, 2n-2, \tag{2.4}$$

where

$$p(\lambda) = (\beta_{n-1}\lambda + \delta_{n-1})^2 q_{2n-4}(\lambda). \tag{2.5}$$

The polynomial $p(\lambda)$ has degree of $2n-2$ and a double root at $\lambda = r$, where

$$r = -\delta_{n-1}/\beta_{n-1}. \tag{2.6}$$

Let

$$\mathbf{V} = \begin{bmatrix} 1 & p(\mu_1) & p^2(\mu_1) & \ldots & p^{2n-3}(\mu_1) \\ 1 & p(\mu_2) & p^2(\mu_2) & \ldots & p^{2n-3}(\mu_2) \\ \vdots & \vdots & \vdots & & \vdots \\ 1 & p(\mu_{2n-2}) & p^2(\mu_{2n-2}) & \ldots & p^{2n-3}(\mu_{2n-2}) \end{bmatrix}, \tag{2.7}$$

$\mathbf{y} = (\mu_1, \mu_2, \ldots, \mu_{2n-2})^T$, $\mathbf{w} = (p^{2n-2}(\mu_1), p^{2n-2}(\mu_2), \ldots, p^{2n-2}(\mu_{2n-2}))^T$, and denote

$$\mathbf{g} = \mathbf{V}^{-1}\mathbf{y}, \quad \mathbf{h} = \mathbf{V}^{-1}\mathbf{w}. \tag{2.8}$$

It has been shown in [5] that the double root r of $p(\lambda)$ is determined by the roots of

$$\pi_1(\lambda)\frac{d}{d\lambda}\pi_2(\lambda) - \pi_2(\lambda)\frac{d}{d\lambda}\pi_1(\lambda) = 0, \tag{2.9}$$

where

$$\pi_1(\lambda) = \sum_{i=1}^{2n-2} g_i\lambda^{i-1}, \quad \pi_2(\lambda) = -\lambda^{2n-2} + \sum_{i=1}^{2n-2} h_i\lambda^{i-1}. \tag{2.10}$$

The polynomial $p(\lambda)$ can thus be determined by its $2n-1$ data points: the $2n-2$ points defined by (2.4) and the point $(r, 0)$ obtained by (2.9)-(2.10). Once $p(\lambda)$ is found, β_{n-1}^2 is determined by the leading coefficient of $p(\lambda)$. Then δ_{n-1} is obtained by (2.6). Substituting $\lambda = r$ in (2.2) gives

$$\gamma_n = \frac{\pi_{i=1}^{2n}(r - \lambda_i)}{\pi_{i=1}^{2n-2}(r - \mu_i)} - r^2 - r\alpha_n, \tag{2.11}$$

which completes the construction of the last raw and column of **C** and **K**. Using (2.2) again, the characteristic polynomial $q_{2n-4}(\lambda)$ may be determined by a synthetic division.

Thus, knowing $q_{2n-2}(\lambda)$ and $q_{2n-4}(\lambda)$, we may reapply the construction procedure to the de-escalated problem of dimension $n-1$ and construct $\alpha_{n-1}, \beta_{n-2}, \gamma_{n-1}$ and δ_{n-2}. Continuing in this manner allows the complete construction of **C** and **K**, successively.

3 Partial Pole Assignment

The system modeled by (1.1) may be controlled by the force $\mathbf{b}u(t)$

$$\mathbf{M}\ddot{\mathbf{x}} + \mathbf{C}\dot{\mathbf{x}} + \mathbf{K}\mathbf{x} = \mathbf{b}u(t). \tag{3.1}$$

Using the *state feedback* control

$$u(t) = \mathbf{f}^T\dot{\mathbf{x}}(t) + \mathbf{g}^T\mathbf{x}(t), \tag{3.2}$$

(3.1) becomes

$$\mathbf{M}\ddot{\mathbf{x}} + (\mathbf{C} - \mathbf{b}\mathbf{f}^\mathbf{T})\dot{\mathbf{x}} + (\mathbf{K} - \mathbf{b}\mathbf{g}^\mathbf{T})\mathbf{x} = \mathbf{o}. \tag{3.3}$$

Let

$$\mathbf{M}\mathbf{V}\mathbf{\Lambda}^2 + \mathbf{C}\mathbf{V}\mathbf{\Lambda} + \mathbf{K}\mathbf{V} = \mathbf{O} \tag{3.4}$$

be the spectral decomposition of (1.4), where $\mathbf{\Lambda} = diag\{\lambda_1, \lambda_2, \ldots, \lambda_{2n}\}$, λ_i distinct, and $\mathbf{V} \in \mathcal{C}^{2\times 2n}$. It has been shown in [2] that the following *bi-orthogonality* relations hold:

$$\mathbf{\Lambda}\mathbf{V}^\mathbf{T}\mathbf{M}\mathbf{V}\mathbf{\Lambda} - \mathbf{V}^\mathbf{T}\mathbf{K}\mathbf{V} = \mathbf{D_1}, \tag{3.5}$$

$$\mathbf{\Lambda}\mathbf{V}^\mathbf{T}\mathbf{C}\mathbf{V}\mathbf{\Lambda} + \mathbf{\Lambda}\mathbf{V}^\mathbf{T}\mathbf{K}\mathbf{V} + \mathbf{V}^\mathbf{T}\mathbf{K}\mathbf{V}\mathbf{\Lambda} = \mathbf{D_2}, \tag{3.6}$$

$$\mathbf{\Lambda}\mathbf{V}^\mathbf{T}\mathbf{M}\mathbf{V} + \mathbf{V}^\mathbf{T}\mathbf{M}\mathbf{V}\mathbf{\Lambda} + \mathbf{V}^\mathbf{T}\mathbf{C}\mathbf{V} = \mathbf{D_3}, \tag{3.7}$$

where $\mathbf{D_1}$, $\mathbf{D_2}$ and $\mathbf{D_3}$ are diagonal matrices. Moreover,

$$\mathbf{D_1} = \mathbf{D_3}\mathbf{\Lambda}, \tag{3.8}$$

$$\mathbf{D_2} = -\mathbf{D_1}\mathbf{\Lambda}, \tag{3.9}$$

$$\mathbf{D_2} = -\mathbf{D_3}\mathbf{\Lambda}^2. \tag{3.10}$$

Partition **V** and **Λ** in the form

$$\mathbf{V} = (\mathbf{V_1}, \mathbf{V_2}), \quad \mathbf{\Lambda} = \mathbf{diag}\{\mathbf{\Lambda_1}, \mathbf{\Lambda_2}\}, \tag{3.11}$$

$\mathbf{V_1} \in \mathcal{C}^{\mathbf{n}\times\mathbf{m}}$ and $\mathbf{\Lambda_1} = diag\{\lambda_1, \lambda_2, \dots, \lambda_m\}$. For an arbitrary vector $\mathbf{h} \in \mathcal{C}^m$ define

$$\mathbf{f} = \mathbf{M}\mathbf{V_1}\mathbf{\Lambda_1}\mathbf{h} \quad \textit{and} \quad \mathbf{g} = -\mathbf{K}\mathbf{V_1}\mathbf{h}. \tag{3.12}$$

Then, using (3.4) we have

$$\begin{aligned} &\mathbf{M}\mathbf{V_2}\mathbf{\Lambda_2^2} + (\mathbf{C} - \mathbf{b}\mathbf{f^T})\mathbf{V_2}\mathbf{\Lambda_2} + (\mathbf{K} - \mathbf{b}\mathbf{g^T})\mathbf{V_2} \\ &\quad = -\mathbf{b}\mathbf{h^T}(\mathbf{\Lambda_1}\mathbf{V_1^T}\mathbf{M}\mathbf{V_2}\mathbf{\Lambda_2} - \mathbf{V_1^T}\mathbf{K}\mathbf{V_2}). \end{aligned}$$

But the first bi-orthogonal relation (3.5) implies that

$$\mathbf{\Lambda_1}\mathbf{V_1^T}\mathbf{M}\mathbf{V_2}\mathbf{\Lambda_2} - \mathbf{V_1^T}\mathbf{K}\mathbf{V_2} = \mathbf{O}. \tag{3.13}$$

It is therefore concluded that for an arbitrary $\mathbf{h}$ the choice of $\mathbf{f}$ and $\mathbf{g}$ as in (3.12) ensures that $\lambda_{m+1}, \lambda_{m+2}, \dots, \lambda_{2n}$ are eigenpairs of *both* the *open loop* system (3.4) and the *closed loop* system (3.3).

We have shown in [2] that if $\mathbf{h} = (h_1, h_2, \dots, h_m)^T$ is chosen such that

$$h_j = \frac{1}{\mathbf{b^T v_j}} \frac{\sigma_j - \lambda_j}{\lambda_j} \prod_{\substack{i=1 \\ i \neq j}}^{m} \frac{\sigma_i - \lambda_j}{\lambda_i - \lambda_j}, \quad j = 1, 2, \dots, m, \tag{3.14}$$

then the m eigenvalues $\lambda_1, \lambda_2, \dots, \lambda_m$ are assigned under (3.3) to the poles $\sigma_1, \sigma_2, \dots, \sigma_m$. Moreover, if $\lambda_1, \lambda_2, \dots, \lambda_m$ and $\sigma_1, \sigma_2, \dots, \sigma_m$ are chosen to be self-conjugate sets then $\mathbf{f}$ and $\mathbf{g}$ obtained by (3.12) are real. The control is thus realisable.

The explicit solution of the partial pole assignment problem, (3.12) and (3.14), shows that the partial assignment can be carried out provided that:

(a) no λ_j, vanishes,

(b) the λ_j are distinct, and

(c) $\mathbf{b}$ must not be orthogonal to $\mathbf{v_j}$,

for $j = 1, 2, \dots, m$.

The conditions **(b)** and **(c)** ensures the partial controllability of the system with respect to the assigned eigenvalues. Hence, they are *necessary* and *sufficient* conditions for the partial pole assignment. The condition **(a)** may be removed by an appropriate shift of origin, as shown in [2]. We note, however, that in vibration a zero eigenvalue of (1.1) has multiplicity of at least two. A rigid body mode of motion thus cannot be removed by partial pole assignment in virtue of **(b)**.

We note that only m eigenpairs, associated with the m shifted poles, are needed to be known. Also, the damping matrix $\mathbf{C}$ does not appear explicitly in the assignment procedure. In applications the required eigenpairs can be measured from experiments by using modal analysis techniques. At the present state of the art there is no reliable method of modeling the damping matrix $\mathbf{C}$. Hence avoiding the explicit use of $\mathbf{C}$ may be advantageous in applications.

4 Pole-Zero Assignment for a Linear Pencil

We now consider the problem of assigning both, poles and zeros, by state feedback control. For simplicity of exposition we assume that the system is undamped; i.e.. the equation of motion (3.1) is reduced to

$$\mathbf{M}\ddot{\mathbf{x}} + \mathbf{K}\mathbf{x} = \mathbf{b}u(t), \tag{4.1}$$

where, as before, $u(t)$ is given by (3.2). We further assume that the poles and zeros are assigned to be real. Following the derivations of Section 3 we find that under these assumptions the velocity control vanishes, i.e..

$$\mathbf{f} = \mathbf{o}. \tag{4.2}$$

The open loop linear pencil is

$$\mathbf{Q_o}(\lambda) = \mathbf{K} - \lambda\mathbf{M}, \tag{4.3}$$

and the closed loop pencil is thus

$$\mathbf{Q_c}(\lambda) = \mathbf{K} - \mathbf{b}\mathbf{g}^{\mathbf{T}} - \lambda\mathbf{M}. \tag{4.4}$$

We may assign m eigenvalues $\lambda_1, \lambda_2, \ldots, \lambda_m$ of the open loop pencil (4.3) to the real set $\sigma_1, \sigma_2, \ldots, \sigma_m$, keeping the other $n-m$ eigenvalues unchanged, by applying the control (4.4) with

$$\mathbf{g} = -\mathbf{K}\mathbf{V_1}\mathbf{h}, \tag{4.5}$$

where $\mathbf{h}$ is given by (3.14) and $\mathbf{V_1} = [\mathbf{v_1}, \mathbf{v_2}, \ldots, \mathbf{v_m}]$ is the truncated eigenvector matrix associated with (4.3). Let $\mathbf{v_1}$ be normalised such that

$$\mathbf{b}^{\mathbf{T}}\mathbf{v_1} = 1. \tag{4.6}$$

If we assign only one eigenvalue, say λ_1 to σ, then by (3.14) and (4.5) the control vector $\mathbf{g}$ takes the form

$$\mathbf{g} = \frac{\lambda_1 - \sigma}{\lambda_1}\mathbf{K}\mathbf{v_1}. \tag{4.7}$$

Note that with $\mathbf{g}$ chosen as in (4.7) the eigenvalues of the closed loop pencil (4.4) are $\sigma, \lambda_2, \ldots, \lambda_n$ for *all* vectors $\mathbf{b}$ which satisfy (4.6). We now show

how to choose the position vector $\mathbf{b}$ which satisfies (4.6) and which assigns the $n-1$ zeros of (4.4) to the real set $\theta_1, \theta_2, \ldots, \theta_{n-1}$.

Let $\hat{\mathbf{Q}}_{\mathbf{o}}(\lambda)$ be the leading $(n-1) \times (n-1)$ principal subpencil of (4.3) and denote its eigenvalues by $\mu_1, \mu_2, \ldots, \mu_{n-1}$ and its eigenvectors by $\mathbf{z_1}, \mathbf{z_2}, \ldots, \mathbf{z_{n-1}}$. Let $\hat{\mathbf{Q}}_{\mathbf{c}}(\lambda)$ be the leading $(n-1) \times (n-1)$ principal subpencil of (4.4). Partition

$$\mathbf{b} = \begin{pmatrix} \hat{\mathbf{b}} \\ b_n \end{pmatrix} \quad \mathbf{g} = \begin{pmatrix} \hat{\mathbf{g}} \\ g_n \end{pmatrix} \quad \mathbf{K} = \begin{bmatrix} \hat{\mathbf{K}} & \hat{\mathbf{k}} \\ \hat{\mathbf{k}}^T & k_{nn} \end{bmatrix}. \tag{4.8}$$

Then, noting that the eigenvalues are invariant under transposition, we may assign the eigenvalues of $\hat{\mathbf{Q}}_{\mathbf{c}}$ to $\theta_1, \theta_2, \ldots, \theta_{n-1}$ by choosing

$$\hat{\mathbf{b}} = -\hat{\mathbf{K}}\mathbf{Z}\mathbf{r}, \tag{4.9}$$

where $\mathbf{Z} = [\mathbf{z_1}, \mathbf{z_2}, \ldots, \mathbf{z_{n-1}}]$ and $\mathbf{r} = (r_1, r_2, \ldots, r_{n-1})^T$ is defined by its coefficients

$$r_j = \frac{1}{\hat{\mathbf{g}}^{\mathbf{T}}\mathbf{z_j}} \frac{\theta_j - \mu_j}{\mu_j} \prod_{\substack{i=1 \\ i \neq j}}^{n-1} \frac{\theta_i - \mu_j}{\mu_i - \mu_j}, \quad j = 1, 2, \ldots, n-1. \tag{4.10}$$

In order to satisfy the condition (4.6) we choose

$$b_n = \frac{1}{a_n}(1 - \hat{\mathbf{b}}^{\mathbf{T}}\hat{\mathbf{g}}), \tag{4.11}$$

which completes the determination of the position vector $\mathbf{b}$.

We have thus shown how to determine the position vector $\mathbf{b}$ and the control vector $\mathbf{g}$ such that one pole and $n-1$ zeros of the open loop pencil (4.3) are assigned to the prescribed values $\sigma, \theta_1, \ldots, \theta_{n-1}$, leaving the other $n-1$ poles of (4.3) unchanged. One may wonder whether it is possible to assign some other poles by reapplying the method described successively. Unfortunately, after the first assignment is made the stiffness matrix $\mathbf{K} - \mathbf{b}\mathbf{g}^T$ is generally no longer symmetric. Hence successive assignment of the other poles cannot be achieved by this method.

Given a symmetric matrix $\mathbf{A}$, the interesting problem of how to choose $\mathbf{b}$ and $\mathbf{g}$ such that the n eigenvalues of $\mathbf{A} - \mathbf{b}\mathbf{g}^T$ and the $n-1$ eigenvalues of its leading $(n-1) \times (n-1)$ principal submatrix are all prescribed, is still open.

5 Summary

We have shown that a symmetric definite tridiagonal quadratic pencil may be constructed with prescribed poles and zeros. We may use this result in applications to construct a passively controlled vibratory system.

Then we have shown how to assign part of the spectrum of a symmetric quadratic pencil in such a way that the complementary spectrum remain unchanged. This has been achieved by the use of incomplete set of eigenpairs and without the explicit knowledge of the damping matrix. This result may thus be used in application to actively reduce the oscillations in structures.

The problem of assigning both poles and zeros of a linear symmetric pencil has been also addressed. It has been shown that certain zeros and one pole may be assigned by state feedback control, leaving the other poles unchanged. This has been done by selecting appropriate position and control vectors.

References

[1] D. Boley and G. H. Golub. A survey of matrix inverse eigenvalue problems. *Inverse Problems* **3** (1987), 595-622.

[2] B.N. Datta, S. Elhay and Y.M. Ram. Orthogonality and partial pole assignment for the symmetric definite quadratic pencil. To appear *Linear Algebra Appl.*

[3] G.M.L. Gladwell. *Inverse Problems in Vibration.* Dordrecht, the Netherlands: Martinus Nijhoff, 1986.

[4] P. Lancaster and J. Maroulas. Selective perturbation of spectral properties of vibrating systems using feedback. *Linear Algebra Appl.* **98** (1988), 309-330.

[5] Y.M. Ram and S. Elhay. An inverse eigenvalue problem for the symmetric tridiagonal quadratic pencil with application to damped oscillatory systems. *SIAM J. Appl. Math.* **56**(1) (1996), 232- 244.

[6] Y. Saad. Projection and deflation methods for partial pole assignment in linear state feedback. *IEEE Trans. Automat. Contr.* **33** (1988), 290-297.

Department of Mechanical Engineering, University of Adelaid, Adelaid, SA 5005, Australia

Inverse Eigenvalue Problems for Multivariable Linear Systems

J. Rosenthal[1] and X.A. Wang[2]

1 Introduction and Motivational Examples

Many inverse eigenvalue problems appearing in control theory and other areas are characterized by polynomial equations. Algebraic geometry is the mathematical theory which deals with polynomial equations. It is extending the theory of linear algebra and it deals with the study of zero sets of systems of polynomial equations, i.e. algebraic sets and varieties, and with polynomial morphisms between varieties.

Historically, algebraic geometry has been mainly developed as a pure theory suitable for the study of geometric objects in vector spaces. Over the years this theory accumulated a wealth of ideas together with increasing abstraction, such as the beautiful language of schemes developed by Grothendieck in the 60th.

Although it is our strong belief that algebraic geometry is potentially very important for many applications (as is, for example, linear algebra) it is clear that a researcher without a background in pure mathematics will have to overcome many obstacles to acquire this knowledge. Fortunately there were in recent years several books written for non-specialists of this field and we would like to mention [1, 7, 9, 13].

This article, which covers some of the material presented in a mini-course at the MTNS, will survey some recent advances in the area of inverse eigenvalue problems. Those advances were mainly based on several strong theorems of algebraic geometry. In order to explain this connection and in order to make the article self contained, we will summarize in the next section the notions and results from algebraic geometry needed. In the sequel we give a series of examples which will explain the kind of problems which we would like to treat in a unified way.

Example 1.1 (Static Pole Placement Problem) Let A, B, C be real matrices of size $n \times n$, $n \times m$ and $p \times n$ respectively. Consider the plant:

$$\frac{d}{dt}x = Ax + Bu, \qquad y = Cx \tag{1.1}$$

[1]Supported in part by NSF grant DMS-9400965.
[2]Supported in part by NSF grant DMS-9500594.

and the compensator $u = Ky$. Identify the set of all monic polynomials of degree n with the real vector space $\mathbb{R}^n$. The static pole placement problem asks for conditions which guarantee that the pole placement map

$$\begin{array}{rccc} \psi: & \mathbb{R}^{mp} & \longrightarrow & \mathbb{R}^n \\ & K & \longmapsto & \det(sI - A - BKC) \end{array} \tag{1.2}$$

is surjective for a generic set of matrices (A, B, C). Note that ψ is a polynomial, but in general not a linear map.

Example 1.2 (Dynamic Pole Placement Problem) Consider as in the last example a linear system, "the plant," described by matrices (A, B, C). The dynamic pole placement problem asks for the construction of a dynamic compensator of the form:

$$\dot{z} = Fz + Gy \qquad u = Hz + Ky,$$

such that the closed loop system

$$\dot{\begin{bmatrix} x \\ z \end{bmatrix}} = \left(\begin{bmatrix} A & 0 \\ 0 & 0 \end{bmatrix} + \begin{bmatrix} BKC & BH \\ GC & F \end{bmatrix} \right) \begin{bmatrix} x \\ z \end{bmatrix}$$

has its pole at a specific location.

On the side of the class of pole placement problems there is a large linear algebra literature on so called matrix extension problems. Probably the main reference for this active area of research is the recent book by Gohberg, Kaashoek and van Schagen [12]. The following example illustrates one of the main problems in this research area:

Example 1.3 (Matrix Extension Problems) Consider the set of all $n \times n$ matrices having complex entries. We will identify this set with the vector space $\mathbb{C}^{n^2}$. Let $\mathcal{L} \subset Mat_{n\times n}$ be linear sub-space and let $A \in Mat_{n\times n}$ be a particular fixed matrix. Again identify the set of monic polynomials of degree n with the vector space $\mathbb{C}^n$. In this situation we are interested in conditions which guarantee that the characteristic map

$$\chi_A : \mathcal{L} \longrightarrow \mathbb{C}^n, \quad L \longmapsto \det(sI - A - L)$$

is surjective or at least "generically" surjective. Since both $\{BKC\}$ and $\left\{ \begin{bmatrix} BKC & BH \\ GC & F \end{bmatrix} \right\}$ define linear subspaces of $Mat_{n\times n}$ and $Mat_{(n+q)\times(n+q)}$ respectively, this last example generalizes the previous two examples.

We would like to illustrate Example 1.3 in the situation, where $n = 2$ and in the situation where $\mathcal{L}$ consists of the set of diagonal matrices. To be specific we will assume:

$$A := \begin{pmatrix} 0 & 1 \\ 1 & 0 \end{pmatrix}, \qquad \mathcal{L} := \begin{pmatrix} x & 0 \\ 0 & y \end{pmatrix}, x, y \in \mathbb{C}. \tag{1.3}$$

In this particular situation χ_A is the polynomial map

$$\chi_A : \mathbb{C}^2 \longrightarrow \mathbb{C}^2, \quad (x, y) \longmapsto (-x - y, xy - 1).$$

One readily verifies that this map is surjective of mapping degree 2. Note that "over the reals" this map fails to be surjective.

2 Basic Concepts from Algebraic Geometry

In this section we summarize the most important concepts from algebraic geometry which will be needed to understand the results presented in the rest of the paper.

2.1 Algebraic Sets, Affine Varieties and Generic Sets

Let $\mathbb{K}$ be an arbitrary field, such as the field $\mathbb{K} = \mathbb{R}$ of real numbers or the field $\mathbb{K} = \mathbb{C}$ of complex numbers. Let $R := \mathbb{K}[x_1, \ldots, x_n]$ be the polynomial ring in n variables and let $S \subset R$ be a set of polynomials.

Definition 2.1 $V(S) := \{x \in \mathbb{K}^n \mid f(x) = 0, \quad \forall f \in S\}$ is called an *algebraic subset* of the vector space $\mathbb{K}^n$.

We would like to remark, that if I is the ideal generated by S then $V(S) = V(I)$.

In general both S and I may have infinite many elements. We have, however, the following important result of Hilbert which essentially states that every algebraic set can be described as the zero locus of finitely many polynomials:

Theorem 2.1 (Hilbert Basis Theorem) *Every ideal $I \subset R$ is finitely generated; i.e. there are polynomials $f_1, \ldots, f_s$ such that $I = < f_1, \ldots, f_s >$.*

Corollary 2.1 If $S \subset R$ is a subset then the algebraic set $V(S) \subset \mathbb{K}^n$ is an intersection of a finite number of hypersurfaces; i.e. there are polynomials $f_1, \ldots, f_s \in R$ such that

$$V(S) = V(f_1, \ldots, f_s) = V(f_1) \cap \ldots \cap V(f_s). \tag{2.1}$$

Example 2.1 Let $\mathbb{K}^2 = \mathbb{R}^2$ be the real 2-plane. Then $V\left((y-x^2)x\right) \subset \mathbb{R}^2$ defines the union of a parabola and the y-axis. In particular this union is an algebraic set.

Example 2.2 Let $Mat_{n\times n}$ denote the set of all $n \times n$ matrices with entries in $\mathbb{K}$. We can identify $Mat_{n\times n}$ with $\mathbb{K}^{n^2}$. Let

$$Sl_n := \{B \in Mat_{n\times n} \mid \det B = 1\}$$

be the special linear group. Then $Sl_n = V(\det -1) \subset \mathbb{K}^{n^2}$ is an algebraic set.

The following theorem states that the algebraic sets in $\mathbb{K}^n$ are the closed sets of a topology.

Theorem 2.2 *The empty set $\emptyset$ and $\mathbb{K}^n$ are algebraic sets and one has:*

1. *If Γ is an index set and I_γ, $\gamma \in \Gamma$ is a set of ideals then:*

$$\bigcap_{\gamma\in\Gamma} V(I_\gamma) = V(\sum_{\gamma\in\Gamma} I_\gamma). \tag{2.2}$$

2. *If I_1, I_2 are ideals then*

$$V(I_1)\bigcup V(I_2) = V(I_1 \bigcap I_2) = V(I_1 I_2). \tag{2.3}$$

Example 2.3 The Zariski closed sets in $\mathbb{C}$ are the finite set of points together with the whole plane $\mathbb{C}$. In this case the Zariski topology coincides with the so called cofinite topology. This example also shows that in general the Zariski topology is not Hausdorff.

Remark 2.1 The nonempty Zariski open sets in $\mathbb{R}^n$ (respectively in $\mathbb{C}^n$) are open and dense in the usual Euclidean topology.

Recall, that in a topological space a nonempty closed subset X is called irreducible if it cannot be expressed as a union of two proper closed subsets. Based on this we define:

Definition 2.2 An irreducible algebraic subset of $\mathbb{K}^n$ is called an affine variety.

Example 2.4

$$V\left((y-x^2)x\right) = V(x) \cup V\left(y-x^2\right)$$

is a decomposition in irreducible components.

Definition 2.3 An affine variety $X \subset \mathbb{K}^n$ is said to have dimension d if there is a (maximal) composition series of affine varieties with

$$X_0 \underset{\neq}{\subset} X_1 \underset{\neq}{\subset} \dots \underset{\neq}{\subset} X_d = X.$$

Definition 2.4 A subset $G \subseteq \mathbb{K}^n$ is called a generic set if there is a nonempty Zariski open set $X \subseteq \mathbb{K}^n$ having the property that $X \subseteq G$.

Note that generic sets of $\mathbb{R}^n$ are necessarily open and dense in the Euclidean topology. Using the above definition we immediately can define:

Definition 2.5 A set of linear system of the form

$$\frac{d}{dt}x = Ax + Bu, \qquad y = Cx \tag{2.4}$$

having a fixed number of inputs, outputs and a fixed McMillan degree will be called a generic set of systems if the corresponding parameter set in $\mathbb{R}^{n^2+mn+mp}$.

There is a more abstract way to define a generic set of linear systems. Since an affine variety is irreducible by definition we can in a natural way extend the definition of genericity to the concept of a "generic subset of a variety." Note also that the systems of the form (2.4) modulo state space equivalence can be naturally identified with a Zariski open subset of a (projective) variety, a so called *Quot Scheme.* (See [24, 27, 28]). This gives an alternative way to speak about a generic set of linear systems. It is not difficult to verify that for the purpose of pole placement problems, the two concepts do coincide.

Remark 2.2 In order to verify that a subset $S \subset \mathbb{K}^n$ is generic it is enough to show:

- The elements of the complement set $\mathbb{K}^n \setminus S$ satisfy a polynomial condition $f(x) = 0$.
- There is one element $\bar{x} \in S$ having the property that $f(\bar{x}) \neq 0$.

Example 2.5 The generic $n \times n$ matrix is invertible; i.e. the set Gl_n of invertible matrices is a generic set in the set of $n \times n$ matrices $Mat_{n\times n} \cong \mathbb{K}^{n^2}$.

Indeed the complement set, i.e. the set of singular matrices, is contained in (actually coincides with) the algebraic set $V(\det)$ and obviously there are matrices with a nonzero determinant.

Example 2.6 The set of semisimple complex $n\times n$ matrices is a generic set. [Consider the resultant $\mathrm{Res}(\chi(\lambda), \frac{d}{d\lambda}\chi(\lambda)) = 0$, where $\chi(\lambda) = \det(\lambda I - A)$.]

The next theorem is the "inverse function theorem" of algebraic geometry. We state it for the field of complex numbers. In general one has to require that the base field is algebraically closed. For a proof see e.g. [14].

Theorem 2.3 (Dominant Morphism Theorem) *Let $\varphi : \mathbb{C}^n \to \mathbb{C}^m$ be a polynomial map. If there is a point $x \in \mathbb{C}^n$ such that the Jacobian $d\varphi_x$ is surjective then* $\mathrm{Im}(\varphi)$ *forms a generic set.*

Example 2.7 Consider once more the polynomial map

$$\chi : \mathbb{C}^2 \longrightarrow \mathbb{C}^2, \quad (x, y) \longmapsto (-x - y, xy - 1).$$

The Jacobian $d\chi$ is

$$d\chi = \begin{bmatrix} -1 & -1 \\ y & x \end{bmatrix}$$

and $d\chi$ is nonsingular as soon as $x \neq y$. It follows that $\mathrm{Im}(\chi)$ is dense in $\mathbb{C}^2$. (A direct computation shows that χ is even surjective.)

2.2 Projective Algebraic Geometry

It is a well known fact in linear algebra, that two subspaces $X, Y \subset \mathbb{K}^n$ do intersect nontrivially as soon as the "dimensions add up." If the base field $\mathbb{K}$ is algebraically closed (e.g. $\mathbb{K} = \mathbb{C}$) and if one accounts for situations, where the varieties intersect "at infinity" then a similar result holds true for two algebraic varieties. In order to make this precise we will have to introduce the concept of a projective variety.

Definition 2.6 The set of all 1-dimensional linear subspaces of the vector space $\mathbb{K}^{n+1}$ is called the n-dimensional projective space and will be denoted by $\mathbb{P}^n_{\mathbb{K}}$ or $\mathbb{P}(\mathbb{K}^{n+1})$.

Remark 2.3 A topology on $\mathbb{K}^{n+1}$ induces a topology on $\mathbb{P}^n_{\mathbb{K}}$ in a natural way. For this, consider the canonical projection

$$\begin{array}{rccc} pr: & \mathbb{K}^{n+1} \setminus \{0\} & \longrightarrow & \mathbb{P}^n_{\mathbb{K}} \\ & x & \longmapsto & p_x. \end{array} \tag{2.5}$$

Example 2.8 $\mathbb{P}^1_{\mathbb{K}}$ is sometimes called the projective line over $\mathbb{K}$. As a set $\mathbb{P}^1_{\mathbb{K}}$ can be viewed as the disjoint union of the sets

$$X_1 := \{(x, 1) \in \mathbb{P}^1 \mid x \in \mathbb{K}\} \text{ and } X_2 := \{(1, 0) \in \mathbb{P}^1\}.$$

Set theoretic X_1 to be isomorphic to the base field $\mathbb{K}$ and X_2 to represent a point which is sometimes called the point at infinity. In particular one has:

$$\mathbb{P}^1_{\mathbb{R}} = S^1 := \{e^{it} \mid t \in [0, 2\pi]\},$$
$$\mathbb{P}^1_{\mathbb{C}} = S^2 := \{(x, y, z) \in \mathbb{R}^3 \mid x^2 + y^2 + z^2 = 1\}.$$

In general we always can view $\mathbb{P}^n_{\mathbb{K}}$ as the set $\mathbb{K}^n$ together with some points at infinity. In a suitable topology $\mathbb{P}^n_{\mathbb{K}}$ forms a "compactification" of the vector space $\mathbb{K}^n$.

Definition 2.7 Let $f_1, \ldots, f_s \in R$ be homogeneous polynomials of degree $d_1, \ldots, d_s$ respectively. Then

$$V(f_1, \ldots, f_s) := \{x \in \mathbb{P}^n_{\mathbb{K}} \mid f_1(x) = \cdots = f_s(x) = 0\}$$

is called an algebraic subset of $\mathbb{P}^n_{\mathbb{K}}$.

It is not hard to show that every algebraic set of $\mathbb{K}^n$ can be embedded in an algebraic set of $\mathbb{P}^n_{\mathbb{K}}$. In analogy to the affine situation one has:

Lemma 2.1 *The algebraic sets in $\mathbb{P}^n_{\mathbb{K}}$ form the closed sets of a topology, called the Zariski topology of $\mathbb{P}^n_{\mathbb{K}}$.*

Definition 2.8 An irreducible algebraic subset of $\mathbb{P}^n_{\mathbb{K}}$ is called a projective variety.

In the same way as for affine varieties, one defines the dimension of a projective variety as through the length of a maximal composition series. The following theorem is a major existence results for the zero sets of polynomial equations:

Theorem 2.4 (Projective Dimension Theorem [14]) *Let X and Y be projective varieties of dimensions r and s in $\mathbb{P}^n_{\mathbb{C}}$. Then every irreducible component of $X \cap Y$ has dimension $\geq r+s-n$. Furthermore, if $r+s-n \geq 0$, then $X \cap Y$ is nonempty.*

By the projective dimension theorem, the intersection of an r dimensional projective variety X and $n-r$ dimensional subspace L in $\mathbb{P}^n$ is always nonempty. There is a well defined number d called the *degree of the variety X* which is defined as the number of points in $X \cap L$ for "almost all" $n-r$ dimensional linear subspaces L. The following theorem gives a precise answer about the number of solutions if two projective varieties are complementary in dimension and if they lie in "general position."

Theorem 2.5 (Bézout's Theorem [14, 26]) *Let X and Y in $\mathbb{P}^n$ be varieties of dimension r and $n-r$. If $X \cap Y$ has finitely many points, then*

$$\#(X \cap Y) = \deg(X)\deg(Y)$$

when counted with multiplicity.

Example 2.9 The equations

$$y - x^2 = 0 \text{ and } x - 1 = 0$$

define projective varieties (curves) in $\mathbb{P}^2_{\mathbb{K}}$ of degree 2 respectively 1 through the homogeneous equations

$$yz - x^2 = 0 \text{ and } x - z = 0.$$

The two points of intersection predicted by Bézout's Theorem are

$$(x, y, z) = (1, 1, 1) \text{ and } (x, y, z) = (0, 1, 0).$$

2.3 Central Projection

The set of projective varieties form the objects of a category. The morphisms between varieties are essentially the polynomial maps. A particular simple and important class of morphisms are the so called central projections which we will now define. For this let M be an $(n+1) \times N$ full rank matrix, $z = (z_0, z_1, \ldots, z_N)^t \in \mathbb{P}^N$ and

$$E = \{z \in \mathbb{P}^N | Mz = 0\}.$$

Then the projection with center E is the map $\rho : \mathbb{P}^N - E \longrightarrow \mathbb{P}^n$ defined by $\rho(z) = Mz$; i.e. a central projection is a linear map restricted on an open set of a projective space.

More generally assume that $X \subset \mathbb{P}^N$ is a projective variety. Then any central projection $\rho : \mathbb{P}^N - E \longrightarrow \mathbb{P}^n$ induces a central projection $\rho : X - E \longrightarrow \mathbb{P}^n$ through restriction of the domain. At this point we would like to mention that the pole placement map studied in the next section has the structure of a central projection.

The geometric meaning of a central projection is the following: take any projective n-subspace $H \subset \mathbb{P}^N$ disjoint from E. Then $\rho : H \longrightarrow \mathbb{P}^n$ is one-to-one and onto. So one can identify H with $\mathbb{P}^n$ and simply say that $\mathbb{P}^n \subset \mathbb{P}^N$. Through any point $z \in X - E$ and E, there passes a unique projective $(N-n)$-subspace $L_z := \text{span}(z, E)$. L_z intersects $\mathbb{P}^n$ in a unique point, namely $\rho(z)$. In other words we have a well defined map

$$\begin{array}{rccc} \rho : & X - E & \longrightarrow & H \\ & z & \longmapsto & \text{span}(z, E) \cap H. \end{array} \tag{2.6}$$

The name central projection results from the following particular situation: let E be a point. To find the image of any other point z one can draw a line through E and z, and then the intersection of the line with H will be the image of z.

The following proposition shows that if X is a projective variety of dimension n not intersecting the center E, then the central projection from X with center E is a finite morphism:

Theorem 2.6 *Let $\rho : \mathbb{P}^N - E \to \mathbb{P}^n$ be a central projection with*

$$\dim E = N - n - 1$$

and $X \subset \mathbb{P}^N - E$ be a projective variety of dimension n. Then $\rho : X \to \mathbb{P}^n$ is onto over $\mathbb{C}$ and there are $\deg X$ complex solutions (counted with multiplicity) of $\rho(z) = b$ in X for each $b \in \mathbb{P}^n$. In particular ρ is a finite morphism.

Proof. (Compare also with [13, p. 235] and [26, Corollary (2.29) and Corollary(5.6)].) Let $b \in \mathbb{P}^n$. Then $z \in \rho^{-1}(b) \cap X$ if and only if the $(N-n)$ linear subspace L_b of $\mathbb{P}^N$ spanned by E and b intersects X at z. Note that X and L_b intersect properly, i.e. $\dim X \cap L_b = 0$, because otherwise $X \cap E$, which is the intersection of $X \cap L_b$ and a hyperplane, would be nonempty by the projection dimension theorem (see e.g. [14, p. 48]). Now, by Bézout's theorem (see e.g. [13, p. 227]), every $(N-n)$ dimensional linear subspace of $\mathbb{P}^N$ intersects X in $\deg X$ points counted with multiplicity.

2.4 The Grassmann Variety and its Plücker Embedding.

Grassmannians represent a class of extremely important varieties. The importance of those varieties lies mainly in the fact that many problems involving matrices and linear systems have a geometric reformulation in terms of Grassmannians.

This translation is usually achieved through an identification of the vector space of $p \times m$ matrices with the "thick open cell" of a certain Grassmann variety. This process will be made precise later. This geometric approach has so far been applied for several different seemingly "applied problems" and we refer to [4, 15, 25, 36, 37, 38] where also further references to the literature can be found.

Formally Grassmannians are defined as follows:

Definition 2.9 Let V be a n-dimensional $\mathbb{K}$-vector space. The set of all p dimensional linear subspaces of V is called the Grassmann variety and will be abbreviated with Grass(m, V) or with Grass(m, n).

If V has a basis $\{e_1, \ldots, e_n\}$ then an element of Grass(m, n) can also be represented by the rowspace of a $m \times n$ matrix A. Such a representation is not unique, namely for any element $T \in Gl_m$, TA represents the same

m–plane. On the other hand, if A and B define the same m–plane, there is a $T \in Gl_m$, such that $TA = B$.

So far we have introduced the Grassmann variety set theoretically. As already indicated through the name, each Grassmannian is also a (projective) variety. In order to verify this we have to embed each Grassmannian into a projective space. This is achieved through the classical Plücker embedding.

For this, consider the vector space of alternating m–tensors $\wedge^m V$. As a vector space $\wedge^m V$ is linearly generated by all vectors of the form

$$v_1 \wedge \ldots \wedge v_m, \qquad v_i \in V, i = 1, \ldots, m$$

subject to the relation

$$v_1 \wedge \ldots \wedge v_i \wedge \ldots \wedge v_j \wedge \ldots \wedge v_m = 0 \qquad \text{if } v_i = v_j \text{ for } i \neq j$$

and subject to the multilinearity in each component. Formally the Plücker embedding is now defined through:

$$\begin{array}{rccl} \varphi : & \text{Grass}(m, V) & \longrightarrow & \mathbb{P}(\wedge^m V) \\ & \text{span}(v_1, \ldots, v_m) & \longmapsto & \mathbb{K} v_1 \wedge \ldots \wedge v_m. \end{array} \tag{2.7}$$

In order to verify that (2.7) is indeed a well defined embedding we will represent (2.7) in terms of coordinates. For this note first that a basis $\{e_1, \ldots, e_n\}$ in V induces a basis in $\wedge^m V$ given by

$$\{e_{i_1} \wedge \ldots \wedge e_{i_m} \mid 1 \leq i_1 < \ldots < i_m \leq n\}.$$

Assume now that $v_1, \ldots, v_m$ are m linearly independent vectors of V. Furthermore assume that

$$v_i = \sum_{j=1}^{n} v_{ij} e_j, \; i = 1, \ldots, m.$$

Using the multilinearity properties and the alternating properties of the vector space $\wedge^m V$ one can express the vector $v_1 \wedge \ldots \wedge v_m$ with respect to the basis $\{e_{i_1} \wedge \ldots \wedge e_{i_m} \mid 1 \leq i_1 < \ldots < i_m \leq n\}$:

$$v_1 \wedge \ldots \wedge v_m = \sum_{1 \leq i_1 < \cdots < i_m \leq n} x_{i_1, \ldots, i_m} \cdot e_{i_1} \wedge \ldots \wedge e_{i_m} \tag{2.8}$$

The coordinates $x_{\underline{i}} := x_{i_1, \ldots, i_m}$ are called the Plücker coordinates of the alternating m-vector $v_1 \wedge \ldots \wedge v_m$. A direct calculation verifies that $x_{\underline{i}}$ is also equal to the $m \times m$ minor of the matrix (v_{ij}) obtained when extracting the columns i_1 to i_m.

In terms of coordinates, the Plücker map (2.7) is therefore described through

$$\begin{array}{rccc} \varphi \;: & Mat_{m\times n} & \longrightarrow & \mathbb{P}(\wedge^m \mathbb{K}^n) \\ & \text{rowspace}(A) & \longmapsto & \sum_{1\le i_1<\cdots<i_m\le n} x_{i_1,\ldots,i_m}\cdot e_{i_1}\wedge\ldots\wedge e_{i_m}. \end{array} \tag{2.9}$$

Now we show that the map φ is indeed well defined. For this assume that A and B define the same m–plane. As already observed earlier there is a $T \in Gl_m$, such that $TA = B$. But this just means that $\varphi(TA) = \det T \cdot \varphi(A)$, in other words a m–plane in $\mathbb{K}^n$is identified with a line in $\wedge^m\mathbb{K}^n$, i.e. a point in $\mathbb{P}(\wedge^m\mathbb{K}^n)$.

On the other hand if A and B define two different m–planes, then it is well known that A and B have different row reduced echelon forms; i.e. the $m \times m$ minors define different points in the projective space $\mathbb{P}(\wedge^m\mathbb{K}^n)$. It therefore follows that (2.7) is an embedding.

The fact that the image of φ is a variety is nontrivial and we refer to the survey article by Kleiman and Laksov, [20].

2.5 An Introduction to Schubert Calculus

Hannibal Schubert, a German high school teacher, studied in the last century a large number of so called enumerative problems in geometry. One of the problems he was dealing with was as follows:

Problem 2.7 Consider n linear subspace $V_i, i = 1,\ldots,n$ in $\mathbb{C}^{m+p}$ having dimension p each. Under what conditions exists a m-dimensional linear subspace W intersecting all subspaces V_i nontrivially?

Schubert treated this problem in [34, 35] using the "Principle of Conservation of Numbers" and several heuristic arguments. Using this principle he derived the following surprising result which reads in modern language:

Theorem 2.8 (Schubert) *If $n \le mp$ then Problem 2.7 has always a solution. Moreover if $n = mp$ then generically there are exactly*

$$d(m,p) = \frac{1!2!\cdots(p-1)!(mp)!}{m!(m+1)!\cdots(m+p-1)!} \tag{2.10}$$

solutions when counted with multiplicity.

Schubert's theorem has been independently verified in this century and the problem is now dealt with in the frame work of a general intersection theory on the Grassmann variety. (See [11, 19] for more details). Schubert's number $d(m,p)$ can also be understood in the following way: define so called Schubert varieties

$$S_i := \{W \in \mathrm{Grass}(m, \mathbb{C}^{m+p}) \mid W \cap V_i \neq \{0\}\}.$$

Then one immediately verifies that in terms of Plücker coordinates, the Schubert variety $S_i \subset \mathbb{P}(\wedge^m \mathbb{K}^n)$ is equal to the intersection of a hyper plane and the Grassmann variety $\mathrm{Grass}(m, \mathbb{C}^{m+p})$. From this one readily establishes:

Corollary 2.2 The degree of the Grassmann variety $\mathrm{Grass}(m, \mathbb{C}^{m+p})$ is $d(m,p)$.

3 The Static Pole Placement Problem

3.1 Results Over the Complex Numbers.

Consider the complex pole placement map

$$\psi: \ \mathbb{C}^{mp} \longrightarrow \mathbb{C}^n, \qquad K \longmapsto \det(sI - A - BKC). \tag{3.1}$$

Hermann and Martin [17] computed the Jacobian $d\psi_0$ at the origin and concluded using the dominant morphism theorem:

Theorem 3.1 *If $n \leq mp$ then the pole placement map ψ is almost surjective for a generic set of complex matrices (A, B, C).*

Let $D^{-1}(s)N(s)$ be a left coprime factorization of the transfer function $C(sI_n - A)^{-1}B$. Then up to a constant factor the closed loop characteristic polynomial $\det(sI_n - A - BKC)$ is also equal to:

$$\det \begin{bmatrix} D(s) & N(s) \\ K & I_m \end{bmatrix}. \tag{3.2}$$

The last equation can be viewed in the following way. Assume that the closed loop characteristic polynomial $\det(sI - A - BKC)$ has (desired) roots at n different locations $s_i \in \mathbb{C}$. Then necessarily we will require that

$$\mathrm{rowsp}\,(D(s_i)\ N(s_i)) \cap \mathrm{rowsp}\,(K\ I_m) \neq \{0\},\ i = 1, \ldots, n.$$

If $n \leq mp$ we know by Schubert's Theorem 2.8 that there exists a full rank matrix $(K_1\ K_2)$ of size $m \times (m+p)$ such that

$$\mathrm{rowsp}\,(D(s_i)\ N(s_i)) \cap \mathrm{rowsp}\,(K_1\ K_2) \neq \{0\},\ i = 1, \ldots, n.$$

Note that Schubert's theorem predicts an m-plane $\mathrm{rowsp}\,(K_1\ K_2)$. On the other hand such a geometric solution is only desirable if the $m \times m$

matrix K_2 is invertible and if this solution does not represent a so called "dependent compensator," i.e., a compensator having the property that

$$\det \begin{bmatrix} D(s) & N(s) \\ K_1 & K_2 \end{bmatrix} = 0 \in \mathbb{C}[s].$$

There are plants $(D(s)\ N(s))$ which do not have dependent compensators at all. Such plants were called *nondegenerate* by Brockett and Byrnes [4] since geometrically the associated Hermann Martin curve [17] represents a nondegenerate curve. The main technical result stated in [4] is as follows:

Lemma 3.1 (Brockett and Byrnes) *If $n \geq mp$, then the generic m-input, p-output plant of McMillan degree n is nondegenerate.*

If $n = pm$ this lemma immediately implies that for a generic set of m input, p output systems n different complex closed loop poles can be achieved.

So far our geometric translation requires that the roots $s_i \in \mathbb{C}$ of the closed loop characteristic polynomial are pairwise different. This difficulty can be overcome. For this, observe that the determinant in (3.2) can be expanded in terms of the $m \times m$ minors of the matrix $(K\ I_m)$. The associated cofactors are the $p \times p$ full size minors of the polynomial matrix $(D(s)\ N(s))$. Let $(k_0, \ldots, k_N)$ be the full size minors of $(K\ I_m)$ and let $(p_0(s), \ldots, p_N(s))$ denote the corresponding cofactors of $(D(s)\ N(s))$. Note that the numbers $(k_0, \ldots, k_N)$ are exactly the Plücker coordinates of rowsp$(K\ I_m)$ in the projective space

$$\mathbb{P}^N = \mathbb{P}(\wedge^m \mathbb{C}^{m+p}) = \mathbb{P}(\wedge^p \mathbb{C}^{m+p}).$$

The crucial observation is now that the pole placement map ψ extends to a central projection in $\mathbb{P}^N$, where the center is defined through

$$E := \{(k_0, \ldots, k_N) \in \mathbb{P}^N \mid \sum_{i=0}^{N} k_i p_i(s) = 0\}.$$

The following result, which follows from the projective dimension theorem and from Schubert's Theorem 2.8 describes the geometric situation.

Theorem 3.2 *The pole placement map ψ induces a central projection*

$$\begin{aligned} \rho_P : \quad \mathbb{P}^N \setminus E &\longrightarrow \mathbb{P}^n \\ (k_0, \ldots, k_N) &\longmapsto \sum_{i=0}^{N} k_i p_i(s). \end{aligned}$$

If $n = mp$ and if the plant $(D(s)\ N(s))$ is nondegenerate, then

$$E \cap \mathrm{Grass}(m, \mathbb{C}^{m+p}) = \emptyset.$$

Moreover if $x \in \mathbb{P}^n$ then $\rho_P^{-1}(x)$ is a linear subspace of dimension $N - mp$ and it intersects the Grassmann variety Grass$(m, \mathbb{C}^{m+p})$ in exactly $d(m,p)$ points when counted with multiplicity.

The above result predicts a "geometric solution" $(K_1\ K_2)$ for every closed loop characteristic polynomial of degree at most n. Since we assumed that $D^{-1}(s)N(s)$ is a proper transfer function, it follows that a compensator $(K_1\ K_2)$ which results in a closed loop characteristic polynomial of degree n is either a dependent compensator or has the property that the $m \times m$ matrix K_2 is invertible. But this argument gives the following major existence result:

Theorem 3.3 (Brockett and Byrnes [4]) *If $n = mp$, then for a generic set of m-input, p-output systems of McMillan degree n the (complex) pole placement map ψ introduced in (3.1) is surjective. Moreover if $p(s) \in \mathbb{C}[s]$ is any monic polynomial of degree n, then there exist $d(m,p)$ different compensators having the property that $\psi(K) = p(s)$.*

3.2 The "Real Static Problem"

Unfortunately some of the strongest results in algebraic geometry require that the base field is algebraically closed, a property which the complex numbers have but the reals do not have. It is therefore probably not surprising that the strongest sufficiency results over the reals did not come close to Theorem 3.3 for a long time. On the other hand the results reported in [4, 25] did sidestep the "real static problem" as formulated in the engineering literature (see e.g. [8, 18]). Still Theorem 3.3 predicted in cases when the degree $d(m,p)$ is odd also some real solutions since solutions of polynomials with real coefficients necessarily have to be invariant under complex conjugation. It was therefore very surprising when in 1992 Wang [38] derived the following strong existence result:

Theorem 3.4 (Wang [38]) *If $n < mp$ then the pole placement map*

$$\psi: \ \mathbb{R}^{mp} \longrightarrow \mathbb{R}^n, \qquad K \longmapsto \det(sI - A - BKC) \tag{3.3}$$

is surjective for a generic set of real matrices (A, B, C) of sizes $n \times n$, $n \times m$ and $p \times n$ respectively.

Before we outline the details of the proof of this major theorem we would like to say a few words about the proof history. The proof originally given by Wang [38] was very geometric in nature and it also considered some situations which were not necessary. Seemingly independently, Leventides [22] (compare also with [23]), Wang [39] and Ariki [2] recognized, that the proof is based on a so called *linearization around a dependent compensator.* This simple proof method was used by Rosenthal, Schumacher and Willems [32] to accomplish a direct elementary proof. In the next subsection we outline the proof.

3.3 Outline of Proof

Step 1: Extension of the pole placement map.

As it was done by Brocket and Byrnes [4] for the complex problem, it will be necessary to consider generalized compensators of the form $F := [K_1\ K_2]$. This time we will allow also compensators which do not have full row rank. Of particular importance will be so called dependent compensators, i.e. compensators which satisfy

$$\det \begin{bmatrix} D(s) & N(s) \\ K_1 & K_2 \end{bmatrix} = 0 \in \mathbb{R}[s].$$

The extended pole placement map is defined as:

Definition 3.1

$$\begin{array}{rccc} \chi: & \mathbb{R}^{m(p+m)} & \longrightarrow & \mathbb{R}^{n+1} \\ & [K_1\ K_2] & \longmapsto & \det \begin{bmatrix} D(s) & N(s) \\ K_1 & K_2 \end{bmatrix}. \end{array} \tag{3.4}$$

Note that χ is a homogeneous map of degree m, i.e.

$$\chi(\lambda [K_1\ K_2]) = \lambda^m \chi([K_1\ K_2]) \text{ for } \lambda \in \mathbb{R}.$$

Step 2: Linearization of χ around a dependent compensator.

As it was done by Martin and Hermann [25] we do linearize the pole placement map. If we linearize the extended pole placement map around a dependent compensator we get:

Lemma 3.2 *If there exists a dependent compensator whose Jacobian $d\chi_F$ is surjective for a particular system (A, B, C), then both the extended pole placement map χ and the map ψ are surjective.*

Proof. Direct consequence from the fact that the map χ is homogeneous.

Step 3: Construction of a polynomially dependent compensator.

In a last step we will have to construct a polynomial map which assigns to every system (A, B, C) a dependent compensator $F = [K_1\ K_2]$.

Every such polynomial map induces an algebraic set

$$S \subset (A, B, C)$$

through the condition that $d\chi_F$ is not surjective (see [32]). If we can exhibit one system (A, B, C) whose dependent compensator results in a surjective Jacobian, then we have shown that a generic set of systems have a surjective pole placement map. An explicit polynomial map and a particular plant (A, B, C) was constructed in [32] under the condition that $n < mp$.

4 The Dynamic Pole Placement Problem

In this section we explain how the techniques developed for the static pole placement problem do extend to the dynamic pole placement problem. In the sequel it will turn out to be the easiest to work with polynomial models of systems.

Consider a $p \times (m + p)$ polynomial matrix. First note the following. Every observable state space system described through

$$\dot{x} = Ax + Bu, y = Cx + Du; \quad x \in \mathbb{R}^n, u \in \mathbb{R}^m, y \in \mathbb{R}^p \tag{4.1}$$

can be equally well described (through elimination of the state variable x) by a system of autoregressive equations of the form

$$P\left(\frac{d}{dt}\right)\begin{pmatrix} u(t) \\ y(t) \end{pmatrix} = 0. \tag{4.2}$$

Here $P(s)$ is a polynomial matrix of size $p \times (m + p)$ and the full size $p \times p$ minors have maximal degree n. As the first order representation (4.1) is only unique up to a change of basis in the state space, there are different ways to describe the system through a matrix $P(s)$. Indeed performing elementary row operations on the matrix $P(s)$ certainly has no effect on the solution set, i.e. the "behavior" in the sense of Willems [40]. On the other hand it is well known that $P_1(s)$ and $P_2(s)$ describe the same behavior if and only if $P_1(s)$ and $P_2(s)$ are row equivalent over the polynomial ring $\mathbb{R}[s]$, i.e., if the rows of $P_1(s)$ and $P_2(s)$ span the same row module.

If the $p \times (m + p)$ matrix $P(s)$ represents such a system, we will assume that $P(s)$ is row reduced having row degrees

$$\nu_1 \geq \cdots \geq \nu_p.$$

For the input-output system (4.1), the ordered indices $\nu = \{\nu_1, \ldots, \nu_p\}$ are the observability indices of the system and $\sum_{i=1}^{p} \nu_i$ is the McMillan degree of the system.

Next we would like to introduce feedback. For this consider a system Σ_1 described through an autoregressive system $P(s)$ and a system Σ_2 described through an autoregressive system $Q(s)$. The *interconnected system* $\Sigma_1 \wedge \Sigma_2$ is then the system described by the combined behavioral equations

$$\begin{bmatrix} P(\frac{d}{dt}) \\ Q(\frac{d}{dt}) \end{bmatrix} \begin{pmatrix} u(t) \\ y(t) \end{pmatrix} = 0. \tag{4.3}$$

In the sequel we will always assume that $P(s)$ has size $p \times (m+p)$, rank p and McMillan degree n and that $Q(s)$ has size $m \times (m+p)$, rank m and McMillan degree q. We call (4.3) a *regular interconnection* (see [21, 41] for more details) if $\det \binom{P(s)}{Q(s)} \neq 0$. If $\det \binom{P(s)}{Q(s)} = 0$, we say $Q(s)$ is a *dependent compensator* of the plant $P(s)$. For a regular interconnection the poles of the closed loop system are given by the roots of the characteristic polynomial

$$\det \begin{pmatrix} P(s) \\ Q(s) \end{pmatrix}.$$

We refer to [21, 31, 33, 41] to connect to the classical formulation of the pole placement problem.

The problem which we study is now as follows. Under what condition is it true that for a generic set of $p \times (m+p)$ polynomial matrices having McMillan degree n the following holds. For every polynomial $\phi \in \mathbb{R}[s]$ of degree $n+q$ there exists a $m \times (m+p)$ polynomial matrix $Q(s)$ of McMillan degree q such that $\det \binom{P(s)}{Q(s)} = \phi$.

In the sequel we will show how the techniques, developed for the static problem and presented in the last section, can be extended to the dynamic pole placement problem.

In a first step we will compactify the total compensator space. For this, let $A^q_{m,p}$ denote the set of all $m \times (m+p)$ polynomial systems of rank m and McMillan degree at most q. Consider the sequence of polynomial maps:

$$\begin{array}{rcl} A^q_{m,p} & \stackrel{Her.-Mar.}{\longrightarrow} & Rat_q(\mathbb{P}^1, Grass(m, \mathbb{C}^{m+p})) \\ & \stackrel{Plücker}{\longrightarrow} & Rat_q(\mathbb{P}^1, \mathbb{P}(\wedge^m \mathbb{C}^{m+p})) \\ & \stackrel{\tau}{\longrightarrow} & \mathbb{P}(\mathbb{C}^{q+1} \otimes \wedge^m \mathbb{C}^{m+p}). \end{array} \tag{4.4}$$

In simple terms, the above sequence of maps computes the $m \times m$ full size minors of a $m \times (m+p)$ polynomial matrix $Q(s)$ and writes down in an ordered way the $\binom{m+p}{m}(q+1)$ polynomial coefficients. A result of Rosenthal [31] now states:

Theorem 4.1 ([31]) *The image describes a projective variety $K^q_{m,p}$ of dimension $q(m+p)+mp$.*

Note that for $q = 0$ the variety $K^q_{m,p}$ is nothing else than the Grassmann variety $\mathrm{Grass}(m, m+p)$.

Let E_P denote the set of $m \times (m+p)$ dependent compensators corresponding to the plant $P(s)$. In analogy to the static situation we can define:

Definition 4.1 The pole placement map for the plant $P(s)$ is defined as the algebraic map:

$$\begin{array}{rcl} \rho_P \ : K^q_{m,p} - E_P & \longrightarrow & \mathbb{P}^{n+q} \\ Q(s) & \longmapsto & \phi(s) = \det \begin{bmatrix} P(s) \\ Q(s) \end{bmatrix}. \end{array} \tag{4.5}$$

The following theorem extends the result of Brockett and Byrnes [4] from the static to the dynamic situation:

Theorem 4.2 (Rosenthal [31]) *The pole placement map ρ_P defines in $\mathbb{P}(\mathbb{C}^{q+1} \otimes \wedge^m \mathbb{C}^{m+p})$ a central projection. If*

$$q(m+p) + mp \geq n + q, \tag{4.6}$$

then the pole placement map ρ_P is surjective for the generic $p \times (m+p)$ system $P(s)$ of McMillan degree n. If

$$q(m+p) + mp = n + q, \tag{4.7}$$

then $E_P = \emptyset$ for the generic system and for every $x \in \mathbb{P}^{n+q}$, $\rho_P^{-1}(x)$ consists of $d(m,p,q)$ compensators, where $d(m,p,q)$ denotes the degree of the variety $K^q_{m,p}$.

In [30] Ravi, Rosenthal and Wang did compute this degree of the variety $K^q_{m,p}$. The result is as follows:

Theorem 4.3 *The degree of the variety $K^q_{m,p}$, i.e., the number of different (complex) feedback compensators in the critical dimension is equal to*

$$(mp + q(m+p))! \left| \sum_{n_1 + \cdots + n_m = q} \frac{\prod_{k<j} (j - k + (n_j - n_k)m + p))}{\prod_{j=1}^{m} (p + j + n_j(m+p) - 1)!} \right|. \tag{4.8}$$

The techniques for the pole placement problem over the reals can be extended as well. In particular the method of linearizing around some dependent compensators can be carried through as for the static pole placement problem, although the technical difficulties are much larger. The following result has been obtained in this way and it constitutes one of the strongest sufficiency results:

Theorem 4.4 (Rosenthal–Wang [33]) *Let $r_m = q - m[q/m]$ and $r_p = q - p[q/p]$ be the remainders of q divided by m and p, respectively. If*

$$q(m+p-1) + mp > n + \min(r_m(p-1), r_p(m-1)), \tag{4.9}$$

then the generic $p \times (m+p)$ plant of McMillan degree n can be arbitrary pole assigned using $m \times (m+p)$ compensators of McMillan degree q only. In particular when q is a multiple of either m or p, one has the sufficient condition

$$q(m+p) + mp > n + q.$$

5 General Matrix Extension Problems

Matrix extension problems have been studied by many authors and we refer to the recent book by Gohberg, Kaashoek and van Schagen [12]. In this section we would like to elaborate on the example of the introduction and show that many of the techniques which we present actually extend to this larger class of problems.

Consider once more a linear subspace $\mathcal{L} \subset Mat_{n\times n}$ and let $A \in Mat_{n\times n}$ be a particular fixed matrix. Again identify the set of monic polynomials of degree n with the vector space $\mathbb{K}^n$. In this situation we are interested in conditions which guarantee that the characteristic map

$$\chi_A : \mathcal{L} \longrightarrow \mathbb{K}^n, \quad L \longmapsto \det(sI - A - L)$$

is surjective or at least "generically" surjective.

The following conditions are obvious necessary conditions:

1. χ_A is almost onto only if $\dim \mathcal{L} \geq n$.

2. There must be at least one element $L \in \mathcal{L}$ whose trace $\operatorname{tr}(L) \neq 0$, i.e. $\mathcal{L} \not\subset sl_n$.

In [16] Helton, Rosenthal and Wang used the dominant morphism theorem to show:

Theorem 5.1 *If the base field is algebraically closed and if there is a linear subspace $\mathcal{L} \subset Mat_{n\times n}$ which satisfies above necessary conditions, then χ_A is generically surjective for a generic set of $n \times n$ matrices A.*

It is of course a natural question if it will be possible to "compactify" the problem in a similar manner as we explain it for the static and dynamic pole placement problem.

In the sequel we outline one way to do this. For this consider the identity

$$\det(sI - A - L) = \det[sI - A, -I] \begin{bmatrix} I \\ L \end{bmatrix}. \tag{5.1}$$

Notice that colsp $\begin{bmatrix} I \\ L \end{bmatrix}$ describes a point in Grass$(n, 2n)$. In terms of the Plücker coordinates of Grass$(n, 2n)$ (compare with Theorem 3.2), we can rewrite (5.1) as:

$$\det(sI - A - L) = \sum_{i=0}^{N} p_i(s) z_i. \tag{5.2}$$

Let E be the linear subspace defined by

$$E = \{z \in \mathbb{P}^N | \sum_{i=0}^{N} p_i(s) z_i = 0\}$$

and let $\bar{\mathcal{L}}$ be the projective closure of $\mathcal{L}$ in Grass$(n, 2n) \subset \mathbb{P}^N$. Then we have:

Theorem 5.2 *The characteristic map χ_A extends to a central projection*

$$\bar{\chi}_A : \text{Grass}(n, 2n) - E \longrightarrow \mathbb{P}^n.$$

In particular if $E \cap \bar{\mathcal{L}} = \emptyset$, then χ_A is surjective of mapping degree equal to the degree of the variety $\bar{\mathcal{L}}$.

We illustrate this result on a situation considered by Friedland [10]. Let $\mathcal{L} := \mathcal{D}_{n \times n}$ be the set of all diagonal matrices, then one verifies that the closure of

$$\begin{bmatrix} I \\ D \end{bmatrix} = \begin{bmatrix} 1 & \cdots & 0 \\ \vdots & \ddots & \vdots \\ 0 & \cdots & 1 \\ x_1 & \cdots & 0 \\ \vdots & \ddots & \vdots \\ 0 & \cdots & x_n \end{bmatrix}$$

in Grass$(n, 2n)$ is equal to the product of n copies of $\mathbb{P}^1$, i.e.

$$\bar{\mathcal{L}} = \mathbb{P}^1 \times \cdots \times \mathbb{P}^1.$$

A calculation of Hilbert polynomials (compare with [6]) shows that the degree of $\bar{\mathcal{L}}$ is $n!$. Finally a direct calculation shows that $E \cap \bar{\mathcal{L}} = \emptyset$ for all $n \times n$ matrices A. We therefore have:

Theorem 5.3 (Friedland [10]) *Let $A \in Mat_{n \times n}$ be arbitrary and let $\mathcal{L} = \mathcal{D}_{n \times n}$ be the set of diagonal matrices. Then χ_A is surjective of mapping degree $n!$.*

References

[1] S. Abhyankar. *Algebraic Geometry for Scientists and Engineers.* Providence, R.I.: AMS, 1990.

[2] S. Ariki. Generic pole assignment via dynamic feedback. Preprint, 1994.

[3] F. M. Brash and J. B. Pearson. Pole placement using dynamic compensators. *IEEE Trans. Automat. Control* **AC-15** (1970), 34-43.

[4] R. W. Brockett and C. I. Byrnes. Multivariable Nyquist criteria, root loci and pole placement: a geometric viewpoint. *IEEE Trans. Automat. Control* **AC-26** (1981), 271-284.

[5] C. I. Byrnes. Pole assignment by output feedback. *Three Decades of Mathematical System Theory.* (H. Nijmeijer and J. M. Schumacher, Eds.). *Lecture Notes in Control and Information Sciences.* Vol. 135. New York: Springer Verlag, 1989. 31-78.

[6] C. I. Byrnes and X. Wang. The additive inverse eigenvalue problem for lie perturbations. *SIAM J. Matrix Anal. Appl.* **14**(1) (1993), 113-117.

[7] D. Cox, J. Little and D.O. O'Shea. *Ideals, Varieties and Algorithms.* New York: Springer Verlag, 1992.

[8] E.J. Davison. On pole assignment in linear systems with incomplete state feedback. *IEEE Trans. Automat. Control* **AC-15** (1970), 348-351.

[9] P. Falb. *Methods of Algebraic Geometry in Control Theory: Part I.* Boston: Birkhäuser, 1990.

[10] S. Friedland. Inverse eigenvalue problems. *Linear Algebra Appl.* **17** (1977), 15-51.

[11] W. Fulton. *Intersection Theory.* New York: Springer Verlag, 1984.

[12] I. Gohberg, M. A. Kaashoek, and F. van Schagen. *Partially Specified Matrices and Operators: Classification, Completion, Applications.* Boston: Birkäuser, 1995.

[13] J. Harris. *Algebraic Geometry, A First Course.* New York: Springer Verlag, 1992.

[14] R. Hartshorne. *Algebraic Geometry.* Berlin: Springer Verlag, 1977.

[15] U. Helmke and J. Rosenthal. Eigenvalue inequalities and Schubert calculus. *Mathematische Nachrichten* **171** (1995), 207-225.

[16] W. Helton, J. Rosenthal, and X. Wang. Matrix extensions and eigenvalue completions, the generic case. To appear *Trans. Amer. Math. Soc.*

[17] R. Hermann and C. F. Martin. Applications of algebraic geometry to system theory. Part I. *IEEE Trans. Automat. Control* **AC-22** (1977), 19-25.

[18] H. Kimura. Pole assignment by gain output feedback. *IEEE Trans. Automat. Control* **AC-20** (1975), 509-516.

[19] S. L. Kleiman. Problem 15: Rigorous foundations of Schubert's enumerative calculus. *Proceedings of Symposia in Pure Mathematics, American Mathematics Society 1976.* 445-482.

[20] S. L. Kleiman and D. Laksov. Schubert calculus. *Amer. Math. Monthly* **79** (1972), 1061-1082.

[21] M. Kuijper. Why do stabilizing controllers stabilize? *Automatica.* **31**(4) (1995), 621-625.

[22] J. Leventides. *Algebrogeometric and Topological Methods in Control Theory.* Ph.D. Dissertation. City University of London. London, England. 1993.

[23] J. Leventides and N. Karcanias. Global asymptotic linearisation of the pole placement map: a closed form solution for the constant output feedback problem. *Automatica.* **31**(9) (1995), 1303-1309.

[24] V. G. Lomadze. Finite-dimensional time-invariant linear dynamical systems: algebraic theory. *Acta Appl. Math.* **19** (1990), 149-201.

[25] C. F. Martin and R. Hermann. Applications of algebraic geometry to system theory: the McMillan degree and Kronecker indices as topological and holomorphic invariants. *SIAM J. Control Optim.* **16** (1978), 743-755.

[26] D. Mumford. *Algebraic Geometry I: Complex Projective Varieties.* New York: Springer Verlag, 1976.

[27] M. S. Ravi and J. Rosenthal. A smooth compactification of the space of transfer functions with fixed McMillan degree. *Acta Appl. Math.* **34** (1994), 329-352.

[28] M. S. Ravi and J. Rosenthal. A general realization theory for higher order linear differential equations. *Systems & Control Letters.* **25**(5) (1995), 351-360.

[29] M. S. Ravi, J. Rosenthal, and X. Wang. On decentralized dynamic pole placement and feedback stabilization. *IEEE Trans. Automat. Contr.*. **40**(9) (1995), 1603-1614.

[30] M. S. Ravi, J. Rosenthal, and X. Wang. Dynamic pole assignment and Schubert calculus. *SIAM J. Control Optim.* **34**(3) (1996), 813-832.

[31] J. Rosenthal. On dynamic feedback compensation and compactification of systems. *SIAM J. Control Optim.* **32**(1) (1994), 279-296.

[32] J. Rosenthal, J. M. Schumacher, and J. C. Willems. Generic eigenvalue assignment by memoryless real output feedback. *Systems & Control Letters.* **26** (1995), 253-260.

[33] J. Rosenthal and X. Wang. Output feedback pole placement with dynamic compensators. *IEEE Trans. Automat. Contr.* **41**(6) (1996), 830-843.

[34] H. Schubert. *Kalkühl der abzählenden Geometrie.* Leipzig: Teubner, 1879.

[35] H. Schubert. Anzahlbestimmung für lineare Räume beliebiger Dimension. *Acta Math.* **8** (1886), 97-118.

[36] M. A. Shayman. Phase portrait of the matrix Riccati equation. *SIAM J. Control Optim.* **24**(1) (1986), 1-65.

[37] R. C. Thompson. High, low, and quantitative roads in linear algebra. *Linear Algebra Appl.* **162/164** (1992), 23-64.

[38] X. Wang. Pole placement by static output feedback. *Journal of Math. Systems, Estimation, and Control.* **2**(2) (1992), 205-218.

[39] X. Wang. Grassmannian, central projection and output feedback pole assignment of linear systems. *IEEE Trans. Automat. Contr.* **41**(6) (1996), 786-794.

[40] J. C. Willems. Paradigms and puzzles in the theory of dynamical systems. *IEEE Trans. Automat. Control.* **AC-36**(3) (1991), 259-294.

[41] J. C. Willems. On interconnections, control and feedback. Submitted to *IEEE AC*, July 1994.

Department of Mathematics, University of Notre Dame, Notre Dame, IN 46556-5683

Department of Mathematics, Texas Tech University, Lubbock, TX 79409-1024

Recursive Designs and Feedback Passivation [1]

Rodolphe Sepulchre, Mrdjan Janković, Petar V. Kokotović

Abstract

The problem of feedback stabilization is reformulated as feedback *passivation* so that the construction of a stabilizing feedback is translated into the construction of a passivating output. This construction is restricted by two geometric requirements of passivity: a relative degree one and a minimum phase property. We show how backstepping and forwarding, the two building blocks of recursive Lyapunov designs, compelement each other by each removing one of the two obstacles to feedback passivation.

1 Introduction

In the recent years, the concept of *passivity* has played an increasing role in a constructive solution to the global stabilization of nonlinear systems. With the problem of feedback stabilization reformulated as feedback passivation, the search for a Lyapunov function $V(x)$ whose time-derivative $\dot{V}$ can be rendered negative by feedback is replaced or accompanied by the search for a passivating output $y = h(x)$ such that the system can be rendered passive by feedback. What is gained in this reformulation is that the search for a passivating output is restricted by two geometric requirements which can be employed to guide the construction: the system must be "relative degree one" and "weakly minimum phase" (a concept introduced in [5] to describe a situation in which the zero-dynamics subsystem is stable but not necessarily asymptotically stable).

As such, the two geometric requirements of feedback passivation are quite restrictive and they often constitute an obstacle rather than a help to a constructive design procedure. As the dimension of the system increases,

[1]This work was completed during the postdoctoral stay of the first author at Santa Barbara. The work was supported in part by the National Science Foundation under Grant ECS-9203491 and in part by the Air Force Office of Scientific Research under Grant F-49620-95-1-049. The first author is Chargé de recherches F.N.R.S., Belgium. He acknowledges partial support from the Belgian Programme on Interuniversity Poles of Attraction, initiated by the Belgian Prime Minister's Office for Science, Technology and Culture.

searching for an output which satisfies these requirements rapidly becomes an unwieldy task. However, combined with recursive designs, which decompose the original problem into a sequence of simpler subproblems, they become geometric tools which guide the construction of a passivating output and of a Lyapunov function for the entire system.

In this paper, we show how the two building blocks of recursive Lyapunov designs, backstepping and forwarding, complement each other to overcome the obstacles of feedback passivation. Backstepping permits a recursive feedback passivation design of systems for which an output is found which satisfies the minimum phase requirement but has a higher relative degree. In a dual manner, forwarding permits a recursive passivation design of systems for which an output is found which has relative degree one but violates the minimum phase requirement. To guarantee global stabilizability, the structural requirements of these two design procedures cannot be relaxed without further assumptions.

2 Feedback passivation for stabilization

Feedback passivation as a tool for stabilization exploits the fundamental connection between ***passivity*** and ***stability***: if a system is passive and zero-state detectable (ZSD), then, with the input $u = 0$, its equilibrium $x = 0$ is stable. Moreover, it is rendered asymptotically stable by closing the loop with the feedback $u = -y$. These statements become global if, in addition, the storage function is radially unbounded (see [7] for details).

The task of stabilization is thus the simplest when an output function $y = h(x)$ can be found such that the system

$$\begin{aligned} \dot{x} &= f(x) + g(x)u \\ y &= h(x) \end{aligned} \tag{2.1}$$

with u as the input and y as the output is passive. However, searching for an output $y = h(x)$ such that the system is passive with a positive definite storage function requires that the system be stable when $u = 0$.

To remove this restriction, we include feedback as a means to achieve passivity. Instead of being stable, the uncontrolled system is assumed to be stabilizable. Therefore, we need to find an output $y = h(x)$ and a feedback transformation $u = \alpha(x) + \beta(x)v$ with $\beta(x)$ invertible, such that the system

$$\begin{aligned} \dot{x} &= f(x) + g(x)\alpha(x) + g(x)\beta(x)v \\ y &= h(x) \end{aligned} \tag{2.2}$$

is passive.

If a feedback transformation can be found to render the system (2.2) passive, we call the original system (2.1) *feedback passive.* The selection of an output $y = h(x)$ and the construction of a passivating transformation is referred to as *feedback passivation.* Under a ZSD assumption, asymptotic stability of the passive system (2.2) is simply achieved with the additional feedback $v = -y$.

The crucial limitation of the feedback passivation design is that the output must have two properties which cannot be modified by feedback. The following result summarizes this geometric characterization of feedback passivity.

Theorem 1 (*Feedback passivity*)
Assume that rank $\frac{\partial h}{\partial x}(0) = m$. Then the system (2.1) is feedback passive with a C^2 positive definite storage function $S(x)$ if and only if it has relative degree one at $x = 0$ and is weakly minimum phase. □

The proof of this (local) result can be found in [1], under an additional technical assumption which was recently removed in [7]. For a definition of the relative degree and the zero dynamics of a nonlinear system, the reader is referred to [2]. A nonlinear system is called weakly minimum phase when the equilibrium $z = 0$ of its zero-dynamics subsystem $\dot{z} = f_{zd}(z)$ is stable and a positive definite Lyapunov function $W(z)$ is known such that $L_{f_{zd}}W \leq 0$ near $z = 0$.

Theorem 1 shows both the interest and the limitation of feedback passivation as a constructive design procedure for stabilization. Because it is restricted by two structural requirements, the search for a passivating output is guided by geometric tools. However, the requirements are so restrictive that, as such, they constitute an obstacle to a systematic construction.

3 Recursive designs

The search for a passivating output can be oriented by physical considerations or by a particular form of the system but many passivation attempts are frustrated by the requirements that the system must have a relative degree one and be weakly minimum phase. Even for a highly structured system such as

$$\begin{aligned}
\dot{z} &= f(z) + \tilde{\psi}(z, \xi_i)\xi_i, \quad i \in \{1, \ldots, n\} \\
\dot{\xi}_1 &= \xi_2 \\
\dot{\xi}_2 &= \xi_3 \\
&\vdots \\
\dot{\xi}_n &= u,
\end{aligned} \tag{3.1}$$

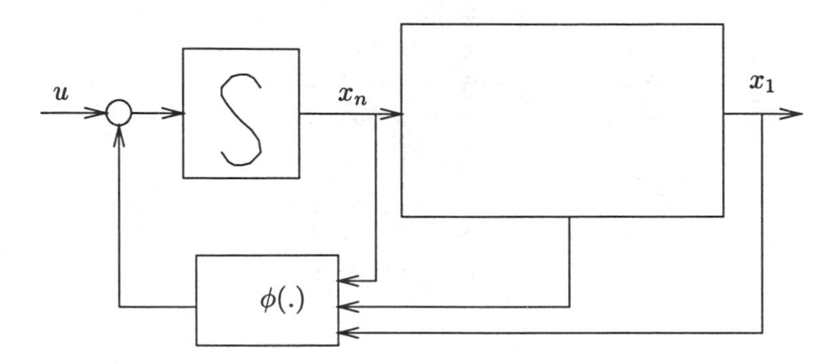

Figure 3.1: A strict-feedback configuration.

with $z = 0$ a globally asymptotically stable equilibrium of $\dot{z} = f(z)$, feedback passivation is difficult because each candidate output $y = \xi_i$ fails to satisfy at least one of the two passivity requirements. Thus, if $y = \xi_1$, the system is minimum phase, but it has a relative degree n. On the other hand, if $y = \xi_n$, the relative degree is one, but the system is not weakly minimum phase because the zero-dynamics subsystem contains an unstable chain of integrators. For all other choices $y = \xi_i$, neither the relative degree one, nor the weak minimum phase requirement are satisfied.

To circumvent the structural obstacles to passivation, *recursive designs* decompose the construction of the passivating output in several steps. At each step, only a subsystem is considered, for which a passivation design is simpler.

The two building blocks of recursive designs are *backstepping* and *forwarding*. These two design procedures complement each other by removing the two obstacles to feedback passivation: backstepping removes the relative degree obstacle, while forwarding removes the minimum phase obstacle. Here we will not consider the most general classes of systems to which these procedures are applicable (see [4, 3, 7]). Instead we will highlight their dual role in a constructive solution of the feedback passivation problem.

Backstepping relies on a system structure illustrated in Figure 3.1 and characterized by the absence of *feedforward* connections from the input u. The block-diagram corresponds to the state equation

$$\begin{aligned} \dot{X}^1_{n-1} &= F^1_{n-1}(X^1_{n-1}, x_n), \quad X^1_{n-1} = [x_1, \ldots, x_{n-1}]^T \\ \dot{x}_n &= \phi(x_1, \ldots, x_n) + u \end{aligned} \tag{3.2}$$

where the $n-1$ first equations do not depend on u. For a feedback passivation design, we must select an output y_n which has relative degree one.

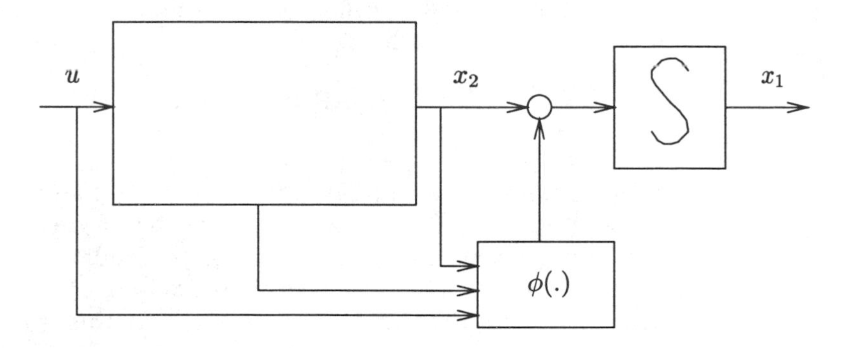

Figure 3.2: A strict-feedforward configuration.

Because of the particular structure (3.2), this implies that y_n depends explicitely on x_n, say in the form

$$y_n = x_n - \alpha_n(x_1, \dots, x_{n-1}) \tag{3.3}$$

which guarantees a global relative degree one. To meet the second passivation requirement, we need to select $\alpha_n(x_1, \dots, x_n)$ in such a way that the system is weakly minimum phase. Setting $y_n \equiv 0$ shows that the zero-dynamics subsystem of (3.2) is

$$\dot{X}^1_{n-1} = F^1_{n-1}(X^1_{n-1}, \alpha_n) \tag{3.4}$$

which means that, to construct α_n, we face a new stabilization problem. However, this stabilization problem is for a system of lower dimension in which the state x_n is treated as an input. If a passivating output y_{n-1} is found for the subsystem (3.2), it is not a passivating output for the entire system (3.2) because it has relative degree one with respect to the virtual "input" x_n, and hence relative degree two with respect to the input u. However, backstepping uses the solution of this $(n-1)$-dimensional subproblem to construct the output y_n which satisfies both passivation requirements. In this sense, backstepping removes the relative degree obstacle to feedback passivation.

Forwarding relies on a system structure illustrated in Figure 3.2 and characterized by the absence of *feedback* connections from the state x_1. The block-diagram corresponds to the state equation

$$\begin{aligned} \dot{x}_1 &= x_2 + \phi(x_2, \dots, u) \\ \dot{X}^2_n &= F^2_n(X^2_n) + G^2_n(X^2_n)u, \quad X^2_n = [x_2, \dots, x_n]^T \end{aligned} \tag{3.5}$$

where the $n-1$ last equations do not depend on x_1. This particular structure allows us to first consider the stabilization of the subsystem

$$\dot{X}_n^2 = F_n^2(X_n^2) + G_n^2(X_n^2)u \tag{3.6}$$

in which the first equation of (3.5) has been ignored. If a passivating output y_2 can be constructed for the subsystem (3.6), it is in general not a passivating output for the entire system (3.5): it satisfies the relative degree one requirement but the zero-dynamics subsystem is no longer stable due to the added equation. A forwarding design will modify the output y_2 into a new output y_1 which satisfies both passivation requirements. In this sense, forwarding removes the minimum phase obstacle to feedback passivation.

4 Backstepping

A family of backstepping designs can be constructed by recursive applications of different versions of the same basic step: the augmentation by one equation of the subsystem already made passive by a "virtual control".

Theorem 2 *(Backstepping as recursive feedback passivation)*
Assume that for the system

$$\dot{z} = f(z) + g(z)u, \tag{4.7}$$

a C^1 feedback transformation $u = \alpha_0(z) + v_0$ and a C^2 positive definite, radially unbounded storage function $W(z)$ are known such that this system is passive from the input v_0 to the output $y_0 = (L_gW)(z)$, that is $\dot{W} \leq y_0 v_0$.

Then the augmented system

$$\begin{aligned} \dot{z} &= f(z) + g(z)\xi \\ \dot{\xi} &= a(z,\xi) + b(z,\xi)u, \end{aligned} \tag{4.8}$$

where $b^{-1}(z,\xi)$ exists for all (z,ξ), is (globally) feedback passive with respect to the output $y = \xi - \alpha_0(z)$ and the storage function $V(z,y) = W(z) + \frac{1}{2}y^2$. A particular control law which achieves passivity of (4.8) is

$$u = b^{-1}(z,\xi)(-a(z,\xi) - y_0 + \frac{\partial \alpha_0}{\partial z}(f(z) + g(z)\xi) + v) \tag{4.9}$$

The system (4.8) with (4.9) is ZSD for the input v if and only if the system (4.7) is ZSD for the input v_0. Moreover, if $W(z(t))$ is strictly decreasing for (4.7) with $u = \alpha_0(z)$, then $W(z(t)) + \frac{1}{2}y^2(t)$ is strictly decreasing for (4.8) with $v = -y$. □

In Theorem 2 a new passivating output y is constructed from the previous passivating control law $\alpha_0(z)$, and the new storage function is obtained by adding y^2 to the old storage function. Moreover, the ZSD property is preserved in the augmented system.

Because Theorem 2 ensures that the augmented system inherits the properties of the original system, we can use it at each step of a recursive design procedure for a system which is an augmentation of the z-subsystem in (4.8) by a lower-triangular ξ-subsystem:

$$\begin{array}{rcl} \dot{z} & = & f(z) + g(z)\xi_1 \\ \dot{\xi}_1 & = & a_1(z,\xi_1) + b_1(z,\xi_1)\xi_2 \\ \dot{\xi}_2 & = & a_2(z,\xi_1,\xi_2) + b_1(z,\xi_1,\xi_2)\xi_3 \\ & \vdots & \\ \dot{\xi}_{n-1} & = & a_{n-1}(z,\xi_1,\ldots,\xi_{n-1}) + b_{n-1}(z,\xi_1,\ldots,\xi_{n-1})\xi_n \\ \xi_n & = & a_n(z,\xi_1,\ldots,\xi_n) + b_n(z,\xi_1,\ldots,\xi_n)u, \quad \xi_i \in I\!R^q, \ i = 1,\ldots,n \end{array} \tag{4.10}$$

The systems in the lower-triangular configuration (4.10) are called *strict-feedback systems*, because every interconnection in the system is a *feedback* connection from the states located farther from the input. Assuming that the z-subsystem satisfies Theorem 2 and that every $b_i(z,\xi_1,\ldots,\xi_i)$ is invertible for all $(z,\xi_1,\ldots,\xi_i)$, the system (4.10) with the output $y_1 = \xi_1 - \alpha_0(z)$ has relative degree n. The relative degree is then recursively reduced to one. For $y_n = \xi_n - \alpha_{n-1}(z,\xi_1,\ldots,\xi_n)$ to be a passivating output for the whole system, the virtual control law $\xi_n = \alpha_{n-1}(z,\xi_1,\ldots,\xi_{n-1})$ must be a passivating feedback for the zero-dynamics subsystem consisting of (4.10) minus the last equation. Likewise, $y_{n-1} = \xi_{n-1} - \alpha_{n-2}(z,\xi_1,\ldots,\xi_{n-2})$ will be a passivating output for this subsystem if α_{n-2} is a passivating feedback for its zero-dynamics subsystem. Continuing this process upward, we end up with the recursive expressions for passivating outputs:

$$y_i = \xi_i - \alpha_{i-1}(z,\xi_1,\ldots,\xi_{i-1})$$

If the $\alpha_i' s$ are constructed employing the passivating transformation (4.9) at each step, then the backstepping recursion is

$$\begin{array}{rcl} y_i & = & \xi_i - \alpha_{i-1}(z,\xi_1,\ldots,\xi_{i-1}) \\ \alpha_i(z,\xi_1,\ldots,\xi_i) & = & b_i^{-1}(-a_i - y_{i-1} + \dot{\alpha}_{i-1} - y_i), \quad i = 2,\ldots,n \end{array} \tag{4.11}$$

In these expressions, the time-derivatives $\dot{\alpha}_i$ are evaluated as explicit functions of the state variables, that is, $\dot{\alpha}_0 = \frac{\partial \alpha_1}{\partial z}(f + g\xi_1)$, $\dot{\alpha}_1 = \frac{\partial \alpha_1}{\partial z}(f + g\xi_1) + \frac{\partial \alpha_1}{\partial \xi_1}(a_1 + b_1\xi_2)$, etc.

The absence of any feedforward connection in the system (4.10) is crucial for recursive backstepping: it guarantees that the relative degree of ξ_i is $r_i = n - i + 1$ for each i. Because of this property, the relative degree one requirement of feedback passivation is met at step i, not with respect to the true input u but rather with respect to the virtual input ξ_{i+1}. Only the output y_n is relative degree one with respect to the true input u.

We stress that, to achieve global feedback passivation via recursive backstepping, it is not sufficient to find a passivating output which satisfies globally the minimum phase requirement and has a global relative degree n, like the output $y_1 = \xi_1 - \alpha_0(z)$ in the system (4.10). In addition, the strict-feedback form excludes feedforward connections, so that the interconnection term $g(z)\xi_1$ in (4.10) cannot depend on the states $\xi_2, \ldots, \xi_n$ and the input u. In general, this restriction is necessary to prevent the peaking obstacle to global stabilization [6].

Example 3 The system

$$\begin{aligned} \dot{z} &= -z + (\xi_1^2 + \xi_2^2)z^2 \\ \dot{\xi}_1 &= \xi_2 \\ \dot{\xi}_2 &= u, \quad y = \xi_1 \end{aligned} \tag{4.12}$$

is globally minimum phase because its zero-dynamics subsystem is $\dot{z} = -z$. It also has a globally defined relative degree two. However, the region of attraction of $(z, \xi_1, \xi_2) = (0,0,0)$ cannot be rendered arbitrary large. The explicit solution

$$z(t) = \frac{e^{-t}}{\frac{1}{z(0)} - \int_0^t e^{-s}(\xi_1^2(s) + \xi_2^2(s))ds} \tag{4.13}$$

shows that the condition

$$z(0) > (\int_0^\infty e^{-s}\|\xi(s)\|^2 ds)^{-1}$$

causes the denominator to be zero at some finite time $t_e > 0$, and hence, $z(t)$ escapes to infinity as $t \to t_e$. Setting $\|\xi(0)\| = 1$, the integral $\int_0^\infty e^{-s}\|\xi(s)\|^2 ds$ cannot be rendered arbitrarily small regardless of the control $u(t)$, so that $z(t)$ will escape to infinity if $z(0)$ is large enough.

5 Forwarding

To present the forwarding procedure we start from a system

$$\dot{\xi} = a(\xi) + b(\xi)u \tag{5.14}$$

which, by assumption, is feedback passive with an already constructed storage function $U(\xi)$ and the output $y_0 = L_g U(\xi)$. Using a preliminary passivating transformation, we assume that the system is passive so that, under a ZSD assumption, the equilibrium $\xi = 0$ of $\dot{\xi} = a(\xi) - b(\xi) L_b U(\xi) := \bar{a}(\xi)$ is globally asymptotically stable (GAS). If U is locally quadratic and if the pair $(\frac{\partial a}{\partial \xi}(0), b(0))$ is stabilizable, $\xi = 0$ is also locally exponentially stable (LES). To make the procedure recursive, we want to achieve the same property for the augmented system

$$\begin{aligned} \dot{z} &= \psi(\xi) + g(\xi) u \\ \dot{\xi} &= a(\xi) + b(\xi) u \end{aligned} \tag{5.15}$$

The main tool of forwarding is the construction of a Lyapunov function for the system (5.15) when $u = -y_0$, which we view as the cascade

$$\begin{aligned} \dot{z} &= 0 + \bar{\psi}(\xi) \\ \dot{\xi} &= \bar{a}(\xi) \end{aligned} \tag{5.16}$$

of the stable system $\dot{z} = 0$ with the GAS/LES system $\dot{\xi} = \bar{a}(\xi)$, the interconnection being $\psi(\xi) - g(\xi) L_b U(\xi) := \bar{\psi}(\xi)$. For this cascade, the sum $U(\xi) + \frac{1}{2} z^2$ fails to be a Lyapunov function because the time-derivative yields

$$(\dot{U} + z\dot{z}) \mid_{u=0} = L_a U(\xi) + z\bar{\psi}(\xi) \leq z\bar{\psi}(\xi) \tag{5.17}$$

which is not negative because of the interconnection term $z\bar{\psi}(\xi)$. To eliminate this term in the time-derivative of the Lyapunov function, we introduce a cross-term $\Psi(z, \xi)$ in the Lyapunov function

$$V(z, \xi) = U(\xi) + \Psi(z, \xi) + \frac{1}{2} z^2 \tag{5.18}$$

in such a way that $\dot{\Psi} = -z\bar{\psi}(\xi)$, so that $\dot{V} = L_a U(\xi) \leq 0$. This yields a definition of Ψ as the line-integral

$$\Psi(z, \xi) = \int_0^\infty \tilde{z}(s) \bar{\psi}(\tilde{\xi}(s)) ds \tag{5.19}$$

evaluated along the solution $(\tilde{z}(s), \tilde{\xi}(s))$ of (5.16) for the initial condition $(\tilde{z}(0), \tilde{\xi}(0)) = (z, \xi)$. It is shown in [3] that the Lyapunov function (5.18) is smooth, positive definite, and radially unbounded, and that the same construction applies to more general situations where z is a vector, $\dot{z} = 0$ is replaced by a stable system $\dot{z} = f(z)$, and ψ and g depend on z. It should be clear that the local exponential stability assumed for $\dot{\xi} = \bar{a}(\xi)$ is crucial for the convergence of the integral (5.19).

For the particular cascade (5.16), we can rewrite $\Psi(z,\xi)$ as

$$\begin{aligned}\Psi(z,\xi) &= \int_0^\infty \tilde{z}(s)\dot{\tilde{z}}(s)ds = \int_0^\infty d(\frac{1}{2}\tilde{z}^2(s)) \\ &= \lim_{s\to\infty} \tilde{z}^2(s) - \frac{1}{2}z^2\end{aligned} \tag{5.20}$$

so that the Lyapunov function (5.18) becomes

$$V(z,\xi) = U(\xi) + \frac{1}{2}(z + \int_0^\infty \bar{\psi}(\tilde{\xi}(s))ds)^2 \tag{5.21}$$

The following theorem presents the basic recursive step of forwarding.

Theorem 4 *(Forwarding as a recursive feedback passivation)*
Let $U(x)$ be a positive definite, radially unbounded, locally quadratic, storage function such that the system

$$\dot{\xi} = a(\xi) + b(\xi)u, \quad y_0 = (L_b U)^T(\xi) \tag{5.22}$$

is feedback passive and ZSD. Furthermore, let the pair $(\frac{\partial a}{\partial \xi}(0), b(0))$ be stabilizable.

Then the cascade

$$\begin{aligned}\dot{z} &= \psi(\xi) + g(\xi)u \\ \dot{\xi} &= a(\xi) + b(\xi)u, \quad y = (L_G V)^T(z,\xi)\end{aligned} \tag{5.23}$$

is feedback passive with a positive definite, radially unbounded storage function $V(z,\xi) = W(z) + \Psi(z,\xi) + U(\xi)$ in which the cross-term $\Psi(z,\xi)$ is given by (5.19). Moreover, if (5.22) is passive and if the Jacobian linearization of (5.23) is stabilizable, then the control law $u = -y = -(L_G V)^T$ achieves GAS and LES of $(z,\xi) = (0,0)$.

With a recursive application of the basic forwarding step we now construct a design procedure for systems in the form

$$\begin{aligned}\dot{z}_1 &= \psi_1(z_2,\ldots,z_n,\xi) + g_1(z_2,\ldots,z_n,\xi)u \\ &\vdots \\ \dot{z}_{n-1} &= \psi_{n-1}(z_n,\xi) + g_{n-1}(z_n,\xi)u \\ \dot{z}_n &= \psi_n(\xi) + g_n(\xi)u \\ \dot{\xi} &= a(\xi) + b(\xi)u, \quad z_i \in I\!R^{q_i},\ i = 1,\ldots,n\end{aligned} \tag{5.24}$$

where (after a preliminary transformation), the subsystem $\dot{\xi} = a(\xi) + b(\xi)u$, $y = L_b U(\xi)$ is assumed to be passive and ZSD. If the Jacobian linearization of (5.24) is stabilizable, we can achieve GAS/LES of $(z,\xi) = (0,0)$ in n

recursive forwarding steps. The design is a bottom-up procedure in which a passivating output y_1 and the Lyapunov function V_1 for the entire system are constructed at the final step. Using the notation

$$G(z,\xi) = (g_1(z_1,\ldots,z_n,\xi),\ldots,g_{n-1}(z_{n-1},z_n,\xi),g_n(z_n,\xi),b(\xi))^T$$

we start with $y_0 = L_bU(\xi)$. The first step of forwarding yields

$$\begin{aligned} V_n(z_n,\xi) &= U(\xi) + \frac{1}{2}(z_n - \phi_n(\xi))^2 \\ \phi_n(\xi) &= \int_0^\infty (\psi_n(\tilde{\xi}) - g_n(\tilde{\xi})L_bU(\tilde{\xi}))\, ds \\ y_n &= L_GV_n(z_n,\xi) \end{aligned}$$

where the integral is evaluated along the solutions of $\dot{\xi} = a(\xi) + b(\xi)u$ with the feedback $u = -L_bU(\xi)$. For $i = n-1,\ldots,1$, the recursive expressions are

$$\begin{aligned} V_i(z_i,\ldots,z_n,\xi) &= V_{i+1}(z_{i+1},\ldots,z_n,\xi) + \frac{1}{2}(z_i - \phi_i(z_{i+1},\ldots,z_n,\xi))^2 \\ \phi_i(z_{i+1},\ldots,z_n,\xi) &= \int_0^\infty (\psi_i - g_iy_{i+1})\, ds \\ y_i &= L_GV_i(z_i,\ldots,z_n,\xi), \qquad i = n-1,\ldots,1 \end{aligned}$$

where the integral is evaluated along the solutions of (5.24) with the control law $u = -y_{i+1}(z_{i+1},\ldots,z_n,\xi)$.

The final Lyapunov function is thus

$$V(z_1,\ldots,z_n,\xi) = U + \frac{1}{2}\sum_{i=n}^{1}(z_i + \phi_i(z_{i+1},\ldots,n))^2$$

and GAS/LES of the entire system (5.24) is achieved with the feedback control law $u = -L_GV$.

The above forwarding procedure started with the output $y_0 = L_bU$, which satisfied only the relative degree requirement. The recursive steps consisted of passivation designs for the subsystems of increasing dimensions. Only the output $y_1 = L_GV$ constructed in the final step satisfied both the relative degree one and the weak minimum phase requirements. In all the intermediate steps, the zero-dynamics subsystems for the constructed outputs can be unstable. Forwarding has thus removed the weak minimum phase obstacle to feedback passivation. In this sense, forwarding complements backstepping which has removed the relative degree obstacle.

It should be stressed, however, that the feedforward structure restricts the type of zero-dynamics instability. Instability in the Jacobian linearization can be caused only by repeated eigenvalues on the imaginary axis and no solution can escape to infinity in finite time. Without other assumptions, this restriction is necessary to guarantee global stabilization.

Example 5 The zero dynamics properties of the system

$$\begin{aligned}\dot{z}_1 &= kz_1 + z_2 + \xi^2 \\ \dot{z}_2 &= \xi \\ \dot{\xi} &= u, \quad y = \xi\end{aligned} \tag{5.25}$$

depend on the sign of the parameter k. For $k < 0$, the system is weakly minimum phase and global asymptotic stability is achieved with a (nonrecursive) feedback passivation design. For $k = 0$, the system is not minimum phase because the zero-dynamics subsystem is a double integrator. However, global stabilization is achieved via one step of forwarding. For $k > 0$, the Jacobian linearization of (5.25) has one zero in the right-half plane. Setting $\zeta = z_1 + \frac{1}{k}z_2$, we obtain

$$\dot{\zeta} = k\zeta + \xi + \xi^2 \geq k\zeta - \frac{1}{2}$$

which shows that, regardless of $u(t)$, $\zeta(t)$ is unbounded if $\zeta(0) > \frac{1}{2}$.

6 Recursive construction of a CLF

Backstepping and forwarding can be executed to achieve the construction of a Control Lyapunov Function (CLF, see [8]) for any nonlinear system which can be represented as a stabilizable subsystem augmented by a lower-triangular structure in "strict-feedback" form and an upper-triangular structure in "strict-feedforward" form. The constructed CLF can be employed to design control laws with optimality properties and stability margins (see [7]).

As an illustration, we return to the system (3.1) which, for each value of $i \in \{1, \ldots, n\}$, can be represented as a subsytem

$$\begin{aligned}\dot{z} &= f(z) + \bar{\psi}(z, \xi_i) \\ \dot{\xi}_i &= \xi_{i+1}, \quad (\xi_{n+1} := u),\end{aligned} \tag{6.26}$$

augmented by a chain of integrators at the "input" ξ_{i+1} and/or at the "output" ξ_i. Each integrator at the input leads to a step of backstepping, while each integrator at the output leads to a step of forwarding.

For the feedback passive subsystem (6.26), a Control Lyapunov Function is $W(z) + \frac{1}{2}\xi_i^2$, where $L_f W(z) < 0$, and a stabilizing feedback is $\xi_{i+1} = \alpha_{i+1}(z, \xi_i) = -L_{\bar{\psi}} W - \xi_{i+1}$. In a step of backstepping, we write

$$\dot{\xi}_i = \alpha_{i+1} + y_i = (\dot{\xi}_i)_{des} + y_i \tag{6.27}$$

and the CLF is augmented by the new "penalty"

$$y_i^2 = (\dot{\xi}_i - (\dot{\xi}_i)_{des})^2 \tag{6.28}$$

which permits to control the subsystem (6.26) through an integrator at the input ξ_{i+1}. In a step of forwarding, the new penalty in the CLF is of the form

$$(\xi_{i-1} + \int_0^\infty \tilde{\xi}_i(s)ds)^2$$

which, using $\dot{\xi}_{i-1} = \xi_i$, can be rewritten as

$$[(\int_0^\infty \tilde{\xi}_i(s)ds) - (\int_0^\infty \tilde{\xi}_i(s)ds)_{des}]^2 \tag{6.29}$$

because $(\int_0^\infty \tilde{\xi}_i(s)ds)_{des} = (\tilde{\xi}_{i-1}(\infty))_{des} - \tilde{\xi}_{i-1}(0) = -\xi_{i-1}$. The new penalty (6.29) permits to stabilize the subsytem (6.26) with an additional integrator at the output ξ_i.

References

[1] C.I. Byrnes, A. Isidori, J.C. Willems, "Passivity, feedback equivalence, and global stabilization of minimum phase systems," *IEEE Trans. on Automatic Control*, vol. 36, pp. 1228-1240, 1991.

[2] A. Isidori, *Nonlinear Control Systems.* 3nd. ed., Springer-Verlag, Berlin, 1995.

[3] M. Jankovic, R. Sepulchre, P.V. Kokotović, "Constructive Lyapunov stabilization of nonlinear cascade systems," *IEEE Trans. on Automatic Control*, to appear, 1996.

[4] M. Krstić, I. Kanellakopoulos, P.V. Kokotović, *Nonlinear and Adaptive Control Design.* Wiley, New York, 1995.

[5] A. Saberi, P.V. Kokotović, H.J. Sussmann, "Global stabilization of partially linear composite systems," *SIAM J. Control and Optimization*, vol. 28, pp. 1491-1503, 1990.

[6] H.J. Sussmann, P.V. Kokotović, "The peaking phenomenon and the global stabilization of nonlinear systems," *IEEE Trans. on Automatic Control*, vol. 36, pp. 424-439, 1991.

[7] R. Sepulchre, M. Jankovic, P. Kokotovi̇̀c, *Constructive Nonlinear Control*, Springer-Verlag, 1996.

[8] E.D. Sontag, "A universal construction of Artstein's theorem on nonlinear stabilization," *Systems & Control Letters*, vol. 13, pp. 117-123, 1989.

Center for Systems Engineering and Applied Mechanics, Université Catholique de Louvain, Av. G. Lemaitre, 4, B1348 Louvain-La-Neuve, Belgium, e-mail: `sepulchre@auto.ucl.ac.be`

Ford Motor Company, Scientific Research Laboratories, Dearborn, MI 48121, e-mail: `mjankov1@ford.com`

Center for Control Engineering and Computation, Dept. of Electrical and Computer Engineering, University of California, Santa Barbara, CA 93106, e-mail: `petar@ece.ucsb.edu`

Ergodic Algorithms on Special Euclidean Groups for ATR

A. Srivastava, M.I. Miller, and U. Grenander[1]

1 Introduction

This paper describes a technique for estimating motions of rigid targets based on the deformable template representations of complex scenes. The efficient modeling of representations for variabilities manifested by objects, shapes and scenes supporting *invariant recognition* is crucial. There are several kinds of variabilities fundamental to these representations: (i) the *variability in target pose and placement*, and (ii) *variability in target numbers and identities.* To model the first variability, templates are constructed from two-dimensional CAD surfaces representing the rigid objects. Using the deformable template approach, these templates are varied via the basic rigid transformations involving the translation and rotation groups. Since complex scenes are composed of multiple moving targets, the complete scene transformations are finite Cartesian products of these Lie groups. Given a set of observations of a particular scene the inference constitutes generating minimum mean squared error (MMSE) estimates and, hence, optimizing on the curved geometry of Lie manifolds.

To account for the other variability, namely the variable target numbers and their identities, the algorithm is equipped with jump transitions over the set of discrete parameters modeling those variations. Similar to a Poisson process, these jump transitions are performed at random exponentially separated times but with the transition measure specified by the *a-posteriori* probability associated with the scene. These jumps result in implicit solutions of the classical detection and recognition steps in automated target recognition (ATR) problem as described in our earlier work [20, 11, 16]. The goal of this paper is to demonstrate an intrinsic geometric technique for constructing stochastic flows through the curved manifolds of Lie groups. The target types and numbers are, therefore, assumed known and fixed with the focus of inference being only on the continuum of orientations and translations.

The authors have previously presented a random sampling algorithm based on the jump-diffusion processes for target tracking/recognition

[1]This work is supported by ARO DAAH04-95-1-0494, ARO DAALO3-92-G-0115, and ARL MDA972-93-1-0012.

[11, 16]. This work modeled the flat parameter spaces based on the conventional representation of rigid orientations through the Euler angles (pitch, yaw and roll). In modeling rigid motions, the parameters are more naturally restricted to curved manifolds, see e.g. [14, 13, 19, 4, 6]. Due to the constraints on the rigid shapes and motions of the objects these representations take values in Lie groups such as the special Euclidean group $\mathbf{SE}(n), \equiv \mathbf{SO}(n) \ltimes \mathbb{R}^n$, $n = 2,\ 3$, where $\ltimes$ stands for the semi-direct product (pg. 92 [3]). $\mathbf{SE}(n)$ forms a Lie group with group operation given by

$$(O_1, p_1) \cdot (O_2, p_2) = (O_1 \cdot O_2, O_1 p_2 + p_1)\ , \quad O_1, O_2 \in \mathbf{SO}(n), \quad p_1, p_2 \in \mathbb{R}^n\ . \tag{1.1}$$

This operation can also be expressed via $n+1 \times n+1$ matrix multiplication,

$$\begin{bmatrix} O_1 & p_1 \\ 0 & 1 \end{bmatrix} \begin{bmatrix} O_2 & p_2 \\ 0 & 1 \end{bmatrix} = \begin{bmatrix} O_1 \cdot O_2 & O_1 p_2 + p_1 \\ 0 & 1 \end{bmatrix}. \tag{1.2}$$

In general, the parameterizations consist of multiple targets, both ground- and air-based, requiring estimation in the space $\mathbf{SE}(n)^J$, $n = 2, 3$, J being the total number of observations of all targets. This paper extends the jump-diffusion methodology to the curved geometry of matrix Lie groups, $\mathbf{SE}(n)^J$ in particular. The basic issue is that the curved manifolds are fundamentally different from the flat Euclidean spaces. We will essentially be studying stochastic gradients on the curved manifolds $\mathbf{SE}(n)^J$ to optimize some cost function. To illustrate, define a C^∞ function $H : \mathbb{R}^n \to \mathbb{R}_+$ and examine the ordinary differential equation (ODE) in $\mathbb{R}^n$,

$$\frac{d\xi(t)}{dt} = -\nabla H(\xi(t))\ . \tag{1.3}$$

The solution $\xi\ :\ [0, \infty) \to \mathbb{R}^n$, given by $\xi(t + \Delta) = \xi(t) + \int_t^{t+\Delta} -\nabla H(\xi(s))ds$, can be viewed as a gradient curve in $\mathbb{R}^n$ parameterized by a single parameter t. In preparation for curved manifolds we rewrite this using representations associated with the tangent spaces, the tangent vectors being studied via their actions on smooth functions on the manifold. For example, a vector $a = [a_1\ a_2\ \ldots\ a_n]^\dagger$ in $\mathbb{R}^n$ is described through the directional derivative $Af = \sum_{i=1}^n a_i \frac{\partial f}{\partial x_i}$ for all smooth functions f. In this sense the vector a is said to be equivalent to the derivative operator $A = \sum_{i=1}^n a_i \frac{\partial}{\partial x_i}$. The action of A on a large enough family of functions completely specifies the vector a. In $\mathbb{R}^n$, inherent to its flat geometry is the fact that the basis vectors of the tangent space at each point are identical (left panel of Figure 1), and are given by the direct partial derivatives $\frac{\partial}{\partial x_i}, i = 1, \ldots, n$. In vector notation we denote them as $\frac{\partial}{\partial x_i} \equiv Y_i \in \mathbb{R}^n$

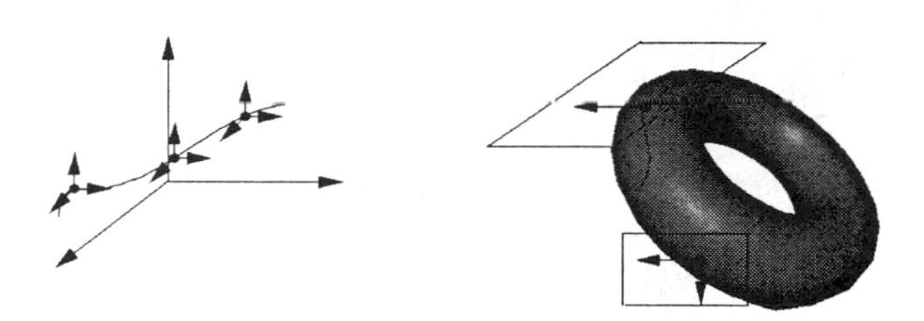

Figure 1: The basis elements of tangent space are identified at all points for Euclidean spaces while for arbitrary manifolds the basis vectors are defined separately at each point.

where Y_i is an n-vector with 1 at the i-th coordinate and 0's elsewhere. In this notation the ODE (1.3) may be rewritten as

$$\frac{d\xi(t)}{dt} = -\sum_{i=1}^{n}(Y_{i,\xi(t)}H)Y_i, \quad \text{where } Y_{i,\xi(t)}H \equiv \frac{\partial H}{\partial x_i}|_{\xi(t)} . \tag{1.4}$$

The vector $\frac{d\xi(t)}{dt}$ is interpreted as the velocity vector of a particle moving along the curve $\xi(t)$ in $\mathbb{R}^n$ and is an element of the tangent space of the manifold with the basis $Y_1, Y_2, \ldots, Y_n$. For small Δ positive it gives,

$$\xi(t+\Delta) = \xi(t) + (-\Delta \sum_{i=1}^{n}(Y_{i,\xi(t)}H)Y_i) + o(\Delta) , \tag{1.5}$$

defining a parameterized curve in $\mathbb{R}^n$. Notice that the incremental translations in $\mathbb{R}^n$ are generated by component-wise addition.

We are concerned with inferences on the curved manifolds $\mathcal{X} = \mathbf{SE}(n)^J$ where due to curvature the usual notations of derivatives and translations have to be modified. Examining 1.5, which is appropriate for $\mathbb{R}^n$, demonstrates two fundamental departures as we move from the flat Euclidean spaces to the curved Lie groups:

1. The first issue is that the tangent vectors at every point on a curved manifold can be different, shown in the right panel of Figure 1 for a taurus. In other words, $Y_i \equiv \frac{\partial}{\partial x_i}$ is not generally tangent to $\mathcal{X}$ at all $x \in \mathcal{X}$. This implies that the tangent spaces at each point must be explicitly tracked. The locally Euclidean property is utilized to define tangent spaces at each point on the manifold. Also, a Lie group has an additional structure, called *parallelizability*, where the tangent spaces can be defined across the whole space in a unified manner. First,

the tangent space at the identity element is defined and the tangent spaces at all other points are just its rotated versions in the sense made precise later.

2. The second aspect fundamental to curved Lie groups is that addition of two elements $(g_1 + g_2)$ does not define translations on $\mathcal{X}$. Instead $+$ is replaced by the group operation, $g_1 \cdot g_2 \in \mathcal{X}$ if $g_1, g_2 \in \mathcal{X}$, to describe translations on Lie groups.

The rigorous mathematical theory for constructing stochastic flows on $\mathbf{SE}(n)^J$ is presented in [14, 21]; herein we focus only on the algorithmic and implementation aspects of these methods illustrated by two specific ATR scenarios: (i) ground-based rigid objects such as tanks, trucks, jeeps, and (ii) flying objects such as airplanes, helicopters, missiles.

Section 2 explains the parameterization of rigid motion using group actions. The geometry of $\mathbf{SE}(n)$ is studied in Section 3 leading to tools for constructing stochastic flows. Section 4 sets up a Bayesian formulation of the problem. The algorithm for solving the motion estimation problem in this Bayesian context is outlined in Section 5, with the implementation results for various specific scenarios presented in Section 6.

3 Representations via Deformable Templates: Group Actions

Representation is the core element of image understanding and therefore recognition. The efficient generation of compact models for representations of objects shapes supporting *invariant recognition* is crucial. The *infinity of poses* manifest by rigid motions during scene evolution should be accommodated.

This is accomplished using the deformable template representations in the following way. Define two target classes:

$$\begin{aligned} \mathcal{A}_{ground} &= \{truck, T72, MT1, jeep, \ldots\}, \\ \mathcal{A}_{air} &= \{F16, X29, 767, cessna, \ldots\}, \end{aligned}$$

with $\mathcal{A} = \mathcal{A}_{ground} \bigcup \mathcal{A}_{air}$. Let $I(\alpha)$ be the complete physical description of the object $\alpha \in \mathcal{A}$ including its shape, size, reflectivity, thermodynamic profile, etc.; i.e. $I(\alpha)$ incorporates all the target-attributes reflected in the sensor outputs. For the sensors considered here, $I(\alpha)$ will correspond to the set of CAD surface models consisting of a finite polygonal patch description, including the vertices and the normals, and their reflectivity properties.

Figure 2 shows the CAD surfaces for some objects used in the simulations later. Individual targets in a given scene are modeled by the action of

Figure 2: CAD surface models for various targets.

transformation groups on the associated rigid surface models; the groups for modeling rigid motion are the rotation and the translation groups. All possible occurrences of targets form a homogeneous space; i.e. for any target attribute (pose and position) there exists a rotation and translation transformation pair whose action models that target occurrence.

The dimensions of these transformation spaces are specified by the allowed degrees of motion for a given object.

1. For ground-based objects, such as tanks, trucks, and jeeps, the motion allows fixed axis rotations and two dimensional translations. The translations are parameterized by the elements of $\mathbb{R}^2$ though the rotation can be equivalently parameterized by elements of the special orthogonal group $\mathbf{SO}(2) = \{g \in \mathbb{R}^{2\times 2} : g^\dagger g = I, det(g) = 1\}$ or the circle $\mathbf{S}^1 = \{g \in \mathbb{R}^2 : g^\dagger g = 1\}$ or the torus $\mathbf{T}^1 = \{g \in [0, 2\pi] : 0, 2\pi \text{ identified }\}$. Note that these groups can all be uniquely identified with each other. We choose $\mathbf{SO}(2)$ to represent the orientations of ground-based objects. Shown in Figure 3 are three orientations of a tank template.

Figure 3: A tank template transformed by three elements of the rotation group $\mathbf{SO}(2)$.

The composite transformation (rotation and translation) is parameterized by elements of the special Euclidean group $\mathbf{SE}(2) \equiv \mathbf{SO}(2) \ltimes \mathbb{R}^2$.

2. For rigid motion in $I\!R^3$, the orientation can be represented in several ways [9], for example by unit quaternions ($\mathbf{S}^3$) or Cayley-Klein parameters ($\mathbf{SU}(2)$) or Euler angles ($\mathbf{S}^2 \times \mathbf{S}^1$) or rotation matrices ($\mathbf{SO}(3)$), not all of them Lie groups. To obtain uniqueness of parameterization and a simple law of composition we employ the special orthogonal group, $\mathbf{SO}(3)$. The motion of a rigid template is given by the special Euclidean group $\mathbf{SE}(3) \equiv \mathbf{SO}(3) \ltimes I\!R^3$, the semi-direct product of $\mathbf{SO}(3)$ and $I\!R^3$. Figure 4 shows a rendering of a transformed template with the transformations parameterized by elements of the translation and rotation groups.

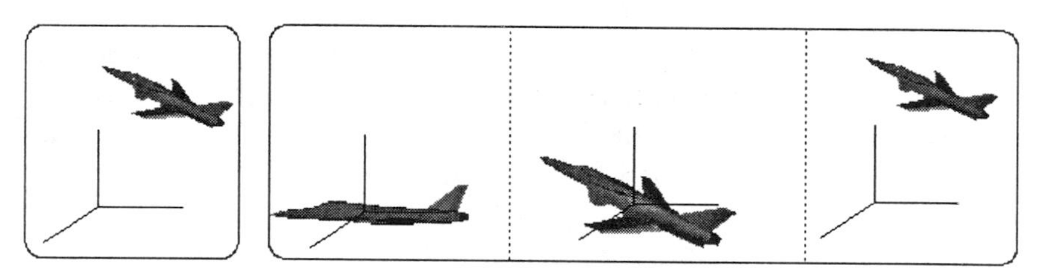

Figure 4: Shown in the left panel is an airplane template transformed by a rotation $O \in \mathbf{SO}(3)$ and a translation $p \in I\!R^3$ as shown in the right panel.

The trajectories of moving targets are parameterized by the finite Cartesian products of the basic unit $\mathbf{SE}(n)$. For a moving object, the object positions in $I\!R^n$, and orientations in $\mathbf{SO}(n)$, parameterize the group transformations at each time along the path. The complete trajectory of object motion observed at J sample times is parameterized by the elements of $\mathbf{SE}(n)^J$. The concatenation, in the natural time sequence, of segments formed during target motion forms a linear graph called a *track*, as shown for an airplane trajectory in Figure 5. In a general scene consisting of multiple moving targets, both on ground and in air, the estimation problem is posed on the parameter space

$$\mathcal{X} = \mathbf{SE}(2)^{J_1} \times \mathbf{SE}(3)^{J_2}, \quad J = J_1 + J_2 \ , \tag{3.1}$$

where J_1 components describe ground-based objects, and J_2 components describe flying objects, J being the total number of motion components.

Remark. More generally, the transformations include some discrete components to accommodate the variability in the complexity of the scene, including the parametric dimension J (model order) as well as the target identities $\alpha \in \mathcal{A}$. To restrict the analysis to inferences via stochastic flows, throughout this paper J and α will be assumed known and fixed.

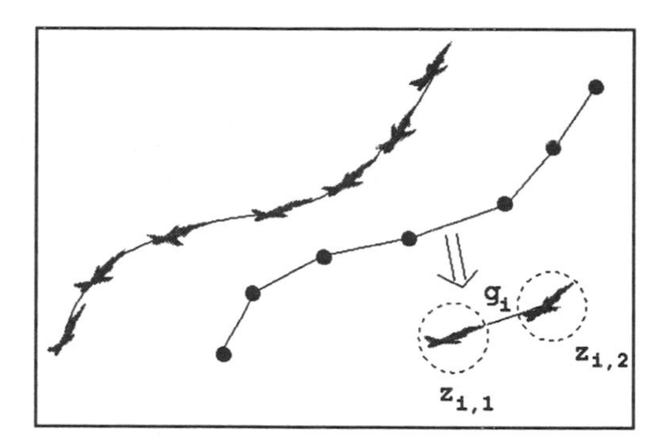

Figure 5: Discrete representation of a flight path: The concatenation of elements from motion group **SE**(3) at sample times forms a linear graph structure.

4 Optimization Via Stochastic Flows

Estimating transformation parameters from the observed images becomes optimization on the constrained manifolds of the representation. In the literature, there exists a large class of constrained optimization techniques (such as conjugate gradient, Newton's, Lagrangian) with solutions posed on Euclidean spaces, $\mathbb{R}^n$. As described in [17], these *extrinsic* techniques depend on embedding the constraint surfaces in bigger Euclidean spaces $\mathbb{R}^n$ and utilizing one of the two standard variational optimization methods: (i) projective methods, where the solutions are evaluated in $\mathbb{R}^n$ and then projected on to the surface, and (ii) Lagrangian methods, where the cost function is the original function plus a constraint term. Often times, not utilizing the underlying geometric structure of the constraint surface these extrinsic methods are not as efficient as compared to the geometry based *intrinsic* methods. We rely most fundamentally on the basic tools from differential geometry for the variational calculus on these manifolds. We start with the basic geometric features of the curved component of $\mathbf{SE}(n)$, the special orthogonal group $\mathbf{SO}(n)$.

4.1 Tangent Vector Fields on SO(n) Generating Flows

The goal is to construct the equivalent of Eqns. 1.4 and 1.5 for the matrix Lie group $\mathbf{SO}(n)$. We utilize the fact that Lie groups are parallelizable; i.e. the tangent vectors $Y_{1,g}, Y_{2,g}, \ldots, Y_{n,g}$ at $g \in \mathbf{SO}(n)$ can all be generated by a rotation on the tangent vectors at the identity $e \in \mathbf{SO}(n)$, $Y_{1,e}, Y_{2,e}, \ldots, Y_{n,e}$. Using the basic results from [3, 18], we now establish this tangent structure on $\mathbf{SO}(n)$. Denote the vector space of tangents to the manifold $\mathbf{SO}(n)$ at a point g by $T_g(\mathbf{SO}(n))$.

The dimension $d(n)$ of $\mathbf{SO}(n)$ is given by $n(n-1)/2$, $d(2) = 1$, $d(3) = 3$, and so on. It can be shown (see pg. 150 [3]) that the tangent space at the identity, $T_e(\mathbf{SO}(n))$, may be identified with the space of $n \times n$ skew-symmetric matrices, with the basis elements $Y_{i,e}$, $i = 1, \ldots, d(n)$, resulting in:

1. For $\mathbf{SO}(2)$, $d(2) = 1$ with one basis element

$$Y_{1,e} = \frac{1}{\sqrt{2}} \begin{bmatrix} 0 & -1 \\ 1 & 0 \end{bmatrix}. \tag{4.1}$$

2. For $\mathbf{SO}(3)$, $d(3) = 3$ with three basis elements

$$Y_{1,e} = \frac{1}{\sqrt{2}} \begin{bmatrix} 0 & -1 & 0 \\ 1 & 0 & 0 \\ 0 & 0 & 0 \end{bmatrix}, Y_{2,e} = \frac{1}{\sqrt{2}} \begin{bmatrix} 0 & 0 & 1 \\ 0 & 0 & 0 \\ -1 & 0 & 0 \end{bmatrix},$$

$$Y_{3,e} = \frac{1}{\sqrt{2}} \begin{bmatrix} 0 & 0 & 0 \\ 0 & 0 & -1 \\ 0 & 1 & 0 \end{bmatrix}. \tag{4.2}$$

To extend this tangent structure to other points, define a left rotation parameterized by $g \in \mathbf{SO}(n)$ as $L_g(h) = g \cdot h$, the $n \times n$ matrix product. Due to continuity of the group operation (matrix product in this case) each vector $Y_{i,e}$ determines uniquely a vector field Y_i on the whole space $\mathbf{SO}(n)$, i.e. evaluated at any point $g \in \mathbf{SO}(n)$ this field gives a tangent vector $Y_{i,g} = L_{g*}(Y_{i,e}) = g \cdot Y_{i,e}$ in $T_g(\mathbf{SO}(n))$. In addition, the fields Y_i's have the property that for any $g \in \mathbf{SO}(n)$ the vectors $Y_{i,g}$ form the orthonormal basis of the tangent space $T_g(\mathbf{SO}(n))$ under the metric

$$< Y_1, Y_2 > = tr(Y_1^\dagger \cdot Y_2), \quad \text{for any } Y_1, Y_2 \in T_g(\mathbf{SO}(n)),$$

where $tr()$ is the matrix trace and the composition rule $(\cdot)$ is $n \times n$ matrix multiplication. As an example, for $g \in \mathbf{SO}(2)$, define the left rotation

$$L_g(h) = g \cdot h = \begin{bmatrix} g_1 & -g_2 \\ g_2 & g_1 \end{bmatrix} \begin{bmatrix} h_1 & -h_2 \\ h_2 & h_1 \end{bmatrix}, \quad \text{for all } \ h \in \mathbf{SO}(2) \ .$$

Then, the rotated tangent vector $L_{h*}(Y_{1,e}) = h \cdot Y_{1,e}$ is tangent to $\mathbf{SO}(2)$ at h. Hence, $Y_{1,e}$ generates a vector field over the complete set $\mathbf{SO}(2)$ described by the rotation $Y_{1,h} = L_{h*}(Y_{1,e})$.

The difficulty of forcing the estimation processes to stay on the manifold $\mathcal{X} = \mathbf{SE}(n)^J$ or $\mathbf{SO}(n)^J$, is handled by constructing flows as the solutions of the ODE built from the tangent basis derived above. A vector field Y on a manifold $\mathcal{X}$ is said to generate a flow, $t \to \xi(t) \in \mathcal{X}$ if the velocity vector $(\frac{d\xi(t)}{dt})$ at any point on the curve is equal to vector field evaluated at that point. We use the fact that every smooth vector field Y generates a smooth flow $\xi(t)$ on the manifold (see Pg. 61 [22]). Let $H_1 : \mathbf{SO}(n) \to I\!R_+$ and $H_2 : I\!R^n \to I\!R_+$ be two C^∞ functions and let $Y_1, \ldots, Y_{d(n)}$ be the orthonormal vector fields on $\mathbf{SO}(n)$ derived above ($Y_{i,g} = g \cdot Y_{i,e}$). The vector fields $-\sum_{i=1}^{d(n)}(Y_{i,g}H_1)Y_{i,g}$, $-\sum_{i=1}^{n}\frac{\partial H_2}{\partial x_i}\frac{\partial}{\partial x_i}$ generate flows satisfying the equations,

$$\frac{d\xi_1(t)}{dt} = -\sum_{i=1}^{d(n)}(Y_{i,\xi(t)}H_1)Y_{i,\xi(t)} \,, \quad \frac{d\xi_2(t)}{dt} = -\nabla H(\xi_2(t)) \,. \tag{4.3}$$

Call $\xi_1(t)$ and $\xi_2(t)$ the deterministic gradient flows on $\mathbf{SO}(n)$ and $I\!R^n$, respectively. For matrix Lie groups, these flows are given by exponential maps; i.e. for $A \in T_e(\mathbf{SO}(n))$, an $n \times n$ skew-symmetric matrix, the parameterized curve $\xi : t \to g \cdot exp(At)$ is a flow on $\mathbf{SO}(n)$ with its generator given by the vector field $Y_g = g \cdot A$. The matrix exponential is given by the infinite series (see [5]), $exp(A) = e + A + \frac{A^2}{2!} + \frac{A^3}{3!} + \ldots$, and for a skew-symmetric matrix A, $exp(A)$ is an orthogonal matrix ([3, 5]).

4.2 Stochastic Flows

The same framework can be extended to generate a diffusion process as a stochastic flow by adding noise terms in Eqns 4.3 [10, 1, 14]. Basically, a diffusion process is simulated through randomization via addition of independent random perturbations to the directional derivative terms in the directions given by the associated tangent vectors.

Define stochastic flows on the $(3J_1 + 6J_2)$-dimensional product group $\mathcal{X} = \mathbf{SE}(2)^{J_1} \times \mathbf{SE}(3)^{J_2}$ as follows. Consider a C^∞-function $H : \mathcal{X} \to I\!R_+$ then the resulting flow has J_1 components on $\mathbf{SE}(2)$ and J_2 components on $\mathbf{SE}(3)$, i.e.

$$\xi(t) = [\xi^{(1)}(t),\ \xi^{(2)}(t),\ \ldots \xi^{(J_1)}(t),\ \xi^{(J_1+1)}(t),\ \xi^{(J_1+2)}(t),\ \ldots \xi^{(J_1+J_2)}(t)] \,.$$

For $j = 1, 2, \ldots, J_1$ the components $\xi^{(j)}(t) = [\xi_1^{(j)}(t),\ \xi_2^{(j)}(t)] \in \mathbf{SE}(2)$ satisfy the equations, obtained by particularizing Eqn 4.3 for $n = 2$,

$$\begin{aligned} \xi_1^{(j)}(t) &= \xi_1^{(j)}(t_0) + \int_{t_0}^{t}[-(Y_{1,\xi_1(s)}^{(j)}H)Y_{1,\xi_1(s)}^{(j)}ds + Y_{1,\xi_1(s)}^{(j)} \circ dW_1^{(j)}(s)] \\ \xi_2^{(j)}(t) &= \xi_2^{(j)}(t_0) + \int_{t_0}^{t}[-\nabla_{p^{(j)}}H(\xi_2(s))ds + dW_2^{(j)}(s)] \,, \end{aligned} \tag{4.4}$$

where $W_1^{(j)}(t) \in \mathbb{R}$, $W_2^{(j)}(t) \in \mathbb{R}^2$ are standard Wiener processes and $\circ$ denotes the Stratonovich interpretation of the stochastic integral. Here $Y^{(j)}_{1,\xi_1(t)}H$ is the directional derivative of H in the direction tangent to the j-th orthogonal group component of $\mathcal{X}$; $\nabla_{p^{(j)}}H$ is the gradient with respect to the position vector in the j-th group component of $\mathcal{X}$. Similarly for $j = J_1 + 1, \ldots, J_1 + J_2$, the components $\xi^{(j)}(t) = [\xi_1^{(j)}(t),\ \xi_2^{(j)}(t)] \in \mathbf{SE}(3)$ are generated by the SDE's

$$\xi_1^{(j)}(t) = \xi_1^{(j)}(t_0) + \int_{t_0}^{t} [-\sum_{i=1}^{3} (Y^{(j)}_{i,\xi_1(s)}H) Y^{(j)}_{i,\xi_1(s)} ds + \sum_{i=1}^{3} Y^{(j)}_{i,\xi_1(s)} \circ dW_i^{(j)}(s)]$$

$$\xi_2^{(j)}(t) = \xi_2^{(j)}(t_0) + \int_{t_0}^{t} [-\nabla_{p^{(j)}} H(\xi_2(s))ds + dW_2^{(j)}(s)]\ , \quad (4.5)$$

where $Y^{(j)}_{i,\xi_1(t)}H$, $i = 1, 2, 3$ are the directional derivatives of H in the three basis directions tangent to the j-th orthogonal group component of $\mathcal{X}$.

4.3 Reference Measures on $\mathbf{SO}(n)$

To reference probability measures and to evaluate expectations one needs to define some base (or reference) measure on the underlying space, $\mathcal{X}$. For the flat component, $\mathbb{R}^n$, the Lebesgue measure provides the reference; in geometry it is expressed as the reference volume element $dx_1 \wedge dx_2 \wedge \ldots \wedge dx_n$. For the curved component $\mathbf{SO}(n)$ the reference measure needs to be specified. Being a compact, connected Lie group $\mathbf{SO}(n)$ has a unique bi-invariant volume element which forms the base measure for Bayesian formulation of the inference problem. This volume element on $\mathbf{SO}(n)$ can be expressed in the desired local coordinates as follows.

1. In $\mathbf{SO}(2)$, the invariant volume form in terms of rotation angle θ is given by $d\theta$. To evaluate the invariant form in the Cartesian coordinates, $\begin{bmatrix} x_1 & -x_2 \\ x_2 & x_1 \end{bmatrix} \in \mathbf{SO}(2)$, such that $x_1^2 + x_2^2 = 1$, substitute $x_1 = cos(\theta)$ and $x_2 = sin(\theta)$, resulting in $-x_2 dx_1 + x_1 dx_2$.

2. In $\mathbf{SO}(3)$, there are several choices for local charts, i.e. Euler angles, exponential coordinates, quaternions. For use in deriving dynamics based prior measure on rigid motions, we are interested in the volume form on $\mathbf{SO}(3)$ in terms of local coordinates given by the exponential map. Rigid body dynamics are naturally expressed in these exponential coordinates associated with the body-frame angular velocities of the rotating object. The chart $exp : \mathbb{R}^3 \rightarrow \mathbf{SO}(3)$ relating the

exponential coordinates to the elements of $\mathbf{SO}(3)$ is given by

$$(q_1, q_2, q_3) \rightarrow exp(Q), \text{ for } Q \text{ skew-symmetric with elements } q_1, q_2, q_3. \tag{4.6}$$

The invariant volume element on $\mathbf{SO}(3)$ in term of exponential coordinates is given by (see [21] for proof),

$$\frac{1}{\pi^2} \frac{sin^2(|q|)}{|q|^2} dq_1 \wedge dq_2 \wedge dq_3 \ , \quad |q| = \sqrt{q_1^2 + q_2^2 + q_3^2} \ . \tag{4.7}$$

5 Posterior for Object Tracking

Taking a Bayesian approach, the variability and uncertainty of the transformation parameters is represented by the observed data ensemble contributing to the posterior measure μ on $\mathcal{X} = \mathbf{SE}(n)^J$. We begin by defining a prior density π_0 on $\mathcal{X}$. The underlying ideal scene identified with its pose $x \in \mathcal{X}$ cannot be observed directly; then $x \in \mathcal{X}$ is sensed through the remote sensor as $y \in \mathcal{Y}$, $\mathcal{Y}$ being the observation space. Since the data are assumed random we characterize them via a statistical transition law, called the *likelihood function* $L(\cdot|\cdot) : \mathcal{X} \times \mathcal{Y} \rightarrow \mathbb{R}$, describing completely the mapping from the input x to the output y. The posterior measure μ with density π becomes the product of the prior density $\pi_0(x)$, the density of the underlying true scene x and the likelihood of the data y according to

$$\pi(x|y) = \frac{\pi_0(x) L(y|x)}{\int_{\mathcal{X}} \pi_0(x) L(y|x) \gamma(dx)} = \frac{e^{-H(x)}}{\mathcal{Z}} \ , \qquad x \in \mathcal{X} \ , \tag{5.1}$$

where H is called the posterior energy, $\mathcal{Z}$ is the normalizer and γ is the base measure on the $\mathcal{X}$.

To make inferences from a given observed data set the posterior distribution is simulated by solving a stochastic flow equation on the manifolds, and non-local jumps to cover $\mathcal{X}$, extending to Lie manifolds the *jump-diffusion* processes as described in [14, 21]. These stochastic flows in $\mathcal{X}$ are constructed so that their stationary measure has the density π on $\mathcal{X}$ with respect to the base measure. This allows for the generation of several classical estimators such as MMSE and MAP as well as the mean and covariance statistics. Define the posterior energy

$$H(x) = E(x) + P(x), \qquad x \in \mathcal{X} \ , \tag{5.2}$$

where $E(x)$ reflects the energy associated with the sensor data likelihood, $e^{-E(x)} \propto L(y|x)$ in 5.1, and P is the energy associated with the prior on $\mathcal{X}$, $e^{-P(x)} \propto \pi_0(x)$ in 5.1.

Prior. The prior on rigid motions is based on the formulation of target dynamics using standard rigid body analysis (neglecting the earth's curvature, motion and wind effects), through the Newtonian differential equations [7]. The prior is induced via the non-linear differential operator associated with these equations of motion, assuming suitable statistical models for the forcing functions, as described in [11, 21, 14].

Data Likelihood. In ATR there are generally several possible remote sensors providing simultaneous data for the inference. The two examples are: (i) low resolution trackers used for global detection and azimuth-elevation tracking of unresolved targets, and (ii) high resolution optical sensors, for collecting detailed information on pose and identity of the target.

1. For pose sensing and target identification a high resolution optical imaging system (see Figure 6) is used. It registers a two-dimensional projection of target profiles on the camera focal plane. For typical

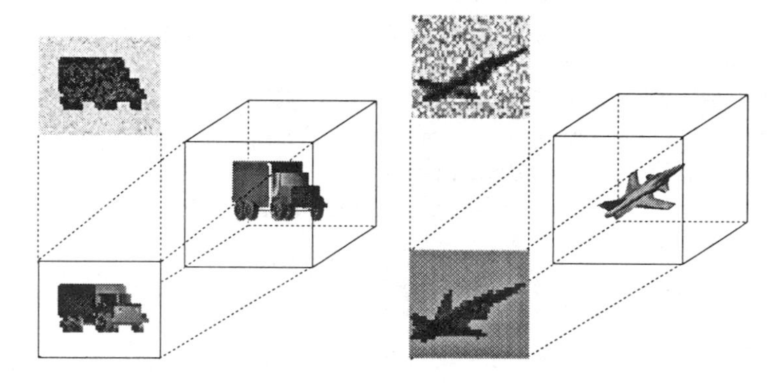

Figure 6: The far-field orthographic imaging system for observing the targets at a high resolution taking the target scene and projecting onto the 2-D lattice with additive noise.

   optical imaging systems [15], the observation is given by the orthographic projection convolved with the point spread function of the camera, plus an additive noise. Under additive white Gaussian noise models, the imaged data is assumed to be a Gaussian random field with mean field given by the target's projection profile. Some data samples are shown in Figure 7.

2. For position tracking, a cross array of 64 narrowband, isotropic sensors (32 elements in each direction at half-wavelength spacing) is assumed as in [11, 16] using the standard narrowband signal model

Figure 7: Sample data sets for the ATR problem: The left panel shows the high-resolution video data for a truck, the middle panel displays an airplane and the right panel shows a tank image.

developed in [12]. The phase lags of signals received at the sensor elements provide information about the source locations. For the additive white Gaussian noise model, the data samples are complex Gaussian vectors with the mean given by the signal component.

6 Ergodic Algorithm on $\mathcal{X} = \mathbf{SE}(2)^{J_1} \times \mathbf{SE}(3)^{J_2}$

Having obtained a posterior measure $\mu(dx) = \frac{e^{-H(x)}}{Z}\gamma(dx)$ on the scene representations taking values in $\mathcal{X} = \mathbf{SE}(2)^{J_1} \times \mathbf{SE}(3)^{J_2}$, a jump-diffusion process is constructed to generate the MMSE estimates. This Markov process $X(t)$ satisfies *jump-diffusion dynamics* through $\mathcal{X}$ in the sense that (i) on random exponential times the process jumps across $\mathcal{X}$, and (ii) between jumps it follows SDEs of the type Eqns. 4.4,4.5 generating a diffusion process. Formally stated, let t_0, t_1, ... be independent and exponentially distributed with parameter λ (a constant). Let $\{\tilde{X}(i),\ i = 1, \ldots\}$ be a Markov chain in $\mathcal{X}$ with some transition function $Q(x, F)$, i.e. $P\{\tilde{X}(i+1) \in F | \tilde{X}(1), \ldots, \tilde{X}(i)\} = Q(\tilde{X}(i), F),\ \ F \subset \mathcal{X}$. Then define a process $X(t)$ on $\mathcal{X}$ by, for $i = 1, 2, \ldots$,

$$X(t) = \begin{cases} \text{solution of the SDEs 4.4,4.5,} & \text{for } t \in [t_i, t_{i+1}) \\ \text{with initial condition } \tilde{X}(i) & \\ \tilde{X}(i+1), & \text{for } t = t_{i+1}. \end{cases}$$

The flows resulting from Eqns. 4.4, 4.5 have infinitesimal drifts given by $Y_{i,X(t)}H$, the directional derivatives of the posterior energy H with respect to the basis element Y_i. The diffusion process taking values in $\mathcal{X}$, results in a smooth transformations on object templates. Shown in Figure 8 is an airplane template transformed to match a given target. The target is shown in light with the template in dark being rotated via a diffusion transformation.

Figure 8: Continuous transformation through the diffusion process: These panels show an airplane template (in dark) being transformed by a diffusion process in **SO**(3) to match a particular observation shown in light.

A jump-diffusion process $\{X(t) \in \mathcal{X}, t \geq 0\}$ jumping on random times $t_1, t_2, ..$ is constructed by the following algorithm:

Algorithm 1 Let $i = 0, t_0 = 0$, $X(0) = x_0$ for some $x_0 \in \mathcal{X}$. Let $Y_{i,g}$ form basis of the tangent space $T_g(\mathcal{X})$ for all $g \in \mathcal{X}$.

1. Generate a sample u of an exponential random variable with mean λ, a constant.

2. Follow the SDEs Eqns. 4.4, 4.5 generating diffusions for time interval $t \in [t_i, t_{i+1})$, $t_{i+1} = t_i + u$:

$$dX(t) = -\sum_{i=1}^{m}(Y_{i,X(t)}H)Y_{i,X(t)}dt + \sum_{i=1}^{m} Y_{i,X(t)} \circ dW_i(t). \qquad (6.1)$$

   Here $m = 3J_1 + 6J_2$, the dimensions of $\mathcal{X}$, and Y_i, $i = 1, 2, \ldots, m$ are the basis vector fields on $\mathcal{X}$ as constructed earlier. More explicitly these are given by Eqns. 4.4, 4.5.

3. At $t = t_{i+1}$, perform a jump move, from $X(t_{i+1})$ to an element of $\mathcal{X}$, according to the transition probability measure

$$Q(x, dy) = \frac{e^{-[E(y)-E(x)]_+}e^{-P(y)}}{\int_{\mathcal{X}} e^{-[E(y)-E(x)]_+}e^{-P(y)}\gamma(dy)}\gamma(dy) \ .$$

4. Set $i \leftarrow i + 1$, go to step 1.

6.1 Implementation Aspects

The core computations in this algorithm are (i) evaluating the derivatives of the posterior energy H by operating tangent vectors, and (ii) computing flows along the given gradient directions. Additionally, we need to compute the MMSE estimate from the sample values $x_i = X(i\Delta)$, $i = 1, 2, \ldots$, described in the next section.

1. In general, the directional derivative of the posterior energy function H in the direction of a tangent vector Y_g is given by

$$Y_g H = \lim_{t \to 0} \frac{1}{t} (H(\psi_g(t)) - H(g)) , \qquad (6.2)$$

where ψ is the flow generated by the field Y such that $\psi_g(0) = g$. In practical situations, the derivative $Y_i H$ in the diffusion SDE is difficult to evaluate analytically due to complicated expressions for H. It can be approximated numerically by, for $\epsilon > 0$ small enough,

$$Y_g H \approx \frac{1}{\epsilon} (H(\psi_g(\epsilon)) - H(g)) . \qquad (6.3)$$

As mentioned earlier, in the case of the orthogonal groups, ψ is given by the exponential map, i.e $\psi_g(t) = g \cdot exp(Y_e t)$. The formulas to evaluate the matrix exponentials for $\mathbf{SO}(n)$, $n = 2, 3$ are given below.

2. On a computer the SDE in 6.1 is approximated by a stochastic difference equation as follows: define $\xi_i(t), -\infty < t < \infty, i = 0, 1, .., m$ as the flows generated by the vector fields Y_i used in 6.1. Let $w_1, w_2, .., w_m$ be independent standard Brownian motions. Define a composite flow,

$$\Gamma(t_1, t_2)x = \xi_m(w_m(t_2) - w_m(t_2))...\xi_1(w_1(t_2) - w_1(t_1))\xi_0(t_2 - t_1)x .$$

Choose $\Delta > 0$ and consider the discrete time Markov process

$$X(k\Delta) = X((k-1)\Delta)\Gamma((k-1)\Delta, k\Delta) , \qquad (6.4)$$

where $X(0) = x_0$. It is shown in [2] that this approximation approaches the diffusion in 6.1 when $\Delta \to 0$, over finite time intervals. As mentioned earlier, in the case of matrix Lie groups the flow is given by the exponentiation of the corresponding tangent vectors, i.e. in $\mathbf{SO}(n)$, $\psi_{i,g}(t) = g.exp(Y_i t)$, $i = 1, 2, 3$. For $\mathbf{SO}(2)$, for all $A \in \mathbb{R}^{2\times 2}$ skew-symmetric,

$$exp(A) = \begin{bmatrix} cos(a) & -sin(a) \\ sin(a) & cos(a) \end{bmatrix} , \quad a = A_{2,1} , \qquad (6.5)$$

while for $\mathbf{SO}(3)$, this matrix exponential can be evaluated by the formula, for all $A \in \mathbb{R}^{3\times 3}$ skew-symmetric,

$$exp(A) = e + \frac{sin(a)}{a} A + \frac{cos(a)}{a^2} A^2 , \quad a = \frac{1}{2} \sum_{i,j} A_{i,j}^2 . \qquad (6.6)$$

6.2 Computation of the Conditional Mean

The process $X(t)$ constructed by the algorithm has the ergodic property (see [14, 21] for details) that the posterior μ is its unique invariant measure and for any measurable function f,

$$\lim_{t\to\infty} \frac{1}{t}\int_0^t f(X(s))ds = \int_{\mathcal{X}} f(x)d\mu(x) .$$

Define $||\cdot||$ to be the Hilbert-Schmidt norm on $\mathbf{SO}(n)$ (by embedding it in $\mathbb{R}^n$), i.e. $||O|| = \sum_{i,j=1}^n O_{ij}^2$. Using the regular Euclidean norm on $\mathbb{R}^n$, we can extend it to $(p, O) \in \mathbf{SE}(n)$ by $||(p, O)|| = ||p|| + ||O||$ and to $\mathbf{SE}(n)^J$ by taking sums over the components. The ergodic result dictates that for posterior μ the sample average of the distance function $f(\cdot) = ||\cdot - x||^2$, $x \in \mathcal{X}$, converges to its expectations

$$\lim_{t\to\infty} \frac{1}{t}\int_0^t ||X(s) - x||^2 ds = \int_{\mathcal{X}} ||x - y||^2 \mu(dy), \quad \forall x \in \mathcal{X} . \tag{6.7}$$

Approximate the minimum mean squared error (MMSE) estimate, $\tilde{x}_N$ for N assumed large, by

$$\tilde{x}_N = arg \min_{x\in\mathcal{X}} \frac{1}{N}\sum_{i=1}^N ||x_i - x||^2 , \tag{6.8}$$

for $x_i = X(i\Delta)$. In other words, given the samples x_i's, the MMSE estimate corresponds to the point $\tilde{x}_N \in \mathcal{X}$ having the least total distance from the samples x_i, $i = 1, 2, \ldots, N$. In $\mathbb{R}^n$ with regular Euclidean norm this is just the sample average, $\tilde{x}_N = \frac{1}{N}\sum_{i=1}^N x_i$, but on curved manifolds the answer is different. For $\mathbf{SO}(n)$ with the Hilbert-Schmidt norm it reduces to,

$$\tilde{x}_N = arg \max_{x\in\mathbf{SO}(n)} tr(a^\dagger x), \quad \text{where } a = \frac{1}{N}\sum_{i=1}^N x_i .$$

From [8], it can be shown that, if $a = u\sigma v^\dagger$ is the singular value decomposition of a then

$$\tilde{x}_N = \begin{cases} uv^\dagger, & \text{if } determinant(a) \geq 0 \\ uLv^\dagger, \quad L = \begin{bmatrix} 1 & 0 & \ldots & 0 \\ 0 & 1 & \ldots & 0 \\ \vdots & & & \\ 0 & 0 & \ldots & -1 \end{bmatrix}, & \text{if } determinant(a) < 0 . \end{cases} \tag{6.9}$$

In the case of **SO**(2) this formula simplifies further. The samples generated on **SO**(2) are of the type: $\begin{bmatrix} cos(\theta_i) & -sin(\theta_i) \\ sin(\theta_i) & cos(\theta_i) \end{bmatrix}$ and hence the average matrix a also has the structure $a = \begin{bmatrix} a_1 & -a_2 \\ a_2 & a_1 \end{bmatrix}$. Both the singular values of matrix a are $\sqrt{a_1^2 + a_2^2}$ and the orthogonal matrix closest to a (in Hilbert-Schmidt distance) is

$$\tilde{x}(N) = \frac{1}{\sqrt{a_1^2 + a_2^2}} a \, . \tag{6.10}$$

7 Results

Now we examine three specific examples to illustrate the methodology: (i) estimate the orientation of a truck from an image, (ii) estimate the orientation of an airplane from an image, and (ii) estimate the trajectory of a flying airplane from a sequence of images.

7.1 Estimating Orientation of a Ground-Based Object

Assume a ground-based target with unknown orientation in **SO**(2). The posterior measure is the product of data likelihood and Haar measure on **SO**(2). The algorithm is as follows:

Algorithm:

Let $i = 0$ and $X(0) = g_0 \in$ **SO**(2) be any initial condition.

1. Generate a sample u of an exponential random variable with constant mean λ.

2. Diffusion: follow the approximation Eqn 6.4 of the SDE, i.e. let $l = 0$.

   (a) For $\epsilon > 0$ small enough, numerically approximate the directional derivative $Y_{i,X(i\Delta)}H$ by α_1 where

   $$\alpha_1 = \frac{1}{\epsilon}\left(H(X(i\Delta) \cdot e^{\epsilon Y_{1,e}}) - H(X(i\Delta))\right) \, .$$

   (b) Generate a Gaussian random variable, $w_1 \sim N(0, 1)$.

   (c) Update the process by $X((i+1)\Delta) = X(i\Delta)exp((\sqrt{\Delta}w_1 + \Delta\alpha_1)Y_{1,e})$. The exponential of a 2×2 skew-symmetric matrix is given in 6.5.

   (d) $i = i + 1$, $l = l + 1$; if $l\Delta < u$ then go to (a).

3. Metropolis jump move: generate a uniform random variable $\theta \in [0, 2\pi]$. Evaluate $g = \begin{bmatrix} cos\theta & -sin\theta \\ sin\theta & cos\theta \end{bmatrix}$. Calculate $H(g)$.
If $H(X(i\Delta)) > H(g)$, set $X(i\Delta) = g$.
Else set $X(i\Delta) = g$ with probability $e^{-(H(g)-H(X(i\Delta)))}$.

4. $i = i + 1$, go to 1.

This algorithm generates a sequence, $\{X(i\Delta) \in \mathbf{SO}(2), i = 1, 2, \ldots\}$, of samples from the posterior distribution from which the MMSE estimate is generated in the following way. The Hilbert Schmidt (H-S) norm on $\mathbf{SO}(2)$ is given by, $||g_1 - g_2||^2 = 4 - 2 * trace(g_1 \cdot g_2^\dagger)$. For the samples $x(i) = X(i\Delta), i = 1, 2, \ldots, N$, the MMSE estimate $\tilde{x}_N$ under H-S norm is given by 6.10.

This algorithm was implemented on a SGI Onyx workstation using its graphics engine for rendering the three-dimensional objects as well as generating the orthographic projection profiles for simulating the video camera image as shown in Figure 6. The projection of a truck, rendered at true orientation x_{true} in the left panel of Figure 9, was sampled on a 64×64 lattice with *i.i.d.* Gaussian noise was added at each pixel to simulate noisy observation shown in the middle panel. The estimation algorithm outlined above was run to generate 100 samples from the posterior with the estimated orientations $\tilde{x}_{100}$ are shown in the right panel of the Figure 9.

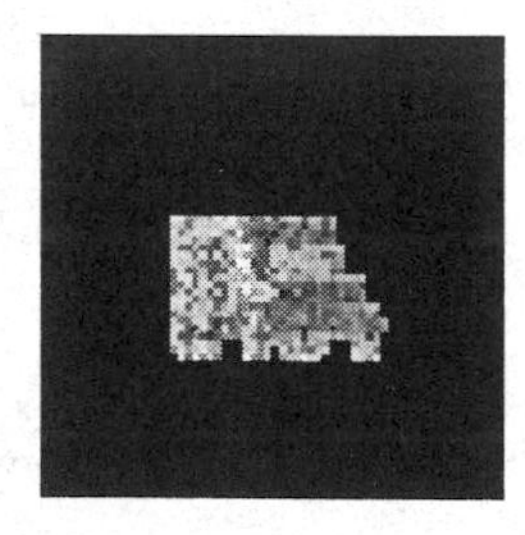

Figure 9: The middle panel show the noisy image of a truck rendered at the true orientation shown in the left panel. The truck is rendered at MMSE estimate $\tilde{x}_{100}$ in the right panel.

The plot in Figure 10 describes the evolution of the algorithm. The two curves correspond to the H-S distance of the samples x_N and the estimate $\tilde{x}_N$ from the true orientation as function of N. The dotted curve displaying $||x_{true} - x_N||^2$ shows the algorithm jumping at times $N = 1, 2, 8, 9, 21, 42$ with diffusions at other times. while the distance profile of the estimated orientation $\tilde{x}_N$, $||x_{true} - \tilde{x}_N||^2$, is plotted by the dark line.

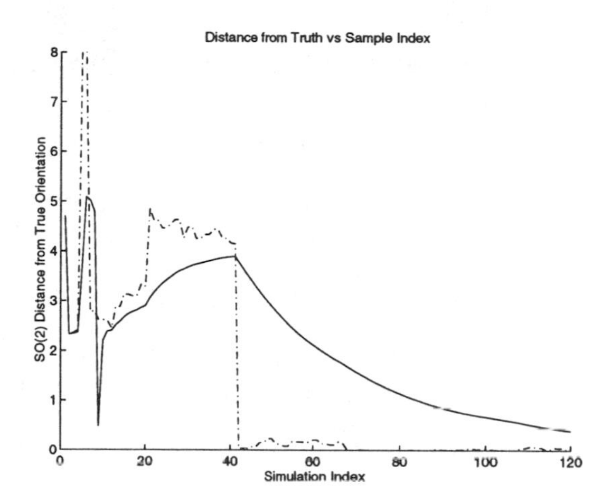

Figure 10: Evolution of the sampling process: the broken line shows the H-S distance of the process $X(s)$ from the reference point x_{true} while the regular line shows the distance function for the sample averages $\tilde{x}_N$ evolving over time.

7.2 Estimating Orientation of an Airplane

Now expand the space to $\mathbf{SO}(3)$ to estimate the orientation of an airplane from its noisy images. A jump-diffusion algorithm is used to search in $\mathbf{SO}(3)$ with the image data likelihood and Haar measure on $\mathbf{SO}(3)$ contributing to the posterior measure. The jump here corresponds to moves within $\mathbf{SO}(3)$, with the diffusions being the solutions of SDE on $\mathbf{SO}(3)$. The H-S norm is given by $||g_1 - g_2||^2 = 6 - 2 * trace(g_1 g_2^\dagger)$ and the corresponding MMSE estimate $\tilde{x}_N$ is given in 6.9. The algorithm becomes:

Algorithm: Let $i = 0$ and $X(0) \in \mathbf{SO}(3)$ be any initial condition.

1. Generate a sample u of an exponential random variable with constant mean λ.

2. Diffusion: follow the approximation Eqn 6.4 of the SDE in step 2 of Algorithm 6 for u cycles; i.e. let $l = 0$ and $Y_{1,e}, Y_{2,e}, Y_{3,e}$ be the three orthonormal basis of the space of skew-symmetric matrices given in 4.2.

   (a) For $\epsilon > 0$ small enough, numerically approximate the directional derivatives $\alpha_i = Y_i H$ using 6.3.

(b) Generate w_1, w_2, w_3 i.i.d Gaussian random variables with mean zero and variance 1.

(c) Update the process according to $X((i+1)\Delta) = X(i\Delta)\Gamma(X(i\Delta))$ where

$$\Gamma(X(i\Delta)) = exp(\sqrt{\Delta}w_3 Y_{3,e})$$

$$exp(\sqrt{\Delta}w_2 Y_{2,e})exp(\sqrt{\Delta}w_1 Y_{1,e})exp(\Delta\sum_{i=1}^{3}\alpha_i Y_{i,e}) .$$

The exponentiation of 3×3 skew-symmetric matrix is given by 6.6.

(d) $i = i+1$, $l = l+1$; if $l\Delta < u$ then go to (a).

3. Metropolis jump move: generate g uniformly over $\mathbf{SO}(3)$, calculate $H(g)$.
If $H(X(i\Delta)) > H(g)$, set $X(i\Delta) = g$.
Else set $X(i\Delta) = g$ with probability $e^{-(H(g)-H(X(i\Delta)))}$.

4. $i = i+1$, go to 1.

The jump step (step 3) involves generating samples from a uniform measure on $\mathbf{SO}(3)$ in the following way. There exists a diffeomorphism between the upper half of the unit 3-sphere, $\mathbf{S}^3_+ = \{(q_0, q_1, q_2, q_3) \in \mathbf{S}^3 | q_0 > 0\}$ and $\mathbf{SO}(3)$ given by, $(q_0, q_1, q_2, q_3) \in \mathbf{S}^3_+ \leftrightarrow Q \in \mathbf{SO}(3)$ where

$$Q = \begin{bmatrix} q_0^2+q_1^2-q_2^2-q_3^2 & -2(q_0q_3-q_1q_2) & 2(q_0q_2+q_1q_3) \\ 2(q_0q_3+q_1q_2) & q_0^2-q_1^2+q_2^2-q_3^2 & -2(q_0q_1-q_2q_3) \\ -2(q_0q_2-q_1q_3) & 2(q_0q_1+q_2q_3) & q_0^2-q_1^2-q_2^2+q_3^2 \end{bmatrix} . \quad (7.1)$$

The uniform measure on the sphere $\mathbf{S}^3_+$ in polar coordinates is given by $sin^2(\theta_1)sin(\theta_2)d\theta_1 d\theta_2 d\theta_3$ (see [18] for example) where for $0 \leq \theta_1 \leq \pi/2$, $0 \leq \theta_2, \theta_3 \leq 2\pi$. Set

$$q_0 = cos\theta_1 \text{ , } q_1 = sin\theta_1 cos\theta_2 \text{ , } q_2 = sin\theta_1 sin\theta_2 cos\theta_3 \text{ , } q_3 = sin\theta_1 sin\theta_2 sin\theta_3 . \quad (7.2)$$

A sample from uniform measure on $\mathbf{SO}(3)$ can be generated by the following steps: (i) Generate a uniform random variable $x \in [0, \pi/4]$ and solve the transcendental equation $\theta_1/2 - sin(2\theta_1)/4 = x$ for θ_1. (ii) Generate a uniform random variable $x \in [-1, 1]$, evaluate $\theta_2 = cos^{-1}(x)$. (iii) Generate a uniform random variable $\theta_3 \in [0, 2\pi]$. (iv) Evaluate $(q_0, q_1, q_2, q_3) \in \mathbf{S}^3_+$ from 7.2. (v) Evaluate $Q \in \mathbf{SO}(3)$ using the 7.1.

The algorithm was implemented in the same simulation environment as earlier. The plot in Figure 11 illustrates the state of the algorithm, the thin

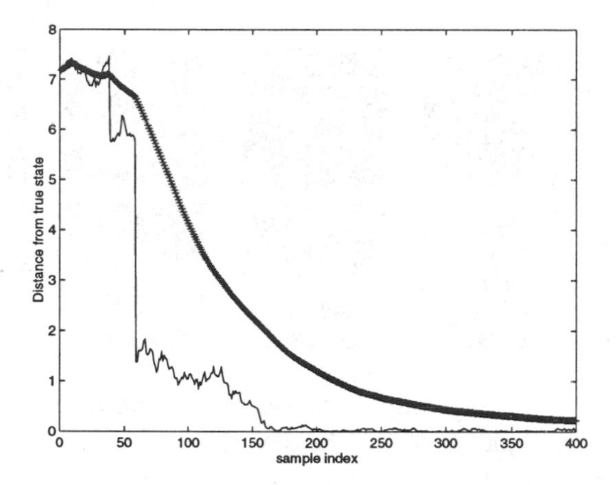

Figure 11: Sampling on **SO**(3): the thin line plots the HS-distance of x_N from x_{true} while the thick line plots the HS of $\tilde{x}_N$ from x_{true}.

line plots $||x_{true} - x_N||^2$ and the thick line plots $||x_{true} - \tilde{x}_N||^2$ against the index N. The jumps are made at $N = 1, 13, 39, 59$ reflecting the random times when the candidates are better matched to the data image than the present state x_N. Shown in the upper panels of Figure 12 is the target rendered at $x_N \in \mathbf{SO}(3)$ for $N = 1, 41, 81, 121, 401$. The middle panels show the data match generated by removing the hypothesized contribution from target at the current sample x_N, from the data. The lower panels display the evolution of sample averages $\tilde{x}_N$ for the same times.

7.3 Estimating Trajectories in SE$(3)^J$

Simultaneous tracking and recognition of multiple targets requires additional jump moves to account for target detection and identification. Examine the tracking of a single target motion via diffusions on the motion components forming a target trajectory in $\mathbf{SE}(3)^J$. Shown in Figure 13 is a diffusion transforming a track in $\mathbf{SE}(3)^4$. The position and orientation of the target at each sample point is modified using the SDEs as described in the last section. The motion components forming a track are related through a dynamics based prior as described in [14]. For detailed algorithm and implementation results please refer to [14].

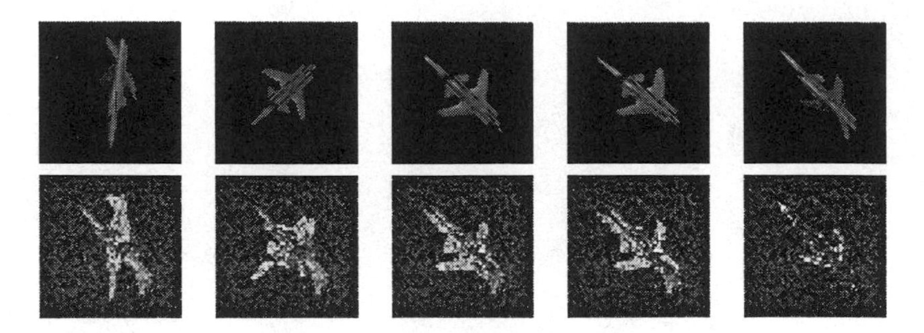

Figure 12: Samples from Jump-Diffusion process on $\mathbf{SO}(3)$ evolving over time: (i) the upper panels show the samples at the simulation index 1, 41, 81, 121, 401, and (ii) the lower panels show the difference images between the observed data and the synthesized data corresponding to that orientation state at these times.

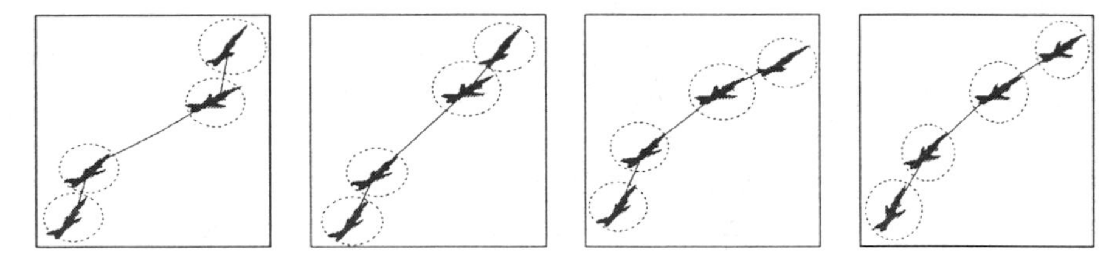

Figure 13: These panels describe a sequence of similarity transformations deforming a 4-length track in $\mathbf{SE}(3)^4$.

8 Conclusion

We have presented algorithms for automated target recognition with parameterizations on Cartesian products of Matrix Lie groups, $\mathbf{SE}(n)$, $n = 2, 3$ in particular. Tools have been developed for intrinsic optimization of Bayesian cost functions on these parametirc manifolds. Stochastic flows with desired statistical properties are constructed to simulate the posterior distribution on the representation space. The algorithmic details are illustrated through specific applications in rigid object tracking and recognition. The algorithms are presented for two specific scenarios: (i) estimating orientation of a ground-based object in $\mathbf{SO}(2)$, (ii) estimating orientation of an airborne-objects in $\mathbf{SO}(3)$, and (iii) estimating the target trajectory for a given target in $\mathbf{SE}(3)^{40}$.

References

[1] Y. Amit and M.I. Miller. Ergodic properties of jump-diffusion processes. Monograph of Electronic Signals and Systems Research Laboratory, Washington University, St. Louis. 1992.

[2] Y. Amit. A multiflow approximation to diffusions. *Stochastic Processes and their Applications* **37**(2) (1991), 213-238.

[3] W.M. Boothby. *An Introduction to Differential Manifolds and Riemannian Geometry.* New York: Academic Press, 1986.

[4] R.W. Brockett. System theory on group manifolds and coset spaces. *SIAM Jounral on Control* **10**(2) (1972), 265-84.

[5] M.L. Curtis. *Matrix Groups.* New York: Springer Verlag, 1979.

[6] T.E. Duncan. An estimation problem in compact lie groups. *Systems and Control Letters* **10**(4) (1990), 257-63.

[7] B. Friedland. *Control System Design: An Introduction To State-Space Methods.* New York: McGraw-Hill, 1986.

[8] G.H. Golub and C.F. Vanloan. *Matrix Computations.* Baltimore: Johns Hopkins University Press, 1989.

[9] K. Kanatani. *Group-Theoretical Methods in Image Understanding.* New York: Springer-Verlag, 1990.

[10] H. Kunita. Stochastic differential equations and stochastic flows of diffeomorphisms. *École d' Été de Probabilités de Saint-Flour, XII -1982.* Vol. 1097. New York: Springer-Verlag, 1984.

[11] M. I. Miller, A. Srivastava, and U. Grenander. Conditional-expectation estimation via jump-diffusion processes in multiple target tracking/recognition. *IEEE Transactions on Signal Processing* **43**(11) (1995), 2678-2690.

[12] R. Schmidt. *A Signal Subspace Approach to Multiple Emitter Location and Spectral Estimation.* Ph.D. Dissertation. Stanford University. Palo Alto, California. 1981.

[13] S. Satto, R. Frezza, and P. Perona. Motion estimation via dynamic vision. To appear *IEEE Transactions on Automatic Control.*

[14] A.Srivastava, U.Grenander, G.R. Jensen, and M.I. Miller. Inferences via jump-diffusion processes on matrix lie groups. Submitted to *Advances in Applied Probability.* 1996.

[15] D.L. Snyder, A.M. Hammoud, and R.L. White. Image recovery from data acquired with a charge-coupled-device camera. *Journal of the Optical Society of America* **10**(5) (1993), 1014-1023.

[16] A. Srivastava, M.I. Miller, and U. Grenander. Multiple target direction of arrival tracking. *IEEE Transactions on Signal Processing* **43**(5) (1995), 1282-85.

[17] S.T. Smith. *Geometric Optimization Methods for Adaptive Filtering.* Ph. D. Dissertation. Harvard University. Cambridge, Massachusetts. 1993.

[18] M. Spivak. *A Comprehensive Introduction to Differential Geometry, Vol I & II.* Berkeley: Publish or Perish, 1979.

[19] S. Soatto and P. Perona. Recursive 3-d visual motion estimation using subspace constraints. To appear *International Journal of Computer Vision.*

[20] A. Srivastava. *Automated Target Tracking and Recognition Using Jump Diffusion Processes.* M.S. Thesis. Washington University. St. Louis, Missouri. 1993.

[21] A. Srivastava. *Inferences on Tranformation Groups Generating Patterns on Rigid Motions.* D.Sc. Thesis. Washington University. St. Louis, Missouri. 1996.

[22] D. H. Sattinger and O. L. Weaver. *Lie Groups and Algebras with Applications to Physics, Geometry and Mechanics.* New York: Springer Verlag, 1986.

Department of Electrical Engineering, Washington University, St. Louis, Missouri 63130

Department of Electrical Engineering, Washington University, St. Louis, Missouri 63130

Division of Applied Mathematics, Brown University, Providence, Rhode Island 02912

Some Recent Results on the Maximum Principle of Optimal Control Theory

H.J. Sussmann[1]

1 Introduction

The maximum principle of optimal control theory (also known as the "Pontryagin maximum principle") was announced in the 1958 International Congress of Mathematicians, and presented in detail in the book [17]. The results of [17], however, are only a *version* of the principle, i.e. one of several possible ways of translating the statement of the principle into rigorous, precise mathematics.

For us, a "principle" is a somewhat vague assertion —such as, for example, the statement that "if the derivative of a function is positive then the function is increasing"— that one feels ought to become true once the terms involved are defined with care and precise technical hypotheses are specified. Since this can usually be done in more than one way, a principle typically has several different *versions*. In modern mathematics, it is customary to be careful about precision and to state technical hypotheses in full, so the discoverer of a new principle will often present it in the form of a rigorous first version, thereby failing to do full justice to the true power of the principle as a potential source of many other, equally rigorous, theorems.

This is what happened to the maximum principle. The rigorous version of [17] required that the right-hand side of the differential equation $\dot{x} = f(x, u, t)$ defining the control system to be of class C^1 with respect to the state variables, with appropriate conditions on the dependence on u and t, and yielded as a conclusion the existence of a nontrivial "adjoint vector" satisfying the Hamiltonian minimization condition. Since then, new and more general versions have been proved by various techniques. These versions strengthen the original one by either weakening the technical hypotheses or yielding stronger conclusions. Examples of the former type of generalization are (a) the "nonsmooth analysis" version (cf. [9, 10]), in which the right-hand side is only required to be Lipschitz continuous with respect to x, and the adjoint equation is replaced by an adjoint inclusion, (b) the maximum principle for differential inclusions, in which seemingly more general dynamical behaviors of the form $\dot{x} \in F(x, t)$ are considered, and (iii) the recent version due to S. Lojasiewicz Jr. ([16]), in which only the reference vector field is required to be Lipschitz with respect to x, and the other ones are only asked to be continuous. As an example of the latter

[1]NSF Grant DMS95-00798 and AFOSR Grant 0923.

kind of extension, we mention various versions of the "high-order maximum principle" (cf. [1, 2, 3, 13, 14, 15]), in which f is usually assumed to be very smooth with respect to x, and this is used to construct "high-order variations" involving Lie brackets of the vector fields of the given system, in addition to the "needle variations" used in the classical proof of the maximum principle. This results in extra necessary conditions for the adjoint vector ζ, in the form of a new set of inequalities $\langle\zeta(t), v\rangle \geq 0$, for vectors v other than the ones arising from the Hamiltonian minimization condition.

The purpose of this note is to announce a new result that contains, generalizes and combines most of the existing versions. We use the word "combines" because the new version applies to "hybrid systems," i.e. systems made of pieces of different kinds, each one of which falls under the scope of one of the existing versions. For example, suppose we have a system whose evolution is governed by a differential inclusion in some region $\mathcal{R}_1$ of the state space X, by a very smooth control system in some other region $\mathcal{R}_2$, and by a controlled O.D.E. with a Lipschitz right-hand in some other region $\mathcal{R}_3$. Suppose a trajectory $\xi_* : [a,b] \mapsto X$ visits all three regions, and we want to test ξ_* for optimality. If ξ_* is optimal, then the maximum principle for differential inclusions will give a nontrivial, Hamiltonian minimizing adjoint vector on any subinterval I of $[a,b]$ such that $\xi_*(I) \subseteq \mathcal{R}_1$, the high-order maximum principle will yield another nontrivial adjoint vector, satisfying Hamiltonian minimization as well as the high-order conditions, on any interval I such that $\xi_*(I) \subseteq \mathcal{R}_2$, and, finally, the nonsmooth maximum principle will produce a nontrivial Hamiltonian minimizing adjoint vector on any interval I such that $\xi_*(I) \subseteq \mathcal{R}_3$. But none of these three results will give a nontrivial adjoint vector ζ defined on the whole interval $[a,b]$ and satisfying Hamiltonian minimization everywhere as well as the high-order conditions at those times t such that $\xi_*(t) \in \mathcal{R}_2$. The reason for this is that the adjoint vectors obtained from the various versions of the maximum principle, by applying each version to a subtrajectory of ξ_* that stays in the appropriate region, need not match. Since the proofs of the different versions are carried out by different techniques, it does not seem possible to combine them into a single construction yielding a global adjoint vector.

Our version takes care of the combination problem and yields a single, global adjoint vector in hybrid situations. In addition, this version

(I) applies to problems where none of the other versions can be used, and

(II) gives stronger results even in some cases where one of the older versions is applicable.

To illustrate the first point, consider the following three examples.

Example 1.1 Find the minimum-time trajectories for the control system in $\mathbb{R}^2$ given by $\dot{x} = u\varphi(x,y)$, $\dot{y} = v\varphi(x,y)$, where the control vector (u,v) is subject to the constraint $u^2+v^2=1$, and $\varphi:\mathbb{R}^2\to\mathbb{R}$ is the function given by $\varphi(x,y)=c_1$ if $y\geq 0$, $\varphi(x,y)=c_2$ if $y<0$, where c_1, c_2 are positive constants such that $c_1\neq c_2$.

Example 1.2 Same question as in Example 1.1, except that now $\varphi(x,y)=\sqrt{|y|}$.

Example 1.3 Let α be such that $1<\alpha<2$. Consider the optimal control problem in $\mathbb{R}$, with the control u subject to $|u|\leq 1$, in which the state variable x evolves according to $\dot{x}=u+|x|^\alpha\sin\frac{1}{x}$. Suppose we want to minimize the integral $\int_0^1 u(t)e^t\,dt$, among all control-trajectory pairs $(u(\cdot),x(\cdot))$ such that $x(0)=x(1)=0$. Is $u(t)\equiv 0$, $x(t)\equiv 0$ a minimizer.

In Examples 1.1 and 1.2, we are dealing with the problem of finding the "light rays" in a "medium" where the speed of light φ is variable. In Example 1.1, the speed of light is discontinuous. It is clear that an optimal path will be a straight line as long as it is contained in one of the half-planes $y\geq 0$, $y\leq 0$. A path joining points A, B in different half-planes will be a broken line, with a corner at the point C where it crosses the x axis. This still leaves a one-parameter family of paths as possible candidates for the solution. To find the true optimal path we need an extra condition that will determine C. Naturally, C can be found by elementary calculus, and the result is the well known Snell law of refraction. However, *this problem does not fit within the framework of the existing versions of the maximum principle, because the right-hand side is discontinuous as a function of the state.*

In Example 1.2, if we restrict ourselves to one of the half-planes $y\geq 0$, $y\leq 0$, we find the famous *brachystochrone problem*, proposed by John Bernoulli in 1696 as a challenge to the mathematical community, and solved the following year, separately, by several eminent mathematicians. (The May 1697 issue of *Acta Eruditorum* contains solutions by John Bernoulli himself, his brother Jacob, Leibniz, Newton, Tschirnhaus, and l'Hôpital.) The solutions can be found by classical calculus of variations or optimal control methods, and turn out to be cycloids. If, however, we want to join points A, B lying in different half-planes, we run into a difficulty similar to that of Example 1.1: one can easily prove that an optimal path will consist of a cycloid going from A to a point C in the axis, followed by another cycloid going from C to B, but none of the existing versions of the maximum principle will narrow down the choice further by determining C. (Notice that the right-hand side of the system is continuous but not Lipschitz with respect to the state variable.)

In Example 1.3, the right-hand side is not Lipschitz with respect to x, even for the "reference control" $u(t)\equiv 0$, so none of the nonsmooth versions apply. Notice, however, that if we proceed naïvely, and observe that the right-hand side is differentiable with respect to x at $x = 0$, and its derivative is 0, we get the "adjoint equation" $\dot{z} = 0$ for the multiplier $t \mapsto z(t)$. On the other hand, the Hamiltonian is

$$H(x, z, z_0, u, t) = z.\Big(u + |x|^\alpha \sin \frac{1}{x}\Big) + z_0 u.e^t \,, \tag{1.1}$$

where z_0 is the abnormal multiplier. If $z_0 = 0$, then the only way that the minimization condition could hold would be to have $z(t) \equiv 0$, which would contradict the nontriviality of the pair (z, z_0). So $z_0 \neq 0$, and we may assume that $z_0 = 1$. Then, for the Hamiltonian minimization condition to hold, it is necessary that $z(t) + e^t = 0$, i.e. that $z(t) = -e^t$, which contradicts the equation $\dot{z} = 0$. So the necessary conditions of the maximum principle are not satisfied, and the trajectory-control pair $(0, 0)$ is not optimal, *provided that we can rigorously justify the application of the "maximum principle" with the "classical" adjoint equation $\dot{z} = 0$, even though the right-hand side is not of class C^1 or even Lipschitz.*

It will turn out that *our version of the maximum principle gives extra necessary conditions that determine C in Examples 1.1 and 1.2, and provides a rigorous justification for the argument used above to exclude the optimality of the pair* $(0, 0)$ *for Example 1.3.*

To illustrate Point (II), we consider

Example 1.4 Same as Example 1.3, but with $\alpha = 2$.

Now the situation is different, because the right-hand side is Lipschitz, so the nonsmooth version of Clarke [9] *does* apply. It turns out, however, that this version gives the adjoint differential inclusion $-\dot{z} \in z.[-1, 1]$, since the Clarke generalized gradient of the function $x \to x^2 \sin \frac{1}{x}$ at $x = 0$ is the interval $[-1, 1]$. The function $z(t) = -e^t$ is a solution of this equation, so *the nonsmooth maximum principle fails to exclude the possibility that the control $u(t) \equiv 0$ might be optimal.* By contrast, the formal argument that we used above for the case when $1 < \alpha < 2$ applies without change when $\alpha = 2$, and establishes the nonoptimality of $u(t) \equiv 0$, provided that we can justify it rigorously. It will turn out that *our version of the maximum principle provides the necessary justification*, so in particular *this version yields a strictly stronger conclusion than the nonsmooth one in some cases where both versions apply.*

Since our version contains the nonsmooth one as a special case, it is clear that in general *our version cannot possibly give rise to a unique adjoint equation.* For instance, in Example 1.4 it is essential that our version implies the conclusion derived from the nonsmooth maximum principle — with the adjoint inclusion $-\dot{z} \in z[-1, 1]$— as well as the different conclusion involving the adjoint equation $\dot{z} = 0$. This nonuniqueness is not a disadvantage of the new version but, quite to the contrary, it makes it even better, since the possibility of choosing among different adjoint equations means that several necessary conditions have to hold, making the new conclusions stronger, as we saw in our analysis of Example 1.4.

We now proceed to outline the statement of our version of the maximum principle, and some of the ideas going into its proof. The sketch given here will of necessity be rather incomplete, due to space limitations. For outlines covering other aspects of the proof, the reader is referred to our papers [19, 20, 21]. A complete account [22], with full details, is in preparation.

The basic strategy for the proof is exactly the same as that of the classical argument given in [17]. One looks at the flow defined by the reference vector field, and constructs "point variations" using the other vector fields of the system. These variations can be combined to yield "endpoint maps," and one applies a suitable open mapping theorem to these maps. Open mapping theorems usually say that if the differential of a map at a point is onto, then the map itself is open at the point. To apply such theorems to our situation, one needs to compute the differential of the endpoint maps, and this is done using the chain rule, since the endpoint map is a composite of simpler maps arising from the reference flow and the variations.

In the classical proof, all this is done in the class of continuously differentiable maps. The main point of our generalization is that exactly the same argument works within much more general classes of possibly set-valued maps (or *map germs*, to be precise), provided only that for these maps there is a concept of "generalized differential" for which the chain rule and a reasonable open mapping theorem hold. In our previous papers [19, 20, 21] we introduced one such class, and called its members "semidifferentiable maps." This class suffices to handle Lojasiewicz-type extensions of the nonsmooth maximum principle. Here we propose to use a still larger class, whose members will be called "multidifferentiable maps." This class still has a good chain rule as well as a suitable open mapping theorem, and in addition contains many maps that are not semidifferentiable. For example, if $F : \mathbb{R}^n \mapsto 2^{\mathbb{R}^n}$ is a locally Lipschitz set-valued map with compact (but not necessarily convex) values, then the flow of the differential inclusion $\dot{x} \in F(x)$ is multidifferentiable along every trajectory. (This follows by looking at the flows of suitable Lipschitz selections of the convexification of F, and approximating their trajectories by trajectories of the nonconvexified inclusion, using the uniform approximation results of [11, 12]. More generally, for a differential inclusion $\dot{x} \in F(x,t)$, and a trajectory ξ such that F is integrably pseudolipschitz along ξ, the flow is multidifferentiable along ξ, as will be explained below.)

It turns out that, in the setting of multidifferentiable set-valued maps, one can rewrite the classical arguments of the book [17] for general *systems* $\{F_u : u \in U\}$ *of differential inclusions*, thereby obtaining a maximum principle that includes the case of a system of vector fields as well as that of a single differential inclusion. All that is needed is for the F_u to be "almost lower semicontinuous" and "locally integrably lower bounded," as defined below, and for the reference trajectory ξ_* to be generated by a reference inclusion $\dot{x} \in F_{u_*}(x,t)$ such that F_{u_*} is integrably pseudolipschitz along ξ_*. We handle this general situation by turning upside-down the conventional wisdom that says that "differential inclusions are more general than systems of vector fields, because a system of vector fields always gives rise to a differential inclusion, but the converse is not true, because set-valued maps, even when they are very smooth, typically do not admit nice single-valued selections." Our rejoinder to this statement is that it is true if "nice" is taken to mean "continuous," but there is no good reason for interpreting the word "nice" so narrowly. It has been shown by A. Bressan, in a series of papers (cf. [4, 5, 6, 7, 8]) that lower semicontinuous set-valued maps admit "conically continuous" selections. By slightly modifying and extending Bressan's results, one can show that, for a general differential inclusion with an almost lower semicontinuous right-hand side $(x,t) \mapsto F(x,t)$, there exist sufficiently many "nice" single-valued selections f of F. Here "nice" is taken to mean, essentially, that the indefinite integral $t \mapsto \int_a^t f(\xi(s),s)ds$ depends continuously on the continuous function ξ. (A precise definition will be given later.) This property suffices for the ordinary differential equation $\dot{x} = f(x,t)$ to be as good as an O.D.E. whose right-hand side is continuous with respect to x. Using these "nice" selections, we can carry out the entire program of the classical proof in the differential inclusions setting. As a byproduct of this work, we feel entitled to claim that *the "vector field approach" has been vindicated, and is in fact* **more general** *than that based on a single differential inclusion.* (We contend that the first part of our summary of the "conventional wisdom," according to which every system of vector fields gives rise to a differential inclusion, is also misleading. It is true that, given a control system $\dot{x} = f(x,u)$, $u \in U$, one can always define a differential inclusion $\dot{x} \in F(x)$ by letting $F(x) = \{f(x,u) : u \in U\}$. But *the passage from the control system to the differential inclusion often involves a loss of information.* For example, the control system $\dot{x} = ux$, with $x \in \mathbb{R}$, $u \in \mathbb{R}$, gives rise to the differential inclusion $\dot{x} \in F(x)$, where $F(x) = \{0\}$ if $x = 0$, $F(x) = \mathbb{R}$ of $x \neq 0$. However, if we specify that an admissible control is an integrable function $t \mapsto u(t)$, then the reachable set from 0 for the control system is just $\{0\}$, whereas the one for the differential inclusion is $\mathbb{R}$. This happens because not every trajectory of $\dot{x} \in F(x)$ is an admissible trajectory of $\dot{x} = ux$).

2 Multidifferentials of Set-valued Maps

If X, Y are sets, we use $SVM(X,Y)$ to denote the set of all set-valued maps from X to Y, so the elements of $SVM(X,Y)$ are exactly the maps from X to the set 2^Y of all subsets of Y. If $F \in SVM(X,Y)$, then the *domain* of F is the set $\mathrm{Dom}(F) = \{x \in X : F(x) \neq \emptyset\}$, and the *graph* of F is the set $\mathrm{Graph}(F) = \{(x,y) \in X \times Y : y \in F(x)\}$. If $F \in SVM(X,Y)$ is a set-valued map, and S a set, the *restriction* of F to S is the map $F\lceil S$ whose graph is $\mathrm{Graph}(F) \cap (S \times Y)$.

If X is a metric space, we use d_X to denote the distance function on X. Then d_X is a function from $X \times X$ to $[0, \infty[$, but we will also use d_X to refer to the distance between subsets of X, given by

$$d_X(S,T) = \inf\{d_X(s,t) : s \in S, t \in T\}\,, \tag{2.2}$$

so d_X takes values in $[0,\infty]$. (The equality $d_X(S,T) = \infty$ holds iff one of the sets S, T is empty). We use Δ_X to denote the function on $2^X \times 2^X$ given by

$$\Delta_X(S,T) = \max\Big(0, \sup\{d_X(s,T) : s \in S\}\Big)\,. \tag{2.3}$$

Then Δ_X takes values in $[0,+\infty]$. (We have $\Delta_X(S,T) = 0$ iff $S \subseteq \mathrm{Clos}(T)$, and $\Delta_X(S,T) = +\infty$ iff $T = \emptyset \neq S$). The Hausdorff distance $d_X^h(S,T)$ between two subsets S, T of X is given by $d_X^h(S,T) = \Delta_X(S,T)+\Delta_X(T,S)$.

If X, Y are metric spaces, then the product $X \times Y$ is a metric space with distance defined by $d_{X\times Y}\Big((x,y),(x',y')\Big) = d_X(x,x') + d_Y(y,y')$.

If K is a compact metric space, and X is a metric space, a *regular set-valued map* from K to X is a set-valued map $F \in SVM(K,X)$ with a compact graph, having the property that there exist a sequence $\{F_j\}$ of continuous (single-valued) maps $F_j : K \mapsto X$ and a compact subset K' of X such that $F_j(K) \subseteq K'$ for all j, and

$$\lim_{j\to\infty} \Delta_{K\times X}(\mathrm{Graph}(F_j), \mathrm{Graph}(F)) = 0\,. \tag{2.4}$$

We use $\mathrm{REG}(K,X)$ to denote the set of all $F \in SVM(K,X)$ that are regular, and identify the space $C^0(K,X)$ in an obvious way with a subset of $\mathrm{REG}(K,X)$. Clearly, if $F \in \mathrm{REG}(K,X)$ then (i) $\mathrm{Dom}(F) = K$, and (ii) $F \in C^0(K,X)$ iff F is single-valued.

If S is a closed subset of $\mathbb{R}^n$, a *map of class C^1 from S to $\mathbb{R}^m$ in the Whitney sense* is a pair $F = (F^0, F^1)$ such that $F^0 \in C^0(S,\mathbb{R}^m)$, $F^1 \in C^0(S,\mathbb{R}^{m\times n})$, and, for each $x \in S$,

$$^0(y) - F^0(x) = F^1(x).(y-x) + o(\|y-x\|) \quad \text{as } y \to x \text{ via values in } S\,. \tag{2.5}$$

We use $C^1(S,\mathbb{R}^m)$ to denote the set of all maps of class C^1 in the Whitney sense from S to $\mathbb{R}^m$. The well known *Whitney extension theorem* (cf. [23, 28]) says that a pair (F^0, F^1) belongs to $C^1(S,\mathbb{R}^m)$ iff there exists a map $\tilde{F} : \mathbb{R}^n \mapsto \mathbb{R}^m$ of class C^1 in the usual sense such that $\tilde{F}\lceil S = F^0$ and $D\tilde{F}\lceil S = F^1$. In many cases —e.g. if S is a closed convex subset of $\mathbb{R}^n$ with nonempty interior— one can identify $C^1(S,\mathbb{R}^m)$ with the space of maps $G \in C^1(\mathrm{Int}(S),\mathbb{R}^m)$ such that both G and DG have continuous extensions to S, but this identification is not valid for general closed sets S, even if $S = \text{Clos}(\text{Int}(S))$. If $F = (F^0, F^1) \in C^1(S, \mathbb{R}^m)$, we will write $F(x)$ for $F^0(x)$, and define $DF = F^1$.

If $\mathbf{L}$ is a subset of $\mathbb{R}^{m\times n}$, and $S \subseteq \mathbb{R}^n$ is closed, we use $C^1_{\mathbf{L}}(S, \mathbb{R}^m)$ to denote the set of all maps $F \in C^1(S, \mathbb{R}^m)$ such that $DF(S) \subseteq \mathbf{L}$ (i.e. $DF(x) \in \mathbf{L}$ for each $x \in S$).

If S is a subset of a metric space X, and $p \in X$, we say that S is *closed near p* if the set

$$S_{\varepsilon,p} = \{x \in S : d_X(x,p) \le \varepsilon\} \tag{2.6}$$

is closed in X for some $\varepsilon > 0$.

Let $F \in SVM(\mathbb{R}^n, \mathbb{R}^m)$ and let $p \in \mathbb{R}^n$, $q \in \mathbb{R}^m$. Let $S \subseteq \mathbb{R}^n$ be such that $p \in S$. Let $\mathbf{L}$ be a compact subset of $\mathbb{R}^{m\times n}$. We say that $\mathbf{L}$ is a *multidifferential of F at (p,q) along S* if for every neighborhood $\mathbf{L}'$ of $\mathbf{L}$ in $\mathbb{R}^{m\times n}$ there exist an $\bar\varepsilon > 0$, families $\{G_\varepsilon\}_{\varepsilon\in]0,\bar\varepsilon]}$, $\{H_\varepsilon\}_{\varepsilon\in]0,\bar\varepsilon]}$, and a function $\theta :]0,\bar\varepsilon] \to]0,\infty[$ such that $\lim_{\varepsilon\downarrow 0}\theta(\varepsilon) = 0$, with the property that, for each $\varepsilon \in]0,\bar\varepsilon]$, if we define $S_{\varepsilon,p}$ by (2.6), then

- (MD.1) $S_{\varepsilon,p}$ is compact,
- (MD.2) $G_\varepsilon \in \text{REG}(S_{\varepsilon,p}, \mathbb{R}^m)$,
- (MD.3) $\text{Graph}(G_\varepsilon) \subseteq \text{Graph}(F)$,
- (MD.4) $H_\varepsilon \in C^1_{\mathbf{L}'}(S_{\varepsilon,p}, \mathbb{R}^m)$,
- (MD.5) $H_\varepsilon(p) = q$,
- (MD.6) $\Delta_{\mathbb{R}^n}(G_\varepsilon(x), H_\varepsilon(x)) \le \varepsilon\theta(\varepsilon)$ for all $x \in S_{\varepsilon,p}$.

We use $MD_{p,q,S}(F)$ to denote the set of all multidifferentials of F at (p,q) along S, and call F *multidifferentiable at (p,q) along S* if $MD_{p,q,S}(F) \neq \emptyset$. It is easy to see that, if F and F' are two set-valued maps in $SVM(\mathbb{R}^n, \mathbb{R}^m)$ such that $F = F'$ near (p,q) (that is, there exists a neighborhood U of (p,q) in $\mathbb{R}^n \times \mathbb{R}^m$ such that $\text{Graph}(F) \cap U = \text{Graph}(F') \cap U$), and S, S' are such that $S = S'$ near p, then $MD_{p,q,S}(F) = MD_{p,q,S'}(F')$. This means that multidifferentiability at (p,q) along S is really a property of germs $\mathbf{F}$ at (p,q) of set-valued maps F from $\mathbb{R}^n$ to $\mathbb{R}^m$ and germs $\mathbf{S}$ of sets S at p, and the multidifferential $M_{p,q,\mathbf{S}}(\mathbf{F})$ of a germ $\mathbf{F}$ at (p,q) along a germ $\mathbf{S}$ at p is well defined.

If F is multidifferentiable at (p,q) along S, then S is necessarily closed near p, and the germ of S at p is contained in the germ of $\text{Dom}(F)$ at p (that is, there exists $\varepsilon > 0$ such that $x \in S$, $d_X(x,p) \le \varepsilon$ implies $x \in \text{Dom}(F)$).

If $F \in SVM(\mathbb{R}^n, \mathbb{R}^m)$, we will write $MD_{p,q}(F)$ for $MD_{p,q,\mathbb{R}^n}(F)$.

Multidifferentials are never unique. (For example, if $\mathbf{L} \in M_{p,q}(F)$, $\mathbf{L}'$ is compact, and $\mathbf{L} \subseteq \mathbf{L}'$, then obviously $\mathbf{L}' \in MD_{p,q}(F)$ as well). It would be nice if among all the members of $MD_{p,q}(F)$ there was one which is a subset of all the others —in which case it would be reasonable to call this set *the* multidifferential of F at p, q. It turns out, however, that in general such a "canonical" multidifferential does not exist.

Example 2.1 Consider the map $\mathbb{R}^2 \ni (x,y) \mapsto F(x,y) = (0,0) \in \mathbb{R}^2$. For $\alpha \geq 0$, let $\mathbf{L}_\alpha$ be the set of all 2×2 matrices (a_{ij}) such that $\sum_{ij} a_{ij}^2 = \alpha^2$. Let $G_\varepsilon(x,y) = (0,0)$, and

$$H_\varepsilon(x,y) = \alpha\varepsilon^2(\sin\frac{x}{\varepsilon^2} + \sin\frac{y}{\varepsilon^2}, \cos\frac{x}{\varepsilon^2} + \cos\frac{y}{\varepsilon^2}). \tag{2.7}$$

Then $DH_\varepsilon(x,y) \in \mathbf{L}_\alpha$ for all (x,y). If we let $p = q = (0,0)$, it follows easily that $\mathbf{L}_\alpha \in MD_{p,q}(F)$. Clearly, the sets $\mathbf{L}_\alpha$ are pairwise disjoint.

So we have to accept the fact that multidifferentials in general are not unique, and there is no canonical way to single out a distinguished member of $MD_{p,q}(F)$.

An important example of multidifferentials is provided by the *semidifferentials*, introduced in our previous papers [19, 20, 21]. A set-valued map $F \in SVM(\mathbb{R}^n, \mathbb{R}^m)$ is *semidifferentiable* at $p \in \mathbb{R}^n$ if there is a compact neighborhood U of p such that (a) $\mathrm{Dom}(F) \cap U$ is compact, (b) $F\lceil U \in \mathrm{REG}(\mathrm{Dom}(F) \cap U, \mathbb{R}^m)$, (c) $p \in \mathrm{Dom}(F)$, and (d) there exist a Lipschitz map $G : \mathbb{R}^n \mapsto \mathbb{R}^m$ and a function $\theta :]0,\infty[\mapsto [0,\infty]$ such that $\lim_{s\downarrow 0}\theta(s) = 0$ and

$$\Delta_{\mathbb{R}^n}(F(x), G(x)) \leq ||x-p||\,\theta(||x-p||) \text{ whenever } x \in \mathrm{Dom}(F). \tag{2.8}$$

(It then follows that $F(p) = \{G(p)\}$, since $\Delta_{\mathbb{R}^n}(F(p), G(p)) = 0$). Any G for which the above conditions hold is called a *first-order Lipschitz approximation of F at p*. A *semidifferential* of F at p is any subset $\mathbf{L}$ of $\mathbb{R}^{m\times n}$ which is a Warga derivate container of G at p for some first-order Lipschitz approximation G of F at p. (A *Warga derivate container* of G at p is a compact subset $\mathbf{L}$ of $\mathbb{R}^{m\times n}$ such that for every neighborhood $\mathbf{L}'$ of $\mathbf{L}$ in $\mathbb{R}^{m\times n}$ there exist a neighborhood V of p and a sequence $\{H_j\}$ of C^1 maps $V \mapsto \mathbb{R}^m$ such that $H_j \to G$ uniformly on V and $DH_j(x) \in \mathbf{L}'$ for all j and all $x \in V$). We use $SD_p(F)$ to denote the set of all semidifferentials of F at p.

Clearly, if F is semidifferentiable at p, and $q = F(p)$, then $\mathrm{Dom}(F)$ is closed near p, F is multidifferentiable at (p,q) along $\mathrm{Dom}(F)$, and the inclusion $SD_p(F) \subseteq MD_{p,q,\mathrm{Dom}(F)}(F)$ holds.

Multidifferentiable maps are of interest to us because (i) they have a good calculus, and in particular a chain rule, (ii) they satisfy good open mapping and separation theorems, and (iii) many maps naturally associated to controls systems and differential inclusions are multidifferentiable.

The *chain rule for multidifferentiable maps* says that,

Theorem 2.2 *If $F_1 \in SVM(\mathbb{R}^n, \mathbb{R}^m)$, $F_2 \in SVM(\mathbb{R}^m, \mathbb{R}^\nu)$, $S_1 \subseteq \mathbb{R}^n$, $S_2 \subseteq \mathbb{R}^m$, $p \in S_1$, $q \in F_1(p)$, $q \in S_2$, $r \in F_2(q)$, $F_1(S_1) \subseteq S_2$, $\mathbf{L_1} \in$*

$MD_{p,q,S_1}(F_1)$, and $\mathbf{L_2} \in MD_{q,r,S_2}(F_2)$, then $\mathbf{L_2} \circ \mathbf{L_1} \in MD_{p,r,S_1}(F_2 \circ F_1)$, if one of the following conditions holds:

(CR.1) S_1 is convex near p and S_2 is convex near q,

(CR.2) F_1 is a single-valued map of class C^1 on S_1 and $\mathbf{L_1} = \{DF_1(p)\}$.

If M is a manifold of class C^1, and $p \in M$, we use T_pM to denote the tangent space of M at p. If $S \subseteq M$ and $p \in S$, the *Bouligand tangent cone* (or *contingent tangent cone*) to S at p is the set $T_p^B(S)$ of all vectors $v \in T_pM$ such that either $v = 0$ or there exist a sequence $\{p_j\}$ in S and real numbers $h_j > 0$ such that $h_j \to 0$ and $f(p_j) - f(p) = h_j.vf + o(h_j)$ as $j \to \infty$ for every $f \in C^1(M, \mathbb{R})$. A subset S of M is said to be *convex near p* if there exists a C^1 coordinate chart $\kappa : U \to \mathbb{R}^m$, defined on a neighborhood U of p, such that $\kappa(S \cap U)$ is convex.

It follows easily from the chain rule that, if M, N are manifolds of class C^1, $p \in M$, $q \in N$, and $S \subseteq M$ is such that $p \in S$, S is closed near p and convex near p, and $F \in SVM(M, N)$, then $MD_{p,q,S}(F)$ is intrinsically defined as a collection of compact subsets of $L(T_pM, T_qN)$, the space of linear maps from T_pM to T_qN.

The *open mapping theorem* gives a sufficient condition for a set-valued map to be open at a point. If $F \in SVM(X, Y)$, where X and Y are topological spaces, we call F *open* at a point $(p, q) \in \text{Graph}(F)$ if for every neighborhood U of p the image $F(U)$ is a neighborhood of q. The theorem then says that

Theorem 2.3 *If M, N are C^1 manifolds, $F \in SVM(M, N)$, $S \subseteq M$, $p \in S$, S is closed and convex near p, $q \in N$, $\mathbf{L} \in MD_{p,q,S}(F)$, and*

(2.3.) every $L \in \mathbf{L}$ maps $T_p^B(S)$ onto T_qN,*

then F is open at (p, q).

More generally, there is a *directional open mapping theorem*: if M, N are C^1 manifolds, $p \in M$, $q \in F(p)$, $v \in T_qN$, we say that F is *directionally open at (p, q) in the direction of v* if for every neighborhood U of p the image $F(U)$ is a v-directional neighborhood of q. (If $q \in N$ and $v \in T_qN$, a *v-directional neighborhood* of q is a subset V of N with the property that, for some chart κ mapping a neighborhood U of q to $\mathbb{R}^n$, the set $\kappa(V \cap U)$ contains a convex set W such that $\kappa(q) \in W$ and $\kappa(q) + r.d\kappa(q).v \in \text{Int}(W)$ for some $r > 0$. It is easy to see that if this property holds for some chart κ defined near p then it holds for every such chart). Then Theorem 2.3 has the following generalization:

Theorem 2.4 *If M, N, F, S, p, q, $\mathbf{L}$ are as in Theorem 2.3, $v \in T_qN$, and instead of (2.4.*) we assume*

(2.4.)* $v \in \mathrm{Int}(LT_p^B(S))$ *for every* $L \in \mathbf{L}$,

then F *is directionally open at* (p,q) *in the direction of* v.

Clearly, Theorem 2.3 is a special case of Theorem 2.4, corresponding to $v = 0$.

3 A Separation Theorem

Theorem 2.4, together with standard separation properties of convex sets, imply the following rudimentary version of the maximum principle.

Theorem 3.1 *Suppose that* M_1, M_2, N *are* C^1 *manifolds,* $q \in N$, *and, for* $i = 1, 2$, $p_i \in S_i \subseteq M_i$, $F_i \in SVM(M_i, N)$, $q \in F_i(p_i)$, S_i *is closed and convex near* p_i, *and* $\mathbf{L_i} \in MD_{p_i,q,S_i}(F_i)$. *Suppose that the sets* $F_1(S_1)$ *and* $F_2(S_2)$ *are separated near* q, *in the sense that*

$$F_1(S_1) \cap F_2(S_2) = \{q\}\,. \tag{3.9}$$

Assume, in addition, that

(i) $T_{q_2}^B S_2$ *is not a linear subspace of* $T_{q_2} M_2$,

(ii) $q \notin F_2(x)$ *whenever* $x \in M_2$ *is close enough to* p_2 *and* $x \neq p_2$.

Then there exist a nontrivial linear functional $z : T_q N \mapsto \mathbb{R}$ *and* $L_i \in \mathbf{L_i}$ *for* $i = 1, 2$ *such that* $z \circ L_1$ *is* ≥ 0 *on* $T_{p_1}^B S_1$ *and* $z \circ L_2$ *is* ≤ 0 *on* $T_{p_2}^B S_2$.

Theorem 3.1 is proved as follows. We apply Theorem 2.4 to the set-valued map $Q \in SVM(M_1 \times M_2, N \times \mathbb{R})$ such that

$$Q(x_1, x_2) = (F_1(x_1) - F_2(x_2), \mu(x_2))\,, \tag{3.10}$$

where we identify a neighborhood W of p_2 in M_2 with a neighborhood of 0 in $T_{p_2} M_2$ using a coordinate chart, and $\mu : T_{p_2} M_2 \mapsto \mathbb{R}$ is a linear functional which is ≥ 0 but not identically 0 on $T_{p_2}^B S_2$. Here

$$F_1(x_1) - F_2(x_2) \stackrel{\text{def}}{=} \{y_1 - y_2 : y_1 \in F_1(x_1), y_2 \in F_2(x_2)\}\,. \tag{3.11}$$

The existence of μ is guaranteed by the fact that $T_{p_2}^B S_2$ is a closed convex cone in $T_{p_2} M_2$ —since S_2 is convex near p_2— which is not a linear subspace. Letting $\mathcal{N} = N \times \mathbb{R}$, $\bar{q} = (q, 0)$, we pick $v = (0, 1) \in T_{\bar{q}} \mathcal{N} \sim T_q N \times \mathbb{R}$. For $L_i \in MD_{p_i,q,S_i}(F_i)$, $i = 1, 2$, let

$$C(L_1, L_2) = \Big\{ \Big(L_1 u_1 - L_2 u_2, \mu(u_2) \Big) : (u_1, u_2) \in T_{p_1}^B S_1 \times T_{p_2}^B S_2 \Big\}\,. \tag{3.12}$$

Theorem 2.4 implies that if $v \in \mathrm{Int}(C(L_1, L_2))$ for all L_1, L_2, then the pair $(0, r)$ belongs to $Q(S_1 \times S_2)$ for small enough r. Pick a small $r > 0$, and write

$$(0, r) = (y_1 - y_2, \mu(x_2)), \tag{3.13}$$

with $x_1 \in S_1$, $x_2 \in S_2$. Then $y_1 = y_2$. Also, $\mu(x_2) = r \neq 0$, and $y_2 \in F_2(x_2)$. Therefore $y_2 \neq q$ by Assumption (ii). So $F_1(S_1) \cap F_2(S_2)$ contains a point of N other than q, contradicting the assumption that $F_1(S_1) \cap F_2(S_2) = \{q\}$. Therefore there must exist $L_1 \in \mathbf{L}_1$, $L_2 \in \mathbf{L}_2$, such that $v \notin \mathrm{Int}(C(L_1, L_2))$. But then there exists a nontrivial linear functional $\bar{z}$ on $T_{\bar{q}}\mathcal{N}$ such that $\bar{z}(v) \leq 0$ and $\bar{z}$ is ≥ 0 on $C(L_1, L_2)$. Write $\bar{z}(x, r) = z(x) + \rho r$, where $z : T_q N \mapsto \mathbb{R}$ is linear. Taking $x = 0$, $r = 1$, we get $\rho \leq 0$. If $u_1 \in T^B_{p_1} S_1$, $u_2 \in T^B_{p_2} S_2$, we have $(L_1 u_1 - L_2 u_2, \mu(u_2)) \in C(L_1, L_2)$, so $z(L_1 u_1) - z(L_2 u_2) + \rho\mu(u_2) \geq 0$. Taking $u_2 = 0$ we find $z(L_1 u_1) \geq 0$. Taking $u_1 = 0$ we find $z(L_2 u_2) \leq \rho\mu(u_2)$. Since $\rho \leq 0$ and $\mu(u_2) \geq 0$, we have $z(L_2 u_2) \leq 0$. Finally, if $z \equiv 0$ we would get $\rho\mu(u_2)) \geq 0$ for all $u_2 \in T^B_{p_2} S_2$. Since $\rho \leq 0$, and μ is ≥ 0 but not $\equiv 0$ on $T^B_{p_2} S_2$, we must have $\rho = 0$. But then $\bar{z} = 0$, which is a contradiction. This completes the proof of Theorem 3.1.

4 Multidifferentials of Flows

A *flow* is a triple $\mathcal{F} = (T, \mathbf{X}, \mathbf{F})$, where T is a totally ordered set, $\mathbf{X} = \{X_t\}_{t \in T}$ is a family of sets parametrized by T, and —using $\preceq$ to denote the order relation in T— $\mathbf{F} = \{F_{t,s}\}_{s,t \in T, s \preceq t}$ is a family of set-valued maps such that

(FL.1) $F_{t,s} \in SVM(X_s, X_t)$ whenever $s, t \in T$ and $s \preceq t$,

(FL.2) $F_{t_3,t_2} \circ F_{t_2,t_1} = F_{t_3,t_1}$ whenever $t_1, t_2, t_3 \in T$ and $t_1 \preceq t_2 \preceq t_3$,

(FL.3) $F_{t,t} = \mathrm{id}_{X_t}$ for every $t \in T$.

A *trajectory* of a flow $\mathcal{F} = (T, \mathbf{X}, \mathbf{F})$ is a single-valued map ξ such that $\mathrm{Dom}(\xi) = T$, $\xi(t) \in X_t$ for every $t \in T$, and $\xi(t) \in F_{t,s}(\xi(s))$ whenever $s, t \in T$ and $s \preceq t$.

If $\mathcal{F} = (T, \mathbf{X}, \mathbf{F})$ is a flow such that each X_t is a C^1 manifold, and ξ is a trajectory of $\mathcal{F}$, a *multidifferential of $\mathcal{F}$ along ξ* is a family $\mathcal{L} = \{\mathbf{L}_{t,s}\}_{s,t \in T, s \preceq t}$ such that

(MDF.1) $\mathbf{L}_{t,s} \in MD_{\xi(s),\xi(t)}(F_{t,s})$ whenever $s, t \in T$ and $s \preceq t$,

(MDF.2) $\mathbf{L}_{t,t} = \{\mathrm{id}_{T_{\xi(t)} X_t}\}$ for all $t \in T$,

(MDF.3) $\mathbf{L}_{t_3,t_2} \circ \mathbf{L}_{t_2,t_1} = \mathbf{L}_{t_3,t_1}$ whenever $t_1, t_2, t_3 \in T$ and $t_1 \preceq t_2 \preceq t_3$.

(Naturally, if $\mathbf{L}$, $\mathbf{L}'$ are subsets of $L(X,Y)$ and $L(Y,Z)$, respectively, where X, Y, Z are linear spaces, $\mathbf{L}' \circ \mathbf{L}$ denotes the set $\{L' \circ L : L \in \mathbf{L}, L' \in \mathbf{L}'\}$).

We call $\mathcal{F}$ *multidifferentiable along* ξ if there exists a multidifferential of $\mathcal{F}$ along ξ. We remark that multidifferentiability of a flow $\mathcal{F} = (T, \mathbf{X}, \mathbf{F})$ along a trajectory ξ of $\mathcal{F}$ implies in particular that the domain of each map $F_{t,s}$ is a full neighborhood of $\xi(s)$. (Indeed, according to our general conventions, $MD_{\xi(s),\xi(t)}(F_{t,s})$ stands for $MD_{\xi(s),\xi(t),X_s}(F_{t,s})$, which cannot be nonempty unless $\xi(s) \in \mathrm{Int}_{X_s}(\mathrm{Dom}(F_{t,s})))$. Moreover, even if $\xi(s) \in \mathrm{Int}_{X_s}(\mathrm{Dom}(F_{t,s}))$ for all t, s, the requirement that $\mathcal{F}$ be multidifferentiable along ξ is in general stronger than the condition that $F_{t,s}$ be multidifferentiable at $(\xi(s), \xi(t))$ for all (s,t), since it is possible to construct examples where $MD_{\xi(s),\xi(t)}(F_{t,s}) \neq \emptyset$ whenever $s, t \in T$ and $s \preceq t$, but a familiy $\mathcal{L}$ satisfying (MDF.1) and (MD.2) as well as the compatibility condition (MDF.3) does not exist.

Given finite-dimensional linear spaces Y_t for t in a totally ordered set T, and a family $\mathbf{L} = \{\mathbf{L}_{t,s}\}_{t,s \in T, s \preceq t}$ such that each $L_{t,s}$ is a subset of the space $L(Y_s, Y_t)$ of linear maps from Y_s to Y_t, a *compatible selection* of $\mathbf{L}$ is a family $L = \{L_{t,s}\}_{t,s \in T, s \preceq t}$ such that $L_{t,s}$ belongs to $\mathbf{L}_{t,s}$ for each t, s, and $L_{t_3,t_2} L_{t_2,t_1} = L_{t_3,t_1}$ whenever $t_1, t_2, t_3 \in T$ and $t_1 \preceq t_2 \preceq t_3$.

The following result is proved by an elementary argument using Zorn's Lemma.

Theorem 4.1 *If all the $\mathbf{L}_{t,s}$ are compact and nonempty, and the compatibility condition (MDF.3) holds, then $\mathbf{L}$ has a compatible selection. More generally, any compatible selection $\ddot{L}$ defined on a subset $\ddot{T}$ of T can be extended to a compatible selection on T.*

5 Point Variations of a Flow

Let M be a manifold of class C^1, let $\bar{x} \in M$, and let k be a positive integer. A *k-parameter point variation on M with base point $\bar{x}$* is a set-valued map $V \in SVM(\mathbb{R}^k_+ \times M, M)$ such that

(i) $\mathrm{Dom}(V)$ is a neighborhood of $(0, \bar{x})$ in $\mathbb{R}^k_+ \times M$,

(ii) $V(0, x) = x$ whenever $(0, x) \in \mathrm{Dom}(V)$.

We use $\mathrm{VAR}_k(M, \bar{x})$ to denote the set of all k-parameter point variations on M with base point $\bar{x}$, and write $\mathrm{VAR}(M, \bar{x}) \stackrel{\mathrm{def}}{=} \cup_{k=1}^{\infty} \mathrm{VAR}_k(M, \bar{x})$.

We call a $V \in \mathrm{VAR}_k(M, \bar{x})$ *nicely multidifferentiable* if there exists a k-tuple $(v_1, \ldots, v_k)$ of vectors in $T_{\bar{x}}M$ such that, if $L : \mathbb{R}^k \times T_{\bar{x}}M \mapsto T_{\bar{x}}M$ is the linear map given by $L(r_1, \ldots, r_k, u) = r_1 v_1 + \ldots + r_k v_k + u$, then $\{L\}$ belongs to $MD_{(0,\bar{x}),\bar{x},\mathbb{R}^k_+ \times M}(V)$. We use $NMD(V)$ to denote the set of all such k-tuples $(v_1, \ldots, v_k)$.

Now suppose $\mathcal{F} = (T, \mathbf{X}, \mathbf{F})$ is a flow such that T has a minimum and a maximum. Let ξ be a trajectory of $\mathcal{F}$. A *combined point variation* of $\mathcal{F}$ along ξ is a map $\mathcal{V}$, defined on a finite subset $|\mathcal{V}|$ of T, such that $\mathcal{V}(t) \in \mathrm{VAR}(X_t, \xi(t))$ for each $t \in |\mathcal{V}|$. (The set $|\mathcal{V}|$ is the *carrier* of $\mathcal{V}$). Given a combined point variation $\mathcal{V}$ of $\mathcal{F}$ along ξ, the *endpoint map* $\mathcal{E}_{\mathcal{V}}$ is defined as follows. Write $|\mathcal{V}| = \{t_1, \ldots, t_m\}$, where $t_1 \prec t_2 \prec \ldots \prec t_m$. Let $a = \min T$, $b = \max T$. Let $\mathcal{V}(t_j) = V_j$, and let k_j be such that $V_j \in \mathrm{VAR}_{k_j}(X_{t_j}, \xi(t_j))$. Let $k = k_1 + \ldots + k_m$, and write $\kappa_j = k_1 + \ldots + k_j$ for $j = 1, \ldots, m$. For $a \preceq t \prec t_1$, let $E_t(x) = F_{t,a}(x)$. Then define

$$E_{t_1}(\varepsilon_1, \ldots, \varepsilon_{k_1}, x) = V_1\Big(\varepsilon_1, \ldots, \varepsilon_{k_1}, F_{t_1,a}(x)\Big). \tag{5.14}$$

For $j = 1, \ldots, m-1$, $t_j \prec t \prec t_{j+1}$, define

$$E_t(\varepsilon_1, \ldots, \varepsilon_{\kappa_j}, x) = F_{t,t_j}\Big(E_{t_j}(\varepsilon_1, \ldots, \varepsilon_{\kappa_j}, x)\Big). \tag{5.15}$$

Then let

$$E_{t_{j+1}}(\varepsilon_1, \ldots, \varepsilon_{\kappa_{j+1}}, x) =$$
$$V_{j+1}\Big(\varepsilon_1, \ldots, \varepsilon_{k_{j+1}}, F_{t_{j+1},t_j}(E_{t_j}(\varepsilon_{k_j+1}, \ldots, \varepsilon_{\kappa_{j+1}}, x))\Big). \tag{5.16}$$

We then let

$$E_t(\varepsilon_1, \ldots, \varepsilon_k, x) = F_{t,t_m}\Big(E_{t_m}(\varepsilon_1, \ldots, \varepsilon_k, x)\Big) \text{ for } t_m \preceq t \preceq b. \tag{5.17}$$

Finally, $\mathcal{E}_{\mathcal{V}}$ is, by definition, the map E_b. Then $\mathcal{E}_{\mathcal{V}} \in SVM(\mathbb{R}_+^k \times X_a, X_b)$.

If $\mathcal{F} = (T, \mathbf{X}, \mathbf{F})$ is a flow such that each X_t is a C^1 manifold, a *point variational vector* for $\mathcal{F}$ along a trajectory ξ of $\mathcal{F}$ is a pair (t, v) such that $t \in T$ and $v \in T_{\xi(t)}X_t$. Let W be a set of variational vectors for $\mathcal{F}$ along ξ. Let $\mathcal{R}$ be a subset of X_b, where $b = \max T$. Let $a = \min T$. We say that W is *$\mathcal{R}$-compatible* if, whenever $t_1 \prec t_2 \prec \ldots \prec t_m$, and $W_1, \ldots, W_m$ are finite subsets of W such that $W_j \subseteq \{t_j\} \times T_{\xi(t_j)}X_{t_j}$ for $j = 1, \ldots, m$, it follows that there exists a combined variation $\mathcal{V}$ of $\mathcal{F}$ along ξ with carrier $\{t_1, \ldots, t_m\}$, such that

(i) each $\mathcal{V}(t_j)$ is nicely multidifferentiable, and there is a $\mathbf{v}_j \in NMD(V(t_j))$ in which all the vectors v for which $(t_j, v) \in W_j$ occur, and

(ii) the image of $\mathcal{E}_{\mathcal{V}}$ is contained in $\mathcal{R}$.

6 The Maximum Principle

If M is a C^1 manifold, $S \subseteq M$ and $q \in S$, an *approximating cone* for S at q is a closed convex cone C in the tangent space T_qM such that there exist a neighborhood U of 0 in T_qM and a continuous map $\varphi : C \cap U \mapsto S$ with the property that $\varphi(0) = q$ and $\varphi(v) = q + v + o(||v||)$ as $v \to 0$, $v \in C$. (The condition that $\varphi(v) = q+v+o(||v||)$ is in principle coordinate-dependent, but one can easily show that it is in fact independent of the choice of coordinates. It can be stated invariantly as follows: for every $\psi \in C^1(M, \mathbb{R})$, $\psi(\varphi(v)) - \psi(q) - v\psi = o(||v||)$ as $v \to 0$).

If X, Y are linear spaces and $A : X \mapsto Y$ is a linear map, we use $A^\dagger$ to denote the transpose of A, so $A^\dagger$ maps the dual space of Y to the dual space of X.

Theorem 6.1 *Let $\mathcal{F} = (T, \mathbf{X}, \mathbf{F})$ be a flow such that T has a minimum a and a maximum b, and each X_t is a C^1 manifold. Let ξ be a trajectory of $\mathcal{F}$, and let $\mathbf{L}$ be a multidifferential of $\mathcal{F}$ along ξ. Let $\mathcal{R}$ be a subset of X_b, and let W be an $\mathcal{R}$-compatible set of variational vectors for $\mathcal{F}$ along ξ. Let S be a subset of X_b such that $\xi(b) \in S$. Assume that $\mathcal{R} \cap S = \{\xi(b)\}$. Let C be an approximating cone for S at $\xi(b)$, such that C is not a linear subspace of $T_{\xi(b)}X_b$. Then there exist a nontrivial linear functional $z : T_{\xi(b)}X_b \mapsto \mathbb{R}$ and a compatible selection $\{L_{t,s}\}_{s,t\in T, s\preceq t}$ of $\mathbf{L}$ such that, if we define*

$$\zeta(t) = L_{b,t}^\dagger z \text{ for } t \in T, \tag{6.18}$$

then

$$z(v) \leq 0 \text{ for all } v \in C, \tag{6.19}$$

and

$$\zeta(t)(w) \geq 0 \text{ whenever } t \in T, (t, w) \in W. \tag{6.20}$$

7 Applications

Having stated the maximum principle in a rather abstract form, we now have to show how all the well known versions, as well as many new results, follow from our result. A detailed account is now in preparation. Here we shall limit ourselves to discussing a few examples.

To begin with, we observe that minimization problems of the kind usually studied in optimal control theory can be reduced to separation problems of the type discussed in Theorem 6.1. For example, suppose the variable x takes values in $\mathbb{R}^n$, and we want to minimize an integral $\int_a^b L(x(t), u(t), t)dt$, subject to a dynamical equation $\dot{x} = f(x, u, t)$, and boundary conditions $x(a) = \bar{x}$, $x(b) = \hat{x}$. We can then consider the new control system in $\mathbb{R}^{n+1}$ with dynamics $\dot{x} = f(x, u, t)$, $\dot{y} = L(x, u, t)$, in which the "running cost" has been added as an extra state variable. A trajectory-control pair $(\xi, u(\cdot))$ of the original system is a minimizer iff the "augmented" trajectory $\tilde{\xi} : [a, b] \mapsto \mathbb{R}^{n+1}$ given by $\tilde{\xi}(t) = (\xi(t), \int_a^t L(\xi(s), u(s))ds)$ satisfies a separation condition. Precisely, let $\mathcal{R}$ be the reachable set over the interval $[a, b]$ for the new system from the point $(\bar{x}, 0)$, and let S be the set

$\{\hat{x}\}\times]-\infty,c]$, where $c=\int_a^b L(\xi(t),u(t))dt$. Then $(\xi,u(\cdot))$ is a minimizer iff $\mathcal{R}\cap S=\{(\hat{x},c)\}$.

A similar reduction is possible for more general optimal control problems, with endpoint constraints of the form $x(a)\in A$, $x(b)\in B$, and even for constraints of the type $(x(a),x(b))\in Q$, where Q is a subset of $\mathbb{R}^n\times\mathbb{R}^n$. As in the simple example of fixed endpoints. This reduction *always* leads to a set S whose approximating cone —if it exists— is not a linear subspace.

From now on, we take the reduction for granted, and concentrate on separation problems.

Consider first the "classical" case of a system $\dot{x}=f(x,u,t)$, $x\in\mathbb{R}^n$, $u\in U$, in which it is assumed that, for the reference control $u_*:[a,b]\mapsto U$ and corresponding trajectory $\xi_*:[a,b]\mapsto\mathbb{R}^n$, the time-varying vector field $(x,t)\mapsto f_*(x,t)\stackrel{\text{def}}{=}f(x,u_*(t),t))$ satisfies C^1-Carathéodory conditions in a tube $\mathcal{T}=\{(x,t):a\le t\le b,||x-\xi_*(t)||\le\varepsilon\}$. (This means that $f_*(x,t)$ is of class C^1 in x for each fixed t, measurable in t for each fixed x, and such that $||\frac{\partial f_*}{\partial x}(x,t)||$ is bounded by an integrable function of t). Then f_* gives rise to a flow $\mathcal{F}$ consisting of C^1 maps, and its is then obvious that $\mathcal{F}$ is multidifferentiable along ξ_*, with a multidifferential $\mathbf{L}=\{\mathbf{L}_{t,s}\}$ consisting of singleton sets $\mathbf{L}_{t,s}=\{L_{t,s}\}$. Moreover, the maps $L_{t,s}$ are obtained by solving the variational equation

$$\dot{M}(t)=\frac{\partial f_*}{\partial x}(\xi_*(t),t).M(t)\,. \tag{7.21}$$

Precisely, $L_{t,s}$ is the value at t of the solution $M(\cdot)$ of (7.21) such that $M(s)=$ identity. The condition $z(t)=L_{b,t}^{\dagger}z(b)$, when differentiated, becomes the familiar adjoint equation. We can define a set of point variational vectors as follows: for each $(t_0,u_0)\in[a,b]\times U$, let

$$v(t_0,u_0)=f(\xi_*(t_0),u_0,t_0)-f_*(\xi_*(t_0),t_0)\,. \tag{7.22}$$

Then we would like to define a variation V_{t_0,u_0} by letting $V_{t_0,u_0}(\varepsilon,x)$ be the point obtained by starting at x at time t_0, following "the" trajectory of $\dot{\xi}=f(\xi,u_0,t)$ up to time $t_0+\varepsilon$, and then going backwards along the trajectory of f_* up to time t. This will easily yield a nicely multidifferentiable variation with variational vector $(t_0,v(t_0,u_0))$, if f_* and the vector field $(x,t)\mapsto f(x,u_0,t)$ are continuous at (x_0,t_0). (Notice, however, that even in this case the map V_{t_0,u_0} can be set-valued, because we are not assuming that $f(x,u_0,t)$ is Lipschitz in x, so the trajectories need not be unique. The regularity of V_{t_0,x_0} follows from standard properties of solutions of ordinary differential equations, together with the fact that continuous vector fields can be approximated by smooth ones, so the corresponding flow maps can be approximated by single-valued ones). So, in the special case when $f(x,u,t)$ is jointly continuous in (x,t) for each fixed u, and f_* is also jointly continuous, we get the Hamiltonian minimization condition in the strong form that $z(t).f(\xi_*(t),u,t) \geq z(t).f(\xi_*(t),u_*(t),t)$ for all (u,t). In the more general case when each vector field $(x,t) \mapsto f(x,u,t)$ is only assumed to be continuous in x, measurable in t, and bounded by an integrable function of t, then well known techniques, based on the Scorza-Dragoni theorem, make it possible to establish the nice multidifferentiability of V_{t_0,u_0} for $t_0 \in E \cap E(u_0)$, where E is a subset of $[a,b]$ of full measure, and $E(u_0)$ is a u_0-dependent subset of full measure. This yields the Hamiltonian minimization condition in the "weak" form:

$$(\forall u \in U)\Big(z(t).f(\xi_*(t),u,t) \geq z(t).f_*(\xi_*(t),t) \text{ for a.e. } t \in [a,b]\Big). \quad (7.23)$$

Under suitable separability conditions (e.g. if $f(x,u,t)$ is independent of t, or if there exists a countable subset U_0 of U such that $\{f(\xi_*(t),u,t) : u \in U_0\}$ is dense in $\{f(\xi_*(t),u,t) : u \in U\}$) we can go from this to the "strong" minimization condition

$$\min\{z(t).f(\xi_*(t),u,t) : u \in U\} = z(t).f_*(\xi_*(t),t) \text{ for a.e.} t \in [a,b]. \quad (7.24)$$

If the reference vector field f_* is only assumed to be Lipschitz with respect to x, with a Lipschitz constant bounded by an integrable function of t, then all the above considerations apply, except only that now the flow maps $F_{t,s}$ arising from f_* are just Lipschitz. A theorem of J. Warga ([24, 25, 26, 27]) allows us to find a multidifferential of this flow. All we have to do is to rewrite (7.21) in the form $\dot{M}(t) = A(t).M(t)$, where A is an arbitrary selection of the map $t \mapsto \partial f_*(\xi_*(t),t)$, and $\partial f_*(x,t)$ is the Clarke generalized Jacobian of $f_*(\cdot,t)$ at x. We then define $\mathbf{L}_{t,s}$, for $s \leq t$, to be the set of all $M(t)$, ranging over all measurable selections $A(\cdot)$, and all solutions $M(\cdot)$ for which $M(s) =$ identity. The resulting family $\mathbf{L}$ is a multidifferential of $\mathcal{F}$ along ξ_*. The needle variations are treated exactly as in the classical case, and we get the Lojasiewicz version of the nonsmooth maximum principle, in which the reference vector field is Lipschitz and the other ones are just continuous.

We now discuss Examples 1.3 and 1.4 in the light of Theorem 6.1. The reference vector field is given by $f(x) = |x|^\alpha \sin \frac{1}{x}$, where $\alpha > 1$, and the reference trajectory is $\xi_*(t) \equiv 0$. Given any trajectory $[0,1] \ni t \mapsto x(t)$ of f, Gronwall's inequality yields, to begin with, $|x(t)| \leq e^{t-s}|x(s)|$ for $0 \leq s \leq t \leq 1$, as long as $|x(t)| \leq 1$ for all $t \in [0,1]$. This implies that $|x(t)| \leq e^{t-s}|x(s)|$ whenever $0 \leq s \leq t \leq 1$, if $|x(s)| \leq e^{-1}$. But then $|x(t) - x(s)| \leq C|x(s)|^\alpha$, where $C > 0$ is a constant. This means that the flow maps $F_{t,s}$ satisfy $F_{t,s}(x) = x + o(|x|)$. In other words, the flow $\mathcal{F}$ generated by our reference vector field is multidifferentiable along our reference trajectory, with multidifferential $\mathbf{L}$ given by $\mathbf{L}_{t,s} = \{\text{identity}\}$. This gives the "adjoint equation" $z(t) \equiv$ constant, providing a rigorous justification to the formal arguments of the introduction.

So far, we have only considered examples where the integrated form $z(s) = L^{\dagger}_{t,s}z(t)$ of the adjoint equation can be differentiated to yield an adjoint equation of the standard form. It should be clear from our discussion that in our approach there is no need for this differentiated form to exist. To illustrate this, we consider Examples 1.1 and 1.2. In Example 1.1, we can write the dynamics in control form by taking as control space the set U of all ordered pairs of unit vectors, and defining, for each $u = (u_1, u_2) \in U$, $f(x,y,u) = c_1u_1$ if $y \geq 0$, and $f(x,y,u) = c_2u_2$ if $y < 0$. Given A, B lying on opposite sides of the x axis, and a C on the x axis, the broken line ξ obtained by concatenating the segments from A to C and from C to B is an integral curve of $f(\cdot,u)$ for a unique constant control $u \in U$. Even though the vector field $f(\cdot,u)$ is discontinuous, the corresponding flow maps $F_{s,t}$ are well defined. If we choose the zero of the time axis so that $\xi(0) = A$, it is easy to see that, if τ denotes the time such that $\xi(\tau) = C$, then every map $F_{t,s}$ is affine linear near $\xi(s)$, as long as $s \neq \tau \neq t$. So the adjoint equation in integrated form makes perfect sense, and yields an adjoint vector $t \mapsto z(t)$ that is actually constant for $t < \tau$ and also for $t > \tau$, but jumps at $t = \tau$. A somewhat long but straightforward calculation yields the expected conclusion, namely, that for ξ to be optimal it has to satisfy the condition of Snell's law of refraction.

As for Example 1.4, the situation is similar, although the calculations are more complicated and the behavior of the adjoint vector is rather different. We omit the details, which are rather straightforward, and limit ourselves to pointing out that the main step is to choose a good way to realize the cycloids that are optimal trajectories in the upper and lower half planes in terms of flows of vector fields depending on a parameter u, so that the broken cycloids described in the introduction become trajectories of a vector field corresponding to a pair (u_1, u_2) of u-values, similarly to what we did in the case of the broken lines. It turns out that the adjoint equation in integrated form still makes sense, and has a solution $t \mapsto z(t)$ that goes to infinity as $t \to \tau$, where τ is the time when our broken cycloid ξ crosses the x axis. If, however, one takes the totally ordered set T of Theorem 6.1 to be the *complement* of $\{\tau\}$ in $\mathrm{Dom}(\xi)$, then z is perfectly well behaved there. The final result of the application of Theorem 6.1 is an extra necessary condition for an optimum: the optimal cycloids in the upper half plane, given as graphs of functions $x \mapsto y(x)$, satisfy $y(1 + \frac{dy}{dx}^2) = \mathit{constant} > 0$, and those in the lower half plane satisfy the same condition, except that now the constant is negative. The extra condition that links the two parts of an optimal broken cycloid $\xi : [0,b] \mapsto \mathbb{R}^2$ crossing the x axis is that the two constants have the same absolute value, i.e. that $|y|(1 + \frac{dy}{dx}^2)$ remain constant throughout the interval $[0,b]$. (Notice that, as $t \to \tau$, $y(t) \to 0$ but $\frac{dy}{dx}^2 \to \infty$, since ξ approaches the x axis vertically).

To conclude our list of examples, we outline the results, described in more detail in our 1996 CDC paper [21], about a maximum principle for systems of differential inclusions. The setting here is a family $\mathbf{F} = \{F_u\}_{u \in U}$ of set-valued maps from $\mathbb{R}^n \times [a, b]$ to $2^{\mathbb{R}^n}$. We assume that each F_u has closed nonempty values and is almost lower semicontinuous (ALSC) and locally integrably lower bounded (LILB). (A map $F : \mathbb{R}^n \times [a, b] \mapsto 2^{\mathbb{R}^n}$ is LILB if for every compact subset K of $\mathbb{R}^n$ there is an integrable function $\rho : [a, b] \mapsto [0, \infty]$ such that $d(0, F(x, t)) \leq \rho(t)$ whenever $x \in K, t \in [a, b]$. We call F ALSC if $[a, b]$ can be expressed as a union $\cup_{j=0}^{\infty} J_j$ of subsets such that each J_j for $j > 0$ is compact, J_0 has measure zero, and for each $j > 0$ the restriction $F \lceil \mathbb{R}^n \times J_j$ is lower semicontinuous. Recall that if X, Y are metric spaces and $F \in SVM(X, Y)$, then F is *lower semicontinuous* if $F^{-1}(\Omega)$ is open in X whenever Ω is open in Y). Also, we assume that $\mathbf{F}$ is closed under interval substitutions, i.e. that if I is a subinterval of $[a, b]$, $u_1, u_2 \in U$, and we define G by $G(x, t) = F_{u_1}(x, t)$ if $t \in I$, and $G(x, t) = F_{u_2}(x, t)$ if $t \notin I$, then $G = F_u$ for some $u \in U$.

Suppose $\xi_* : [a, b] \mapsto \mathbb{R}^n$ is a trajectory of F_{u_*} for some $u_* \in U$. Assume, moreover, that F_{u_*} is *integrably pseudolipschitz* along ξ_*. (This means that there exist an integrable function $k(\cdot) : [a, b] \mapsto [0, \infty]$ and an $\varepsilon > 0$ such that, for almost all t, the inclusion

$$F(x, t) \cap \bar{B}(\dot{\xi}_*(t), \varepsilon) \subseteq F(x', t) + \bar{B}(0, k(t)||x' - x||) \tag{7.25}$$

holds whenever $||x - \xi_*(t)|| \leq \varepsilon$ and $||x' - \xi_*(t)|| \leq \varepsilon$. Here $\bar{B}(x, r)$ denotes the closed ball with center x and radius r).

Under these hypotheses, one can show that the flow generated by the differential inclusion $\dot{x} \in F_{u_*}(x, t)$ is multidifferentiable along ξ_*, and one can describe a way to compute a multidifferential. (This requires, to begin with, the construction of an integrably Lipschitz vector field $(x, t) \mapsto f(x, t)$ which is a selection of the convexification of F_{u_*} and is such that $\dot{\xi}_*(t) = f(\xi_*(t), t)$ for a.e. t. Once this is done, the flow of f is clearly semidifferentiable, and a semidifferential $\mathbf{L}$ of this flow is given by Warga's theorem quoted above. Finally, to prove that the flow of F_{u_*} is multidifferentiable with multidifferential $\mathbf{L}$, one approximates the trajectories of f by trajectories of F_{u_*}, using a variant of the results of [11, 12] on uniform approximation of relaxed trajectories by non-relaxed ones).

Having shown the multidifferentiability of the reference flow, the next step is to construct needle variations. For this purpose, we use a modification of the theory of "conically continuous" selections of lower semicontinuous set-valued maps, due to A. Bressan (cf. [4, 5, 6, 7, 8] and also Pucci [18]). The main point is that, if F is ALSC, then for almost every t_0 the following is true: for each $x_0 \in \mathbb{R}^n$ and each $y_0 \in F(x_0, t_0)$, there exists a Borel$\times$Lebesgue-measurable selection f of F such that $f(x_0, t_0) = y_0$, f satisfies integral bounds $||f(x, t)|| \leq \psi_K(t)$, with $\psi_K \in L^1([a, b])$, on each compact subset K of $\mathbb{R}^n$, and f has the following continuity property: for each K, the map $\xi \mapsto \int_a^{\cdot} f(\xi(s), s)ds$ is continuous from the space A_K of all absolutely continuous $\xi : [a,b] \mapsto K$ such that $||\dot{\xi}(t)|| \leq \psi_K(t)$ for a.e. t to the space $C^0([a,b],\mathbb{R}^n)$ of continuous maps from $[a,b]$ to $\mathbb{R}^n$. (Both spaces are endowed with the sup norm). It turns out that this continuity property suffices for the O.D.E. $\dot{x} = f(x,t)$ to have the same properties as O.D.E.'s with a continuous right-hand side. In particular, this O.D.E. has local existence of solutions, the flow maps $\Phi_{t,s}$ are upper semicontinuous and compactly valued for small enough $|t-s|$, and they can be approximated by continuous single-valued maps, obtained by regularizing f. Finally, f can also be chosen so that f is "almost continuous" at (x_0,t_0), in the sense that there exists a measurable subset E of $[a,b]$ such that t_0 is a point of density of E and $f\lceil\mathbb{R}^n \times E$ is continuous at (x_0,t_0). This suffices to construct, for every $u \in U$, and for almost all $t \in [a,b]$, needle variations generating the variational vectors (t,y) for all $y \in F_u(\xi_*(t),t)$. (The order of the quantifiers matters here. The bad set of measure zero may depend on u). The compatibility condition is easily verified. The conclusion is that, if $\mathcal{R}$ denotes the reachable set from $\xi_*(a)$ over $[a,b]$, and $S \subseteq \mathbb{R}^n$ is such that $\xi_*(b) \in S$ and S has an approximating cone C at $\xi_*(b)$ which is not a linear subspace, then $\mathcal{R} \cap S = \{\xi_*(b)\}$ implies that for every locally integrably Lipschitz selection f of the convexification of F_{u_*} such that ξ is an integral curve of f there exists a nontrivial solution $t \mapsto z(t)$ of the adjoint inclusion $-\dot{z}(t) \in z(t)\partial_x f(\xi_*(t),t)$ such that, for each $u \in U$, we have

$$\inf\Big\{z(t).y : y \in F_u(\xi_*(t),t)\Big\} \geq z(t).\dot{\xi}_*(t) \ \text{ for a.e. } t \in [a,b]\,. \tag{7.26}$$

Under suitable separability conditions the exceptional set of measure zero for which the conclusion of (7.26) fails can be taken to be independent of u, and one then gets the stronger conclusion that

$$\inf\Big\{z(t).y : y \in \bigcup_{u\in U} F_u(\xi_*(t),t)\Big\} \geq z(t).\dot{\xi}_*(t) \ \text{ for a.e. } t \in [a,b]\,. \tag{7.27}$$

References

[1] A.A. Agrachev and R. V. Gamkrelidze. A second-order optimality principle for a time-optimal problem. *Math. Sbornik* **100** (1976), 142.

[2] A.A. Agrachev and R. V. Gamkrelidze. The exponential representation of flows and the chronological calculus. *Math. Sbornik* **109** (1978), 149.

[3] A.A. Agrachev and R. V. Gamkrelidze. The Morse index and the Maslov index for smooth control systems. *Doklady Akad. Nauk USSR* **287** (1986).

[4] A. Bressan. On differential relations with lower continuous right-hand side. An existence theorem. *J. Diff. Equations* **37** (1980), 89-97.

[5] A. Bressan. Solutions of lower semicontinuous differential inclusions on closed sets. *Rend. Sem. Mat. Univ. Padova* **69** (1983), 99-107.

[6] A. Bressan. Directionally continuous selections and differential inclusions. *Funk. Ekvac.* **31** (1988), 459-470.

[7] A. Bressan. On the qualitative theory of lower semicontinuous differential inclusions. *J. Diff. Equations* **77** (1989), 379-391.

[8] A. Bressan. Upper and lower semicontinuous differential inclusions: a unified approach. *Nonlinear Controllability and Optimal Control.* (H.J. Sussmann, Ed.). New York: M. Dekker Inc., 1990. 21-31.

[9] F.H. Clarke. The Maximum Principle under minimal hypotheses. *S.I.A.M. J. Control Optim.* **14** (1976), 1078-1091.

[10] F.H. Clarke. *Optimization and Nonsmooth Analysis.* New York: Wiley, 1983.

[11] R.M. Colombo, A. Fryszkowski, T. Rzežuchowski and V. Staicu. Continuous selections of solution sets of Lipschitzean differential inclusions. *Funkcialaj Ekvacioj* **34** (1991), 321-330.

[12] A. Fryszkowski and T. Rzežuchowski. Continuous version of the Filippov-Ważewski relaxation theorem. *J. Diff. Equations* **94** (1993), 254-265.

[13] H.J. Kelley, R. E. Kopp and H. G. Moyer. Singular extremals. *Topics in Optimization.* (G. Leitman Ed.). New York: Academic Press, 1967.

[14] H.W. Knobloch. *High Order Necessary Conditions in Optimal Control.* New York: Springer-Verlag, 1975.

[15] A.J. Krener. The higher order maximum principle and its application to singular extremals. *SIAM J. Control Opt.* **15** (1977), 256-293.

[16] S. Lojasiewicz, Jr. Local controllability of parametrized differential equations. Preprint, 1996.

[17] L.S. Pontryagin, V.G. Boltyanskii, R.V. Gamkrelidze and E.F. Mischenko. *The Mathematical Theory of Optimal Processes.* New York: Wiley, 1962.

[18] A. Pucci. Sistemi di equazione differenziale con secondo membro discontinuo rispetto all'incognita. *Rend. Ist. Mat. Univ. Trieste* **3** (1971), 75-80.

[19] H.J. Sussmann. A strong version of the Maximum Principle under weak hypotheses. *Proc. 33rd IEEE Conference on Decision and Control 1994*, 1950-1956.

[20] H.J. Sussmann. A strong version of the Lojasiewicz Maximum Principle. *Optimal Control of Differential Equations.* (N.H. Pavel, Ed.). New York: M. Dekker, 1994.

[21] H.J. Sussmann. A strong maximum principle for systems of differential inclusions. To appear *Proceedings of the 35th IEEE Conference on Decision and Control 1996.*

[22] H.J. Sussmann. The maximum principle of optimal control theory. Preprint, 1996.

[23] J.C. Tougeron. *Idéaux de Fonctions Différentiables.* New York: Springer-Verlag, 1972.

[24] J. Warga. Fat homeomorphisms and unbounded derivate containers. *J. Math. Anal. Appl.* **81** (1981), 545-560.

[25] J. Warga. Controllability, extremality and abnormality in nonsmooth optimal control. *J. Optim. Theory Applic.* **41** (1983), 239-260.

[26] J. Warga. Optimization and controllability without differentiability assumptions. *SIAM J. Control and Optimization* **21** (1983), 837-855.

[27] J. Warga. Homeomorphisms and local C^1 approximations. *Nonlinear Anal. TMA* **12** (1988) 593-597.

[28] H. Whitney. Analytic extensions of differentiable functions defined in closed sets. *Trans. Amer. Math. Soc.* **36** (1934), 63-89.

Department of Mathematics, Rutgers University, New Brunswick, New Jersey 08903

Nonlinear Input-output Stability and Stabilization

A.R. Teel[1]

1 Introduction

In this paper we focus on the input-output method of stability analysis for nonlinear systems. The fundamental ideas underlying this method were developed in the 1960's by authors such as Zames and Sandberg (see [8], [17], [9], [18] and the references therein). Some notable additional insight was provided by Safonov [7] in 1980. In the last five years, this area has been greeted with renewed enthusiasm by nonlinear control researchers searching for additional tools that can be used as the basis for nonlinear control law synthesis. Our goal here is to illustrate this heightened interest by describing a selected set of recent results on input-output stability and stabilization, especially in the $\mathcal{L}_\infty$ setting.

The input-output method is relevant for the interconnection of two dynamical systems as illustrated in figure 1. This is a generalization of the additive input interconnection, shown in figure 2, which was commonly studied in the 1960's. The problems we will be interested in are 1) stability, e.g., "do bounded inputs produce bounded outputs"?; 2) performance, e.g., "is there an acceptable amplification from the size of the input to the size of the output"?; and, 3) synthesis, i.e., given system 1, design system 2 to induce stability and/or performance.

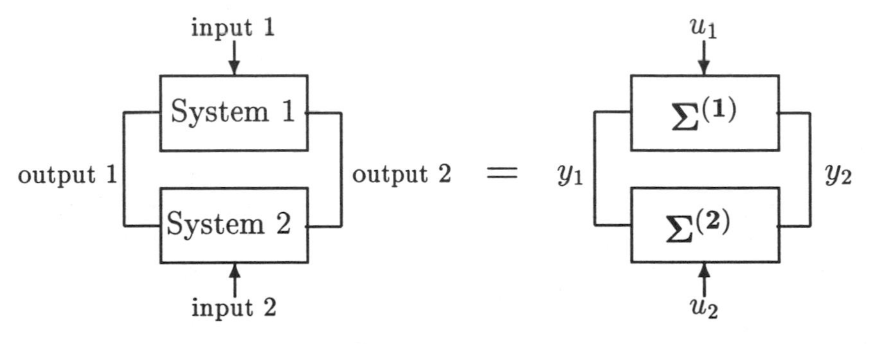

Figure 1: General interconnection.

[1]Research supported in part by the NSF under grants ECS-9309523 and ECS-9502034 and by the AFOSR under grant F49620-96-1-0144.

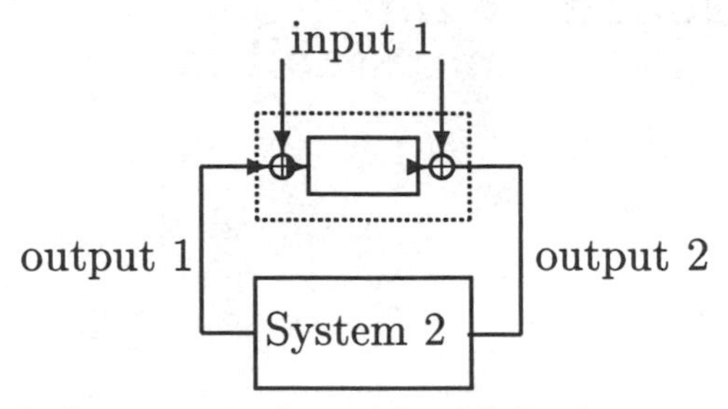

Figure 2: Interconnection with additive inputs.

To quantify these properties, norms (generically represented by the symbol $||\cdot||$) are used. We will use capital, script letters for the associated (extended) normed spaces, e.g., $u \in \mathcal{U}_e$ and if $||u|| < \infty$ then $u \in \mathcal{U}$. A dynamical system is a subset of an extended normed space. We will assume we have a truncation-type operator π_τ, where $\tau \in [0, \infty)$, to extract finite measurement information from elements that might otherwise have infinite norm. We will use $z_\tau := \pi_\tau z$. It is not necessary that $||z_\tau|| < \infty$, i.e., $z_\tau \in \mathcal{Z}$, for all $\tau \in [0, \infty)$.

Stability will be described in terms of so-called class-$\mathcal{G}$ functions. A continuous function $\gamma : \mathbb{R}_{\geq 0} \to \mathbb{R}_{\geq 0}$ is said to be of class $\mathcal{G}$ if it is zero at zero and is nondecreasing. (We will also eventually use class-$\mathcal{G}_+$ functions: γ is nondecreasing; and class-$\mathcal{K}_\infty$ functions: γ is zero at zero, strictly increasing and unbounded, which implies that it has a class-$\mathcal{K}_\infty$ inverse on $\mathbb{R}_{\geq 0}$.)

Definition 1.1 A dynamical system $\mathbf{\Sigma}$ is **stable** if there exists $\gamma \in \mathcal{G}$ such that, for each τ and $z := (u, y) \in \mathbf{\Sigma}$ satisfying $z_\tau \in \mathcal{Z}$, we have $||y_\tau|| \leq \gamma(||u_\tau||)$.

Specific "norms" that we will use frequently when assessing stability are

$$||z||_\infty := \sup_{t \geq 0} |z(t)| \quad , \qquad ||z||_a := \limsup_{t \to \infty} |z(t)| \tag{1.1}$$

where $|\cdot|$ is (equivalent to) the Euclidean norm.

We will study stability for the system in figure 1 by characterizing the properties of its component systems $\mathbf{\Sigma}^{(1)}$ and $\mathbf{\Sigma}^{(2)}$. The composite dynamical system has input $u = (u_1, u_2)$ and output $y = (y_1, y_2)$. By defining the *graphs* of the systems $\mathbf{\Sigma}^{(i)}$ as

$$\mathcal{G}^{(i)}(u_i) := \left\{ (y_1, y_2) \in \mathcal{Y}_{1e} \times \mathcal{Y}_{2e} : \ (u_i, y_1, y_2) \in \mathbf{\Sigma}^{(i)} \right\} ,$$

we see that

$$y \in \mathcal{G}^{(1)}(u_1) \cap \mathcal{G}^{(2)}(u_2) . \tag{1.2}$$

Part I: Input-Output Stability

2 Generalized, Nonlinear Small Gain Theorem

Drawing upon the input-output stability work of the 1960's, Safonov, in 1980, formulated a conceptually simple sufficient condition for input-output stability for the system in figure 1 in terms of the graphs of the systems $\mathbf{\Sigma}^{(i)}$. It was (essentially) the following:

Theorem 2.1 [[7]] *If there exist $\gamma \in \mathcal{G}$ and functionals $\lambda^{(1)}, \lambda^{(2)} : \mathcal{Z} \to \mathbb{R}$ such that, for each τ and $z = (u, y) \in \mathcal{Z}_e$ satisfying $z_\tau \in \mathcal{Z}$,*

$$
\begin{aligned}
& y \in \mathcal{G}^{(1)}(u_1) \Longrightarrow \lambda^{(1)}(z_\tau) \leq 0 \\
& y \in \mathcal{G}^{(2)}(u_2) \Longrightarrow \lambda^{(2)}(z_\tau) \leq 0 \\
& \max\left\{\lambda^{(1)}(z_\tau), \lambda^{(2)}(z_\tau)\right\} \leq 0 \Longrightarrow ||y_\tau|| \leq \gamma\left(||u_\tau||\right),
\end{aligned}
$$

then the interconnection is stable with gain γ.

While this theorem provides a nice theoretical tool for guaranteeing stability, it opens up the question: "how does one construct the functionals $\lambda^{(i)}$"? In [7], Safonov constructed functionals for an ordinary differential equation (system 1 is a multi-channel integrator with initial condition as input and system 2 is the function on the right-hand side of the differential equation) to guarantee stability in the $|| \cdot ||_\infty$ norm under the same assumptions that one makes to guarantee global stability using Lyapunov's method. Safonov also constructed functionals using the idea of a dynamic conic sector, generalizing the static conic sector idea that was put forth in the 1960's. The dynamic conic sectors of [7] can be further generalized using the nonlinear, small gain idea proposed by Mareels and Hill in [5]. We now state such a generalization (for more details see [13].) Let $\mathbf{C}$ and $\mathbf{R}$ be mappings from $\mathcal{Y}$ to some finite norm subspace and suppose there exists $\alpha \in \mathcal{K}_\infty$ such that

$$
\alpha(||y_\tau||) \leq \max\left\{||\mathbf{R}(y_\tau)||, ||\mathbf{C}(y_\tau)||\right\} . \tag{2.1}
$$

Define

$$
\left.
\begin{aligned}
\lambda^{(1)}(z_\tau) &:= ||\mathbf{C}(y_\tau)|| - \max\left\{\phi_1(||u_{1_\tau}||), \gamma^{(1)}(||\mathbf{R}(y_\tau)||)\right\} \\
\lambda^{(2)}(z_\tau) &:= ||\mathbf{R}(y_\tau)|| - \max\left\{\phi_2(||u_{2_\tau}||), \gamma^{(2)}(||\mathbf{C}(y_\tau)||)\right\}
\end{aligned}
\right\} \tag{2.2}
$$

where $\gamma^{(1)}, \gamma^{(2)}, \phi_1, \phi_2 \in \mathcal{G}$. The following result, which we will call the *generalized, nonlinear small gain theorem*, addresses when these two functionals can be used to assert stability with the help of theorem 1.

Theorem 2.2 *If* $\gamma^{(1)} \circ \gamma^{(2)}(s) < s$ *for all* $s > 0$ *then there exists* $\gamma \in \mathcal{G}$ *such that* $\max\left\{\lambda^{(1)}(z_\tau), \lambda^{(2)}(z_\tau)\right\} \leq 0 \Longrightarrow ||y_\tau|| \leq \gamma(||u_\tau||)$.

The analysis used to prove this result is straightforward. First note that $\max\left\{\lambda^{(1)}(z_\tau), \lambda^{(2)}(z_\tau)\right\} \leq 0$ implies

$$\begin{aligned} ||\mathbf{C}(y_\tau)|| &\leq \max\left\{\phi_1(||u_{1_\tau}||), \gamma^{(1)}(||\mathbf{R}(y_\tau)||)\right\} \\ ||\mathbf{R}(y_\tau)|| &\leq \max\left\{\phi_2(||u_{2_\tau}||), \gamma^{(2)}(||\mathbf{C}(y_\tau)||)\right\} . \end{aligned} \tag{2.3}$$

Substituting the second inequality into the first, we get the inequality

$$a \leq \max\{b, \alpha(a)\} \tag{2.4}$$

where $a = ||\mathbf{C}(y_\tau)||$, $b = \max\left\{\phi_1(||u_{1_\tau}||), \gamma^{(1)}\left(\phi_2(||u_{2_\tau}||)\right)\right\}$ and $\alpha = \gamma^{(1)} \circ \gamma^{(2)}$. Assuming $\alpha(s) < s$ for all $s > 0$, we must have $a \leq b$. Otherwise, we would have $a \leq \alpha(a)$ which is a contradiction. We can use a similar argument to bound $||\mathbf{R}(y_\tau)||$ and then we can use (2.1) to bound $||y_\tau||$.

Remark 2.1 For later use we note that, by the same reasoning, if $\alpha(s) < s$ for $s \in (\underline{s}, \overline{s})$ then (2.4) and $a, b \in [0, \overline{s})$ imply $a \leq \max\{\underline{s}, b\}$.

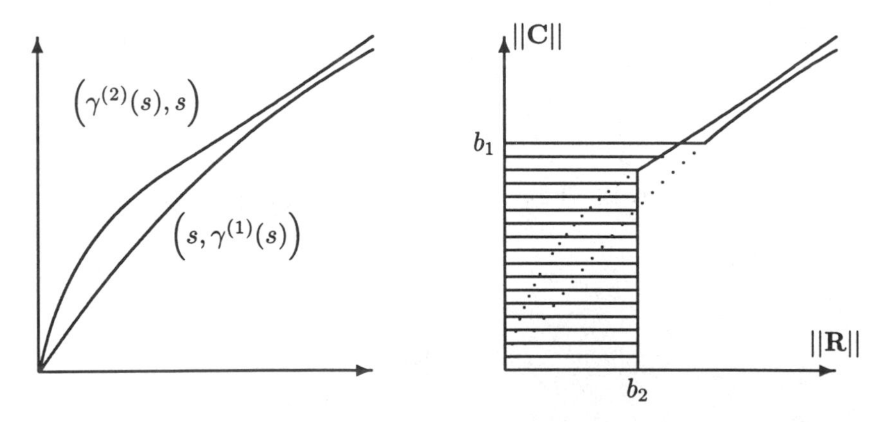

Figure 3: Graphical interpretation of small gain condition.

The small gain condition in Theorem 2.2 also has a straightforward graphical interpretation. Indeed, asking that $\gamma^{(1)} \circ \gamma^{(2)}(s) < s$ for all $s > 0$ is equivalent to asking that the curve $(s, \gamma^{(1)}(s))$ in the first graph of figure

3 is always below the curve $\left(\gamma^{(2)}(s), s\right)$. The latter is sufficient for stability because it guarantees that the region where both inequalities in (2.3) are satisfied is bounded. This is illustrated in the second graph of figure 3 where the overlapping region is shaded ($b_i = \phi_i(||u_{i_\tau}||)$).

Finally, as an aside, we discuss the small gain condition in Theorem 2.2 as it relates to the small gain condition given in [5]. The analogous starting point in [5] would be the inequalities (now using some shorthand notation)

$$\begin{array}{rcl} \mathbf{C} & \leq & \phi_1 + \gamma^{(1)}(\mathbf{R}) \\ \mathbf{R} & \leq & \phi_2 + \gamma^{(2)}(\mathbf{C}) \end{array} \tag{2.5}$$

and the small gain condition is that

$$(\mathrm{Id} + \rho) \circ \gamma^{(1)} \circ (\mathrm{Id} + \rho) \circ \gamma^{(2)}(s) < s \qquad \forall s > 0 \tag{2.6}$$

for some $\rho \in \mathcal{K}_\infty$. The reason that the small gain condition is different is because a sum of two terms is used on the right-hand side in (2.5) rather than the maximum of two terms. To see the connection between the two conditions, note that the sum can be converted to a maximum with the help of any function $\rho \in \mathcal{K}_\infty$ by noting that

$$\phi + \gamma \leq \max\left\{(\mathrm{Id} + \rho^{-1})(\phi), (\mathrm{Id} + \rho)(\gamma)\right\} . \tag{2.7}$$

This follows by considering the two cases $\phi \leq \rho(\gamma)$ and $\rho(\gamma) \leq \phi$. In this form, the small gain conditions are equivalent.

Graphically, the condition (2.6) asks that the distance between the curves in the first graph of figure 3 grows without bound. This stronger condition is needed because, when using the summation on the right-hand side, the extra terms ϕ_i shift the original curves toward each other (in contrast to the case where the maximum is used on the right-hand side so that the extra terms ϕ_i only alter the curves on a bounded domain.)

3 $\mathcal{L}_\infty$ Applications of Small Gain Theorem

The starting point for most of the applications of the small gain theorem in the $\mathcal{L}_\infty$ setting is the following lemma.

Lemma 3.1 *Let $V : [0, \infty) \to \mathbb{R}_{\geq 0}$ be continuous and let $D^+V(t)$ represent its upper right-hand derivative. If $V(t) \geq \mu \geq 0$ implies $D^+V(t) \leq 0$ then $||V||_\infty \leq \max\{V(0), \mu\}$. Moreover, if for each $\epsilon \in (0, 1]$ there exists $\delta > 0$ such that $(V(t) - \mu) \in [\epsilon, 1/\epsilon]$ implies $D^+V(t) \leq -\delta$ then $||V||_a \leq \mu$.*

To briefly illustrate an application of this lemma, consider the case where $V(t) = x^2(t)$ and $\dot{x} = -x^3(t) + u(t)$ where $u(t)$ is bounded. In this case, the conditions of the lemma are satisfied with $\mu = ||u||_\infty^{2/3}$. From the relationship between V and x we can also conclude that

$$\begin{aligned} ||x||_\infty &\leq \max\left\{|x_\circ|, ||u||_\infty^{1/3}\right\} \\ ||x||_a &\leq ||u||_a^{1/3} . \end{aligned} \tag{3.1}$$

We will now use the lemma above and the nonlinear, small gain theorem to derive results on stability for functional differential equations and some "total stability" results.

3.1 A Razumikhin-type Theorem

(A more detailed discussion of the result in this section can be found in [16].) Consider a functional differential equation of the form

$$\dot{x} = f(x_d, u_d) \qquad x_d(0) = \zeta . \tag{3.2}$$

The notation here (which is nonstandard) is analogous to that for an ordinary differential equation $\dot{x} = f(x, u)$. In particular, the time dependence on the right-hand side is suppressed. In this case, however, the objects $x_d(t)$ and $u_d(t)$ are vector-valued functions rather than vectors. They are defined as $x_d(t)(\chi) = x(t - \chi)$ for $\chi \in [0, \delta]$ (similarly for $u_d(t)(\cdot)$) where $x : [0, T) \to \mathbb{R}^n$ is a function satisfying the differential equation (3.2). When $\delta = 0$, equation (3.2) is an ordinary differential equation. We are specifically interested in the case where $\delta > 0$. We assume local (in time) existence of solutions. A common method for analyzing the stability of functional differential equations is to use a so-called Lyapunov functional (an energy function that depends on $x_d(t)$). Sometimes, however, stability for functional differential equations can be determined using a Lyapunov function (an energy function that depends only on $x(t)$). General stability results based on Lyapunov functions have come to be known as Razumikhin-type theorems (see [3, section 5.4]). We can use the nonlinear, small gain theorem and lemma 3.1 to reproduce the classical Razumikhin-type theorem and extend it to the $\mathcal{L}_\infty$ input-output stability case. We need the following "norm" definitions for functions:

$$||x_d||_\infty = \sup_{t \geq 0} |x_d(t)| \quad , \qquad ||x_d||_a = \limsup_{t \to \infty} |x_d(t)| \tag{3.3}$$

where $|x_d(t)| = \max_{\chi \in [0,\delta]} |x(t - \chi)|$. Then we can state the following result:

Theorem 3.1 *If there exists a C^1 function V, functions $\underline{\alpha}$, $\overline{\alpha} \in \mathcal{K}_\infty$ and $\gamma_v, \gamma_u \in \mathcal{G}$ such that*

1. $\underline{\alpha}(|x|) \leq V(x) \leq \overline{\alpha}(|x|)$,

2. $V(x(t)) > \max\{\gamma_u(|u_d(t)|), \gamma_v(|V_d(t)|)\} \Longrightarrow \frac{\partial V}{\partial x}(x) f(x_d, u_d) < 0,$

3. $\gamma_v(s) < s$ *for* $s > 0$,

then

$$\left\{ \begin{array}{lcl} ||x_d||_\infty & \leq & \underline{\alpha}^{-1}\left(\max\left\{\overline{\alpha}(|\zeta|), \gamma_u(||u_d||_\infty)\right\}\right) \\ ||x_d||_a & \leq & \underline{\alpha}^{-1} \circ \gamma_u(||u_d||_a) \ . \end{array} \right.$$

The proof of this result starts by applying lemma 3.1. In the case where $||u_d||_\infty$ and $||V_d||_\infty$ are finite we obtain

$$\begin{array}{lcl} ||V||_\infty & \leq & \max\left\{V(x(0)), \gamma_u(||u_d||_\infty), \gamma_v(||V_d||_\infty)\right\} \\ ||V||_a & \leq & \max\left\{\gamma_u(||u_d||_a), \gamma_v(||V_d||_a)\right\} \ . \end{array} \tag{3.4}$$

(Even though we don't know that $||V_d||_\infty$ is finite *a priori*, causality allows us to recognize that the first inequality in (3.4) holds for all standard truncations of V and V_d within the maximal interval of definition for x.) On the other hand, from the definition of V_d, we also have

$$\begin{array}{lcl} ||V_d||_\infty & \leq & \max\{|V_d(0)|, ||V||_\infty\} \\ ||V_d||_a & \leq & ||V||_a \ . \end{array} \tag{3.5}$$

We are now in a position to apply the generalized nonlinear, small gain theorem ($\mathbf{C} = V, \mathbf{R} = V_d$), first using the measure $||\cdot||_\infty$ on truncated signals. From the continuation property of the differential equation we can conclude that x is defined on $[0, \infty)$ and $||V_d||_\infty$ is finite. So, we can also apply the nonlinear, small gain theorem using the measure $||\cdot||_a$. The result of the theorem follows after using $|x_d(t)| \leq \underline{\alpha}^{-1}(|V_d(t)|)$ and $|V_d(0)| \leq \overline{\alpha}(|\zeta|)$.

3.2 Total Stability: Persistent Disturbances

(This section's result was reported in [11].) We consider the system

$$\dot{x} = f(x, u) \qquad x(0) = x_\circ \tag{3.6}$$

under the assumption that, when $u(t) \equiv 0$, the origin is globally asymptotically stable. By "total stability" it is usually meant that if $||u||_\infty$ is small then $||x||_a$ is small. Such a property can be established in a fairly straightforward manner with the help of Lemma 3.1 and the nonlinear, small gain theorem. First, from the assumption that the origin is globally asymptotically stable when $u(t) \equiv 0$, it follows that there exists a smooth function $V(x)$ and class-$\mathcal{K}_\infty$ functions $\underline{\alpha}$ and $\overline{\alpha}$ such that

$$\underline{\alpha}(|x|) \leq V(x) \leq \overline{\alpha}(|x|) \tag{3.7}$$

and

$$\frac{\partial V}{\partial x}(x)f(x,0) < 0 \qquad \forall x \neq 0 \,. \tag{3.8}$$

From continuity, it follows that there exist $\gamma, \kappa \in \mathcal{G}$ such that

$$\left.\begin{array}{rcl} |x| & > & \gamma(|u|) \\ |u|\kappa(|x|) & < & 1 \end{array}\right\} \Longrightarrow \frac{\partial V}{\partial x}(x)f(x,u) < 0 \tag{3.9}$$

(cf. [11, Lemma 3.1]). Let ρ be a strictly positive real number with $\rho < 1$. Let $\beta_\rho \in \mathcal{G}$ be such that $\beta_\rho(s) = 0$ for $s \leq \rho < 1$ and $\beta_\rho(s) = 1$ for $s \geq 1$. Define $d = x\beta_\rho\left(|u|\kappa(|x|)\right)$. Note that $|x| > |d|$ implies $|u|\kappa(|x|) < 1$. Then it follows that

$$|x| > \max\{\gamma(|u|), |d|\} \Longrightarrow \frac{\partial V}{\partial x}(x)f(x,u) < 0 \,. \tag{3.10}$$

Using (3.7) and lemma 3.1, it follows that

$$\begin{array}{rcl} ||x||_\infty & \leq & \underline{\alpha}^{-1} \circ \overline{\alpha}\left(\max\left\{|x_\circ|, \gamma(||u||_\infty), ||d||_\infty\right\}\right) \\ ||x||_a & \leq & \underline{\alpha}^{-1} \circ \overline{\alpha}\left(\max\left\{\gamma(||u||_a), ||d||_a\right)\right\} \,. \end{array} \tag{3.11}$$

In turn,

$$\begin{array}{rcl} ||d||_\infty & \leq & ||x||_\infty \beta_\rho\left(||u||_\infty \kappa(||x||_\infty)\right) \\ ||d||_a & \leq & ||x||_a \beta_\rho\left(||u||_\infty \kappa(||x||_a)\right) \,. \end{array} \tag{3.12}$$

If we define $\epsilon := ||u||_\infty$, the small gain condition becomes

$$\underline{\alpha}^{-1} \circ \overline{\alpha}\left(s\beta_\rho\left(\epsilon\kappa(s)\right)\right) < s \,. \tag{3.13}$$

From the properties of β_ρ, (3.13) is satisfied for $s \in [0, \bar{s}]$ (at least) where

$$\bar{s} = \arg\min_s \{\epsilon\kappa(s) = \rho\} \,. \tag{3.14}$$

Then, using remark 2.1 as it applies to the small gain theorem and taking the limit as $\rho \to 1$ we get the following result.

Theorem 3.1 *If* $||u||_\infty \cdot \kappa \circ \underline{\alpha}^{-1} \circ \overline{\alpha}\left(\max\{|x_\circ|, \gamma(||u||_\infty)\}\right) < 1$ *then*

$$\left\{\begin{array}{rcl} ||x||_\infty & \leq & \underline{\alpha}^{-1} \circ \overline{\alpha}\left(\max\left\{|x_\circ|, \gamma(||u||_\infty)\right\}\right) \\ ||x||_a & \leq & \underline{\alpha}^{-1} \circ \overline{\alpha} \circ \gamma(||u||_a) \,. \end{array}\right.$$

Remark 3.1 In the case where $\kappa \in \mathcal{K}_\infty$, as $|x_\circ| \to \infty$ the constraint imposes the requirement that $||u||_\infty \to 0$.

3.3 Total Stability: Small Input Time Delays

(The generalization of the result in this section for the case of $\mathcal{L}_\infty$ input-output stability is given in [16].) We consider the system

$$\dot{x} = f\Big(x, k\left(x(t-\delta)\right)\Big) \qquad x_d(0) = \zeta \tag{3.15}$$

under the assumption that k is differentiable and, when $\delta = 0$, the origin is globally asymptotically stable. Again we can establish total stability, this time with respect to the time delay δ. We begin by rewriting (3.15) as

$$\dot{x} = f(x, k(x) + u) \qquad x_d(0) = \zeta \tag{3.16}$$

where $u = k(x(t-\delta)) - k(x(t))$. Using the same argument as in the preceding section, we can find class-$\mathcal{G}$ functions $\gamma_\circ$, γ, κ and β_ρ such that, with $d = x\beta_\rho\Big(|u|\kappa(|x|)\Big)$, we have

$$\begin{aligned} ||x_d||_\infty &\leq \max\Big\{\gamma_\circ(|\zeta|), \gamma(||u||_\infty), \gamma_\circ(||d||_\infty)\Big\} \\ ||x_d||_a &\leq \max\Big\{\gamma(||u||_a), \gamma_\circ(||d||_a)\Big\} \ . \end{aligned} \tag{3.17}$$

Moreover, since u can be rewritten as

$$u = \int_{t-\delta}^{t} \frac{\partial k}{\partial x}(x(s)) f\Big(x, k\left(x(s-\delta)\right)\Big)\, ds \tag{3.18}$$

and $f(0, k(0)) = 0$, there exists $\gamma_u \in \mathcal{G}$ such that

$$\begin{aligned} ||u||_\infty &\leq \delta\gamma_u(||x_d||_\infty) \\ ||u||_a &\leq \delta\gamma_u(||x_d||_a) \ . \end{aligned} \tag{3.19}$$

The small gain condition then becomes

$$\max\Big\{\gamma\Big(\delta\gamma_u(s)\Big), \gamma_\circ\Big(s\beta_\rho\Big(\delta\gamma_u(s)\kappa(s)\Big)\Big)\Big\} < s \ . \tag{3.20}$$

Since the left-hand side of (3.20) is identically zero when $\delta = 0$, it follows that there exist functions φ and ψ in class-$\mathcal{G}$ such that (3.20) holds for all s satisfying $\varphi(\delta) < s < \frac{1}{\psi(\delta)}$. Again from remark 2.1 and the small gain theorem, we have the following result:

Theorem 3.2 *If* $\max\{\gamma_\circ(|\zeta|), \varphi(\delta)\} < \dfrac{1}{\psi(\delta)}$ *then*

$$\left\{\begin{aligned} ||x_d||_\infty &\leq \max\Big\{\gamma_\circ(|\zeta|), \varphi(\delta)\Big\} \\ ||x_d||_a &\leq \varphi(\delta) \ . \end{aligned}\right.$$

Part II: Input-output Stabilization

4 Gain Assignment: Pass 1

With the nonlinear, small gain theorem available as an analysis tool it is natural to ask, for control problems, when it is possible to assign a certain gain from an input to an output. As we will see, such a question can be motivated by certain robust stabilization problems. For an initial result that addresses this question, we consider the system

$$\dot{x} = f(x) + g(x)[u + d] \tag{4.1}$$

where it is assumed that the origin is globally asymptotically stabilizable by (sufficiently) smooth feedback when $d(t) \equiv 0$. To simplify the statement of the theorem below, we will say that when

$$\begin{aligned} ||y||_\infty &\leq \max\left\{\gamma_0(|x_\circ|), \gamma_1(||d||_\infty)\right\} \\ ||y||_a &\leq \gamma_1(||d||_a) \end{aligned} \tag{4.2}$$

the mapping from $(x_\circ, d)$ to y is $\mathcal{L}_\infty$ input-output stable with gain (γ_0, γ_1). Also, Id stands for the $\mathcal{K}_\infty$ function $\mathrm{Id}(s) = s$. The following gain assignment result was established in [6]:

Theorem 4.1 [6] *Given any class-$\mathcal{K}_\infty$ functions γ_x and γ_u, there exist a continuous function $k(x)$ and a function $\gamma_\circ \in \mathcal{K}_\infty$ such that the following mappings for the system (4.1) with $u = k(x)$ are $\mathcal{L}_\infty$ input-output stable with the given gains:*

$$\begin{aligned} (x_\circ, d) &\mapsto x: \quad && (\gamma_\circ, \gamma_x) \\ (x_\circ, d) &\mapsto u: \quad && (\gamma_\circ, \mathrm{Id} + \gamma_u)\ . \end{aligned}$$

Remark 4.1 If the inverse of the functions γ_x and γ_u are suitably smooth at the origin then the control law will be smooth.

This result is analogous to the $\mathcal{L}_2$ ($\mathcal{H}_\infty$) result for the linear system

$$\begin{aligned} \dot{x} &= Ax + B[u + d] \\ y &= \begin{bmatrix} \ell x \\ u \end{bmatrix} \end{aligned} \tag{4.3}$$

where (A, B) is stabilizable: the $\mathcal{L}_2$-gain from d to y can be made arbitrarily close to the value 1 regardless of the value for ℓ. This follows from the fact that the corresponding $\mathcal{H}_\infty$ algebraic Riccati equation in this case,

$$A^T P + PA + (\frac{1}{\gamma^2} - 1)PBB^T P + \ell^2 I = 0\ , \tag{4.4}$$

always has a stabilizing, positive semi-definite solution as long as $\gamma > 1$.

In the next section we will use the gain assignment theorem (theorem 6) to solve some robust stabilization problems.

5 Application: Robust Stabilization

We will say that a system $\dot{x} = f(x,u), y = h(x,u)$ is $\mathcal{L}_\infty$-*detectable* if there exist functions $\gamma_\circ$, γ_u and γ_y of class-$\mathcal{G}$ such that

$$\begin{aligned} ||x||_\infty &\leq \max\left\{\gamma_\circ(|x_\circ|), \gamma_u(||u||_\infty), \gamma_y(||y||_\infty)\right\} \\ ||x||_a &\leq \max\left\{\gamma_u(||u||_a), \gamma_y(||y||_a)\right\} . \end{aligned} \tag{5.1}$$

Consider the system

$$\begin{aligned} \dot{x}_1 &= f_1(x_1, x_2, u) \\ \dot{x}_2 &= f_2(x_2) + g_2(x_2)[u + \phi(x_1, x_2, u)] . \end{aligned} \tag{5.2}$$

Suppose that

1. The x_1 subsystem is $\mathcal{L}_\infty$-detectable.
2. The mapping $(x_{1_\circ}, x_2, u) \mapsto \phi$ is $\mathcal{L}_\infty$ input-output stable with gain $(\gamma_0, \gamma_1, \gamma_2)$.
3. The origin of the x_2 subsystem is globally asymptotically stabilizable when $\phi \equiv 0$.

Then, the next result follows from the gain assignment theorem of the previous section and the nonlinear, small gain theorem.

Theorem 5.1 *If there exists $\gamma_u \in \mathcal{K}_\infty$ such that $\gamma_2 \circ (\mathrm{Id} + \gamma_u)(s) < s$ for all $s > 0$ then there exists a feedback $u = k(x_2)$ such that the origin of (5.2) is globally asymptotically stable.*

This result also extends to the case where there are exogenous disturbances and then can be used recursively as a means for "backstepping" the $\mathcal{L}_\infty$ input-output stability property. For more details see [4]. Another possible application is the following:

Robust dynamic inversion for aircraft dynamics: Consider a (simplified) model of the longitudinal dynamics of an aircraft:

$$\begin{aligned} \dot{y} &= b(y,z) + a(y,z)u \\ z &= \Sigma(z^\circ, y, u) \end{aligned} \tag{5.3}$$

where $y \in \mathbb{R}$ is the variable to be controlled and Σ is a dynamical system with initial state z°. In the ideal case, the functional form of b and a would be known exactly, the variables y and z would be available for feedback, and the system

$$z = \Sigma\left(z^\circ, y, a^{-1}(y,z)\left[b(y,z) + v\right]\right) =: \Sigma'\left(z^\circ, y, v\right) \tag{5.4}$$

would be stable with respect to y and v. This dynamical system represents the so-called inverse dynamics induced by the choice of the controlled variable y. The standard dynamic inversion algorithm would have us pick

$$\begin{aligned} u &= a^{-1}(y,z)\,[b(y,z)+v(y,x_c)] \\ \dot{x}_c &= C(x_c,y) \end{aligned} \tag{5.5}$$

with $x_c \in \mathbb{R}^{n_c}$, where the origin of the system

$$\dot{x}_c = C(x_c,y) \quad , \quad \dot{y} = v(y,x_c) \tag{5.6}$$

is globally asymptotically stable. The compensator dynamics $\dot{x}_c$ may be in place to meet some additional performance objectives; e.g., they may include integral action. Since the functional form of a and b is not known exactly, and perhaps not all of z is measurable, the control (5.5) cannot be implemented. So we proceed in the following fashion : we first choose u to have the form given in (5.5) where a and b are replaced by a model of the actual a and b and the models only include those parts containing variables that are measured. The function v is left unspecified and the form of the compensator dynamics x_c is left unchanged. Then we can write the system (5.3) together with the compensator dynamics as

$$\begin{aligned} \dot{x}_c &= C(x_c,y) \\ \dot{y} &= v+\phi(y,z,v) \\ z &= \tilde{\Sigma}(z^\circ,y,v) \end{aligned} \tag{5.7}$$

where ϕ represents the disturbance to the $\dot{y}$ equation due to uncertainty and/or unmodeled or unmeasured dynamics and $\tilde{\Sigma}$ has the same form as Σ' but with the new a and b. The (x_c,y) subsystem with control v and disturbance ϕ has the form considered in theorem 6 and the hypothesis of theorem is satisfied (in particular, the function $v(y,x_c)$ given in (5.5) globally asymptotically stabilizes the origin of the (x_c,y) subsystem when $\phi \equiv 0$.) So, given any class-$\mathcal{K}_\infty$ functions γ_x and γ_v, we can find a new feedback function $v = k(y,x_c)$ that assigns the gain γ_x from ϕ to (x_c,y) and the gain $\mathrm{Id}+\gamma_v$ from ϕ to v. Now, for stability it remains to analyze the interconnection of the (x_c,y) subsystem, having input ϕ and outputs $(y,k(y,x_c))$, and the $\tilde{\Sigma}$ subsystem, having inputs $(y,k(y,x_c))$ and output ϕ. Stability, at least based on the small gain theorem, depends strongly on stability of $\tilde{\Sigma}$ and the associated gains. In particular, according to theorem 7 we can achieve stability if the gain γ from v to ϕ is sufficiently small (such that we can find a function $\gamma_v \in \mathcal{K}_\infty$ so that $\gamma \circ (\mathrm{Id}+\gamma_v)(s) < s$ for all $s > 0$.) Since, as discussed above, $\tilde{\Sigma}$ is closely related to the inverse dynamics induced by choosing y as the controlled variable, this further illustrates why it is important to make a good choice for y and carefully analyze the behavior of its inverse dynamics.

6 Gain Assignment: Pass 2

Thus far, the philosophy has been to use "high gain" controllers to induce low gains from disturbances to states. It is appreciated in the control community that this idea has its pitfalls. Consider the control system

$$\dot{x} = u + d \tag{6.1}$$

where $x \in \mathbb{R}$. We will consider a tradeoff between assigning a small gain (from d to x) "at infinity" and robustness to small time delays at the input. Consider the two control strategies $u_1 = -x$ and $u_2 = -x - x^3$. When there is no input time delay, the second input assigns an $\mathcal{L}_\infty$-gain that looks like $s^{1/3}$ (sublinear) at infinity and s at the origin while the first input assigns a gain s for both small and large disturbances. So, for example, the second input will guarantee that the state remains bounded for all disturbances d that satisfy $||d||_\infty = k||x||_\infty$ while the first input will only guarantee that the state remains bounded when $k < 1$. Now, consider the situation when there is a small time delay at the input and $d \equiv 0$, i.e., (6.1) becomes

$$\dot{x} = u(t - \delta) \tag{6.2}$$

where $\delta > 0$. It can be verified that the origin is asymptotically stable using the first input for $\delta < 1.5$ seconds (even if the time-delay is time-dependent; see [3, Ch.5, Theorem 5.2]). On the other hand, it can be established that if $|x(-\chi)| \geq \bar{x}$ for all $\chi \in [0, \delta]$ where (conservatively) $\delta(1 + \bar{x}^2) \geq 4$ and the initial data is continuous then the solution does not converge to the origin. This is seen by integrating and noting that there exist $t_\circ \geq 0$ and $t_1 \in \left(0, \dfrac{2\bar{x}}{\bar{x} + \bar{x}^3}\right] \subset (0, \delta/2]$ such that

$$\begin{array}{rcll} |x(t)| & \leq & \bar{x} & \forall t \in [t_\circ, t_\circ + t_1] \\ |x(t)| & \geq & \bar{x} & \forall t \in [t_\circ + t_1, t_\circ + \delta] \\ |x(t_\circ + \delta)| & \geq & (\bar{x} + \bar{x}^3)\delta - \bar{x} \geq 3\bar{x} & . \end{array} \tag{6.3}$$

Using the first and third inequalities in (6.3), integrating and using the upper bound on t_1 we get that for $t \in [t_\circ + \delta, t_\circ + \delta + t_1]$

$$|x(t)| \geq 3\bar{x} - (\bar{x} + \bar{x}^3)t_1 \geq \bar{x} \; . \tag{6.4}$$

We now have that $|x(t)| \geq \bar{x}$ for all $t \in [t_\circ + t_1, t_\circ + t_1 + \delta]$ and we can repeat the argument with the starting time $t_\circ + t_1 + \delta$. It follows that $x(t)$ does not converge to the origin.

The problem we encounter trying to balance making the gain from d to x small in (6.1) while still being stable in the presence of small time delays is analogous to that encountered for the system

$$\dot{x}_1 = x_2(t - \delta) + x_1 \sin(x_1) + d \quad , \quad \dot{x}_2 = x_3 \quad , \quad \dot{x}_3 = u \tag{6.5}$$

when trying to make the gain from d to x_1 small. When $\delta = 0$, this problem is well-understood and can be solved with the tools of the previous section. When $\delta > 0$, it is not clear what global gain from d to x_1 can be assigned. The system (6.5) can be written down equivalently as

$$\begin{aligned} \dot{x}_1 &= x_2 + x_1 \sin(x_1) + \bar{d} \quad , \quad \dot{x}_2 = x_3 \quad , \quad \dot{x}_3 = u \\ \bar{d}(t) &= \int_{t-\delta}^{t} x_3(s)ds + d(t) \ . \end{aligned} \tag{6.6}$$

So, one sees that the goal of achieving a small gain from $\bar{d}$ to x_1 needs to be balanced with the goal of achieving a small gain from $\bar{d}$ to x_3. Most gain assignment problems for nonlinear systems with this type of tradeoff are not well-understood at this point. In [12], a formalism for addressing such problems was presented. For example, consider the system

$$\dot{x} = f(x) + g(x)u + w(x)d \tag{6.7}$$

and let $\gamma \in \mathcal{G}$ be the target $\mathcal{L}_\infty$ input-output gain from d to x.

Theorem 6.1 [12] *If there exist class-$\mathcal{K}_\infty$ functions $\underline{\alpha}$, $\overline{\alpha}$ and ϕ and a smooth function $V(x)$ such that*

1. $\underline{\alpha}(|x|) \leq V(x) \leq \overline{\alpha}(|x|)$,
2. $\underline{\alpha}^{-1} \circ \overline{\alpha} \circ \phi^{-1}(s) \leq \gamma(s)$ *for all* $s \geq 0$,
3. $\inf_u \left\{ L_f V + L_g V u + |L_w V| \phi(|x|) \right\} < 0 \ \forall x \neq 0$,

then there exists a function $k(x)$ that is smooth on $\mathbb{R}^n \setminus \{0\}$ such that, with $u = k(x)$, the system (6.7) has $\mathcal{L}_\infty$ input-output gain γ from d to x.

Work on constructive algorithms for establishing related properties is ongoing (see [2] for example.)

7 Conclusion

In this paper we have tried to illustrate how the input-output method has been used recently in nonlinear stability analysis and control synthesis. We have focused primarily on results in the $\mathcal{L}_\infty$ setting. Of course, there are many other interesting results in other settings, most notably the flurry of recent results in the area of nonlinear "$\mathcal{H}_\infty$" control. Other novel, small gain theorem-based results are also appearing (see [14] and [15] for example.) We would be remiss not to point to the 1989 paper of Eduardo Sontag [10] as the genesis for much of the renewed interest in $\mathcal{L}_\infty$ input-output methods for nonlinear systems.

The presentation here is intended mainly to give a flavor for what results exist in the literature. There are many other results that we have not discussed, e.g., local small gain theorems (see, for example, [14, Theorem 1]), other "norm" measures such as distances to sets (see, for example, [12]), time-varying systems, and total stability with respect to stable singular perturbations [1].

Many interesting open problems remain, most notably those associated with general gain assignment and related computational issues. We look forward to seeing such problems addressed in the literature in the near future.

References

[1] P.D. Christofides and A.R. Teel. Singular Perturbations and input-to-state stability. To appear *IEEE Transactions on Automatic Control.*

[2] I.J. Fialho. *Worst-case Analysis and Design of Nonlinear Feedback Systems.* Ph.D. Dissertation. University of Minnesota. Minneapolis, Minnesota. 1996.

[3] J.K. Hale and S.M. Verduyn Lunel. *Introduction to Functional Differential Equations.* New York: Springer-Verlag, 1993.

[4] Z.P. Jiang, A.R. Teel and L. Praly. Small gain theorem for ISS systems and applications. *Mathematics of Control, Signals and Systems* **7** (1994), 95-120.

[5] I.M.Y. Mareels and D.J. Hill. Monotone stability of nonlinear feedback systems. *Journal of Mathematical Systems, Estimation and Control* **2**(3) (1992), 275-291.

[6] Y. Wang and L. Praly. Stabilization in spite of matched unmodeled dynamics and an equivalent definition of input-to-state stability. To appear *Mathematics of Control, Signals and Systems.*

[7] M.G. Safonov. *Stability and Robustness of Multivariable Feedback Systems.* Cambridge, MA: MIT Press, 1980.

[8] I.W. Sandberg. On the $\mathcal{L}_2$-boundedness of solutions of nonlinear functional equations. *Bell Sys. Tech. J.* **43** (1964), 1581-1599.

[9] I. W. Sandberg. Bell Labs in the 60's and input-output stability. *Proceedings of the 34th Conference on Decision and Control 1995* 2790-2791.

[10] E.D. Sontag. Smooth stabilization implies coprime factorization. *IEEE Transactions on Automatic Control* **AC-34** (1989), 435-443.

[11] E.D. Sontag. Further facts about input to state stabilization. *IEEE Trans. on Automatic Control* **35**(4) (1990), 473-476.

[12] E.D. Sontag and Y. Wang. On characterizations of input-to-state stability with respect to compact sets. *Proceedings of the Nonlinear Control Systems Design Symposium (NOLCOS) 1995*. 226-231.

[13] A.R. Teel. On graphs, conic relations, and input-output stability of nonlinear feedback systems. *IEEE Transactions on Automatic Control* **41**(5) (1996), 702-709.

[14] A.R. Teel. A nonlinear small gain theorem for the analysis of nonlinear control systems with saturation. *IEEE Transactions on Automatic Control* **41**(9) (1996), 1256-1270.

[15] A.R. Teel. On $\mathcal{L}_2$ induced by feedbacks with multiple saturations. *ESAIM: Control, Optimisation and Calculus of Variations* **1** (1996), 225-240.

[16] A.R. Teel. Connections between Razumikhin-type theorems and the ISS nonlinear small gain theorem. Submitted to *IEEE Transactions on Automatic Control.*

[17] G. Zames. On the input-output stability of time-varying nonlinear feedback systems. Part I: Conditions using concepts of loop gain, conicity, and positivity. *IEEE Transactions on Automatic Control* **11** (1966), 228-238.

[18] G. Zames. Input-output feedback stability and robustness. *Control Systems Magazine* **16**(3) (1996), 61-66.

Electrical Engineering Department, University of Minnesota, 4-174 EE/CS Building, 200 Union St. SE, Minneapolis, Minnesota 55455

Repetitive Control Systems: Old and New Ideas

G. Weiss

1 What is a Repetitive Control System?

Many signals in engineering are periodic, or at least they can be well approximated by a periodic signal over a large time interval. This is true, for example, for most signals associated with engines, electrical motors and generators, converters, or machines performing a task over and over again. Thus, it is a natural control problem to try to track a periodic signal with the output of a plant, or (what is almost the same), to try to reject a periodic disturbance acting on a control system. We examine this problem in Sections 1 and 3 of this paper (Section 2 is for background). In Section 4 we shall indicate a way of generalizing these ideas to cope with superpositions of periodic signals of arbitrary periods.

We assume that the plant to be controlled is linear, time-invariant and finite-dimensional. Then one possible approach to the tracking and/or disturbance rejection problem described above is to use the internal model principle of Francis and Wonham [1]. This leads to the control system shown in Figure 1, where **P** is the transfer function of the plant, **C** is the transfer function of the compensator and **M** is the transfer function of the internal model. In the same diagram, r is the reference signal, d is the disturbance, and e is the error signal which should be made small.

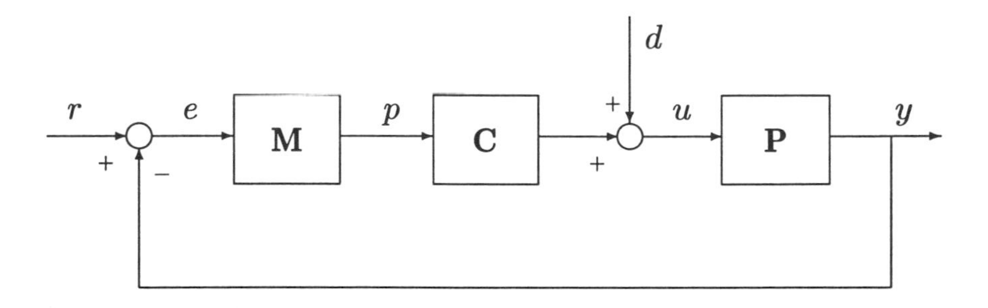

Figure 1: Repetitive control system. **M** is the internal model, **C** is the stabilizing compensator and **P** is the plant. The reference r and the disturbance d are periodic with period τ, and the error e should be kept small.

The internal model and the compensator are also linear and time-invariant systems. The internal model is infinite-dimensional and it is capable of generating signals which are similar to r and d, that is, periodic signals of a given period τ. The compensator is finite-dimensional. Neither of the three subsystems is stable, in general, but the compensator must be designed such that the whole feedback system shown in Figure 1 should be exponentially stable. The internal model and the compensator are the two components of the controller. Such a feedback system is called a *repetitive control system.* It has been studied by Inoue, Nakano and Iwai [6], by Inoue, Nakano, Kubo, Matsumoto and Baba [7] and by others. Probably the main references about this subject are the papers by Hara, Yamamoto, Omata and Nakano [4] and by Yamamoto [17]. A closely related and active area of control theory is *iterative learning control*, see for example the books by Rogers and Owens [12] and by Moore [10].

For the sake of simplicity, in this paper we shall consider that all signals are scalar, so that the three subsystems in Figure 1 are SISO. Our arguments will be in the frequency domain and (again to simplify the exposition) the connections to the state space theory will be mentioned briefly and without proof. We shall not strive to state the results in their greatest generality.

We assume that the internal model has the following transfer function:

$$\mathbf{M}(s) = \frac{1}{1 - e^{-\tau s}\mathbf{W}(s)}, \tag{1.1}$$

where $\mathbf{W}$ is a real-rational stable transfer function with

$$\|\mathbf{W}\|_\infty \leq 1. \tag{1.2}$$

In (1.2), $\|\mathbf{W}\|_\infty$ denotes the H^∞ norm (the supremum of $|\mathbf{W}(s)|$ over all $s \in \mathbb{C}$ with Re $s > 0$). Such a transfer function $\mathbf{M}$ can be obtained by connecting a delay line into a feedback loop, as shown in Figure 2.

We explain very briefly how such a control system works, and what the role of $\mathbf{W}$ is (the rigorous details will be given in Section 3).

First we consider the case $\mathbf{W} = 1$. Then $\mathbf{M}$ has infinitely many poles on the imaginary axis, namely, $k\nu i$, where $k \in \mathbb{Z}$ and $\nu = \frac{2\pi}{\tau}$. Thus, the transfer function $\mathbf{G}$ from r to e has zeros in the same points, because

$$\mathbf{G} = (1 + \mathbf{PCM})^{-1}, \tag{1.3}$$

and there are no unstable pole-zero cancellations in the product $\mathbf{PCM}$ (since the whole system is stable). The same is true for the transfer function from d to e, which is $-\mathbf{GP}$. These zeros of $\mathbf{G}$ imply that if $r, d \in L^2_{loc}[0, \infty)$ are periodic with period τ, then $\hat{e}$ (the Laplace transform of e) will *not* have poles at the points $k\nu i$. This, combined with the stability of the system,

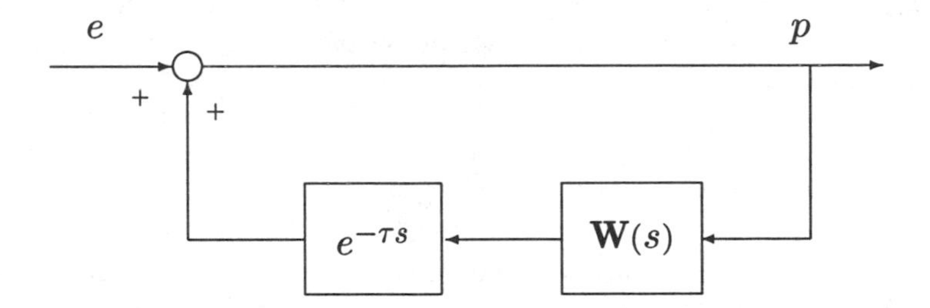

Figure 2: The structure of the internal model **M**. Usually, **W** is a low-pass filter, needed both for technological and for robustness reasons.

implies that for some $\delta > 0$ we have

$$\int_0^\infty e^{\delta t} |e(t)|^2 dt < \infty, \tag{1.4}$$

which in particular is much stronger than $e \in L^2[0, \infty)$.

In practice, we cannot choose $\mathbf{W} = 1$, for three reasons:

(1) The first reason is technological: we cannot realize a delay line with infinite bandwidth, and thus any delay line that we can build has to be modelled as an ideal delay line in series with a low-pass filter.

(2) The conditions that **C** must satisfy in order to stabilize the whole system are the following: it must stabilize **P** (as if **M** would not exist, meaning $\mathbf{M} = 1$), and moreover

$$\|\mathbf{W}(1 + \mathbf{PC})^{-1}\|_\infty < 1. \tag{1.5}$$

If $\mathbf{W} = 1$, then (1.5) can be satisfied if and only if $\mathbf{P}^{-1} \in H^\infty$, which is almost never the case. But if **W** is a genuine filter, then the chances of finding a **C** satisfying (1.5) become much better.

(3) If $\mathbf{W} = 1$, then the resulting feedback system, even if it is stable, is not robustly stable with respect to delays. This means that arbitrarily small delays at any point in the feedback loop shown in Figure 1 will destabilize the system. We can overcome this problem by imposing

$$|\mathbf{W}(\infty)| = 0 \quad \text{and} \quad |\mathbf{P}(\infty)\mathbf{C}(\infty)| < 1. \tag{1.6}$$

(These conditions can be relaxed, see Section 3 for details, but in any case we need that $|\mathbf{W}(\infty)| < 1$.)

The above considerations show that **W** must be a nonconstant filter. How should we choose **W** and what tracking and disturbance rejection performances can we hope for? The answer, roughly, goes like this: In every practical situation, r and d are confined to a certain frequency band

$[\omega_l, \omega_u]$ (the lower bound ω_l might be zero, but the upper bound ω_u cannot be infinity). This means that r can be written as a finite Fourier series (recall that $\nu = \frac{2\pi}{\tau}$):

$$r(t) = \sum_{|k\nu| \in [\omega_l, \omega_u]} r_k \, e^{ik\nu t} .$$

A similar formula holds for d, of course. Choose $\mathbf{W}$ such that $\mathbf{W}(i\omega)$ is very close to 1 for $\omega \in [\omega_l, \omega_u]$. Then $\mathbf{M}$ will have poles very close to $k\nu i$, for those $k \in \mathbb{Z}$ for which $|k\nu| \in [\omega_l, \omega_u]$. Now $\mathbf{G}$ will have zeros at the poles of $\mathbf{M}$, and hence it will be close to zero (but not quite) at the points $k\nu i$, for the relevant values of k. The error signal can be decomposed into two components:

$$e = e_{ss} + e_{tr} \tag{1.7}$$

where the *steady-state error* $e_{ss} \in L^2_{loc}[0, \infty)$ is periodic and small, and the *transient error* e_{tr} behaves as in (1.4). The "size" of e_{ss} is measured by its L^2-norm over one period (of length τ). The closer $\mathbf{W}(i\omega)$ is to 1 in the relevant frequency band, the smaller is e_{ss}. Precise statements and formulas will be given in Section 3.

2 Some Background on Regular Transfer Functions and the Feedback Connection

In this section we recall some known facts about well-posed and regular transfer functions, stability, stabilizing compensators, and robustness with respect to small delays. As already mentioned, we consider only SISO systems and transfer functions, without repeating this assumption.

We make the following convention: if a meromorphic function is defined on some right half-plane and can be extended meromorphically to a greater right half-plane, we will not make any distinction between the initial function and its extension. This will not lead to confusions, since the extension (if it exists) is unique.

For each $\alpha \in \mathbb{R}$, H^∞_α denotes the space of bounded and analytic functions on the right half-plane

$$\mathbb{C}_\alpha = \{s \in \mathbb{C} \mid \text{Re } s > \alpha\} .$$

It is well-known that H^∞_α is a Banach space with the supremum norm. With the convention of the previous paragraph, we have that

$$H^\infty_\alpha \subset H^\infty_\beta \quad \text{if} \quad \alpha \leq \beta .$$

For $\alpha = 0$ we use also the notation H^∞ instead of H^∞_0.

A *well-posed transfer function* is an element of one of the spaces H^∞_α. The well-posed transfer functions form an algebra (i.e., we can add and multiply them). Any well-posed transfer function $\mathbf{G}$ defines a shift-invariant and continuous operator $\mathbb{F}$ on $L^2_{loc}[0,\infty)$. If $u \in L^2_{loc}[0,\infty)$ has a Laplace transform $\hat{u}$, then $y = \mathbb{F}u$ is given via its Laplace transform $\hat{y}$, as follows:

$$\hat{y} = \mathbf{G}\hat{u}. \tag{2.1}$$

Let $\mathbf{G}$ be a well-posed transfer function. We say that $\mathbf{G}$ is *exponentially stable* if $\mathbf{G} \in H^\infty_\alpha$ for some $\alpha < 0$. For example, $e^{-\tau s}$ and $\frac{1}{s}(e^{-\tau s} - 1)$ (with $\tau > 0$) are exponentially stable. For each $\beta \in \mathbb{R}$, we denote by $L^2_\beta[0,\infty)$ the space of functions of the form $e^{\beta t}v(t)$, where $v \in L^2[0,\infty)$. If $\mathbf{G} \in H^\infty_\alpha$ and $u \in L^2_\beta[0,\infty)$ for some $\beta \geq \alpha$, then the function y defined by (2.1) is also in $L^2_\beta[0,\infty)$. For this reason, in view of the particular case $\alpha = \beta = 0$, we say that $\mathbf{G}$ is *L^2-stable* if $\mathbf{G} \in H^\infty$.

A well-posed transfer function $\mathbf{G}$ is called *regular* if the limit

$$\lim_{\lambda \to +\infty} \mathbf{G}(\lambda) = D$$

exists. In this case, the number D is called the *feedthrough value* of $\mathbf{G}$. For example, any well-posed transfer function obtainable from rational functions and delays by finitely many algebraic operations is regular (this includes all the transfer functions which arise in feedback systems of the type encountered in the previous sections). For a detailed discussion of well-posed and of regular transfer functions we refer to Weiss [14] and [15].

We shall not explain here what a well-posed linear system is: the reader may consult, for example, [14]. The transfer function of any well-posed linear system is well-posed. Conversely, for any well-posed transfer function $\mathbf{G}$ we can find many well-posed linear systems whose transfer function is $\mathbf{G}$, as follows from results in Salamon [13]. A well-posed linear system is called *regular* if its transfer function is regular. Such systems have a simple description via their generating operators A, B, C, D, which are the analogues of the matrices appearing in the usual representation $\dot{x} = Ax + Bu$, $y = Cx + Du$ of finite-dimensional linear systems. The operators A, B and C are unbounded in general. For details we refer again to [14].

If a well-posed linear system is exponentially stable, then its transfer function is also exponentially stable. The converse is false (also in finite dimensions), but the following result has been proved by Rebarber [11]: if a regular linear system is stabilizable, detectable and its transfer function is L^2-stable, then the system is exponentially stable. For the precise definition of stabilizability and detectability (at the level of generality needed here) we refer to [11] and to Weiss and Curtain [16].

Let $\mathbf{P}$ and $\mathbf{C}$ be well-posed transfer functions. We say that $\mathbf{C}$ *stabilizes* $\mathbf{P}$ if the matrix $\mathbf{L} = \begin{bmatrix} 1 & \mathbf{P} \\ -\mathbf{C} & 1 \end{bmatrix}^{-1}$ is L^2-stable, i.e., each of its four entries is in H^∞. This means that if we connect $\mathbf{P}$ and $\mathbf{C}$ in a feedback loop with two external inputs, as in Figure 3, then the four transfer functions from r and d to e and u are L^2-stable. If both $\mathbf{P}$ and $\mathbf{C}$ are L^2-stable, then $\mathbf{C}$ stabilizes $\mathbf{P}$ if and only if $(1+\mathbf{PC})^{-1}$ is L^2-stable. For details and for further references on this subject we refer to the survey paper of Logemann [8] and to Georgiou and Smith [3].

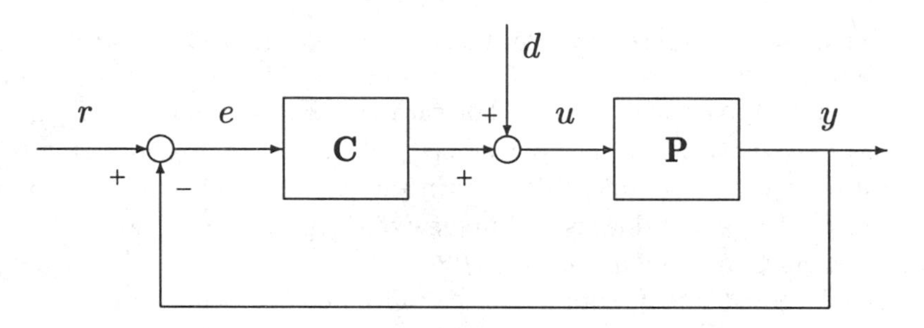

Figure 3: The standard feedback connection of two systems. We may think of $\mathbf{P}$ as the plant, $\mathbf{C}$ as the compensator, r as the reference, d as the disturbance and e as the tracking error.

Using the result of Rebarber mentioned earlier, the following proposition was proved in Section 4 of [16] (here we give the simplified version corresponding to SISO systems).

Proposition 2.1. *Let Σ_p and Σ_c be stabilizable and detectable SISO regular linear systems, with transfer functions $\mathbf{P}$ and $\mathbf{C}$, respectively. Then the feedback connection shown in Figure 3 defines an exponentially stable regular linear system if and only if $\mathbf{C}$ stabilizes $\mathbf{P}$.*

Let $\mathbf{P}$ and $\mathbf{C}$ be well-posed transfer functions. For each $\varepsilon \geq 0$, we define the transfer function $\mathbf{C}_\varepsilon$ by

$$\mathbf{C}_\varepsilon(s) = e^{-\varepsilon s}\mathbf{C}(s).$$

We say that $\mathbf{C}$ *stabilizes* $\mathbf{P}$ *robustly with respect to delays* if there exists an $\varepsilon_0 > 0$ such that for each $\varepsilon \in [0, \varepsilon_0]$, $\mathbf{C}_\varepsilon$ stabilizes $\mathbf{P}$. The intuitive meaning of this is that the introduction of sufficiently small delays into the feedback loop shown in Figure 3 does not destroy its L^2-stability. This concept has been introduced in Logemann, Rebarber and Weiss [9] and it is closely related (in the SISO case, almost equivalent) to the concept of w-stability, introduced by Georgiou and Smith in [2] and further studied by them in [3]. The following theorem is taken from Section 8 of [9].

Theorem 2.2. *Let* **P** *and* **C** *be regular transfer functions and suppose that* **C** *stabilizes* **P**. *Let* γ *be defined by*

$$\gamma = \limsup_{|s|\to\infty} |\mathbf{P}(s)\mathbf{C}(s)| . \tag{2.2}$$

(*i*) *If* $\gamma < 1$, *then* **C** *stabilizes* **P** *robustly with respect to delays.*

(*ii*) *If* $\gamma > 1$, *then* **C** *does not stabilize* **P** *robustly with respect to delays.*

In the simple particular case when **P** and **C** are rational, the condition $\gamma < 1$ becomes $|\mathbf{P}(\infty)\mathbf{C}(\infty)| < 1$, which we have encountered in Section 1.

Let **P** and **C** be regular transfer functions which are meromorphic on $\mathbb{C}_\alpha$ for some $\alpha < 0$. It follows from the last theorem that if **C** stabilizes **P** and **P** has infinitely many poles in the closed right half-plane (where Re $s \geq 0$), then **C** does not stabilize **P** robustly with respect to delays. (This statement is the SISO version of Theorem 1.2 in [9].)

Suppose that the systems Σ_p and Σ_c are as in Proposition 2.1, and their feedback connection (shown in Figure 3) is exponentially stable. This feedback system is called ***robustly stable with respect to delays*** if there exists an $\varepsilon_0 > 0$ such that for each $\varepsilon \in [0, \varepsilon_0]$, any feedback system obtained from the previous one by introducing a delay of ε into the feedback loop, is still exponentially stable. From Proposition 2.1 it follows that this is the case if and only if **C** stabilizes **P** robustly with respect to delays, so that we can use Theorem 2.2 to verify if the condition is satisfied.

3 The Main Theorems for a Single Period

In this section we state our results concerning tracking and disturbance rejection of periodic signals. Thus, we are dealing with a single period τ, the repetitive control system is as shown in Figure 1 and the internal model is as in Figure 2. We give only a short outline of the proofs.

As in Section 1, we assume that the plant is a linear time-invariant finite-dimensional SISO system. Moreover, the plant is assumed to be stabilizable and detectable (for example, this is the case if the plant is minimal). Thus, its transfer function, denoted by **P**, is a rational proper scalar function. Exactly the same assumptions are made about the filter in cascade with the delay line, whose transfer function is denoted by **W**, and about the stabilizing compensator, whose transfer function is denoted by **C**. Moreover, as in Section 1, we assume that $\mathbf{W} \in H^\infty$ and $\|\mathbf{W}\|_\infty \leq 1$. The delay line with transfer function $e^{-\tau s}$ is realized as in [14], so that its state space is $L^2[-\tau, 0]$, and it is exponentially stable (its growth bound is $-\infty$).

The main stability and robustness result for the repetitive control system shown in Figures 1 and 2 is the following.

Theorem 3.1. *Assume that the compensator transfer function* $\mathbf{C}$ *satisfies the following three conditions:*

(1) $(1+\mathbf{PC})^{-1} \in H^\infty$;

(2) *there are no unstable pole-zero cancellations in the product* $\mathbf{PC}$;

(3) $\|\mathbf{W}(1+\mathbf{PC})^{-1}\|_\infty < 1$.

Then the feedback system shown in Figures 1 *and* 2 *is exponentially stable. If, moreover, the conditions*

$$|\mathbf{W}(\infty)| < 1 \quad \textit{and} \quad |\mathbf{P}(\infty)\mathbf{C}(\infty)| < 1 - |\mathbf{W}(\infty)| \tag{3.1}$$

hold, then this feedback system is robustly stable with respect to delays.

The first part of this theorem (the exponential stability in state space) is due to Hara, Yamamoto, Omata and Nakano [4] (see their Corollary 1, with $C_2 = 0$). The robustness with respect to delays, claimed in the second part of the theorem, also refers to exponential stability in the state space, as introduced at the end of Section 2.

Note that the above theorem does not make any mention of the reference or disturbance signals appearing in Figure 1, or of any periodic signals: these will make their entry only in the next theorem. We sketch our proof for both parts because our reasoning is different from the one in [4], and it is instructive for the multi-periodic case which we want to handle later.

Sketch of the Proof. First we prove the exponential stability. By simple transformations, the feedback system (with no inputs, only with an initial state) is equivalent to the one shown in Figure 4, where $\mathbf{S}$ is the transfer function of a finite-dimensional system Σ_s containing everything except the delay line.

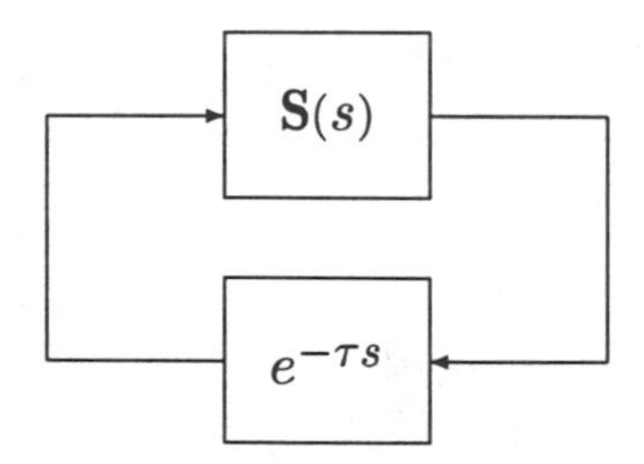

Figure 4: A feedback system which, when there are no inputs, is equivalent to the one in Figures 1 and 2. The subsystem Σ_s with transfer function $\mathbf{S}$ contains everything but the delay line.

A simple finite-dimensional argument starting from conditions (1) and (2) yields that Σ_s is stable (in the state-space) and its transfer function is

$$\mathbf{S} = \mathbf{W}(1+\mathbf{PC})^{-1}.$$

It now follows from Proposition 2.1 that the system in Figure 4 is exponentially stable iff $(1 - e^{-\tau s}\mathbf{S}(s))^{-1} \in H^\infty$. Since $\|e^{-\tau s}\|_\infty = 1$, it follows from condition (3) that this is indeed the case.

Now we turn to the robustness part of the theorem. It is not difficult to show that both the internal model (with transfer function $\mathbf{M}$) and the cascade connection of the compensator and the plant (with transfer function $\mathbf{PC}$) are stabilizable and detectable regular linear systems. Hence, as explained at the end of Section 2, their feedback connection is robustly stable with respect to delays if

$$\limsup_{|s|\to\infty} |\mathbf{P}(s)\mathbf{C}(s)\mathbf{M}(s)| < 1. \tag{3.2}$$

Since (by (3.1)) $|\mathbf{W}(\infty)| < 1$, the left-hand side of (3.2) can be computed to be $|\mathbf{P}(\infty)\mathbf{C}(\infty)|/(1 - |\mathbf{W}(\infty)|)$, so that (3.1) implies (3.2).

Given a rational proper $\mathbf{P}$, the problem of finding a rational proper $\mathbf{C}$ such that conditions (1), (2) and (3) of Theorem 3.1 should be satisfied is called the *weighted sensitivity* H^∞ *problem.* This has been extensively studied in the recent robust control literature and good algorithms (and programs) for its solution are available, see for example Green and Limebeer [5]. It may happen that the problem has no solution, in which case we should try to modify the filter $\mathbf{W}$.

Note that (1.6) is more restrictive than (3.1), but in most practical applications we expect the conditions (1.6) to be satisfied. The condition $|\mathbf{W}(\infty)| < 1$ cannot be eliminated from (3.1) (by allowing $|\mathbf{W}(\infty)| = 1$ with $\mathbf{P}(\infty)\mathbf{C}(\infty) = 0$). Indeed, if we had $|\mathbf{W}(\infty)| = 1$, then $\mathbf{M}$ would have infinitely many poles on the imaginary axis. In this case, according to the comments after Theorem 2.2, $\mathbf{PC}$ would not stabilize $\mathbf{M}$ robustly with respect to delays.

The next lemma concerns decompositions of signals into steady-state and transient parts, as in (1.7). The reader is requested to recall the notations H^∞_α and $L^2_\beta[0,\infty)$ from Section 2.

Lemma 3.2. *Let* $\mathbf{G} \in H^\infty_\alpha$ *for some* $\alpha < 0$. *Assume that* $u \in L^2_{loc}[0,\infty)$ *has the decomposition* $u = u_{ss} + u_{tr}$, *where* $u_{tr} \in L^2_\beta[0,\infty)$ *with* $\alpha \leq \beta \leq 0$, *and* u_{ss} *is periodic with period* τ. *If* y *is the output function corresponding to the input function* u *and the transfer function* $\mathbf{G}$, *as in* (2.1), *then* y *has a similar decomposition:* $y = y_{ss} + y_{tr}$, *with* $y_{tr} \in L^2_\beta[0,\infty)$ *and* y_{ss} *is periodic with period* τ.

Moreover, if the sequences (u_k) *and* (y_k) *are the Fourier coefficients of* u_{ss} *and* y_{ss}, *i.e., denoting* $\nu = \frac{2\pi}{\tau}$,

$$u_{ss}(t) = \sum_{k\in\mathbb{Z}} u_k e^{ik\nu t}, \qquad y_{ss}(t) = \sum_{k\in\mathbb{Z}} y_k e^{ik\nu t},$$

then

$$y_k = \mathbf{G}(ik\nu)u_k. \tag{3.3}$$

We remark that the sequences (u_k) and (y_k) are in l^2, of course, and the two Fourier series appearing above converge in the L^2-sense. Note that in the particular case when $u_{tr} = 0$, we can take $\beta = \alpha$, obtaining that $y_{tr} \in L^2_\alpha[0,\infty)$ (this case will be used later).

We do not give the proof of this lemma, only mention the fact that, in spite of this being a purely "frequency domain" statement, the only way the author knows to prove that $y_{tr} \in L^2_\beta[0,\infty)$ is by state space methods, invoking a realization of $\mathbf{G}$.

The following theorem concerns the situation when the feedback system shown in Figures 1 and 2 is exponentially stable and the external signals r (reference) and d (disturbance) are periodic with period τ, i.e.,

$$r(t) = \sum_{k\in\mathbb{Z}} r_k e^{ik\nu t}, \qquad d(t) = \sum_{k\in\mathbb{Z}} d_k e^{ik\nu t}, \tag{3.4}$$

where, as usual, $\nu = \frac{2\pi}{\tau}$. We assume that these signals are in $L^2_{loc}[0,\infty)$, so that their Fourier coefficients (r_k) and (d_k) are in l^2. We show that the decomposition (1.7) of the error signal holds and we give a formula for the Fourier coefficients of the steady-state error.

Theorem 3.3. *With the notation of Theorem* 3.1, *assume that the conditions* (1), (2) *and* (3) *are satisfied. Suppose that r and d are in $L^2_{loc}[0,\infty)$ and are periodic with period τ, as in* (3.4). *Then the error e can be decomposed as*

$$e = e_{ss} + e_{tr},$$

where $e_{tr} \in L^2_\alpha[0,\infty)$ for some $\alpha < 0$ and e_{ss} is periodic, with period τ. Let (e_k) be the sequence of Fourier coefficients of e_{ss}, similarly as in (3.4). *Then, denoting $\mathbf{S}_0 = (1+\mathbf{PC})^{-1}$,*

$$e_k = \frac{(1-\mathbf{W}(ik\nu))\mathbf{S}_0(ik\nu)}{1-\mathbf{W}(ik\nu)\mathbf{S}_0(ik\nu)}\,[r_k - \mathbf{P}(ik\nu)d_k]. \tag{3.5}$$

Sketch of the Proof. By Theorem 3.1, the feedback system of Figure 1 is exponentially stable, so that its sensitivity $\mathbf{G} = (1+\mathbf{PCM})^{-1}$ belongs to H^∞_α, for some $\alpha < 0$ which is greater then the growth bound of the operator semigroup of the system. If the initial state of the system is zero, then

$$\hat{e} = \mathbf{G}(\hat{r} - \mathbf{P}\hat{d}), \tag{3.6}$$

so that, according to Lemma 3.2, e can be decomposed into a steady-state and a transient part, as claimed in the theorem. If the initial state is not zero, then it generates an error signal which belongs to $L^2_\alpha[0,\infty)$, so that it can be absorbed into e_{tr} and it does not affect e_{ss}. A simple computation shows that

$$\mathbf{G}(s) = \frac{(1 - e^{-\tau s}\mathbf{W}(s))\mathbf{S}_0(s)}{1 - e^{-\tau s}\mathbf{W}(s)\mathbf{S}_0(s)}.$$

Applying formula (3.3) to this particular situation, we obtain (3.5).

An important conclusion which we can draw from formula (3.5) is the following: if the Fourier coefficients of r and d are concentrated in a frequency band (i.e., they are very small outside this band) and $\mathbf{W}(ik\nu)$ is very close to 1 in this band, then the coefficients e_k will be very small, meaning that the L^2-norm of e_{ss} over a period will be very small. To illustrate how (3.5) can be used to get estimates, assume that $d = 0$ and r_k vanishes outside the frequency band $|k\nu| \in [\omega_l, \omega_u]$. Denote $\gamma = \|\mathbf{WS}_0\|_\infty$ and $\gamma_0 = \|\mathbf{S}_0\|_\infty$, so that $\gamma < 1$ by condition (3). ¿From (3.5) and the Parseval equality we get

$$\|e_{ss}\|_{L^2[0,\tau]} \leq \frac{\gamma_0}{1-\gamma} \cdot \max_{|k\nu| \in [\omega_l, \omega_u]} |1 - \mathbf{W}(ik\nu)| \cdot \|r\|_{L^2[0,\tau]}.$$

We want to emphasize the strength of the conclusion $e_{tr} \in L^2_\alpha[0, \infty)$ in the last theorem. This implies that

$$\int_t^{t+\tau} |e_{tr}(\sigma)|^2 d\sigma \leq m \cdot e^{-2\alpha t},$$

for some $m > 0$ and for all $t \geq 0$. If the derivatives $\dot{r}$ and $\dot{d}$ belong to $L^2_{loc}[0, \infty)$, then the theorem can be applied to the derivatives, yielding that $\dot{e}_{tr} \in L^2_\alpha[0, \infty)$ holds as well. From this we can deduce that $e_{tr}(t)$ converges to zero at an exponential rate.

4 How to Track (or Reject) a Superposition of Two Periodic Signals

In certain engineering problems, it could be a reasonable assumption that the reference r and/or the disturbance d are superpositions of several periodic functions. For the sake of simplicity, we assume in this paper that there are only two periods involved, τ_1 and τ_2:

$$r = r_1 + r_2, \quad r_1(t + \tau_1) = r_1(t), \quad r_2(t + \tau_2) = r_2(t), \tag{4.1}$$

and similarly for $d = d_1 + d_2$. For example, it might happen that r is periodic with period τ_1 and d is periodic with period τ_2, in which case $r_2 = d_1 = 0$. All these signals are assumed to be in $L^2_{loc}[0, \infty)$. The generalization of our results to any finite number of periods is straightforward.

We are trying to use again the feedback structure from Figure 1, but with a more sophisticated internal model. Since a good internal model should be able to generate functions as in (4.1), it is an obvious guess to try to take $\mathbf{M}$ to be the sum of two partial internal models, one of which generates signals with period τ_1 and the other with period τ_2. This guess turns out to be almost correct, but more precisely, the internal model should be a convex combination of models of the type (1.1):

$$\mathbf{M} = \alpha_1 \mathbf{M}_1 + \alpha_2 \mathbf{M}_2 , \tag{4.2}$$

where $\alpha_1 > 0$, $\alpha_2 > 0$, $\alpha_1 + \alpha_2 = 1$ and for $j = 1, 2$

$$\mathbf{M}_j(s) = \frac{1}{1 - e^{-\tau_j s}\mathbf{W}(s)} . \tag{4.3}$$

For simplicity, we have assumed that the filters used in the partial internal models $\mathbf{M}_1$ and $\mathbf{M}_2$ are identical. Their transfer function $\mathbf{W}$ is such that

$$\|\mathbf{W}\|_\infty \leq 1 \quad \text{and} \quad |\mathbf{W}(\infty)| < 1 . \tag{4.4}$$

The above conditions imply that there are only finitely many unstable poles of $\mathbf{M}_1$ and $\mathbf{M}_2$ and these are on the imaginary axis. Moreover, we require $\mathbf{W}$ to be such that $\mathbf{M}_1$ and $\mathbf{M}_2$ should not have any common unstable poles. The intuitive explanation of this condition is the following: if the two partial internal models had a common unstable pole $i\omega$, then each of them could generate functions of the form $p(t) = p_0 e^{i\omega}$, where p_0 is a constant. These could cancel each other out at the output of the internal model, so that the compensator $\mathbf{C}$ would not "see" them. Then, assuming $r = d = 0$, the compensator would not generate any response, $u = y = e = 0$, and the partial internal models would continue to generate their signals forever, meaning that the feedback system is unstable.

Our assumptions on the plant with the transfer function $\mathbf{P}$ and the compensator with the transfer function $\mathbf{C}$ are the same as in Section 3: stabilizability and detectability. We use the same realization for the delay lines as in Section 3 and we assume that the two filters with transfer function $\mathbf{W}$ are stable and satisfy (4.4). Given such an internal model with a relatively complicated structure, it is somewhat surprizing that our stability and robustness result is almost identical to Theorem 3.1:

Theorem 4.1. *Assume that the compensator transfer function* $\mathbf{C}$ *satisfies the three conditions listed in Theorem* 3.1; $\mathbf{W}$ *satisfies* (4.4) *and* $\mathbf{M}_1$, $\mathbf{M}_2$ *do not have any common unstable poles.*

Then the feedback system shown in Figure 1, *with* $\mathbf{M}$ *given by* (4.2) *and* (4.3), *is exponentially stable. If, moreover, the condition*

$$|\mathbf{P}(\infty)\mathbf{C}(\infty)| < 1 - |\mathbf{W}(\infty)|$$

holds, then this feedback system is robustly stable with respect to delays.

Note that the assumptions of the theorem imply that $\mathbf{W}(0) < 1$.

Sketch of the Proof. First we prove the exponential stability. Like in the proof of Theorem 3.1, by simple transformations we obtain an equivalent system as in Figure 4, but the single delay gets replaced by two delay lines, having as transfer function the 2×2 matrix $\Delta(s) = \text{diag}(e^{-\tau_1 s}, e^{-\tau_2 s})$. The finite-dimensional system which contains everything else is stable (by conditions (1) and (2)) and its transfer function is

$$\mathbf{S}_e = \mathbf{W}\left(\begin{bmatrix} 1 & 0 \\ 0 & 1 \end{bmatrix} + \mathbf{PC}\begin{bmatrix} \alpha_1 & \alpha_2 \\ \alpha_1 & \alpha_2 \end{bmatrix}\right)^{-1},$$

which replaces $\mathbf{S} = \mathbf{W}(1 + \mathbf{PC})^{-1}$ in Figure 4.

We introduce the numbers $\beta_1 = \sqrt{\alpha_1}$, $\beta_2 = \sqrt{\alpha_2}$ and the matrices

$$R = \begin{bmatrix} \beta_2/\beta_1 & 0 \\ 0 & 1 \end{bmatrix}, \qquad V = \begin{bmatrix} \beta_1 & \beta_2 \\ -\beta_2 & \beta_1 \end{bmatrix}.$$

Note that V is unitary, $V^* V = I$. Then

$$\mathbf{S}_e = \mathbf{W} R V^* \begin{bmatrix} (1 + \mathbf{PC})^{-1} & 0 \\ 0 & 1 \end{bmatrix} V R^{-1}.$$

By the matrix-valued version of Proposition 2.1, the feedback system is exponentially stable iff $(I - \Delta\mathbf{S}_e)^{-1} \in H^\infty$. Since $\Delta R = R\Delta$, this condition is equivalent to the invertibility in H^∞ of the matrix-valued function

$$\mathbf{E} = I - \Delta V^* \begin{bmatrix} \mathbf{S} & 0 \\ 0 & \mathbf{W} \end{bmatrix} V.$$

Since $\|\Delta\|_\infty = \|V\| = \|V^*\| = 1$, it is easy to see, using (4.4) and condition (3), that there is a $\lambda > 0$ such that $\mathbf{E}^{-1}$ is uniformly bounded for $|s| > \lambda$ and $\text{Re}\ s \geq 0$. It remains to check the uniform boundedness of $\mathbf{E}^{-1}$ on the compact set $\mathcal{C} = \{s \in \mathbb{C} \mid |s| \leq \lambda,\ \text{Re}\, s \geq 0\}$, which is equivalent to checking that $\mathbf{E}^{-1}$ has no poles in $\mathcal{C}$.

Suppose that $\mathbf{E}^{-1}$ has a pole $p \in \mathcal{C}$, then 1 must be an eigenvalue of $I - \mathbf{E}(p)$. Let $x_0 \neq 0$ be a corresponding eigenvector, then $|\mathbf{S}(p)| < 1$ implies that $|\mathbf{W}(p)| = 1$ and $z_0 = V x_0$ is a scalar multiple of $[0\ 1]^*$. Now by a simple argument $\mathbf{W}(p)\Delta(p)x_o = x_0$, which implies that p is a common pole of $\mathbf{M}_1$ and $\mathbf{M}_2$, in contradiction with the assumptions. Thus, $\mathbf{E}^{-1}$ has no poles in $\mathcal{C}$, and the feedback system is exponentially stable.

The proof of the robustness part is done similarly as for Theorem 3.1, with a few minor extra points to be considered.

As in Section 3, we denote by $\mathbf{G}$ the sensitivity of the multi-periodic repetitive control system discussed above, i.e.,

$$\mathbf{G} = (1 + \mathbf{PCM})^{-1}.$$

We have to point out that, although Theorem 4.1 guarantees the exponential stability of the feedback system, so that $\mathbf{G} \in H^\infty_\alpha$ for some $\alpha < 0$, $\mathbf{G}$ might have poles very close to the imaginary axis. Indeed, if p_1 is a pole of $\mathbf{M}_1$, p_2 is a pole of $\mathbf{M}_2$, and p_1 is very close to p_2, then there will be a pole p of $\mathbf{G}$ very close to p_1 and p_2. If, moreover, p_1 and p_2 are very close to the imaginary axis, then obviously so will be p.

The design procedure for W and C. We describe how the filter and compensator have to be chosen in order to achieve tracking and disturbance rejection. We assume that the signals $r = r_1 + r_2$ and $d = d_1 + d_2$ are as in (4.1). We denote $\nu_1 = \frac{2\pi}{\tau_1}$ and $\nu_2 = \frac{2\pi}{\tau_2}$ (these are the two fundamental frequencies). We assume further that r and d are concentrated in a finite frequency band $[\omega_l, \omega_u]$, which means that the Fourier coefficients of r_1 and d_1 (see (3.4)) are very small if the index $k \in \mathbb{Z}$ is such that $|k\nu_1| \notin [\omega_l, \omega_u]$, and similarly for r_2 and d_2. We choose $\mathbf{W}$ such that (4.4) holds and $\mathbf{W}(i\omega)$ is very close to 1 if $|\omega| \in [\omega_l, \omega_u]$. Then $\mathbf{M}_1$ will have poles very close to $ik\nu_1$ if $|k\nu_1| \in [\omega_l, \omega_u]$, and similarly for $\mathbf{M}_2$. Some of these poles may actually be on the imaginary axis (this can happen where $|\mathbf{W}(i\omega)| = 1$). If a pole of $\mathbf{M}_1$ on the imaginary axis coincides with a pole of M_2, then we have to modify our choice of $\mathbf{W}$ in order to satisfy the assumptions of Theorem 4.1. Now, using H^∞ control theory (and a program package such as the Robust Control Toolbox of MATLAB, a trademark of The MathWorks, Inc.) we design the compensator with transfer function $\mathbf{C}$ such that the three conditions of Theorem 3.1 are satisfied. If this is impossible, then again we have to modify our choice of $\mathbf{W}$, possibly by a compromise in quality: $\mathbf{W}(i\omega)$ might get further away from 1, for $\omega \in [\omega_l, \omega_u]$.

After the above design procedure has been successfully completed, from the formula of $\mathbf{G}$ we see that $|\mathbf{G}(i\omega)|$ is very small if $|\omega| \in [\omega_l, \omega_u]$ and $\omega = k\nu_1$ or $\omega = k\nu_2$, with $k \in \mathbb{Z}$. It follows from Lemma 3.2, Theorem 4.1 and superposition that the error signal e can be decomposed as follows:

$$e = e_{ss1} + e_{ss2} + e_{tr},$$

where $e_{tr} \in L^2_\alpha[0, \infty)$ for some $\alpha < 0$, e_{ss1} is periodic with period τ_1 and e_{ss2} is periodic with period τ_2. (This decomposition is unique iff τ_1/τ_2 is irrational.) Moreover, it follows from (3.3) and (3.6) that e_{ss1} and e_{ss2} will be very small (as measured by their L^2-norms over one period).

Multi-periodic repetitive control as described above would be of questionable value if the gain $|\mathbf{G}(i\omega)|$ could have high peaks for ω lying between the integer multiples of ν_1 and ν_2. Fortunately, this is not the case: good upper bounds for $|\mathbf{G}(i\omega)|$ can be found, for $\omega \in [\omega_l, \omega_u]$, in spite of $\mathbf{G}$ possibly having poles close to the imaginary axis. For lack of space, we do not discuss these bounds here.

References

[1] B.A. Francis and W.M. Wonham. The internal model principle for linear multivariable regulators. *Appl. Math. Optim.* **2** (1975), 170-194.

[2] T. Georgiou and M.C. Smith. w-Stability of feedback systems. *Systems & Control Letters* **13** (1989), 271-277.

[3] T. Georgiou and M.C. Smith. Graphs, causality and stabilizability: linear, shift-invariant systems on $\mathcal{L}_2[0,\infty)$. *Mathematics of Control, Signals, and Systems* **6** (1993), 195-223.

[4] S. Hara, Y. Yamamoto, T. Omata and M. Nakano. Repetitive control system: A new type servo system for periodic exogenous signals. *IEEE Trans. Aut. Contr.* **33** (1988), 659-668.

[5] M. Green and D.J.N. Limebeer. *Linear Robust Control.* Englewood Cliffs, NJ: Prentice-Hall, 1995.

[6] T. Inoue, M. Nakano and S. Iwai. High accuracy control of servomechanism for repeated contouring. *Proc. of the 10th Annual Symp. on Incremental Motion Control, Systems and Devices 1981.* 258-292.

[7] T. Inoue, M. Nakano, T. Kubo, S. Matsumoto and H. Baba. High accuracy control of a proton synchrotron magnet power supply. *Proc. of the IFAC 8th World Congress 1981.* 216-221.

[8] H. Logemann. Stabilization and regulation of infinite-dimensional systems using coprime factorizations. *Analysis and Optimization of Systems: State and Frequency Domain Approaches for Infinite-Dimensional Systems.* (R.F. Curtain, A. Bensoussan and J.L. Lions, Eds.). Vol. 185. *LNCIS.* Berlin: Springer-Verlag, 1993.

[9] H. Logemann, R. Rebarber and G. Weiss. Conditions for robustness and nonrobustness of the stability of feedback systems with respect to small delays in the feedback loop. *SIAM J. Control and Optim.* **34** (1996), 572-600.

[10] K.L. Moore. *Iterative Learning Control for Deterministic Systems. Adv. in Ind. Control.* London: Springer-Verlag, 1993.

[11] R. Rebarber. Conditions for the equivalence of internal and external stability for distributed parameter systems. *IEEE Trans. Aut. Contr.* **38** (1993), 994-998.

[12] E. Rogers and D.H. Owens. *Stability Analysis for Linear Repetitive Processes.* Vol. 175. *LNCIS.* Berlin: Springer-Verlag, 1992.

[13] D. Salamon.Realization theory in Hilbert space. *Mathematical Systems Theory* **21** (1989), 147-164.

[14] G. Weiss. Transfer functions of regular linear systems. Part I: characterizations of regularity. *Trans. Amer. Math. Society* **342** (1994), 827-854.

[15] G. Weiss. Regular linear systems with feedback. *Mathematics of Control, Signals, and Systems* **7** (1994), 23-57.

[16] G. Weiss and R.F. Curtain. Dynamic stabilization of regular linear systems. To appear *IEEE Trans. Automatic Control.*

[17] Y. Yamamoto. Learning control and related problems in infinite-dimensional systems. *Essays on Control: Perspectives in the Theory and its Applications.* (H.L. Trentelman and J.C. Willems, Eds.). Boston: Birkhäuser, 1993. 191-222.

Center for Systems and Control Engineering, School of Engineering, University of Exeter, Exeter EX4 4QF, United Kingdom

Fitting Data Sequences to Linear Systems

Jan C. Willems[1]

1 Dynamical Systems

In this introduction, we review some basic notions and results from the behavioral approach to dynamical systems. We refer to [7, 10, 11] for a more thorough exposition of this theory.

A *dynamical system* is a triple $\Sigma = (\mathbb{T}, \mathbb{W}, \mathfrak{B})$ with $\mathbb{T} \subset \mathbb{R}$ the *time-axis*, $\mathbb{W}$ a set called the *signal space*, and $\mathfrak{B} \subset \mathbb{W}^{\mathbb{T}}$ the *behavior*. The behavior is the central object of interest when we view a dynamical system as a mathematical model: it tells us what time signals (those in $\mathfrak{B}$) are, according to the model, possible, and what time signals are not possible (those not in $\mathfrak{B}$). The theory developed in this paper holds for *discrete–time systems* with time axis $\mathbb{Z}, \mathbb{Z}_+$, or $\mathbb{Z}_-$. For the sake of concreteness, however, we assume at first that $\mathbb{T} = \mathbb{Z}_+$. We assume that the signal space $\mathbb{W} = \mathbb{F}^q$, where $\mathbb{F}$ is a field. The main cases of interest are $\mathbb{F} = \mathbb{R}, \mathbb{C}$, or a Galois field. We will assume throughout that $\mathbb{F}$ is one of these. The dynamical system $\Sigma = (\mathbb{Z}_+, \mathbb{F}^q, \mathfrak{B})$ is said to be *linear* if $\mathfrak{B}$ is a linear subspace of $(\mathbb{F}^q)^{\mathbb{Z}}$, and *time–invariant* if $\sigma\mathfrak{B} \subset \mathfrak{B}$. Here σ denotes the *backward shift*, defined as $(\sigma w)(t) := w(t+1)$.

A typical way of representing a discrete–time dynamical system is by a difference equation. However, we will introduce such representations from a conceptual point of view. We call the dynamical system $\Sigma = (\mathbb{Z}_+, \mathbb{F}^q, \mathfrak{B})$ *complete* if $w : \mathbb{Z}_+ \to \mathbb{F}^q$ belongs to $\mathfrak{B}$ whenever for all $t_0, t_1 \in \mathbb{Z}$ there holds $w\mid_{[t_0,t_1]\cap\mathbb{Z}_+} \in \mathfrak{B}\mid_{[t_0,t_1]\cap\mathbb{Z}_+}$. Denote the polynomials in the indeterminate ξ and with coefficients in $\mathbb{F}$ by $\mathbb{F}[\xi]$ and the polynomial vectors or matrices by $\mathbb{F}^{\bullet}[\xi], \mathbb{F}^{\bullet\times\bullet}[\xi]$. If the number of rows is n, we use the notation $\mathbb{F}^n[\xi], \mathbb{F}^{n\times\bullet}[\xi]$, for columns or rows and columns.

Let $R \in \mathbb{F}^{\bullet\times q}[\xi]$ and consider the system of difference equations

$$\boxed{R(\sigma)w = 0} \qquad (1.1)$$

This defines the system $\Sigma = (\mathbb{Z}_+, \mathbb{F}^q, \ker R(\sigma))$ with $R(\sigma)$ viewed as a map from $(\mathbb{F}^q)^{\mathbb{Z}_+}$ to $(\mathbb{F}^{rowdim(R)})^{\mathbb{Z}_+}$. It is trivial to see that this system is linear, time–invariant and complete. In fact:

Theorem 1.1 *Let $\Sigma = (\mathbb{Z}_+, \mathbb{F}^q, \mathfrak{B})$ be a dynamical system. The following statements are equivalent:*

1. *Σ is linear, time–invariant, and complete;*

---

[1]The research on which this paper is based was done in collaboration with Dr. Margreet Kuijper, presently at the University of Melbourne.

2. *$\mathfrak{B}$ is a linear, shift–invariant ($\sigma\mathfrak{B} \subset \mathfrak{B}$), closed (in the topology of pointwise convergence) subspace of $(\mathbb{F}^q)^{\mathbb{Z}_+}$;*

3. *there exists a polynomial matrix $R \in \mathbb{F}^{\bullet \times q}[\xi]$ such that $\mathfrak{B} = \ker R(\sigma)$.*

We refer to [10, 11] for the proof of this (and the other results whose proof will not be given here).

We denote the family of dynamical systems $\Sigma = (\mathbb{Z}_+, \mathbb{F}^q, \mathfrak{B})$ satisfying any of the equivalent conditions 1, 2, or 3 of the above theorem by $\mathfrak{L}^q_{\mathbb{F}}$.

2 Equivalent Kernel Representations

We call (1.1) (or simply R) a *kernel representation* of the system $\Sigma = (\mathbb{Z}_+, \mathbb{F}^q,$
$\ker R(\sigma)) \in \mathfrak{L}^q_{\mathbb{F}}$. If $R_1, R_2 \in \mathbb{F}^{\bullet \times q}[\xi]$ are both kernel representations of the same system in $\mathfrak{L}^q_{\mathbb{F}}$ then we call R_1, R_2 *kernel equivalent* (denoted as $R_1 \underset{K}{\sim} R_2$). We call (1.1) (or simply R) a *minimal kernel representation* if $R' \underset{K}{\sim} R$ implies $rowdim(R) \leq rowdim(R')$. The following proposition classifies minimality and kernel equivalence.

Proposition 1

1. $R_1 \underset{K}{\sim} R_2$ iff there exists $F_1, F_2 \in \mathbb{F}^{\bullet \times \bullet}[\xi]$ such that $R_1 = F_2 R_2$ and $R_2 = F_1 R_1$;

2. R induces a minimal kernel representation iff R is of full row rank (i.e., its rank must equal its row dimension);

3. Let (1.1) be a minimal kernel representation of $\Sigma \in \mathfrak{L}^q_{\mathbb{F}}$. Then all minimal kernel representations can be obtained by the action of the transformation group $R \underset{U}{\longmapsto} UR$ where $U \in \mathbb{F}^{rowdim(R) \times rowdim(R)}[\xi]$ ranges over the group of unimodular polynomial matrices (i.e., $\det U$ must be a nonzero polynomial of degree 0).

Motivated by the above proposition, we now introduce a canonical form for kernel representations of a given system $\Sigma \in \mathfrak{L}^q_{\mathbb{F}}$. This canonical form is a refinement of the minimal kernel representations, where it was only the number of equations that is minimized. In the canonical form we also minimize the lags in the individual equations. Consider (1.1) and write the polynomial matrix R in terms of its columns, $R = rol(r_1, r_2, \cdots, r_{rowdim(R)})$. Let d_k be the degree of r_k. We call the integers $(d_1, d_2, \cdots, d_{rowdim(R)})$ obtained this way the *lag indices* of the kernel representation (1.1), and call $d_1 + d_2 + \cdots + d_{rowdim(R)}$ the *total lag* of (1.1). Finally, we call (1.1) a *minimal lag representation* if it is a minimal kernel representation and if $R' \underset{K}{\sim} R$

implies that the total lag associated with R does not exceed the total lag associated with R'. Finally, recall the notion of a row proper polynomial matrix. Consider the polynomial matrix $M = col(m_1, m_2, \cdots, m_{rowdim(M)})$, and $M \in \mathbb{F}^{\bullet\times\bullet}(\xi)$ in terms of its columns: let d_k denote the degree of m_k. Write $m_k(\xi) = c_{d_k}\xi^{d_k} + c_{d_{k-1}}\xi^{d_k-1} + \cdots$. Now form the leading coefficient matrix $C_L = col(c_{d_1}, c_{d_2}, \cdots, c_{d_{rowdim(M)}})$. Then M is said to be *row proper* if C_L is a matrix of full row rank (whence the polynomial matrix M itself is also of full row rank). The following theorem analyzes the structure of minimal lag representations.

Theorem 2.1 *Let $\Sigma \in \mathfrak{L}^q_{\mathbb{F}}$.*

1. *(1.1) is a minimal lag representation of Σ iff R is row proper;*
2. *Assume that (1.1) is a minimal lag representation and that its rows are such that its lag indices $d_1, d_2, \cdots, d_{rowdim(R)}$ are ordered as $d_1 \leq d_2 \leq \cdots \leq d_{rowdim(R)}$. Let $R' \in \mathbb{F}^{\bullet\times q}[\xi], R' \underset{K}{\sim} R$, have as lag indices $d'_1 \leq d'_2 \leq \cdots \leq d'_{rowdim(R')}$. Then $rowdim(R') \geq rowdim(R)$, and for all $1 \leq k \leq rowdim(R)$, there holds*
$$d_{rowdim(R)-k} \leq d'_{rowdim(R')-k}.$$
3. *If R' induces a minimal kernel representation, for all $1 \leq k \leq rowdim$ (R), there holds*
$$d_k \leq d'_k.$$

The above theorem shows that a minimal lag representation not only minimizes the total lag but also the individual lags of the difference equations involved in the kernel description of a system $\Sigma \in \mathfrak{L}^q_{\mathbb{F}}$. There are however still many minimal lag representations of a given system $\Sigma \in \mathfrak{L}^q_{\mathbb{F}}$, but similarly as for minimal kernel representations where the non–uniqueness involved pre–multiplication by a unimodular polynomial matrix, it is possible to classify the non–uniqueness of minimal lag representations. We now describe this non–uniqueness. Let (1.1) be a minimal lag kernel representation of a given $\Sigma \in \mathfrak{L}^q$. Denote the lag indices by $d_1, d_2, \cdots$, and assume that they are in increasing order $d_1 \leq d_2 \leq \cdots \leq d_{rowdim(R)}$. Lump the lags that are equal, obtaining

$$\begin{aligned} d_1 = \cdots = d_{n_1} < d_{n_1+1} = \cdots = d_{n_1+n_2} < \cdots \\ < d_{n_1+n_2+\cdots+n_{k-1}+1} = \cdots = d_{rowdim(R)}. \end{aligned} \tag{2.1}$$

Now consider a lower triangular $rowdim(R) \times rowdim(R)$ polynomial matrix of the following form

$$U = \begin{bmatrix} U_{11} & 0 & \cdots & 0 \\ U_{21} & U_{22} & \cdots & 0 \\ \vdots & \vdots & \ddots & \vdots \\ U_{k1} & U_{k2} & \cdots & U_{kk} \end{bmatrix}$$

with

(i) $U_{ij} \in \mathbb{F}^{n_i \times n_j}[\xi]$

(ii) degree $U_{ij} \leq n_i - n_j$

(iii) $U_{11}, U_{22}, \cdots, U_{kk}$ nonsingular.

It is easy to see that the polynomial matrix U is unimodular, and that UR has the same lag indices as R. In fact,

Theorem 2.2 *Let (1.1) be a minimal lag representation of $\Sigma \in \mathfrak{L}^q_{\mathbb{F}}$. Then all minimal lag representations of $\Sigma \in \mathfrak{L}^q_{\mathbb{F}}$ are obtained by premultiplying R by a polynomial matrix U satisfying the above conditions.*

3 The MPUM

In this section we will introduce the notion of more powerful models in the context of the model class $\mathfrak{L}^q_{\mathbb{F}}$. Let $\Sigma_k = (\mathbb{Z}_+, \mathbb{F}^q, \mathfrak{B}_k) \in \mathfrak{L}^q_{\mathbb{F}}, k = 1, 2$. We call Σ_1 *more powerful* than Σ_2 if $\mathfrak{B}_1 \subset \mathfrak{B}_2$: the more a model forbids, the more powerful it is. Let $\mathfrak{D} \subset (\mathbb{F}^q)^{\mathbb{Z}_+}$; $\mathfrak{D}$ stands for *"data"*. We call $\Sigma = (\mathbb{Z}_+, \mathbb{F}^q, \mathfrak{B}) \in \mathfrak{L}^q$ *unfalsified* by $\mathfrak{D}$ if $\mathfrak{D} \subset \mathfrak{B}$, which means that the observations do not contradict the fact that $\mathfrak{B}$ may be the behavior of the dynamical system that produced the observations $\mathfrak{D}$. Finally we call $\Sigma = (\mathbb{Z}_+, \mathbb{F}^q, \mathfrak{B}) \in \mathfrak{L}^q_{\mathbb{F}}$ the *most powerful unfalsified model (MPUM)* in $\mathfrak{L}^q_{\mathbb{F}}$ for $\mathfrak{D}$ if whenever $\Sigma' = (\mathbb{Z}_+, \mathbb{F}^q, \mathfrak{B}') \in \mathfrak{L}^q_{\mathbb{F}}$ is also unfalsified by $\mathfrak{D}$, then $\mathfrak{B} \subset \mathfrak{B}'$.

The MPUM in $\mathfrak{L}^q_{\mathbb{F}}$ indeed exists:

Proposition 2 Let $\mathfrak{D} \in (\mathbb{F}^q)^{\mathbb{Z}_+}$. Then the MPUM in $\mathfrak{L}^q_{\mathbb{F}}$ for $\mathfrak{D}$ exists. It is denoted as $\Sigma^*_{\mathfrak{D}}$.

Proof. The system $\Sigma \in \mathfrak{L}^q_{\mathbb{F}}$ with behavior $\bigcap_{\alpha} \mathfrak{B}_\alpha$, where $\mathfrak{B}_\alpha$ ranges over all the behavior unfalsified models in $\mathfrak{L}^q_{\mathbb{F}}$, is obviously the MPUM.

There is a handy recursive way to obtain a kernel representation of the MPUM for a finite data set $\mathfrak{D} = \{w_1, w_2, \cdots, w_n\} \subset (\mathbb{F}^q)^{\mathbb{Z}_+}$. The recursive step of this algorithm is explained as follows. Assume that $R_k \in \mathbb{F}^{\bullet \times q}[\xi]$ induces a kernel representation for the MPUM in $\mathfrak{L}^q_{\mathbb{F}}$ for the data $\mathfrak{D}_{k-1} = \{w_1, w_2, \cdots, w_{k-1}\}$. First compute the error at stage k: $e_k = R_{k-1}(\sigma) w_k$. Then $e_k \in (\mathbb{F}^{rowdim(R_{k-1})})^{\mathbb{Z}_+}$. Next compute $E_k \in \mathbb{F}^{\bullet \times rowdim(R_{k-1})}[\xi]$ where E_k induces a kernel representation for the MPUM of $\{e_k\}$. The recursive step $R_k = E_k R_{k-1}$ induces with $R_0 = I$ through $\mathbb{R}_n$ a kernel representation in $\mathfrak{L}^q_{\mathbb{F}}$ for the data $\mathfrak{D}_n = \{w_1, w_2, \cdots, w_n\}$.

Summarizing, the algorithm runs as follows:

- Set $R_0(\xi) = I$.
- For $k = 1, 2, \cdots, r$, compute
  - the vector time-series $e_k = R_{k-1}(\sigma)w_k$
  - the polynomial matrix $E_k \in \mathbb{F}^{\bullet \times rowdim(R_{k-1})}[\xi]$ which induces a kernel representation of the MPUM in $\mathfrak{L}_{\mathbb{F}}^{rowdim(R_{k-1})}$ for $\{e_k\}$.
  - the product
$$R_k = E_k R_{k-1}. \tag{3.1}$$
- The algorithm returns $R_n \in \mathbb{F}^{\bullet \times q}[\xi]$; R_n induces a kernel representation of $\Sigma_{\mathfrak{D}}^*$ the MPUM in $\mathfrak{L}_{\mathbb{F}}^q$ for $\mathfrak{D} = \{w_1, w_2, \cdots, w_n\}$.

However, the above algorithm depends completely on our ability of computing E_k, a kernel representation of a single time–series. There are a number of situations in which it can be done easily. For example, if w is an *exponential time–series*: $w(t) = a\rho^t$ with $a \in \mathbb{F}^q$ and $\rho \in \mathbb{F}$. Then a kernel representation for the MPUM for $\{w\}$ is given by $R(\xi) = I$ if $a = 0$, and if $a \neq 0$, for example, if the i–th component of $a, a_i \neq 0$, by

$$R(\xi) = \begin{bmatrix} a_i & & & & -a_1 & & & \\ & a_i & & & -a_2 & & & \\ & & \ddots & & \vdots & & & \\ & & & a_i & -a_{i-1} & & & \\ & & & & \xi - \rho & & & \\ & & & & -a_{i+1} & a_i & & \\ & & & & \vdots & & \ddots & \\ & & & & -a_q & & & a_i \end{bmatrix}. \tag{3.2}$$

The column with the elements $-a_1, \cdots, -a_{i-1}, \xi - \rho, a_{i+1}, \cdots, -a_q$ is the i–th column and the blanks are zero elements. It is easily verified that the above indeed specifies the MPUM for this exponential time series.

The combination of this result with the recursive algorithm yields a rather explicit way of computing the MPUM for $\mathfrak{D} = \{a_1\rho_1^t, a_2\rho_2^t, \cdots, a_n\rho_n^t\}$ with $\rho_1, \rho_2, \cdots, \rho_n \in \mathbb{F}$ and $a_1, a_2, \cdots, a_n \in \mathbb{F}^q$ (of course, $a_k\rho_k^t$ is a careless notation for $t \in \mathbb{Z}_+ \mapsto a_k\rho^t \in \mathbb{F}^q$). In this case $e_k(t) = R_{k-1}(\rho_k)a_k\rho_k^t$. Note that e_k is also an exponential time–series. Therefore E_k can be readily computed, and the recursive algorithm with $R_0 = I, R_k = E_k R_{k-1}$, returns the MPUM R_n in a computationally very efficient way.

A special case of what we have called an exponential time–series is a *pulse*, i.e., a time-series of the form $(a, 0 \cdots, 0, \cdots)$ (set $\rho = 0$ in the above, in particular in formula (3.2)). Pursuing this idea in the case of an arbitrary time-series yields a recursive algorithm for computing the MPUM for an arbitrary time–series $w = (w(0), w(1), \cdots, w(t), \cdots) \in (\mathbb{F}^q)^{\mathbb{Z}_+}$. From w, derive the shifted time–series $w, \sigma w, \cdots, \sigma^t w, \cdots$. Now assume that we have an *initialization*; i.e., assume that somehow we know the MPUM R_n for $\{\sigma^n w\}$. For example, if $w(t) = 0$ for $t \geq n$, then $R_n(\xi) = I$. Now, apply the recursive algorithm for $\mathfrak{D} = \{\sigma^{n-1}w, \sigma^{n-2}w, \sigma^{n-3}w, \cdots, w\}$. For obvious reasons, we will now let the recursive step go from n down to 0. The algorithm now runs as follows.

Initialization. R_n. At stages $k = n, n-1, \cdots, 0$, compute $e_k = R_{k+1}(\sigma)\sigma^k w$ with R_{k+1} an MPUM for $\{\sigma^{n-1}w, \cdots, \sigma^{k+1}w\}$ and observe that e_k is a pulse! Consequently, E_k is readily computed, yielding $R_k = E_k R_{k+1}$. Finally, proceed recursively down to R_0. This yields the MPUM for $\{w\}$.

4 Recursive Computation of the MPUM

It turns out that the recursive algorithm explained in Section 3 can be refined without increase in computational complexity so as to generate a minimal lag kernel representation of the MPUM. The only thing that we have to do is to make sure that the recursive step preserves the minimal lag structure; i.e., in (3.1), we should choose E_k such that if R_{k-1} is row proper, so will be $R_k = E_k R_{k-1}$. This can be done as follows.

Consider $R_{k-1} \in \mathbb{F}^{\bullet \times q}[\xi]$ and assume that it is row proper. Arrange its lag indices as in (2.1). Assume thus that R_k has as *lag index sequence* $L_{k-1} := (L_0^{k-1}, L_1^{k-1}, \cdots, L_{\mathfrak{L}}^{k-1}l, \cdots)$ where $L_{\mathfrak{L}}l^{k-1}$ denotes the number of rows in R_{k-1} with degree $\mathfrak{L}l$. Thus R_{k-1} has L_0^{k-1} lag indices equal to 0, L_1^{k-1} equal to 1, etc. In particular, its number of rows equals $\sum\limits_{\mathfrak{L}l} L_{\mathfrak{L}}^{k-1}l$ and its total lag equals $\sum\limits_{\mathfrak{L}} l\mathfrak{L}lL_{\mathfrak{L}}^{k-1}l$. The recursive update of R_{k-1} will now involve a recursive update of R_{k-1} and of L_{k-1}.

Assume that $e_k(t) = \epsilon_k \rho_k^t$ with $\epsilon_k \in \mathbb{F}^q$ and $\rho_k \in \mathbb{F}$. Now consider the k–th error vector $\epsilon_k = col(\epsilon_k^1, \epsilon_k^2, \cdots, \epsilon_k^{rowdim(R_{k-1})})$. In the applications that we have in mind $rowdim(R_{k-1})$ will usually be equal to q, but that need not concern us here. Let r_k be such that $\epsilon_k = \cdots = \epsilon_k^{r_k-1} = 0$ and $\epsilon_k^{r_k} \neq 0$. Now analyze R_{k-1}. Let $\mathfrak{L}l_{k-1}$ be such that

$$\sum_{\mathfrak{L}l=0}^{\mathfrak{L}l_{k-1}-1} L_{\mathfrak{L}}^{k-1}l < r_k \leq \sum_{\mathfrak{L}l=0}^{\mathfrak{L}l_{k-1}} L_{\mathfrak{L}}l^{k-1} =: m_k. \tag{4.1}$$

The desired E_k which preserves row properness is given by

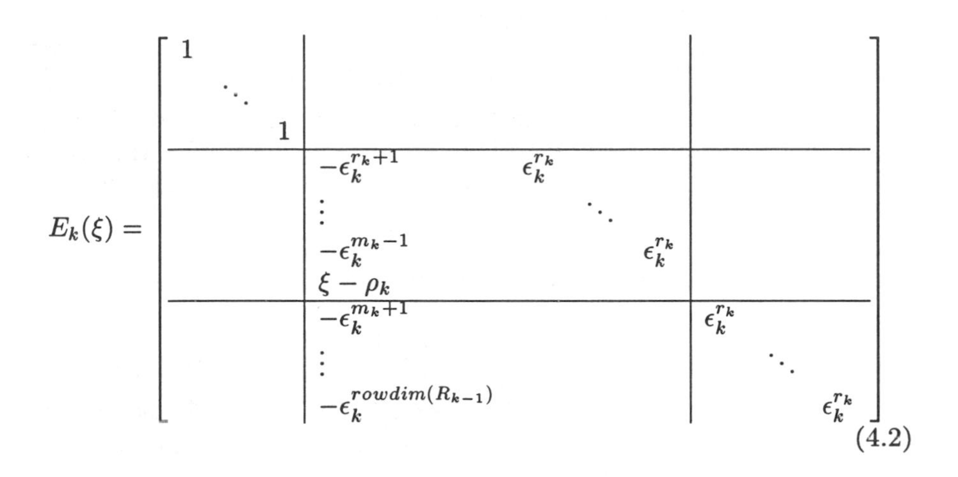

(4.2)

If it happens that $r_k = m_k$, then the middle submatrix in (4.2) is just $(\xi - \rho_k)$. The blanks in (4.2) are again meant to be zeros. The matrix E_k looks more complicated than it is. The underlying idea is as follows. We look down the vector ϵ_k until we meet the first non–zero entry. That determines the lag of R_{k-1} that we has to increased by one. By multiplying the corresponding row of R_k by $(\xi - \rho_k)$ we indeed achieve this. The matrix E_k is further arranged so that this particular row is subsequently transferred to the group with one lag higher, and such that no other lags are increased.

The recursive step consists of putting the k–th kernel representation equal to $R_k = E_k R_{k-1}$ and the k–th lag index sequence equal to $L_k = L_{k-1} + (0, 0, \cdots, 0, 1, 0, \cdots)$ with the 1 in this (infinite) vector the m_k–th entry. Then if R_n induces a minimal lag kernel representation of the MPUM in $\mathfrak{L}^q_{\mathbb{F}}$ for $\mathfrak{D} = \{w_1, w_2, \cdots, w_n\}$.

Obviously by using this construction on $\mathfrak{D} = \{\sigma^{n-1}w, \sigma^{n-2}w, \cdots, \sigma w, w\}$ we obtain a MPUM for the single time–series $\{w\}$ starting with a known MPUM R_n for $\sigma^n w$.

5 Finite Time-Series

In this section we extend the theory of the previous sections, which concerns *infinite* vector time–series, to the modeling of *finite* vector time–series.

Consider the time–series

$$a = (a_1, a_2, \cdots, a_T) \tag{5.1}$$

with $a_i \in \mathbb{F}^q, i = 1, 2, \cdots, T$. We call the sequence

$$c = (c_0, c_1, \cdots c_n) \tag{5.2}$$

with $c_i \in \mathbb{F}^q, i = 0, 1, \cdots, n$, an *annihilator* of a if

$$c_0^T a_t + c_1^T a_{t-1} + \cdots + c_n^T a_{t-n} = 0 \tag{5.3}$$

for $t = n+1, \cdots, T$; n is called the *lag* of the annihilator c. We call c a *shortest lag annihilator* for a if among all annihilators of a, the lag n is as short as possible. The problem that we consider is to describe all annihilators.

The idea followed is to replace a by an infinite time–series, and subsequently apply the theory of the previous sections. Define $\tilde{a} : \mathbb{Z} \to \mathbb{F}^{q+1}$ as follows

$$\tilde{a} = (\cdots \begin{bmatrix} 0 \\ 0 \end{bmatrix} \cdots \begin{bmatrix} 0 \\ 0 \end{bmatrix} \begin{bmatrix} 0 \\ 1 \end{bmatrix} \begin{bmatrix} a_1 \\ 0 \end{bmatrix} \cdots \begin{bmatrix} a_T \\ 0 \end{bmatrix}). \tag{5.4}$$

Note that in $\tilde{a}$, vectors $0 \in \mathbb{F}^q$ have been added before a and that a sequence $(\cdots 0 \cdots 0\ 1\ 0 \cdots 0)$ has been appended to this extension of a.

Call the non–zero sequence $\tilde{c} = (\tilde{c}_0, \tilde{c}_1, \cdots, \tilde{c}_n)$ in $\mathbb{F}^{q+1}$ an annihilator for $\tilde{a}$ if

$$\tilde{c}_0^T \tilde{a}_t + \tilde{c}_1^T \tilde{a}_{t-1} + \cdots + \tilde{c}_n^T \tilde{a}_{t-n} = 0 \tag{5.5}$$

for $t \in \mathbb{Z}, t \leq T$. Partition $\tilde{c}_k$ as $\begin{bmatrix} \tilde{c}'_k \\ \tilde{c}''_k \end{bmatrix}$ with $\tilde{c}'_k \in \mathbb{F}^q$, and $\tilde{c}''_k \in \mathbb{F}$. Writing out (5.5) yields the following relations:

$$(\tilde{c}'_0)^T a_t + (\tilde{c}'_1)^T a_{t-1} + \cdots + (\tilde{c}'_n)^T a_{t-n} = 0$$

for $t = n+1, \cdots, T$

$$(\tilde{c}'_0)^T a_t + (\tilde{c}'_1)^T a_{t-1} + \cdots + \tilde{c}'_{t-1} a_1 = -\tilde{c}''_t$$

for $t = 1, \cdots, n$

$$\tilde{c}''_0 = 0.$$

From these relations it is obvious that c is annihilating for a iff a sequence $*$ can be appended to c such that $\begin{bmatrix} c \\ * \end{bmatrix}$ becomes an annihilating sequence for $\tilde{a}$.

The recursive procedure described in the previous paragraph makes it obvious how an MPUM for $\tilde{a}$ can be computed. Of course, there is a minor difference here since $\tilde{a}$ is defined on $\mathbb{Z}_-$ instead of $\mathbb{Z}_+$. However, this is easily accommodated by working with the forward shift σ^{-1} instead of with the backward shift σ.

More interesting than the problem of finding annihilators as (5.3), is the problem of finding a recursion. Thus we call the sequence

$$G = (G_0, G_1, \cdots, G_n) \tag{5.6}$$

with $G_i \in \mathbb{F}^{q \times q}, i = 1, 2, \cdots, n$, a *recursion* for a if G_0 is non–singular, and if

$$G_0 a_t + G_1 a_{t-1} + \cdots + G_n a_{t-n} = 0 \tag{5.7}$$

for $t = n+1, \cdots, T$. If again we extend a to $\tilde{a}$ as in (5.4), then the problem of finding a minimal lag recursion becomes analogous to the ***partial realization problem***. Actually, for the case at hand, it is the single–input, multiple–output version of it, but by extending the algorithm below to the case of a data set with more than one time–series, the multiple–input version can be treated as well.

Let $\tilde{a}$ be as defined above, and define $\mathfrak{D} = (\tilde{a}_0, \tilde{a}_1, \cdots, \tilde{a}_{T-1}, \tilde{a}_T)$ with $\tilde{a}_k := \sigma^{-T+k}\tilde{a}$. The algorithm is initialized as $R_{-1}(\xi) = I$ where I denotes the identity matrix in $\mathbb{F}^{(q+1)\times(q+1)}$. Define $e_k = R_{k-1}(\sigma^{-1})\tilde{a}_k$. Then $e_k : \mathbb{Z}_- \to \mathbb{F}^{q+1}$ will be a pulse. Denote $e_k(0) = col(\epsilon_k^{\prime}, \epsilon_k^2, \cdots, \epsilon_k^{q+1})$. Now proceed as in Section 4, yielding $E_k(\xi)$ given by

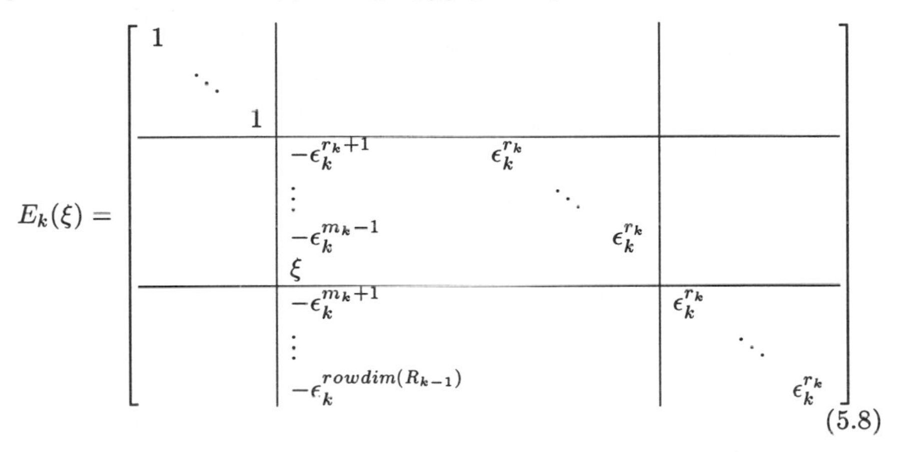

$$E_k(\xi) = \left[\begin{array}{ccc|cccc|ccc} 1 & & & & & & & & & \\ & \ddots & & & & & & & & \\ & & 1 & & & & & & & \\ \hline & & & -\epsilon_k^{r_k+1} & \epsilon_k^{r_k} & & & & & \\ & & & \vdots & & \ddots & & & & \\ & & & -\epsilon_k^{m_k-1} & & & \epsilon_k^{r_k} & & & \\ & & & \xi & & & & & & \\ \hline & & & -\epsilon_k^{m_k+1} & & & & \epsilon_k^{r_k} & & \\ & & & \vdots & & & & & \ddots & \\ & & & -\epsilon_k^{rowdim(R_{k-1})} & & & & & & \epsilon_k^{r_k} \end{array}\right] \tag{5.8}$$

Note that $R_k = E_k R_{k-1}$ now has a common factor ξ in all the elements of its m_k–th row. Define the reduced matrix $R_k^{\prime}$ as the one obtained by deleting the m_k th row and $(q+1)$–th column of R_k. Then

$$R_k^{\prime}(\xi) = G_0^k + G_1^k\zeta + \cdots + G_{n_k}^k\xi^{n_k} \tag{5.9}$$

yields by

$$(G_0^k, G_1^k, \cdots, G_{n_k}^k) \tag{5.10}$$

a minimal–lag recursion for $(a_1, a_2, \cdots, a_k)$, and $R_T^{\prime}$ yields a recursion for a. This procedure is summarized in the diagram

$$\begin{array}{ccccccccccccc} R_{-1} = I & \to & E_0 & \to & R_0 & \to & \cdots & \to & R_k & \to & \cdots & \to & R_T \\ & & & & \downarrow & & & & \downarrow & & & & \downarrow \\ & & & & R_0^{\prime} & & & & R_k^{\prime} & & & & R_T^{\prime} \end{array}.$$

Observe that, if we only desire to find $R_T^{\prime}$, it is not necessary to compute the intermediate $R_k^{\prime}$'s. Note the similarity of the above procedure and that of finding the controllable minimal complexity unfalsified model explained in [11] and [1].

Instead of following the above procedure, one could relay the row with the common factor ξ to the $(q+1)$–th row. This leads to bookkeeping which is more involved than the above. In the case $q = 1$, this is straightforward, and leads to the following algorithm, derived in [4].

The algorithm involves again a recursion in the R_k's. Exploiting the fact that the sum of the lag indices equals the iteration step, and the fact that things can be arranged in such a way that the second element of the error vector ϵ_k always equals one, yields the following algorithm for computing the MPUM R_T in the case $q = 1$.

Initialization: $R_{-1} = I, L_{-1} = 0$. The recursive step requires computing $R_k = E_k R_{k-1}$. Denote the $(1,1)$ element of R_{k-1} by g_{k-1}, $g_{k-1}(\xi) = g_0^{k-1} + g_1^{k-1}\xi + \cdots$. At each stage, compute

$$\epsilon_k = g_0^{k-1} a_k + g_1^{k-1} a_{k-1}. \tag{5.11}$$

Then compute E_k and L_k as follows:
if $\epsilon_k = 0$ or $L_{k-1} > k/2$, set

$$E_k(\xi) = \begin{bmatrix} 1 & -\epsilon_k \\ 0 & \xi \end{bmatrix}; L_k = L_{k-1} \tag{5.12}$$

if $\epsilon_k \neq 0$ and $L_{k-1} \geq k/2$, set

$$E_k(\xi) = \begin{bmatrix} 1 & -\epsilon_k \\ \xi/\epsilon_k & 0 \end{bmatrix}; L_k = k - L_{k-1}. \tag{5.13}$$

Then the $(1,1)$ element of R_T defines the shortest lag recursion for (5.1).

The above algorithm is not new: it is the celebrated Berlekamp–Massey algorithm ([2], [5]), which plays an important role in coding theory and cryptography. Its interpretation as a recursive algorithm for computing an MPUM delivers a great deal of insight, for example in the rationale of setting up a recursion that involves four polynomials. Indeed, the MPUM at each stage is an autonomous system involving two variables, and thus its kernel representation is a 2×2 polynomial matrix.

6 Application to Decoding

The relevance of the Berlekamp–Massey algorithm for decoding Reed–Salomon codes, and, more generally for decoding, alternant codes, is well–known [3, 6]. In this section, we will explain this using system theory language.

Consider the linear system (without inputs)

$$x(t+1) = Fx(t); y(t) = Hx(t) \tag{6.1}$$

with $x \in \mathbb{F}^n, y \in \mathbb{R}^p, F \in \mathbb{F}^{n\times n}, H \in \mathbb{F}^{p\times n}$. Let $T \in \mathbb{N}$ and consider the code in $\mathbb{F}^n$ defined by

$$\mathfrak{C} = \{x(1) \in \mathbb{F}^n | y(1) = \cdots = y(T) = 0\}$$

where the y's are computed using (6.1). Thus we view the code consisting of all initial conditions of a linear system that generate zero outputs for $t = 1, 2, \cdots, T$. The special case that we are interested in, has F diagonal: $F = diag(\lambda_1, \lambda_2, \cdots, \lambda_{kn})$ with $\lambda_i \neq \lambda_j$ for $i \neq j$. The parity check matrix of this code equals

$$\begin{bmatrix} h_1 & h_2 & \cdots & h_n \\ \lambda_1 h_1 & \lambda_2 h_2 & \cdots & \lambda_n h_n \\ \vdots & \vdots & & \vdots \\ \lambda_1^{T-1} h_1 & \lambda_2^{T-1} h_2 & \cdots & \lambda_n^{T-1} h_n \end{bmatrix}. \tag{6.2}$$

In view of the structure of this matrix, we call this code a *Vandermonde code.* It is easy to see that the distance of this code is larger than or equal to $d_H + T - 1$ (assuming this to be less than n), with d_H the distance of the code with parity check matrix H.

Assume the received word equals $r = x + e$ with $x \in \mathfrak{C}$ and e the error. The syndromes may be computed by

$$z(t+1) = \Lambda z(t), s(t) = Hz(t) \tag{6.3}$$

with $z(1) = r$. This algorithm yields the finite time–series

$$(s(1), s(2), \cdots, s(T)) \tag{6.4}$$

with $s(t) \in \mathbb{F}^p$. Denote the error vector by $col(e_1, e_2, \cdots, e_n)$. Then there holds

$$s(t) = \sum_{k=1}^{n} \lambda_k^{t-1} e_k h_k. \tag{6.5}$$

Assume that $j_1, j_2, \cdots, j_{n'}$, are the indices for which $e_k \neq 0$. Thus (6.5) may be viewed as an equation in which $s(t)$ is given for $t = 1, 2, \cdots, T$ and in which $j_1, j_2, \cdots, j_{n'}$ and $e_1, e_2, \cdots, e_{n'}$ are the unknowns. Obviously the difficulty is to find the j_k's. Once these are known, then finding the e_k's is a linear problem.

This problem is readily solved by computing the minimum lag recursion for (6.4). This may be done using the algorithm outlined in the previous section. This algorithm returns the polynomial matrix $R'_T \in \mathbb{F}^{q \times q}[\xi]$. The j_k's and the e_k's are then readily derived from R'_T. Thus finding the error locations $j_1, j_2, \cdots, j_{n'}$ requires finding the λ_k's for which the matrix $R'_T(\lambda_k)$ is singular.

The key point of this brief discussion is to illustrate the application of the recursive MPUM calculation in coding theory. We glossed over many important points, in particular, over the consequences of the use of finite fields.

References

[1] A.C. Antoulas and J.C. Willems. A behavioral approach to linear exact modeling. *IEEE Transactions on Automatic Control* **38** (1993), 1776-1802.

[2] E.R. Berlekamp. *Algebraic Coding Theory.* New York: McGraw-Hill, 1968.

[3] R. Blahut. *Theory and Practice of Error Control Codes.* New York: Addison-Wesley, 1983.

[4] M. Kuijper and J.C. Willems. On constructing a shortest linear recurrence relation. To appear *IEEE Transactions on Automatic Control.*

[5] J.L. Massey. Shift-register synthesis and BCH decoding. *IEEE Transactions on Information Theory* **IT-15** (1969), 122-127.

[6] S. Roman. *Coding and Information Theory.* Berlin: Springer Verlag, 1992.

[7] J.C. Willems. From time series to linear system. Part I: Finite dimensional linear time invariant systems. *Automatica* **22** (1986), 561-580.

[8] J.C. Willems. From time series to linear system. Part II: Exact modeling. *Automatica* **22** (1986), 675-694.

[9] J.C. Willems. From time series to linear system. Part III: Approximate modeling. *Automatica* **23** (1987), 87-115.

[10] J.C. Willems. Models for dynamics. *Dynamics Reported* **2** (1989), 171-269.

[11] J.C. Willems. Paradigms and puzzles in the theory of dynamical systems. *IEEE Transactions on Automatic Control* **36** (1991), 259-294.

University of Groningen, PO Box 800, 9700 AV Groningen, The Netherlands

Fighter Aircraft Control Challenges and Technology Transition

K.A. Wise

1 Introduction

The design of the next generation of fighter aircraft is a multidisciplinary challenge in aerodynamics, control, electromagnetics, and structural design. Numerical prototyping [1] will be used to optimize configurations in order to make these systems affordable. Critical flight control research problems for these kind of aircraft exist in adaptive flight control for reconfiguration, multivariable control (integrated controls), performance optimizing controls, tailless aircraft (unstable), and thrust vectoring for envelope expansion. Applications in these research areas are highly nonlinear and require robust control. The single most important issue is that of affordability. Any problem remains unsolved if the solution is too expensive to implement. Reconfiguration, integrated controls, performance optimizing controls, tailless aircraft, and thrust vectoring control are technologies that can significantly impact affordability and, at the same time, improve aircraft performance. These technologies also reduce aircraft life cycle costs (cost of maintaining and operating). The following list summarizes important issues directed at the control of high performance fighter aircraft. It is important to note that many of these technologies have a strong tie to commercial aircraft.

- Propulsion Controlled Aircraft
- Optimal Control Effector Distribution (Integrated Controls)
- Damage Adaptive Aircraft Control
- Real-time Aircraft Identification
- Real-time Drag Reduction/Signature Reduction
- Optimal Trajectories For Minimum Signature
- Terrain Following/Terrain Avoidance/Threat Avoidance
- Pilot Aids and Control Laws That Improve Flying Qualities
- Load Alleviation/Flutter Suppression
- Optimal Engine/Nozzle Control

- Anti-Pilot Induced Oscillation (PIO) Compensation
- Optimal Aeroservoelastic Filter Design

These topics have been grouped into four technology areas as shown in Figure 1. Payoffs for DoD and industry for solving these problems are also shown in the figure. The following paragraphs discuss in more detail these applied controls research topics. The order of the presentation does not imply any priority ranking.

Technology Area	Payoff
Integrated Controls	ÂLower Operating Costs
ÂAero/Propulsion/Engine	ÂLower Maintenance Costs
ÂPropulsion Controlled Aircraft	ÂLower Design Costs
ÂImproved Flight Safety	ÂEliminate PIO Problems
Reconfigurable Flight Control	ÂIncreased Departure Resistance
ÂOptimal Control Effector Distribution	ÂImproved Agility
ÂDamage Adaptive Aircraft Control	ÂDual Use Capability
ÂReal-time Aircraft Identification	ÂEngine Out Recovery
Aircraft Performance Improvements	ÂDamage Tolerance
ÂTailless Aircraft	ÂSignature Reduction
ÂDrag Reduction/Signature Reduction	ÂReduced Aircraft/Missile Weight
ÂLoad Alleviation/Flutter Suppression	ÂControllable Flight With No Aero Control
ÂEngine/Nozzle Control	ÂMaximize System Reliability
Aircraft/Missile Robust Nonlinear Control	ÂOptimal Control Power Usage
ÂAffordability Issues In Control System Design	ÂIncreased Survivability
ÂAnti-PIO Compensation	
ÂASE Compensation Design Tools	
ÂAsymmetric Vortex Control	
ÂHigh AOA Control Laws	

Figure 1 - Technology areas for controls research

2 Integrated Control Effectors

This area of research is multidisciplinary in nature (aerodynamics, dynamics and control), and includes topics important to reconfigurable / damage adaptive controls, agility/envelope expansion, and robust nonlinear flight control. Many of today's evolving aircraft configurations present a considerable design challenge due to their use of innovative control effectors to increase maneuver capabilities and expand operating envelopes. Thrust vectoring, reaction control systems, pneumatic devices, and vortex control surfaces are commonly being used to augment conventional aerodynamic control surfaces. The design of flight control systems for these configurations is complicated by both the number of control effectors and their highly nonlinear response characteristics. The challenge is to meet stability and flying qualities requirements while fully realizing the performance capabilities of the vehicle. An example of a conceptual high performance tailless fighter aircraft with multiple control effectors is shown in Figure 2. This figure illustrates many of the candidate aerodynamic control effector technologies available. This aircraft model is being used under an AFOSR sponsored tailless aircraft research initiative (begun in May 96) to address critical control technologies for tailless fighters. The institutions/researchers participating are: MDA (Kevin Wise), Univ. of Colorado (John Hauser), and Univ. of Minnesota (Andy Teel). The simulation has also being used at the Univ. of Illinois (Petros Voulgaris) and MIT (Eric Feron, Alexander Megretski, Jim Paduano). Advanced aircraft will require superior mission performance, agility, and stealth levels of operation. Innovative control effectors, as shown in Figure 2,

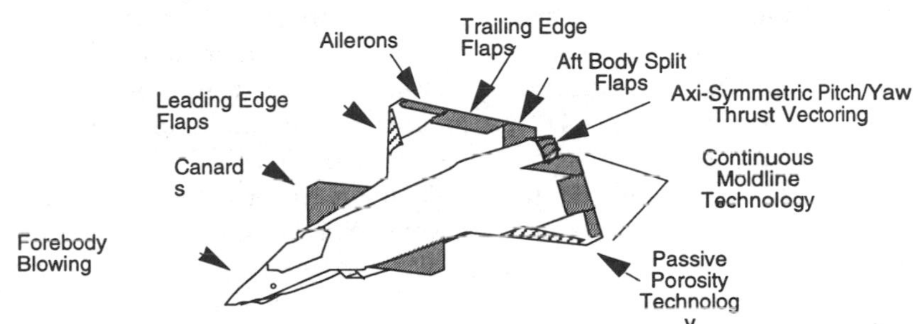

Figure 2 Conceptual high performance tailless fighter.

can help achieve these goals. Advanced effectors that manipulate the aircraft forebody/lex/wing vortex system offer improved high angle-of-attack ( ) pitch and yaw control authority, resulting in increased agility and controllability at high- conditions. The incorporation of advanced low- yaw and roll effectors can enable the reduction or elimination of the vertical tail and rudder control surfaces. This can significantly improve the aircraft survivability by reducing observability. It also results in reduced weight and costs. The use of innovative aerodynamic effectors can decrease the rates and deflections required from main engine thrust vectoring systems, resulting in reduced subsystem weight and cost, as well as increased reliability.

As new aircraft designs progress towards tailless versions, the integration of aerodynamic controls and alternate controls (pneumatic, thrust vectoring, MEMs, passive porosity, active flexible structure control) becomes more critical. In new aircraft designs there are strong incentives for minimizing weight, costs, and maintainability requirements while maximizing system reliability. Conventional control surfaces are ineffective at high 's and also create signature problems. Non-traditional control effectors (such as shown in Figure 2) for achieving flight control need to be optimized and integrated within design constraints of the total weapon system so that the aircraft meet stealth requirements without compromising maneuverability and agility goals.

This problem of control integration is a relatively new problem and has not received adequate attention (because it is so application dependent). Only in the newest fly-by-wire aircraft is this an issue (in older aircraft mechanical linkages between effectors and the pilot stick is used). In fly-by-wire systems, the vehicle management system processes the pilot inputs and commands independent actuators, thus offering new capabilities and new problems.

Methods for control integration include add hoc solutions [2], matrix inverse and psuedoinverse approaches [3], constrained control allocation [4-6], psuedocontrols, and daisy chaining. In industry, during configuration synthesis prior to control law development, an aerodynamicist usually defines (in an add hoc manner) how the control effectors are integrated, and this definition is used in wind tunnel tests to obtain force and moment data. In Speyer [2], the control integration problem was investigated by examining the principal gains of $(j\omega I - A)^{-1}B$ versus frequency, and selecting the control effectors based upon the control power provided throughout the pilot frequencies (0.1 to 1.0 rad/s). In Wise [3] matrix inverse and psuedoinverse methods were used to relate moment commands to control effector commands. Durham [4-6] makes a comparison of constrained control allocation methods, psuedocontrols, and daisy chaining approaches.

The algorithms/logic used to integrate control effectors are not a control law in themselves. The stability, robustness, and flying qualities of the aircraft depend upon how the control laws command the control effectors. Classically, aircraft control effectors are viewed as moment producing devices (roll, pitch, and yaw). Given that the control laws have determined the required moment commands, control distribution logic distributes commands to the individual control effectors to satisfy these requirements.

The problem of integrated controls is one of optimizing the control distribution logic. Different criteria could be used to be optimize/integrate the controls (fuel economy, max agility, max stealth, etc.).

State of the art multivariable flight control system designs usually blend/integrate the control effectors at the elevon δ_e, aileron δ_a, rudder δ_r level. Mode logic and blending logic (on the control effector level) is usually determined in an add hoc fashion. The recent dynamic inversion [7,8] based on control law thrusts [9-19] further complicate this problem. Consider the aircraft dynamics modeled by the equations of the form

$$\dot{x} = f(x) + g(x)u, \tag{1}$$

with n-dimensional state x and m-dimensional control u. A dynamic inversion control law which achieves the desired response characteristics may be formulated as

$$u = g^{-1}[\nu - f(x)] \tag{2}$$

where ν specifies the desired response. The number of controls available ($m > 13$) on the aircraft shown in Figure 2 does not directly (easily) support the use of dynamic inversion.

Typical aircraft applications of dynamic inversion [9-19] use a two time scale approach in which control surface commands $(\delta_a, \delta_e, \delta_r)$ are calculated in order to generate desired angular accelerations $(\dot{p}, \dot{q}, \dot{r})$. This inner loop inversion (fast time scale) requires all m controls surfaces in u to be chained (ganged) together into a form for roll, pitch, and yaw control. Outer loops are then wrapped around these inner loops to achieve aircraft flying qualities (slow time scale).

Snell, et. al. [9] extended the use of dynamic inversion to an aircraft which has more control surfaces than dynamic quantities to control. The control surfaces consisted of aileron, rudder, canard, lateral axis thrust vector control (TVC), and pitch axis TVC. Inversion of the fast dynamics results in a non-square control distribution matrix [$g(x)$ in Eq. (1)]. Equation (1) thus represents an underdetermined set of equations. Since $g(x)$ is right invertible, the required control deflections may still be generated by Eq. (2).

Probably the most common method used to integrate control effectors together is daisy chaining [4]. The basic idea is that a primary control effector (or set of surfaces) is used until it saturates. Upon saturation, another control effector (or set of surfaces) is used to provide additional control power, thus creating the daisy chain. As long as the primary control effector does not saturate the secondary effectors are not used.

A recent study [20] at Wright Laboratory has shown that daisy chaining can add significant phase lag into the control system. Too much phase lag in flight control systems leads to pilot induced oscillations which can lead to instability (YF-22 crash [21]). It is clear that methods are needed to optimize the integration of control effectors.

One benefit of better integration will be to improve the flying qualities of the aircraft (ability to perform precision maneuvers such as landing, refueling, guns on target, etc.). Improved flying qualities will benefit all types of aircraft (fighters, commercial, and transport).

For high performance fighter aircraft, there are additional benefits. For stealth type aircraft, improving the integration of the control effectors can lead an improved stealth mode, where the controls are integrated such that the radar signature is optimized as the aircraft flies. Conventional aerodynamic surfaces, which have edges that cause increased RCS signatures, are not stealthy. For this reason, unconventional control effectors are being considered for stealth aircraft. These effectors include pneumatic devices, passive porosity, and flexible wings.

For pneumatic devices, ports or slots are used to control the boundary layer (sucking or blowing) attaching the flow. Placed in different positions along an aircraft's wing, these devices can provide moderate roll, pitch, and yaw control power.

Passive porosity is used to equalize pressure gradients across an airfoil. A missile forebody example is shown in Figure 3. In the application shown, a control valve (actuator) is placed in the vent pipe. The control action is to open and shut the valve for active vortex control.

Flexible wings, through active control of the shape, can also be used to generate control forces and moments. Conceptually, this allows for thinner, higher aspect ratio wings to be used. The wings would then be deformed aeroelastically into shapes that maximize performance.

Traditional approaches treat wings as rigid structural members, using flaps and deflected surfaces to generate

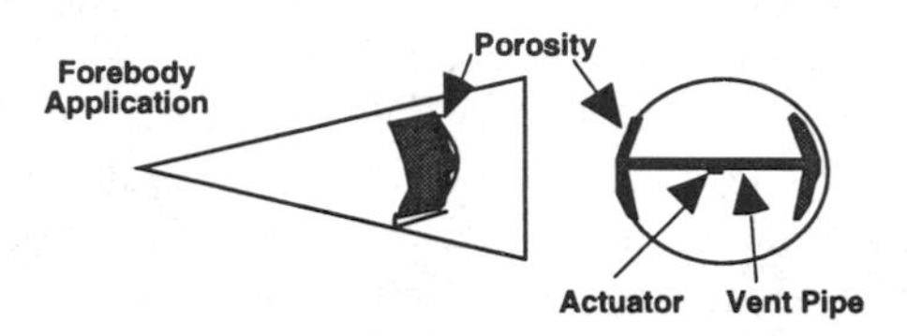

Figure 3 Passive porosity for aero control.

control forces and moments. In this case wing flexibility reduces control effectiveness and is a negative attribute. Wing stiffeners would be added to make the wing less flexible. This significantly increases the weight and the cost.

3 Reconfigurable Flight Control

Three main problem areas need to be addressed through applied research in reconfigurable flight control [22-33]: 1) Identify aircraft parameters in real time (damage assessment); 2) Real time control (optimal control); and 3) Digital implementation (implementation and redundancy). Reconfigurable/damage adaptive (R/DA) flight control must accommodate major changes to the aircraft's stability and control characteristics. This area of research has strong civil and military interest, and contains significant technical risk, with a primary goal to improve pilot/passenger safety. R/DA approaches also support improved fault diagnostics which can lower support costs.

McDonnell Douglas has been heavily involved with the flight testing of R/DA flight control laws using the NASA F-15 ACTIVE aircraft [26, 27, 29, 31, 32]. Under several joint Air Force and NASA sponsored programs [Self Repairing Flight Control System (SRFCS)], damage adaptive control laws were designed, evaluated, and flight tested. During the flight test, a pre-programmed stabilator failure was used to test the approach (the right stabilator was commanded to a fixed position and held). The control laws adapted, giving the pilot full control over the aircraft. The success of this program indicates the potential to use R/DA flight control laws. However, in this case the failure and its characteristics were known a priori. The approach used in the SRFCS program is not applicable when there are large numbers of failures, and does not guarantee that the failures (and/or models of the aircraft) have been properly detected and classified. New thrusts are aimed at real-time identification and optimal control [22-24, 34] in the presence of uncertainty.

Using the F-16 VISTA aircraft, the Air Force is sponsoring Baron Associates [34] in the development of a "Self Designing Controller (SDC)" for R/DA flight control. Under this SBIR program, Barron Associates will be designing a real-time adaptive flight controller taking into account: multiple control effectors; saturation; hard limits; control axis prioritization; command limiting; and adherence to flying qualities requirements

This approach will use an indirect adaptive control methodology that uses explicit parameter identification to characterize the plant dynamics, and standard LQR design techniques on-line for control of the identified plant. the control effectors that are remaining will have to be re-allocated (based upon a control axis prioritization).

It should be clear that real-time control effector integration (and optimal integration) is strongly related to the R/DA flight control problem. For a damaged aircraft, current adaptive control based approaches rely on the identification of the control distribution matrix B to solve this problem. This does not address the fundamental problem of how to best use the controls that are available.

In Bodson [24] a comparison is made of several of the adaptive control approaches being considered for R/DA flight control. Bodson's assessment of the approaches tends to favor the indirect adaptive approach based upon its flexibility in that algorithms that account for actuator saturation [25] can be used. Bodson's papers, and the references therein, provide excellent background material for R/DA control strategies.

4 Aircraft Performance Improvements

There are many unsolved problems in the following areas:

- Aircraft Agility and Envelope Expansion
- Tailless Aircraft
- Drag Reduction/Signature Reduction
- Engine/Nozzle Control
- Load Alleviation/Flutter Suppression

A common theme among these topics is the integrated controls problem. Aircraft agility improvements and envelope expansion can be achieved by optimizing the use of the control effectors, thus maximizing the aircraft's performance. To extend the flight envelope outside the regions accessible by (limited by) aerodynamic controls requires the use of nontraditional controls, such as thrust vectoring, reaction jets, or other vortex control devices, which can be highly nonlinear and further couple the dynamics/aerodynamics (out of plane forces caused by vortex shedding).

Tailless aircraft are operational today, but none achieve the performance and maneuverability levels envisioned for the next generation strike fighters. Tailless aircraft pose new challenges in providing lateral-directional control power for low and moderate angle-of-attacks as well as at low speed conditions (powered approach and carrier suitability).

High bandwidth actuation systems are needed for stability augmentation. With the high bandwidth comes high cost. In addition, the redundancy requirements imposed on these critical subsystems (further increasing costs) makes tailless aircraft a challenge to produce.

Performance optimizing controls for enhanced maneuvering and envelope expansion, and drag reduction to maximize range requires the use of nonlinear controls. Depending upon the aircraft's mode of operation, an aircraft may want to stow its aerodynamic control surfaces in their "zero" deflection state. Control power for maneuvers would be provided by alternate controls (pneumatic, thrust vectoring, MEMs, passive porosity, flexible wing control). Nonlinear robust controls will be needed to provide the pilot with this type of mode selection capability. For large maneuvers, outside the control range of the alternate controls (primary control effectors), the aerodynamic surfaces would be faded in (secondary control effectors). Only when the alternate control effectors fail to provide enough control to satisfy pilot commands would the aerodynamic surfaces be used.

The control laws, through the control effectors, give the pilot control over the aircraft's trajectory. The control augmentation system (CAS) (control effectors, control laws, sensors, flight computers) stabilizes the aircraft and provide what is called flying qualities. An aircraft's flight control system (FCS) must satisfy certain flying quality specifications. When a pilot performs precision tasks (landing, refueling, guns on target), poor integration of the control effectors leads to a poor design of the FCS. This can lead to poor performance, pilot induced oscillations, and crashes.

There are flying qualities specifications for all modes of an aircraft's operation [35]. When designing an aircraft or extending the flight envelope to new regions, pilots fly simulations of the aircraft using dome simulators. During this process, the desired flying qualities are determined [36,37]. Control laws are then designed to shape the aircraft's response such that it meets these criteria. This includes developing control effector mixing logic, control laws, gain schedules, and nonlinear command gradients (stick force shaping). Aircraft flight control research must address flying qualities needs.

5 Aircraft Robust Nonlinear Control

This section describes applied research topics in robust nonlinear control. Candidate topics include:

- Affordability Issues
- Anti-PIO Compensation
- High AOA Control Laws/Vortex Control
- Robust Optimal Nonlinear Control

5.1 Affordability Issues

As stated in the introduction, industry's number one goal is to make affordable systems. Methods for designing robust nonlinear control systems must be automatable, in that full envelope flight control laws can be rapidly designed with limited manpower. In addition, computer aided design tools are required to support the design, analysis, robustness, and simulation of flight control systems.

MDA has been using the commercial control system design product called MATRIXx to reduce the cost in developing flight control systems. Our process is shown in Figure 4. The tailless fighter aircraft shown in Figure 2 has been modeled within this framework.

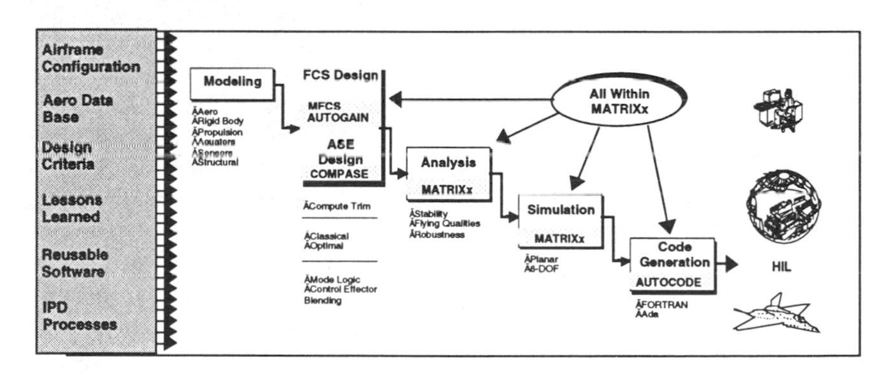

Figure 4 Rapid prototyping control law process.

In the past, aerospace engineers developed the requirements for a flight control system by developing a FORTRAN 6 degree of freedom nonlinear simulation for the application, designing the flight control system by linear and nonlinear analyses, defining the flight control system requirements (software requirements), and delivering the requirements to software engineers for software development. This process was expensive in time and manpower. In addition, the software development process precluded changes in the design (requirements) late in the development cycle as more accurate/higher fidelity models of hardware were obtained.

By using a tool like MATRIXx, the requirements definition process remains the same, but the analyses are all completed within the same tool.

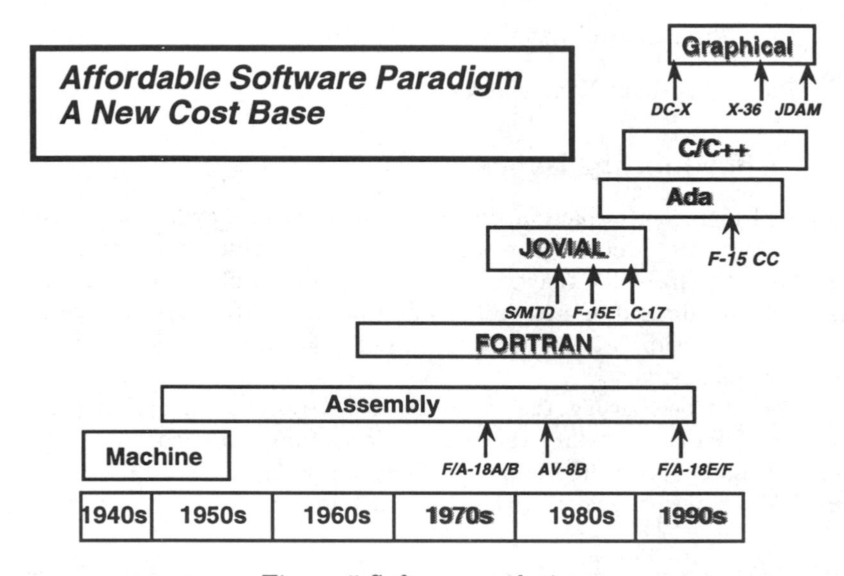

Figure 5 Software evolution.

The big change is that the flight control system is designed graphically, and the autocode feature provided by MATRIXx allows for rapid development of the flight software. This autocoded software is used by the engineer, is used in the manned simulators, is used in the hardware in the loop simulators (HIL), and will fly in the aircraft.

This new process represents a new paradigm in developing flight software. Figure 5 illustrates this paradigm shift. Instead of writing the flight software by hand, where errors could be introduced, the graphical block diagrams are autocoded. This software is then unit tested and transitioned to the software engineers for incorporation into the operation flight program (OFP). The vision behind this new process is to have the same engineer who is developing the requirements also create the software implementing those requirements, all within the same tool, thus minimizing errors and reducing costs.

The use of these tools and processes by industry makes its easier for academia to transition new control technologies. If researchers also use these same tools, then the cost associated with the transition of new control analysis or design approaches is greatly reduced. Also, this makes the transition of accurate models to academia for research purposes feasible.

It is well known that current and future flight control challenges are nonlinear. Feedback linearization methods have been the only tools available for engineers to address the nonlinear flight control problems in a rigorous way. Recent progress in disturbance attenuation for nonlinear systems (nonlinear H_∞) also offers the potential to address these challenges.

Feedback linearization has become the most popular method for designing robust nonlinear flight controls [9-19, 40-43] since these types of controllers, after linearizing the nonlinear dynamics, rely on linear control design methods. This makes it easier for the engineer to design for aircraft flying qualities. Nonlinear H_∞), as a design method, requires the solution of partial differential equations. Very few lessons learned are available using this approach. Other than in missile research [58], only low order academic problems have been solved. When selecting an approach for control system design, industry performs trade studies, with comparisons between approaches made on a level playing field. Recently, new methods for control system design have appeared based upon linear matrix inequalities (LMIs) [38] and ℓ_1-optimal control [39]. These new methods provide alternate ways of addressing many of the nonlinear characteristics in robust nonlinear flight control problems. Comparisons of these approaches with feedback linearization and nonlinear H_∞) will help industry to develop improved flight control systems. Lessons learned are needed that describe which of these methods is best for a particular type of control problem.

5.2 Anti-PIO Compensation

Pilot induced oscillations (PIOs) [40-48] are sustained aircraft oscillations due to an undesirable coupling between the aircraft and pilot. Prior to stability augmentation systems, aircraft only exhibited PIOs (referred to as classical PIOs) early in their development. Today's aircraft (with digital stability augmentation systems) exhibit PIOs when one or more of the following occur: excessive control sensitivity, excessive feedback gains, excessive time delays, or excessive lags in the control system. As the sophistication in flight control system design has grown, so too has the potential for inadvertently designing an aircraft with strong susceptibility to PIOs.

PIOs related to excessive time delays can be traced to neglected higher order dynamics contributing phase lag over the pilot frequencies, and to actuator rate limiting. The need exists for compensating the control system to prevent PIOs. This problem remains unsolved.

Recent progress has been made in the design and implementation of control laws for the problem of actuator saturation [49-57]. Kapasouris [49, 50] introduced this approach as a nonlinear reference governor that attenuates, when necessary, the input commands. Gilbert [56] has extended this to systems with state constraints, which would occur during control rate saturation. Using these new methods, as well as the possibilities from -optimal control [39], algorithms to prevent PIO need to be researched.

5.3 High AOA Control Laws/Vortex Control and Robust Optimal Nonlinear Control

Dynamic inversion has emerged as the most common nonlinear aircraft control design method. This method eases gain scheduling but complicates robustness analysis. At high AOA flight conditions where the aerodynamic surfaces are ineffective, aerodynamic control effectors are augmented with active vortex controls or thrust vectoring. This complicates the inversion of the control distribution matrix. Due to the nonlinear nature of the problem, robust performance is hard to predict and ASE compensation is hard to design and analyze. Additional nonlinear control design methods are needed to design aircraft control laws.

For high performance fighter aircraft (and all aircraft), the amount of control surface deflection used to perform a given maneuver is critical. (To minimize drag and power requirements implies flying with the minimum surface deflections at all times.) Dynamic inversion based controls can be viewed as a deaugmentation - augmentation process. The deaugmentation comes from subtracting off the nonlinearities. The augmentation comes from substituting the desired dynamics. This process does not make use of any of the natural dynamics and aerodynamics to minimize the control effort.

Kokotovic has presented a simple example that demonstrates this principal quite well. The nonlinear scalar plant is $\dot{x} = x - x^3 + u$. Using a feedback linearizing control, with desired dynamics of $\dot{x} = -x$, the dynamic inversion control would be $u = -2x + x^3$, which clearly would saturate any actuator for moderate x.

New methods need to be developed that can address nonlinear aircraft control problems. Possibilities include conventional optimal control methods using linear parameter varying models, controller synthesis using linear matrix inequalities [38], and ℓ_1-optimal control [39] approaches.

6 Technology Transition

In the last few years there has been a slight shift in emphasis by research funding agencies. In the past, the major focus of 6.1 basic research programs was to perform basic research. This is still the case, but included now is the requirement to focus this research on critical technology areas (industry problems) and to transition technology into products. For example, AFOSR is measuring how much funded AFOSR technology is transitioning into

AF products. With the competition for funding becoming greater, it is important for researchers (if they plan to get funding) to address technology transition in their proposals/white papers.

To transition technology a researcher must:

1. identify the customer and focus on their objectives; and

2. establish a mechanism for making the transition happen.

For example, some of the AF Fixed Wing Vehicle Technology Objectives include: 50% reduction in flight control system development costs, a new capability in intelligent and self adaptive control, a 50% reduction in mission critical vulnerability, and a 70% reduction in control related accidents. The first part is fairly easy, because most research objectives usually tie to customer objectives. The hard part is the making the transition happen.

Making the transition happen usually requires someone from industry taking the developed technology and applying it to/on a program. Thus, industry (the producer of AF products) must also be interested in the research and believe that there is a payoff for applying the new technology. Unfortunately this process takes time.

Industry researches/evaluates/matures technologies under Internal Research and Development (IRAD) programs. At MDA, candidate IRAD programs are proposed 6 to 8 months prior to beginning the work (usually starts in June). These candidate IRAD programs are collected and evaluated under a Technology Prioritization Process (TPP). The TPP collects all the proposed programs and sends out customer evaluation forms to our internal customers (like F/A-18, F-15, C-17, etc.), and the internal customers score the IRADs based on how well they address their needs. The scores are collected and funding is allocated based on the IRAD obtaining a high score. Low scoring IRADs are usually not funded.

Once a technology is matured under the IRAD program it is delivered to an internal customer for application to their program. This requires algorithms/design tools to be user friendly and implementable by project engineers. An example of a successful transition is the following.

In the late 1980's, control researchers at the University of Illinois's Coordinated Science Laboratory developed, under AF funding, some projective control theory used to project optimal state feedback control designs into output feedback control designs. The papers published from this research were used by MDA engineers (under MDA IRAD) to develop a design tool (called AUTOGAIN) that would project optimal missile autopilot designs into output feedback designs. The output feedback designs reduced hardware costs (less sensors), were superior to autopilots designed classically, and had roughly the same digital implementation as classical autopilots. However, the design tool that was developed greatly reduced the manpower required to complete the design over the flight envelope, thus improving affordability. AUTOGAIN has been used on MDA's Tomahawk, Standoff Land Attack Missile (SLAM-ER), Joint Direct Attack Munition (JDAM), and Miniaturized Munition Technology (MMT) programs.

7 Summary

In light of today's current funding trends, industry will be relying on universities to support technology development. Topics in integrated controls, reconfigurable flight control, aircraft performance improvements, and aircraft/missile robust nonlinear control (and references) have been presented in the hopes to stimulate academic research in these areas, and to promote the transition of technology from academia to industry.

References

[1] L. Medgyesi-Mitschang. Numerical prototyping: the new paradigm in low observables system development. MDA Briefing on Control Issues for LO High Performance Aircraft. 1995.

[2] R. Douglas, S. Mackler and J. Speyer. Robust hover control for a short takeoff/vertical landing aircraft. *Proc. of the AIAA GNC Conf. 1990.* 146-163.

[3] K. Wise and J. Brinker. Linear quadratic flight control for ejection seats. To appear *AIAA J. Guid., Control, and Dyn.*

[4] W. Durham. Constrained control allocation. *J. Guid., Control, and Dyn.* **16**(4) (1993), 717-725.

[5] K. Bordignon and W. Durham. Closed form solutions to the constrained control allocation problem. *Proc of the AIAA GNC Conf. 1993.* 113-123.

[6] W. Durham. Constrained control allocation: three moment problem. *J. Guid., Control, and Dyn.* **17**(2) (1994), 330-336.

[7] A. Isidori. *Nonlinear Control Systems.* New York: Springer-Verlag, 1989.

[8] M. Vidyasager. *Nonlinear Systems Analysis.* New Jersey: Prentice-Hall, 1993.

[9] S. Snell, D. Enns and W. Garrard. Nonlinear inversion flight control for a supermaneuverable aircraft. *Proc. of the AIAA Guidance, Navigation, and Control Conference 1990.* 808-825.

[10] D. Bugajski, D. Enns and M. Elgersma. A dynamic inversion based control law with application to the high angle-of-attack research vehicle. *Proc. of the AIAA Guidance, Navigation, and Control Conference 1990.* 826-839.

[11] B. Morton. A dynamic inversion control approach for high-mach trajectory tracking. *Proc. of the American Control Conference 1992.* 1332-1336.

[12] S. Snell. Cancellation control law for lateral-directional dynamics of a supermaneuverable aircraft. *Proc. of the AIAA Guidance, Navigation, and Control Conference 1993.* 701-709.

[13] P. Menon. Nonlinear command augmentation system for a high performance aircraft. *Proc. of the AIAA Guidance, Navigation, and Control Conference 1993.* 720-730.

[14] *Multivariable Control Design Guidelines.* Honeywell Technology Center for Air Force. 2 November 1993.

[15] K. Wise, M. Dierks, B. Kerkemeyer and J. Tang. Linear and nonlinear aircraft flight control for the aiaa controls design challenge. *Proc. of the AIAA Guidance, Navigation, and Control Conference 1992.*

[16] J. Buffington, R. Adams and S. Banda. Robust, nonlinear, high angle-of-attack control design for a supermaneuverable vehicle. *Proc. of the AIAA Guidance, Navigation, and Control Conference 1993.* 690-700.

[17] J. Reiner, G. Balas and W. Garrard. Robust dynamic inversion for control of highly maneuverable aircraft. *AIAA Journal of Guidance, Control, and Dynamics* **18**(1) (1995).

[18] D. Enns. Robustness of dynamic inversion vs. m synthesis: lateral-directional flight control example. *Proc. of the AIAA Guidance, Navigation, and Control Conference 1990.* 210-222.

[19] G. Balas, W. Garrard and J. Reiner. Robust dynamic inversion control laws for aircraft control. *Proc. of the AIAA Guidance, Navigation, and Control Conference 1992.* 192-205.

[20] J. Berg, K. Hammett, C. Schwartz and S. Banda. An analysis of the destabilizing effect of daisy chained rate limited actuators. Submitted to *IEEE Trans. on Control Systems Technology.*

[21] M. Dornheim. Report pinpoints factors leading to yf-22 crash. *Aviation Week and Space Technology* (1992), 49-50.

[22] M. Bodson. Identification with Modeling Uncertainty and Reconfigurable Control. *Proc. of the IEEE CDC 1993.* 2242-2243.

[23] M. Bodson, et. al. Control reconfiguration in the presence of software failures. *Proc. of the IEEE CDC 1993.* 2284-2289.

[24] M. Bodson and J.E. Groszkiewicz. Multivariable adaptive algorithms for reconfigurable flight control. *Proc. of the IEEE CDC 1994.* 3330-3335.

[25] P. Chandler, M. Mears and M. Patcher. On-line optimizing networks for reconfigurable control. *Proc. of the IEEE CDC 1993.* 2272-2277.

[26] J. Weiss and S. Hoy. Flight control reconfiguration for structurally damaged aircraft. Presentation. NAECON 1990. Dayton, Ohio.

[27] W. Havern, J. Tromp and M. McCay. Aerodynamic definition and simulation of a battle damaged fighter aircraft. Presentation. NAECON 1990. Dayton, Ohio.

[28] W. Weinstein. Control reconfigurability handbook. Presentation. NAECON 1990. Dayton, Ohio.

[29] J. Urnes, R. Yeager and J. Stewart. Flight demonstration of the self-repairing flight control system in a NASA F-15 aircraft. Presentation. NAECON 1990. Dayton, Ohio.

[30] J. Schroeder and N. Fifield. Airborne fault isolation flight test. Presentation. NAECON 1990. Dayton, Ohio.

[31] J. Valencia, J. Lane and J. Schroeder. Ground station reporting on in flight maintenance diagnostics. Presentation. NAECON 1990. Dayton, Ohio.

[32] P. Chandler and E. Wells. Detection and isolation of control element failures on the NASA HIDEC F-15. Presentation. NAECON 1990. Dayton, Ohio.

[33] B. Migyanko and R. Yeager. Flight test reconfiguration performance and handling qualities results. Presentation. NAECON 1990. Dayton, Ohio.

[34] D. Ward and R. Baron. Self-designing controller (SDC). Presentation. Government and Industry at WL/FIGC. 1994. Wright-Patterson AFB. Dayton, Ohio.

[35] Anon. Military standard - flying qualities of piloted vehicles. MIL-STD-1797. 1987.

[36] D. Moorhouse and W. Moran. Flying qualities design criteria for highly augmented systems. Presentation. NAECON 1985. Dayton, Ohio.

[37] G. Krekeler, D. Wilson and D. Riley. High angle of attack flying qualities criteria. *Proc of the AIAA GNC 1989.*

[38] S. Boyd, L. El Ghaoui, E. Feron and V. Balakrishnan. *Linear Matrix Inequalities in System and Control Theory.* Philadelphia: SIAM, 1994.

[39] M. Dahleh and I. Diaz-Bobillo. *Control of Uncertain Systems: A Linear Programming Approach.* Englewood Cliffs, NJ: Prentice-Hall, 1995.

[40] R. Hess and R. Kalteis. Technique for predicting longitudinal pilot induced oscillations. *J. Guid., Control, and Dyn.* **14**(1) (1991), 198-204.

[41] M. Givens. Evaluation of B-2 susceptibility to pilot-induced oscillations. Northrop Grumman Technical Report. 1994.

[42] K. McKay. Summary of an AGARD workshop on pilot induced oscillation. *Proc. of the AIAA GNC 1994.* 1147-1157.

[43] R. Smith. Predicting and validating fully-developed PIO. *Proc. of the AIAA GNC 1994.* 1158-1166.

[44] D. Mitchell, R. Hoh, B. L. Aponso and D. Klyde. The measurement and prediction of pilot-in-the-loop oscillations. *Proc. of the AIAA GNC 1994.* 1167-1177.

[45] D. Moorhouse. Experience with the R. Smith PIO criterion on the F-15 STOL and maneuver technology demonstrator. *Proc. of the AIAA GNC 1994.* 1178-1184.

[46] E. Rynaski. Multivariable design to directly satisfy flying qualities. *Proc. of the AIAA GNC 1994.* 1185-1193.

[47] R. A'Harrah. An alternate control scheme for alleviating aircraft pilot coupling. *Proc. of the AIAA GNC 1994.* 1194-1201.

[48] D. McRuer. Pilot induced oscillations and human dynamic behavior. Technical Report 2450-111. Systems Technology, Inc. Hawthorne, California. 1994.

[49] P. Kapasouris, M. Athans and G. Stein. Design of feedback control systems for stable plants with saturating actuators. *Proc of the IEEE CDC 1988.* 429-439.

[50] P. Kapasouris and M. Athans. Control systems with rate and magnitude saturation for neutrally stable open loop systems. *Proc. of the IEEE CDC 1990.* 3404-3409.

[51] A. Rodriguez and J. Cloutier. Control of a bank-to-turn missile with saturating actuators. *Proc. of the ACC 1994*. 1656-1664.

[52] F. Mazenc and L. Praly. Adding an integration and global asymptotic stabilization of feedforward systems. *Proc of the IEEE CDC 1994*. 121-126.

[53] Y. Chitour, W. Liu, and E. Sontag. On the continuity and incremental gain properties of certain saturated feedback loops. *Proc of the IEEE CDC 1994*. 1127-132.

[54] A. Teel. Additional stability results with bounded controls. *Proc of the IEEE CDC 1994*. 133-137.

[55] Z. Lin, A. Saberi and A. Stoorvogel. Semi-global stabilization of linear discrete-time systems subject to input saturation via linear feedback – an ARE-based approach. *Proc of the IEEE CDC 1994*. 138-143.

[56] E. Gilbert, I. Kolmanvosky and K. Tan. Nonlinear control of discrete-time linear systems with state and control constraints: a reference governor with global convergence properties. *Proc of the IEEE CDC 1994*. 140-145.

[57] F. Tyan and D. Bernstein. Antiwindup compensator synthesis for systems with saturating actuators. *Proc of the IEEE CDC 1994*. 146-151.

[58] K.A. Wise and J.L. Sedwick. Nonlinear optimal control for agile missiles. To appear *Journal of Guidance, Control, and Dynamics.*

[59] Joint Services Advanced Flight Control Technology Study Briefing. Lockheed, Fort Worth. 2 December 1994.

McDonnel Douglas Aerospace, Mail Code 1067126, PO Box 516, St. Louis, Missouri 63166

Systems & Control: Foundations & Applications

Founding Editor
Christopher I. Byrnes
School of Engineering and Applied Science
Washington University
Campus P.O. 1040
One Brookings Drive
St. Louis, MO 63130-4899
U.S.A.

Systems & Control: Foundations & Applications publishes research monographs and advanced graduate texts dealing with areas of current research in all areas of systems and control theory and its applications to a wide variety of scientific disciplines.

We encourage the preparation of manuscripts in TEX, preferably in Plain or AMS TEX— LaTeX is also acceptable—for delivery as camera-ready hard copy which leads to rapid publication, or on a diskette that can interface with laser printers or typesetters.

Proposals should be sent directly to the editor or to: Birkhäuser Boston, 675 Massachusetts Avenue, Cambridge, MA 02139, U.S.A.

Estimation Techniques for Distributed Parameter Systems
H.T. Banks and K. Kunisch

Set-Valued Analysis
Jean-Pierre Aubin and Hélène Frankowska

Weak Convergence Methods and Singularly Perturbed Stochastic Control and Filtering Problems
Harold J. Kushner

Methods of Algebraic Geometry in Control Theory: Part I
Scalar Linear Systems and Affine Algebraic Geometry
Peter Falb

H^∞-Optimal Control and Related Minimax Design Problems
Tamer Başar and Pierre Bernhard

Identification and Stochastic Adaptive Control
Han-Fu Chen and Lei Guo

Viability Theory
Jean-Pierre Aubin

Representation and Control of Infinite Dimensional Systems, Vol. I
A. Bensoussan, G. Da Prato, M. C. Delfour and S. K. Mitter

Representation and Control of Infinite Dimensional Systems, Vol. II
A. Bensoussan, G. Da Prato, M. C. Delfour and S. K. Mitter

Mathematical Control Theory: An Introduction
Jerzy Zabczyk

H_∞-Control for Distributed Parameter Systems: A State-Space Approach
Bert van Keulen

Disease Dynamics
Alexander Asachenkov, Guri Marchuk, Ronald Mohler, Serge Zuev

Theory of Chattering Control with Applications to Astronautics, Robotics, Economics, and Engineering
Michail I. Zelikin and Vladimir F. Borisov

Modeling, Analysis and Control of Dynamic Elastic Multi-Link Structures
J. E. Lagnese, Günter Leugering, E. J. P. G. Schmidt

First Order Representations of Linear Systems
Margreet Kuijper

Hierarchical Decision Making in Stochastic Manufacturing Systems
Suresh P. Sethi and Qing Zhang

Optimal Control Theory for Infinite Dimensional Systems
Xunjing Li and Jiongmin Yong

Generalized Solutions of First-Order PDEs: The Dynamical Optimization Process
Andreĭ I. Subbotin

Finite Horizon H_∞ and Related Control Problems
M. B. Subrahmanyam

Control Under Lack of Information
A. N. Krasovskii and N. N. Krasovskii

H^∞-Optimal Control and Related Minimax Design Problems
A Dynamic Game Approach
Tamer Başar and Pierre Bernhard

Control of Uncertain Sampled-Data Systems
Geir E. Dullerud

Robust Nonlinear Control Design: State-Space and
Lyapunov Techniques
Randy A. Freeman and Petar V. Kokotović

Adaptive Systems: An Introduction
Iven Mareels and Jan Willem Polderman

Sampling in Digital Signal Processing and Control
Arie Feuer and Graham C. Goodwin

Ellipsoidal Calculus for Estimation and Control
Alexander Kurzhanski and István Vályi

Minimum Entropy Control for Time-Varying Systems
Marc A. Peters and Pablo A. Iglesias

Chain-Scattering Approach to H^∞-Control
Hidenori Kimura

Systems and Control in the Twenty-First Century
*Christopher I. Byrnes, Biswa N. Datta, Clyde F. Martin,
and David S. Gilliam*